Engineering Materials

This series provides topical information on innovative, structural and functional materials and composites with applications in optical, electrical, mechanical, civil, aeronautical, medical, bio- and nano-engineering. The individual volumes are complete, comprehensive monographs covering the structure, properties, manufacturing process and applications of these materials. This multidisciplinary series is devoted to professionals, students and all those interested in the latest developments in the Materials Science field, that look for a carefully selected collection of high quality review articles on their respective field of expertise.

Indexed at Compendex (2021) and Scopus (2022)

Upendra Kumar
Editor

Defects Engineering in Electroceramics for Energy Applications

Springer

Editor
Upendra Kumar
Department of Applied Sciences
Indian Institute of Information Technology
Allahabad
Prayagraj, Uttar Pradesh, India

ISSN 1612-1317 ISSN 1868-1212 (electronic)
Engineering Materials
ISBN 978-981-97-9017-3 ISBN 978-981-97-9018-0 (eBook)
https://doi.org/10.1007/978-981-97-9018-0

This Springer imprint is published by the registered company Springer Nature Singapore Pte Ltd.
The registered company address is: 152 Beach Road, #21-01/04 Gateway East, Singapore 189721, Singapore

Contents

Fundamentals of Solid-State Physics

V. J. Usha Praveena and Shikha Chander

Abstract Solid-state physics is the branch of physics that offers the widest range of phenomena for the application of both classical and quantum physics on macroscopic, microscopic, and atomic scales. Matter exists either in a solid state or a fluid state, and fluid is subdivided into liquid and gaseous states. In modern concept it is divided into condensed or gaseous states. Moreover, condensed matter is subdivided into the solid and liquid states. From the present point of view, these solids are conveniently divided into three classes (1) Crystalline, (2) Semi-crystalline (polycrystalline), and (3) Non-crystalline (Amorphous). Technology has benefited tremendously from the development of solid-state research. In general, instrumentation and measurement techniques incline heavily on methods and gadgetry initially developed for pure solid-state research. Today, much of technology is based on applying stable state phenomena. If that's what it is, then industrial physicists and material scientists should have some idea of the basic origin of these phenomena. Only this way can they expect to follow and direct the development of new materials and techniques. We can specifically mention the factors controlling the mechanical strength and stability of solids and the electrical and magnetic properties of solids. The present chapter, "Fundamentals of Solid-State Physics," is planned with the following sequence of topics. We've already realized that solid-state physics includes a wide spectrum of subjects and materials, which is why one can choose more specific topics, for example, in terms of taste or professional experience. Nonetheless, they are fundamental and technical elements of magnetic and dielectric phenomena that demand to be included. Similarly, the surprising phenomena of superconductivity and unique theoretical interpretation justify its inclusion. This chapter also presents the current research developments in which major advances are still being made.

Keywords Crystalline solids · Condensed matter · Instrumentation · Superconductivity

V. J. U. Praveena (✉)
Department of Physics, St. Francis College for Women, Begumpet, Hyderabad 500016, India
e-mail: usha.praveena@sfc.ac.in

S. Chander
Department of Chemistry, St. Francis College for Women, Begumpet, Hyderabad 500016, India

U. Kumar (ed.), *Defects Engineering in Electroceramics for Energy Applications*,
Engineering Materials, https://doi.org/10.1007/978-981-97-9018-0_1

1 Introduction

Electroceramics are a group of materials, with electrical characteristics that are essential for various electronic devices and systems. These materials can control charges, fields and currents typically reacting to factors like temperature, pressure or electromagnetic fields. The exploration of electroceramics is closely linked to solid state physics, which offers insights, into how they work and how to design them effectively. Researchers can refine materials by delving into their atomic and electronic structures. This can lead to advancements in fields such, as electronics, telecommunications, energy conversion and more. Understanding electroceramics at a level requires exploring the concepts of solid state physics.

'Solid state physics is the branch of Physics which offers widest range of phenomena for the application of both classical and quantum physics, on macroscopic, microscopic and atomic scales'. Matter exists either in solid state and fluid state and fluid subdivided in to liquid and gaseous state. In modern concept it is divided into condensed or gaseous states. Moreover, condensed matter is subdivided in to the solid and liquid states. From the present point of view, these solids are conveniently divided in to three classes as (1) Crystalline (2) Semi crystalline (poly crystalline) and (3) Non crystalline (Amorphous).

The advances in solid-state research have greatly benefited technology. Generally speaking, instruments and quantifying techniques bank on techniques and devices originally created for pure solid-state explore. The use of stable state phenomena is the foundation for a significant portion of modern technology. If that's the case, material scientists and industrial physicists ought to be aware of the fundamental causes of these phenomena. They can only hope to oversee and follow the creation of novel materials and methods in this manner. We can specifically list the elements governing a solid's electrical and magnetic properties, as well as its mechanical strength and stability [1, 2].

Present chapter "Fundamentals of Solid-State Physics" is planned with following sequence of topics. Since solid state physics covers widespread of topics and materials—as we've already established—one can select more specialized subjects based on factors like personal preference or professional background. However, they must be included because they are an essential and technical component of magnetic and dielectric phenomena. In a similar vein, superconductivity's unexpected phenomena and original theoretical explanation support its inclusion. Additionally, the current state of research is presented in this chapter, where significant advancements are still being made.

Solid state phenomena have played a major role in the historical development, regardless of whether interest is focused on fundamental understanding or for exercise of materials. Hence, existence of super conductivity and magnetism piques scientist's interests and captures the imagination of electrical engineers. To comprehend solid state phenomena, problems must be made as simple as feasible. Studying pure metals, alloys, or simple compounds is necessary. However, material scientists who want to exploit solid-state phenomena for practical purposes rarely discover that pure

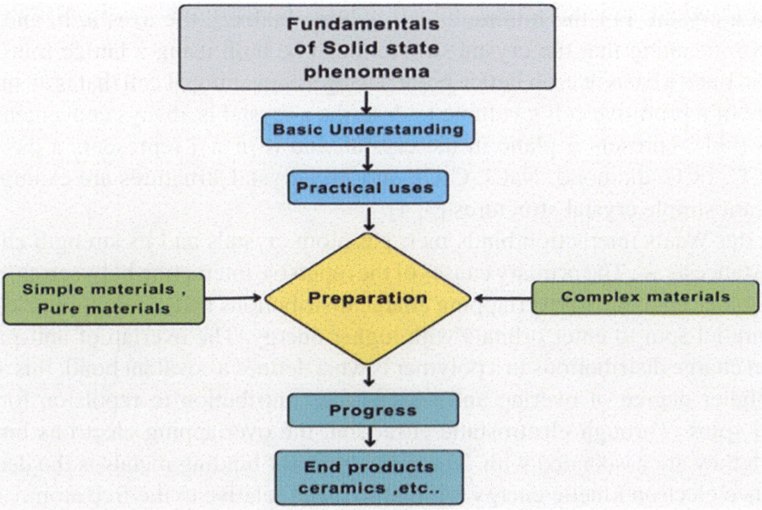

Fig. 1 Block diagram demonstrating materials synthesis

elements are suitable in and of themselves. Instead, they must either develop or create through trial-and-error compounds that, due to their affordability and performance, may meet practical needs. We use a straightforward block diagram to illustrate these two points. Understanding the characteristics of common matter that we encounter in our daily lives is some of the main benefits of delving deeper in to solid state physics. Understanding the qualities of a material, including its structural, mechanical, thermal, electrical, magnetic, and dielectric characteristics, is made easier by a basic understanding of solid state physics (Fig. 1).

It describes the distinctions between copper and mercury. The unique qualities of silicon that contribute to its significance in semiconductor technology. We perform real experiments using typical bulk stuff. Ordinary bulk matter also tends to be polycrystalline, meaning that it is made up of aggregates of tiny crystals known as "grains." The characteristics of grains or crystallites greatly influence their size, shape, and orientation. Bulk matter's polycrystalline nature is frequently detrimental. It is therefore necessary to create large single crystals of metals, alloys, compounds, and salts. In the last 45 years, output has increased significantly at a rapid pace. They can be grown using multiple specialized methods, and scores of compounds are sold commercially as monocrystals. In technical applications as well, these single crystals are frequently required. We will highlight the necessity of doing comparative research in solid state physics and in the creation of new technologies based on their inherent characteristics. Atoms and molecules make up all materials. The regular arrangement of atoms within crystals is what sets them apart from non-crystals. 'A lattice is a collection of points connected by the lattice translation operator $T = n_1a + n_2b + n_3c$, where a, b, and c are referred to as primitives and n_1, n_2 and n_3 are integers'. We must affix an identical basis of n atoms to each lattice point in order to form a crystal. For the minimum cell volume [axbxc], the axes a, b, and c are primitive, meaning that the crystal structure can be built using a lattice translation operator T and a basis at each lattice point. Using a conventional cell that is an integral multiple of a primitive cell's volume to describe a crystal is always convenient. The indices (hkl) represent a plane in the crystal, and $[n_1n_2n_3]$ represents a direction. The BCC, FCC, diamond, NaCl, CsCl, and ZnS crystal structures are examples of significant simple crystal structures [3, 4].

Van der Waals interaction binds inert gas atom crystals and its strength changes with distance as $\frac{1}{R^6}$. The primary causes of the repulsive interaction between atoms are electrostatic repulsion of overlapping charge distributions and overlapping electrons with parallel spin to enter orbitals with higher energy. The overlap of anti-parallel electron charge distributions in a polymer is what defines a covalent bond; this allows for a higher degree of overlap and lessens the contribution to repulsion for anti-parallel spins. Through electrostatic attraction, the overlapping electrons bind the ion path they are associated with. The primary factor binding metals is the decrease in valence electron kinetic energy within the metal relative to the free atom.

Understanding the workings of crystals, in materials is crucial for progressing research in creating materials that have potential applications, across different technological sectors which can be achieved by X ray diffraction. It also provides insights about the properties of different materials such as determining crystal structure identifying phases, analysing defects characterizing texture and tracking changes in structure. The detailed structural information it offers is key to progressing research, development and production of materials.

2 X-Ray Diffraction

As discussed, X-ray diffraction experiments can reveal information about the crystal structure. Bragg's law, which states that "$n\lambda = 2d\sin\theta$," serves as the foundation for three techniques that identify crystal structures illustrated in Fig. 2.

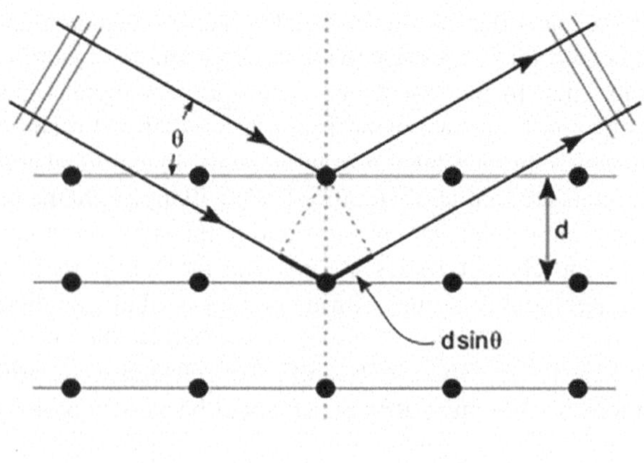

Fig. 2 X-ray diffraction from solids [5]

1. Laue method: An X-ray beam with a constant wavelength is used to hold a single crystal still. The crystal separates and diffracts the discrete values of "λ" that satisfy Bragg's law by having planes with spacing "d" and incidence angle "θ" that match.
2. Rotating crystal method: A monochromatic beam of X-ray rotates a mono crystal around a fixed axis. Different atomic planes are brought into position for reflection by variations in θ.
3. Powder method or Debye–Scherrer method: A crystalline material sample that has been powdered is fixed in place within a monochromatic beam. There might be some crystalline orientations in the distribution where the angle of incidence satisfies Bragg's law.

Of course, the three methods mentioned above are used in the current research with some modifications. The Laue method can be used to quickly determine the orientation and symmetry of crystals as well as to measure how much crystalline imperfection is present after heat and mechanical treatment. When a single crystal specimen is available, the rotating crystal method works best for determining the structure. Because the Powder method does not require a single crystal, it is used in metallurgical and applied work.

Scintillation counters or proportional counters are employed in diffractometers, which are modern instruments, to detect diffracted radiation. These techniques make it possible to gather data automatically and are still more beneficial when studying dislocation or the location of interstitial ions within a single crystal.

With exception of scattering mechanism, the same physical principles that govern X-ray diffraction also govern electron diffraction in neutron beams. Positive nucleus and outer electrons of an atom produce a strong potential that scatters electrons. This potential is orders of magnitude stronger than that of X-rays, and diffracted beams can be easily sighted by photographic plates or seen in fluorescent screen pictures. To produce monochromatic radiation, the accelerating voltage needs to be carefully stabilized, and the wavelength can be changed continuously in theory. The electron beams can only exist in a high vacuum and require the use of magnetic or electrostatic lenses. Moreover, specimens used in transmission experiments need to be extremely thin due to their strong absorption by matter. Studying bulk specimens at glancing incidence is done by reflection. The study of surfaces benefits greatly from electron diffraction, which is frequently thought to originate from two-dimensional structures. Research reactor moderators contain a large concentration of thermally energetic neutrons (about 60 °C), and recently, reactors have been constructed specifically to produce powerful neutron beams. Maxwell's law states that both the energies of neutrons and molecules in a gas are distributed. The equivalent wavelengths, which are approximately 1.55 A° and 1.33 A°, respectively, correlating to the RMS velocities of neutron gas at 0 and 100 °C, are ideal for crystal diffraction. In a large crystal, Bragg reflection from a selected family of planes selects an appropriate wavelength band. This monochromatic beam can be used to examine single crystals or with a powder sample and a Bragg spectrometer.

Unlike the case of atoms scattering X-rays, the neutron scattering length does not regularly vary across the periodic table. Atoms are capable of equally intense scattering as heavy ones. Neutron scattering can divulge the sequence structures of alloys such as Carbonyl zinc or Iron carbonate, which are not visible with X-ray scattering. Neutron diffraction is therefore suitable for studying compounds containing both light and heavy elements. Different isotopes of a particular element scatter variantly, which makes sense since scattering involves the formation of an unstable compound nucleus. The neutron can interact with magnetic atoms, or those with a net electron spin, because it also has a magnetic moment. This enables the identification of magnetic structures with varying degrees of complexity.

Since an atom's magnetic moment originates in its electrons, magnetic scattering follows the same general angular dependence as X-rays.

Not all solids have to be crystalline. Additionally, they can be obtained as glasses or as non-crystalline, amorphous aggregates with a meaningful scale which implies that no translation symmetry and no long-range correlation between atom positions is observed. Depending on the kind of bond that forms, there may be short range order over distances equivalent to few atomic spacings. The level of ordering will vary. One accurate description of amorphous metals is haphazardly densely packed structures. Additionally, simple liquids—that is, liquids made up of separate, comparable atoms—do not exhibit translational symmetry. It is impossible to describe in terms of unit cells or other geometric concepts when there is no symmetry. The atoms' relative positions are always shifting in the liquid's dynamic equilibrium. Amorphous solids have drawn increasing attention in the past few years. The amorphous state can be obtained for alloys and even for some pure metals. Certain metals, such as gallium or beryllium, "freeze" as their atoms strike the substrate and form an ordered structure when they evaporate on very cold substrates, around 2 K. Crystalline order can also be destroyed by ion bombardment. Even at room temperature, many covalently bonded materials remain stable in their amorphous form. The ability to produce metallic alloy glasses through extremely rapid cooling is a relatively recent development. This method is not suitable for preparing amorphous pure metals; instead, stabilizing the amorphous glassy structure seems to require the addition of elements like silicon, boron, or phosphorus. After forming, these alloys require heating to between 300 and 400 °C before they recrystallize. They have much better corrosion resistance than crystalline phases, which makes them potentially useful for commercial applications. Certain magnetic alloys may also find use as transformer cores.

The discovery of X-ray diffraction by crystals revolutionized our comprehension of the structure of materials, enabling profound insights into their properties. However, it became evident that certain properties, such as plasticity, electrolytic conductivity, electrical resistance, and crystal strength, could not be entirely accounted for by structural differences alone. Thus, it became imperative to delve deeper into deviations from ideal crystal structures in order to fully elucidate these properties.

3 Defects and Dislocations

A material is helpful for construction when it combines malleability and ductility, not when it is malleable alone. At the point when stacked past a specific cutoff known to be yield pressure, the stuff shouldn't break as glasses for instance, however ought to yield plastically, turning out to be for all time disfigured without cracking. It is said that glass is fragile while copper and aluminium are ductile. Unadulterated metals are having a more prominent or lower level of ductility, and the material scientists try to configure combinations to increase great strength by keeping up with and permissible proportion of malleability. We portray the mechanical way of behaving of a material with the guide of a pressure strain outline. Preferably this chart shows the pressure expected to keep an expected pace of strain as an element of strain. Practically speaking the capacity of testing machine to answer the way of behaving of the example can cause subtleties of stress strain bend showing up diversely on various machines yet it won't concern us. Significant factors are strain rate, temperature and arrangement of sample. Therefore, it became necessary to investigate the deviation from the ideal flawless crystals in order to explain them. C. G. Darwin proposed in 1914 that the atoms in a pure crystal form a lattice array, but that this array is not continuous throughout, but is instead divided into blocks by discontinuities at the edges of the blocks. Several scientists employed Darwin's hypothetical postulate to explain a variety of solid-state phenomena.

A. Smekel proposed that defects of a different kind might account for characteristics like diffusion and electrolytic conductivity. Only the mobility of atoms allows for electrical conduction. An atomic-scale defect results from an atom's movement.

The idea of a perfect crystal is really helpful. Numerous lattice properties, including density, dielectric capacity, specific heat, and elastic properties, can be explained by using it. We refer to these characteristics as structure insensitivity. All real crystalline materials aren't flawless, though. Although totally regular crystals are extremely rare, regular atomic arrangement is one of the most significant characteristics of crystals. Any variation from periodicity in a solid is referred to as a flaw. There must be some kind of lattice imperfection in every crystal. While lattice faults can be ignored for many purposes, they also play a crucial part in present-day solid state physics. The following are the different means that flaws affect a solid's properties: A metal's electrical resistance may be increased by several percent in many pure metals and significantly more in alloys due to imperfections that scatter conduction electrons in the metal.

1. Some imperfections reduce the crystal's strength, even if they are present occasionally.
2. Color is frequently present in pure salts that have flaws and impurities.
3. Hysteresis loss of ferromagnets: Structure-sensitive properties are those that are highly responsive to the crystal's state of perfection. The presence of dislocations, which are line defects, makes crystals extremely soft. In 1934, Taylor initially introduced the concept within crystal physics to elucidate the factors contributing to the observed lower levels of critical shear stress. "Burgers" later generalized

Fig. 3 Edge dislocation in a cubic crystal [6]

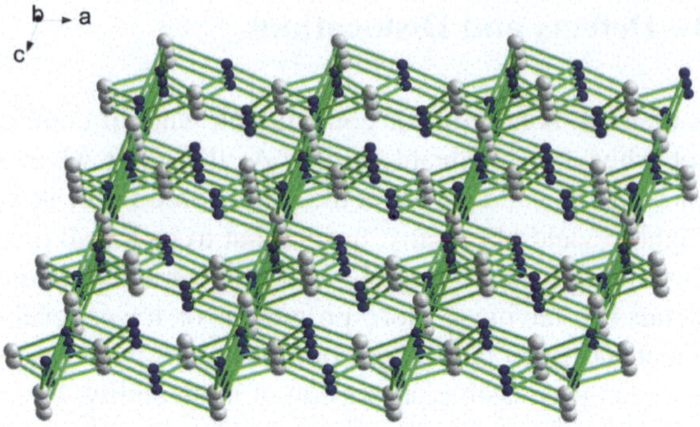

the idea in 1938. In 1948, the growth patterns observed on the surfaces of naturally occurring beryl crystals were found to precisely match the predictions of dislocation theory, marking the initial instance of direct experimental evidence supporting the existence of dislocations. There are now many methods available for studying them, but transmission electron microscopy is certainly the most significant.

The "edge" and the "screw" are the most basic forms of dislocation. Formally, we could consider them to be made of perfect crystal blocks that are envisioned to be cut, distorted, and then rejoined. It is evident that in actual crystals, a dislocation results in a movement equivalent to an interatomic separation for both screw and edge types (Fig. 3). The additional "half plane" added to the otherwise perfect crystal defines the edge dislocation, which is a directed ray. We observe that dislocation line may move very slightly due to atomic rearrangements in its nearby vicinity; if the line exits the crystal, we will have a surface step and have essentially passed above part of the crystal in relation to the below part. We have achieved a certain amount of slip without having to firmly move crystal blocks. Therefore, if crystals have dislocations, we could anticipate that they would move in response to tiny shear stresses, causing slippage. This suggests that about 10^{-1} G of ideal huge critical shear stress should be present in a crystal that has no dislocations and is hard to deform. In actuality, such crystals can be grown frequently and arise as tiny objects known as "Whiskers" that resemble the ideal behavior. The reason why polycrystalline materials have higher yield strengths than single crystal is also understandable. At low stresses, grain boundaries act as barriers, impeding the dislocations from extending beyond the grains and thereby inhibiting slippage. Notably, the yield strength of polycrystalline materials exhibits a proportional relationship to d^{-2}, where d represents the average grain diameter, with an increase observed as the grain size decreases. Excepting glasses and amorphous solids, almost all solids have periodic atom arrays that make up a crystal lattice. Typical vibrations have a medium thanks to the periodic crystal lattice that exists in solids. There are quantized vibrational modes called phonons between every lattice spacing. Since phonons are crucial to the physical characteristics of solids and to the materials' electrical and thermal conductivity, studying them is an essential aspect of solid state physics.

The phonon's long wavelength attribute confers characteristics to sounds in solids as well. Another manifestation of a particular kind of vibrational motion in quantum mechanics is phonons. This kind of vibrational motion, which is characterized by all system components moving in a sinusoidal manner at the same frequency, is known as a normal mode (elementary chunks of lattice vibration). A standing wave, is a continuous form of the normal mode with varying amplitude and oscillating frequencies at the same phase and x, y, and z coordinates, is an instance for a normal mode.

The sound velocity, thermal, elastic, and optical properties of materials can all be explained by lattice vibrations. Atoms in a solid oscillate around their equilibrium position, a phenomenon known as "lattice vibration". Due to the atoms' bonds with one another, equilibrium positions of a crystal form a regular lattice. These nearby atoms' vibrations are not independent of one another. Lattice waves are regular lattices with harmonic forces between atoms and regular modes of vibration. Lattice waves occur at low frequencies and reach high frequencies of up to 1,000 Hz. At very high frequencies, however, the wavelengths are on the order of interatomic spacing. Since these wavelengths are so short, there is no correlation between the motion of neighbouring atoms. Instead, each atom moves about its average position in three dimensions with an average vibrational energy of, typically, 3kT, where 'k' stands for the Boltzmann constant and 'T' for the absolute temperature. Electrical resistance can also result from lattice vibrations interacting with free electrons in a conducting solid.

4 Dielectric Properties

A non-conducting substance, such as glass, rubber, or waxed paper, serves as dielectric. Some of the characteristics of dielectric are as follows:

(i) The capacitance rises by factor K, also named as the dielectric constant of dielectric, when the dielectric material is inserted between the capacitor's plates. $C = KC_0$, where C_0 is the capacitance in the absence of the dielectric
(ii) The dielectric constant K is specific to a given material and lacks a unit.
(iii) $K = \varepsilon/\varepsilon_0$ if ε_o is the permittivity of free space and ε is the permittivity of dielectric material.
(iv) When dielectric materials are added, the energy density changes to $u = 1/2\varepsilon E^2$.
(v) The potential difference V between the plates after dielectric material is inserted is as follows, assuming that the charge stays constant: $V = V_0/K$. where V_o is the potential difference in the absence of a dielectric. The potential difference V between plates decreases by a factor of K, according to the equation.
(vi) $Q = KQ_0$ if the potential difference remains constant after the dielectric material is inserted.

As a result, factor K increases charge on the capacitor with a fixed potential difference between its plates.

(vii) If the material with a dielectric constant of K is pushed into the space in between plates of a parallel plate capacitor, and electric field between plates, E_o, then $E = E_0/K$.

Dielectric constant multiplied by the electric field inside the dielectric reduces it.

(viii) The dielectric's net field E is given as $E = E_0 - E_{ind}$, where $E_{ind} = E_0(1 - 1/K)$.

4.1 Local Electric Field

The following facts must be remembered: every material is made up of a very large number of atoms and molecules; every atom is made up of a positively charged nucleus and negatively charged electrons; every system made up of two equal and opposite charges, q, separated by a certain distance, is an electric dipole; and every dielectric atom or molecule in an electric field causes the positive and negative charges to feel opposite forces and separate, forming a dipole.

4.2 Dipole Relaxation and Dielectric Losses

Dipole relaxation time is a measure of how long it takes for a disturbed system to return to its equilibrium configuration. The relaxation frequency is equal to the relaxation time times its reciprocal. The main reason why the dielectric constants at high and low frequencies differ is because of the orientational contribution to dielectric constant. Temperature and frequency have a significant influence on orientational relaxation frequencies as follows:

$$\alpha = \frac{\alpha_0}{(1 + i\omega\tau)}$$

where 'τ' is called as Debye relaxation time, 'α_0' is static orientational Polarizability and 'ω' is angular frequency.

4.3 Dielectric Breakdown

The insulator becomes conducting and takes on the characteristics of a conductor when stronger electric fields are applied. Dielectric materials lose their dielectric qualities under such circumstances. Dielectric Breakdown is the name given to this phenomenon. The process is irreversible. Dielectric materials eventually fail as a result of this.

4.4 Types of Dielectric Material

Dielectrics are classified according to the kind of molecule that is found in the substance. Dielectrics are classified into two categories: polar and non-polar.

4.4.1 Polar Dielectrics

Positive particle mass centres do not line up with negative particle mass centres in polar dielectrics. The dipole moment is present here. The molecules' shapes are asymmetrical. The molecules align themselves with electric field when it is applied. The net dipole moment in the molecules drops to zero when electric field is removed, resulting in observation of random dipole moment. Few examples are CO_2, H_2O, etc.

4.4.2 Non-polar Dielectrics

The centers of mass of positive and negative particles coincide in non-polar dielectrics. These molecules don't have a dipole moment. The shape of these molecules is symmetrical. Non-polar dielectrics include H_2, N_2, O_2, and so on.

4.5 Examples of Dielectric Material

Materials that act as dielectrics include liquids, solids, gases, and vacuums. In electrical engineering, solid dielectrics are widely used. Porcelain, ceramics, glass, paper, and other materials are a few examples of sold dielectrics. Gaseous dielectrics include nitrogen, dry air, sulphur hexafluoride, and the oxides of different metals. Two typical examples of liquid dielectrics are transformer oil and distilled water.

4.6 Applications of Dielectric Material

The following are a few uses for dielectrics:

- Capacitors employ these to store energy.
- Materials with high permittivity dielectrics are used in semiconductor devices to improve their performance.
- Liquid crystal displays make use of dielectric materials.
- The Dielectric Resonator Oscillator uses ceramic dielectric.
- Strontium Barium Titanate thin films are dielectrics with high tunability and low leakage current that are used in microwave tunable devices.

- Polyethylene serves as a barrier between the substrate and the outside world in industrial coatings.
- Mineral oils help with cooling and are used as a liquid dielectric in electrical transformers.
- To boost the capacitance value of high-voltage capacitors, castor oil is used.
- A specially processed dielectric, Electrets, material acts as electrostatic equivalent to magnets.

5 Electrons in Free Atoms and in Solids

Electrical properties of solids are fixed by electrons in the atoms' outermost orbits. It is possible to apply electron theory to any solid, including non-metals. It also describes the solids' thermal, magnetic, and electrical characteristics. The electron theory of solids uses the electronic structure of solids to explain their structure and properties. There have been three primary phases of development. 1. The theory of classical free electrons 2. Theory of Quantum Free Electrons 3. The Zone Theory.

1. Classical free electron theory: In 1900, Drude and Lorentz created the first theory. This theory states that the electrical conductivity of metal comes from free electrons, which are subject to the laws of classical mechanics.
2. Quantum Free Electron Theory: The theory was created by Sommerfeld in 1928. Sommerfeld states that the potential of free electrons is constant. Quantum laws are followed by this theory. 3. Zone Theory: In 1928, Bloch presented the band theory. This theory states that free electrons travel within a periodic potential that the lattice provides. The "Band Theory of Solids" is another name for this theory. It provides a thorough informative analysis of electrons.

The classical free electron theory of metals was put forth by Lorentz and Drude. This theory states that the metals with free electrons follow the principles of classical mechanics [7].

Salient Features of Classical Free Electron Theory

1. There are a lot of free electrons in metals, and they can move freely in any direction.
2. These free electrons follow the kinetic theory of gases' rules just like gas molecules in a container.
3. The energy associated with each electron for temperature T in absence of a field is given by kT. Since $kT = mv_t^2$, it has to do with kinetic energy. Where 'k' is Boltzmann constant and
1. 'v_t' is thermal velocity.
4. Positive ion cores are fixed in metals, while free electrons roam randomly and collide with boundaries, other free electrons, or other positive ion cores. These collisions are elastic as a result. Thus, only free electrons are responsible for electric conduction.

5. The Maxwell–Boltzmann distribution of velocities governs the electron's velocity in a metal.
6. There is a constant potential field around the free electrons. Because of this, the electrons' potential energy is constant.
7. Free electrons are accelerated in the direction against applied electric field with onset of electric field to any metal. This velocity is known as the drift velocity and is represented by the symbol v_{df}.

The classical free electron theory's success:

(1) It confirms Ohm's law.
(2) It describes how metals conduct heat and electricity.
(3) Widemann-Franz law is derived.
(4) It describes the metals' optical characteristics.

Constraints or shortcomings or merits of classical theory:

(1) Electrical conductivity of insulators and semiconductors is not explained.
(2) It is unable to account for how electrical conductivity varies with temperature at low temperatures.
(3) The idea of a metal's specific heat is not explained.
(4) It is unable to account for the electrons' mean free path.
(5) Black body radiation, the Compton effect, and the photoelectric effect could not be explained by the conventional free electron theory.
(6) It is unable to explain how ferromagnetism and paramagnetic susceptibility are dependent on temperature.

6 Quantum Free Electron Theory

This theory was put forward by Sommerfeld in 1928. Numerous shortcomings of classical theory are addressed by it. By selecting Fermi–Dirac statistics over Maxwell–Boltzmann statistics, Sommerfeld provided an explanation. He used the ideas of quantum mechanics to develop this theory.

6.1 Quantum Free Electron Theory Assumptions

(1) Inside the metal's boundaries, valence electrons travel freely at a fixed potential and are stopped from leaving the metal at the boundaries (high potential). The electron is consequently imprisoned in a potential well.
(2) The Pauli Exclusion Principle dictates how electrons are distributed among the different permitted energy levels.
(3) Ignored are the repulsion between electrons and the attraction between free electrons and lattice ions.

(4) Fermi–Dirac statistics describe how energy is distributed among the free electrons.
(5) Free electron energy values are quantized.
(6) The electron Schrodinger time-independent wave equation is applied to determine potential energy values. The particle present in a potential box problem is similar to this one. Energy of electron is $E_n = \frac{n^2 h^2}{8mL^2}$, where n = 1, 2, 3, ….

6.2 *Advantages of the Free Electron Quantum Theory*

(1) It does a good job of explaining how metals conduct heat and electricity.
(2) The Thermionic phenomenon can be explained by it.
(3) It describes how the conductivity of metals varies with temperature.
(4) It explains why metals have a specific heat.
(5) It describes how metals are magnetically susceptible.
(6) It explains black body radiation, the Compton Effect, and the photoelectric effect, among other things.
(7) It provides the accurate mathematical formula for thermal conductivity of a metal.

6.3 *Quantum Free Electron Theory's Drawbacks*

(1) It cannot account for the metallic characteristics that some crystals only display.
(2) It cannot explain why the metallic crystals' atomic arrays should favour specific structures exclusively.
(3) The distinction between an insulator, a semiconductor, and a metal is not made by this theory and it is also unable to account for the Hall Coefficient's positive value.
(4) It is untrue that conduction is caused by the two electrons that are present at Fermi level.

Origin of energy bands in solids is that distinct electron energies are present in an isolated atom. Assume that when two separate atoms are brought very close to one another, their orbiting electrons will interact. Thus, the electron energies in the combined system will fluctuate and be somewhat greater or lower than their initial values rather than remaining at the same level. Thus, there are two energy levels that are closely spaced apart at each energy level. Bands of allowed energies are created when 'N' number of atoms are combined to form a solid and electrons of these atoms interact to produce 'N' number of nearly spaced energy levels instead of discrete energy levels.

The degree of atom overlap determines the width of the energy band, which is greatest for outermost electrons. Conduction bands are the bands that correspond to the outermost orbits, and valence bands are the next inner band. The term "forbidden energy gap" or "band gap" refers to the space created by these two permitted bands.

Understanding how electrons behave in a solid involves looking at the range of energy levels they have access, to, which is influenced by how atoms are arranged in the crystal structure. The band theory offers a way to comprehend and forecast the characteristics of materials.

7 Classification of Crystalline Solids

Solids are categorised as: conductors, semiconductors, and insulators based on the width of forbidden band.

Conductors: There is no energy gap (E_g) and an overlap between the valence and conduction bands. Free electrons are already present in large quantities in conduction band at room temperature. Because of this, these solids have good thermal and electrical conductivity. At high temperatures, the electrical resistivity rises due to collisions between free electrons.

Example: Al, Cu, Ag, Au etc.

Semiconductors: A small energy band gap, $E_g \approx 1$ eV, separates valence and conduction bands. In conduction band, free electrons are unavailable at low temperatures (0 K). For this reason, at low temperatures, they act like insulators. At high temperatures, free electrons move from valence band to conduction band, increasing electrical conductivity. At high temperatures, these solids act like electrical conductors. Example, Germanium $E_g = 0.7$ eV and Silicon $E_g = 1.1$ eV.

Insulators: A very large energy gap ($E_g \geq 3$ eV) separates the conduction and valance bands. At room temperature, there are electrons in valance band and none in conduction band. These materials are electrical insulators as a result. Valence electrons are unable to enter the conduction band, even at very high temperatures. Example: rubber, glass, mica, ebonite, etc.

Band Energy

Combinations of molecular orbitals close in energy result in continuous bands of energy. Of course, bands with different energies will form because of the mass amounts of various molecular orbital mixings. The band gap is the name given to the difference between these band energies as indicated in Fig. 4.

The electron jump across the band gap is examined by the band theory. Specifically, the transition of electrons across their Fermi energy level from valence band to conduction band. The solid's optical and magnetic properties are determined by this "jump". The Valence Band is energy band that contains all of valence electrons present in highest molecular orbital.

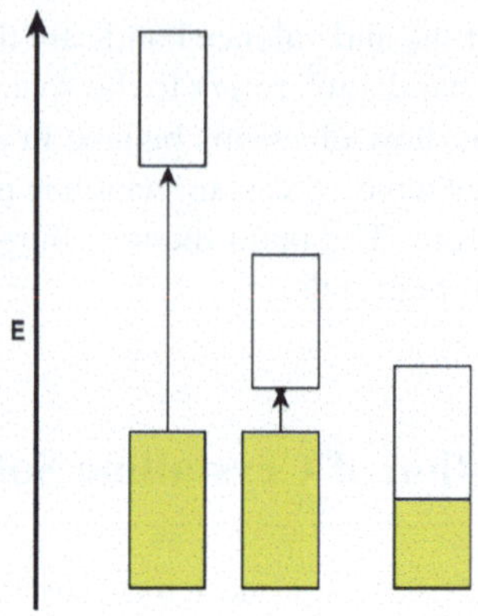

Fig. 4 Conduction bands are represented by white boxes, and valence bands by the yellow boxes

Band of Conduction

Energy band at which there are mobile charge carriers, either positive or negative. Elementary electrons with sufficient energy to leave valence band and enter conduction band are known as negative mobile charge carriers. Here, they are directly involved in semiconductor conductivity and can freely move throughout the crystal lattice. Another name for positive mobile charge carriers is holes. The absence of an electron in conduction band is referred as a hole. Put differently, a hole is the feature that, although an electron (a negative mobile charge carrier) can exist within a band, it does not exist at that specific location. The electron is called a positive mobile charge carrier because it has the potential to exist but does not.

Fermi Level

The giant occupied molecular orbital at absolute zero is referred to as this level. Usually, it can be located in middle of valence and conduction bands. In this state, particles are independent of one another and have their own quantum states. When temperature gets close to rising above absolute zero, particles will take states above Fermi level and leave states below Fermi level empty.

Semiconductor

Conductivity between an insulator and a conductor is what is known as a semiconductor. Because they are unlikely to short circuit like conductors do, semiconductors are widely used in everyday electronics. Their narrow band gap gives them their distinct conductivity. Since the electrons aren't always in conduction band, a band gap prevents short circuits. For a solid to have some conductivity, there must be sufficient electron flow from valence to conduction bands, which is possible by a small band gap.

As an atom's nuclear charge is released, electrons in conduction band are free to travel freely throughout the band. Since electron in this band contributes to solid's electrical conductivity, it is known as a negative charge carrier. The electron is free to move about in this band. The state changes to a positive charge carrier, or hole, when the electron exits valence band.

Pure Semiconductors

Pure semiconductors, whose characteristics are derived only from the substance. In this case, the quantity of holes in valence band and the number of electrons in conduction band are equal. Another name for these semiconductors is i-types.

Independent Semiconductors

Semiconductors that are impure but have been "doped" to increase their conductivity. P-type and n-type extrinsic semiconductors are the two varieties. To extract electrons from the valence band, a "dopant" atom is incorporated into the lattice. An acceptor is the name given to this atom. A p-type (positive type) semiconductor will eventually result from number of holes in lattice exceeding number of negative charge carriers as more acceptors are added. Numerous "dopant" atoms, or donors, contribute electrons to conduction band in N-type semiconductors.

Applications of Semiconductors

One of the major uses of semiconductors was the creation of low frequency alternating current rectifiers. C. E. Fitts created these rectifiers with selenium as early as 1886. Since its introduction by Logrondahl and P. H. Geiger in 1927, the copper oxide rectifier has found widespread application as a low power rectifier in wireless sets, battery charges, and other applications. Much research on copper oxide and selenium has resulted from the development of these rectifiers.

The photoconductive qualities of copper oxide and selenium have been utilized to create exposure meters for photography and photocells, which are commonly employed in the film industry to convert soundtrack markings into electric currents for loudspeaker amplification and reproduction. These cells are also utilized in many automated systems, including train counters and burglar alarms.

Hertz had demonstrated radio waves in 1888, but it wasn't until roughly 1904 that it was realized that semiconductors' rectifying properties could be used to create a detector for high frequency currents that these waves had set up in an electric circuit.

Many different drugs were tried over the next few years. Radio receiver sensitivity increased when it was discovered that a thin wire, similar to a cat's whisker, when it came into contact with a semiconducting material crystal, worked incredibly well as a rectifier for high frequency currents. The two materials that were typically favoured were silicon and lead sulphide, the latter of which was present as natural galena crystals. The question of whether the rectification effects were thermal or electrical in nature gave rise to an intriguing debate. G. W. Pierce's thorough research, which demonstrated beyond a reasonable doubt that these are electrical, helped to resolve this.

Thermionic valves quickly supplanted crystal detectors, and the former was all but rendered obsolete. It was once again used when radar technology developed very short wavelengths because no other device was found to be an effective rectifier of frequency changes at these extraordinarily high frequencies. The most effective material for this purpose turned out to be silicon.

The infrared photo conductor class of semiconductors, which includes lead sulphide, selenide, and telluride, is another group that has seen significant use. Using these, the most sensitive infrared detectors for the crucial 1.5–7μ infrared spectrum region have been produced thus far. Smith has reviewed the characteristics of these detectors and how they are used in infrared spectroscopy.

8 Magnetism

Solid state physics plays an important role for magnetic phenomena. Because magnetic ions at different crystal sites may behave differently depending on where they are in the lattice, many magnetic materials contain multiple types of magnetic species. Thus, a crucial factor in the study of the internal field and lattice is the interaction between nearby magnetic ions.

While there is much room for experimentation and theoretical conjecture due to variety of magnetic properties displayed in widely disparate classes of material, the practical applications of magnetism are crucial both technically and commercially. All conventional matter exhibits magnetism because it develops a magnetic dipole moment when exposed to an external magnetic field, regardless of whether it takes the form of free atoms, ions, molecules, or condensed aggregates. On the other hand, it is also possible to have a permanent magnetic dipole moment. The magnetic dipole moment with respect to either unit volume or unit mass in terms of magnetization per unit mass is known as the magnetization M of a bulk sample by

$$\chi = \mu_o \frac{M}{B_o}$$

B_o represents both the vacuum permeability and the surrounding magnetic field.

The orbital and spin motions of electrons as well as their interactions with one another are the source of magnetism. Describe how materials react to magnetic fields to give an overview of the various types of magnetism. Some people might find this surprising, but matter is magnetic. Simply put, certain materials have a much higher magnetic strength. Primary difference being that while there is a very strong interaction between atomic moments in some materials, there is no collective interaction of atomic magnetic moments in other materials.

The following five major groups can be used to categorize magnetic behaviour of substances (Fig. 5):

1. Diamagnetism
2. Paramagnetism
3. Ferromagnetism
4. Ferrimagnetism
5. Antiferromagnetism

The substances in the first two categories are not magnetically ordered and show no signs of collective magnetic interactions. Below a critical temperature, substances in final three groups show long-range magnetic order. The majority of materials that we think of as magnetic are ferrimagnetic and ferrromagnetic (i.e., behaving like iron). The remaining three are typically regarded as "nonmagnetic" because of how weakly magnetic they are.

1. Diamagnetism

Though it is often very weak, diamagnetism is a fundamental property of all matter. It results from orbiting electrons' obstructive behaviour in the presence of an applied magnetic field. Atoms lacking net magnetic moments-that is, all orbital shells filled and no unpaired electrons-make up diamagnetic substances. On the other hand, the susceptibility is negative due to the production of a negative magnetization upon exposure to a field.

Keep in mind that there is no magnetization when the field is zero. Another distinguishing feature of diamagnetic materials is their temperature-independent susceptibility. Some examples of diamagnetic substances, in units of $10^{-8}\text{m}^3/\text{kg}$, include.

$$\text{Quartz} \ -0.62, \ \text{Water} \ -0.90, \ \text{Calcite} \ -0.48$$

2. **Paramagnetism**

Because of unpaired electrons in partially filled orbitals, some of the atoms or ions in this class of materials have a net magnetic moment. Iron is among the most significant atoms with unpaired electrons. Unlike diamagnetism, the magnetization is zero when

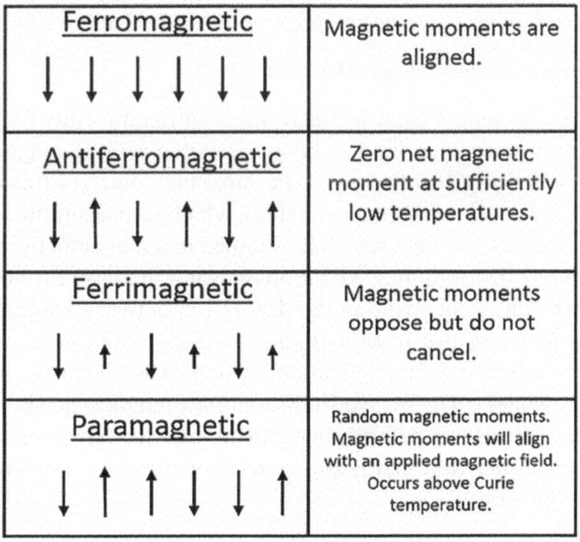

Fig. 5 Types of magnetism [8]

field is removed, and the individual magnetic moments do not interact magnetically. A net positive magnetization and positive susceptibility are result of partial alignment of the atomic magnetic moments in the direction of the field in presence of the field.

Additionally, the random effects of temperature oppose the field's efficiency in aligning the moments. Curie Law, a temperature-dependent susceptibility, is the outcome of this.

The paramagnetic susceptibility is small at room temperature and in moderate fields. When applied field is not too high or the temperature is very low (less than 100 K), paramagnetic susceptibility remains unaffected. In these circumstances, the relationship between paramagnetic susceptibility and total iron content is linear. At room temperature, many minerals that contain iron are paramagnetic.

In units of $10^{-8}m^3/kg$, a few instances are as follows:

$$\text{Clay} - 13,\ \text{Silicate} - 79,\ \text{Siderite} - 100$$

3. Ferromagnetism

Most likely, iron, nickel, or magnetite come to mind when you think of magnetic materials. These materials show very strong interactions at the atomic moments, in contrast to paramagnetic materials. Atomic moments align in parallel or antiparallel directions as a result of these interactions, which are brought about by electronic exchange forces. Exchange forces are extremely strong; they are roughly million times stronger than earth's field, or a field on the order of 10^3 Tesla. The exchange force is a quantum mechanical phenomenon caused by relative orientation of the spins of two electrons.

Ferromagnetic materials show parallel moment alignment in absence of a magnetic field, which results in a large net magnetization. Most cited examples are elements Fe, Ni, and Co, as well as their alloys. Among them two unique properties are their spontaneous magnetization and existence of magnetic ordering temperature.

Spontaneous Magnetization

The net magnetization that results from a uniformly magnetized microscopic volume in absence of a field is known as spontaneous magnetization. The spin magnetic moments of the electrons determine how much of this magnetization there is at 0 K.

The saturation magnetization, which is measurable in the lab is maximum induced magnetic moment which is obtained in a magnetic field (H_{sat}) is known as saturation magnetization; magnetization does not increase further beyond this field. Magnetic domains play a role in the distinction between saturation magnetization and spontaneous magnetization (more about domains later). Particle size has no bearing on saturation magnetization, which is an intrinsic property that is temperature-dependent. The susceptibilities of ferromagnets and paramagnets differ significantly. In contrast to paramagnetic materials, ferromagnetic materials exhibit saturation of magnetization at high (room-temperature) temperatures and moderate magnetic fields (Table 1).

Table 1 Magnetic parameters

	H_{sat} Tesla	T range (K)	χ $10^{-8}m^3/kg$
Paramagnetic	>10	100	~50
Ferromagnetic	~1	~300	10^3–10^4

Curie Temperature

Thermal energy eventually outweighs the electronic exchange forces in ferromagnets, resulting in a randomizing effect, despite the extremely high exchange forces. The Curie temperature (T_C) is specific temperature at which this happens. The ferromagnet is ordered below Curie temperature and disordered above it. At Curie temperature, saturation magnetization becomes zero. Below is a typical plot of magnetization against temperature for magnetite.

Another intrinsic feature that can be utilized to identify minerals is the Curie temperature, which is also a diagnostic parameter. It is not infallible, though, as distinct magnetic minerals could theoretically have the same Curie temperature.

4. **Ferrimagnetism**

More complex forms of magnetic ordering can happen because of crystal structure of ionic compounds. Ferrimagnetism is one kind of magnetic ordering. This figure illustrates magnetic spins in a ferrimagnetic oxide.

Two magnetic sublattices (referred to as A and B) spaced apart by oxygens make up the magnetic structure. The oxygen anions act as a mediator in these exchange interactions. The interactions are referred to as super exchange interactions or indirect interactions when this occurs. The A and B sublattices' spin alignment is antiparallel under the strongest super exchange interactions.

Net magnetic moment in ferrimagnets is caused by unequal magnetic moments of the A and B sublattices. Thus, ferromagnetism and ferrimagnetism are comparable. It displays hysteresis, remanence, Curie temperatures, and spontaneous magnetization—all characteristics of ferromagnetic behaviour. Ferro- and ferrimagnets, on the other hand, have quite distinct magnetic orderings.

One well-known ferrimagnetic substance is magnetite. In fact, until Néel developed the theoretical foundation for ferrimagnetism in the 1940s, magnetite was thought to be a ferromagnet.

Crystal Structure of Magnetite

The spinel structure is crystallized with magnetite, Fe_3O_4. The smaller Fe ions fill in spaces created by the closely packed large oxygen ions in a cubic configuration. Two types of the gaps are found (Fig. 6):

Tetrahedral location: Fe ion is surrounded by four oxygen atoms.

Octahedral Site: Fe ion is surrounded by six oxygen atoms.

The two magnetic sublattices, A and B, are formed by the tetrahedral and octahedral sites, respectively. A sublattice's spins are opposite to those of B sublattice. Due to the significant differences between the two crystal sites, there are intricate

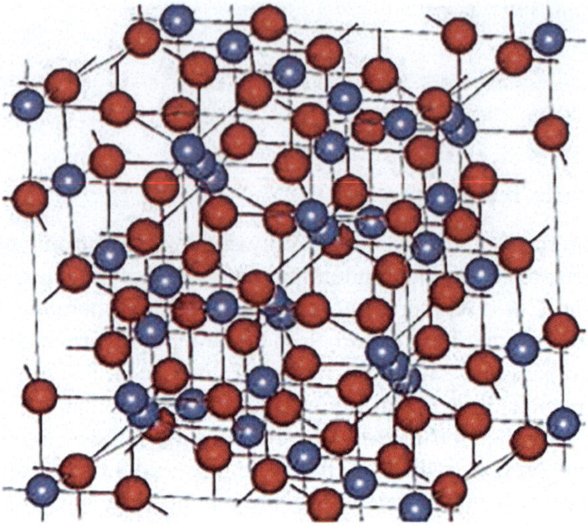

Fig. 6 Crystal structure of magnetite [9]

exchange interactions between and between the iron ions in the two different types of sites.

The structural formula for magnetite is $[Fe^{3+}]$ A $[Fe^{3+}, Fe^{2+}]$ B O_4.

We refer this specific configuration of cations on the A and B sublattice as an inverse spinel structure. The B-site Fe^{2+} is responsible for net magnetic moment of magnetite when there are negative AB exchange interactions.

5. **Antiferromagnetism**

The total moment is zero if sublattice moments of A and B are primarily equal and opposite. Antiferromagnetism is the name given to this kind of magnetic ordering.

The behaviour of susceptibility above a critical temperature, known as the Néel temperature (T_N), provides a clue to antiferromagnetism. The Curie–Weiss law governs the susceptibility of paramagnets above T_N, but a negative intercept denotes negative exchange interactions.

In corundum structure, hematite crystallizes in a close-packed, hexagonal structure with oxygen ions. Fe^{3+} ions' magnetic moments are antiferromagnetically coupled between the planes but ferromagnetically coupled with in particular c-planes.

The Magnetisation Curve: Hysteresis

Apart from saturation magnetization and Curie temperature, ferromagnets can also remember an applied field even after it is removed. A plot of the variation in magnetization with magnetic field is referred to as a hysteresis loop, and this behaviour is known as hysteresis (Fig. 7).

The coercivity of remanence (H_r) is another characteristic of hysteresis. This is reverse field that brings the saturation remanence down to zero when it is applied and then removed. It consistently exceeds the coercive force. The initial susceptibility

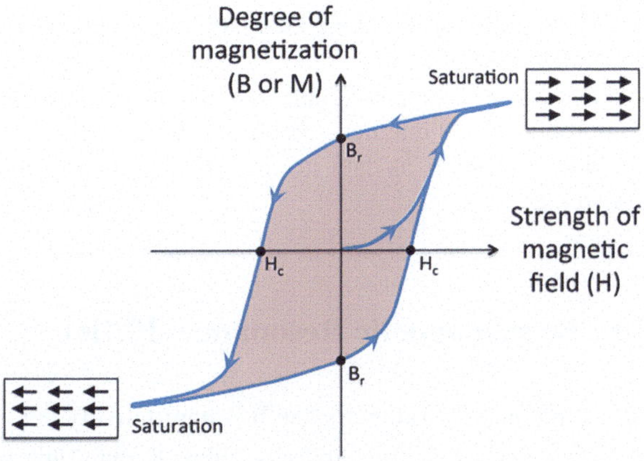

Fig. 7 Hysteresis loop for a ferro- or ferrimagnet [10]

(χ_0) is the magnetization observed in the order of the earth's field (60–110 μT). Different hysteresis parameters depend on temperature, stresses, domain state, grain size, and other factors in addition to being intrinsic properties. Hysteresis variables are helpful for magnetic grain sizing of natural materials because they depend on grain size.

9 Magnetic Resonance

In magnetic systems with both magnetic moment and angular momentum, a phenomenon known as magnetic resonance occurs. Resonance suggests that we are in tune with the natural magnetic system frequency, which is same as the magnetic moment's frequency of gyroscopic precision in an external static magnetic field. Due to the similarities between characteristic frequency of atomic spectra and magnetic resonance frequencies, which are usually in microwave frequency range for electron spins or the frequency region for nuclear spins, respectively which is microwave spectroscopy. The ability to choose from the entire range of magnetic susceptibility is the resonance method's most significant benefit. The observation of iron's weak nuclear paramagnetism against the background of electron ferromagnetism is the most striking example. Resonance is the ability of the materials under study to absorb energy from an incident oscillating electromagnetic field at certain critical frequencies. A variety of experimental methods make use of resonance. The two most well-known ones are electrons in resonance (ESR), which absorbs energy from electrons, and nuclear magnetic resonance (NMR), which absorbs energy from nucleus [11].

Information about the internal local electric fields of alloys and cold-worked materials can be obtained through nuclear magnetic resonance. Electron spin resonance is a valuable method for assessing various properties of solids and is also applied in the study of semiconductor and insulator properties. Exploring magnetic resonance phenomena offers insights, into the electromagnetic characteristics of materials allowing for their analysis, improvement and application, across various technological fields.

Next concept, Ferromagnetic resonance proves to be a method that not deepens our comprehension of magnetic materials but also propels progress, in diverse technological domains, such, as data storage and processing magnetic sensing and spin-based electronics.

10 Ferromagnetic Resonance (FMR)

Static magnetic parameters can be quantitatively determined, microwave losses can be calculated, and the magnetic state of materials can be identified using ferromagnetic resonance (FMR). The opposite of nuclear and paramagnetic resonance absorption is ferromagnetic resonance absorption. Griffiths made the initial discovery of the ferromagnetic effect, and Yager and Bozorth have since reported additional confirmation. C. Kittel has put forth a theory regarding the resonance effect. In high magnetic fields, ferromagnetic resonances are widely used to study magnetic materials at microwave frequencies. The study of resonance, in solids as ceramics is pivotal, for enhancing our comprehension of materials and how they are used.

11 Magnetoresistance

The term "magnetoresistance" describes how a crystal's electrical resistance changes in a magnetic field. This effect results from the electrons' paths becoming curved and not going straight in the direction of superimposed electric field on applying magnetic field. Several researchers with fields up to 50,000 oersted and Kapitza with fields up to 300,000 oersted have thoroughly investigated this effect.

- The ratio of $\Delta R/R$ increases proportionally to H^2 for relatively low field values, but proportionally to capital H for higher field values, if R is the metal's resistance in zero field at a given temperature and ΔR is the increase brought about by the application of a magnetic field. $\Delta R/R_o$, where R_o is the resistance in the zero field at 0 °C, is occasionally utilized because some workers believe that the latter ratio is less reliant on the purity of the metal than the former.
- The transverse magneto resistance is the effect of greatest interest, and it is typically studied in the following configuration: an external power supply is used to establish a DC electric field E_x in a long, thin wire that is directed along the X-axis. Along the Z axis, which is normal to wire's axis, a uniform magnetic field H is applied. A metal's transverse magnetoresistance can typically accomplish one of three very distinct actions in extremely high fields.

(i) At a resistance that is multiple times the zero-field value, the resistance may saturate or become independent of H. Every crystal axis orientation with respect to the measurement axis experiences saturation.
(ii) The resistance may rise to the maximum field for every crystal orientation investigated.
(iii) The resistance might reach saturation in certain directions of the crystal but not in others, which are frequently very close. This behavior is exhibited by an exceptional anisotropy of the resistance in a magnetic field.

While ceramics are typically considered insulating materials, certain ceramic compounds may exhibit magneto-resistance effects under specific conditions.

12 Piezoelectricity

The combined action of mechanical strain and electric fields on a material is known as piezoelectricity; mechanical strain on piezoelectric materials results in polarity in the material, and the application of an electric field causes strain in the material. Piezoelectric sample experiences net polarization and the production of a dipole in the direction of the applied stress when pressure is applied. There are numerous uses for piezoelectricity in signal and electrical transducers. In transducers, which change one form of energy into another—in this case, mechanical energy into electricity—piezoelectricity plays a major role. Sound waves vibrate and are transformed into electrical signals in microphones. On the other hand, loudspeakers transform those electrical signals into sound waves. Ultrasonic wave ranges are where piezoelectric transducers are employed, not audible wave ranges.

Applications using ultrasonic waves benefit from the use of piezoelectric transducers. Finding the ultrasonic velocity of the wave passing through the solid and translating it into an electric signal that can be read on an oscilloscope will yield the solid's Young's modulus. Because the ultrasonic waves preserve the material's structural integrity, they will either reflect or scatter ultrasonic waves that contain cracks and defects. This enables material failure analysis. The Surface Acoustic Wave (SAW) device was created mainly to be used as a filter; it can either increase or decrease specific frequencies produced by electric devices by interacting with the surface waves on circuit boards. Transformers, which raise or lower electric voltages, and pressure sensors, which use the piezoelectric effect to generate an electric response in the event of a pressure or frequency change, are two additional applications for piezoelectrics.

13 Pyroelectricity

The temperature dependence of spontaneous polarization on specific solids, which can be aggregates of single crystals or polycrystalline materials, is manifested as pyroelectric phenomena. A small increase in temperature causes the material's electrical polarization to alter, allowing voltage between specific pyroelectric sample faces to be measured. On the other hand, the same magnitudes of polarization and voltage are found, and the signs are reversed if the same amount lowers the material's temperature. Pyroelectric materials, therefore, offer a convenient way to create infrared (IR) detectors that work at room temperature.

Ferroelectrics are a subgroup of pyroelectrics, and pyroelectric materials are a subgroup of piezoelectric. Important candidate materials and their applications and the relationships between crystal structure and pertinent properties are discussed.

14 Functional Materials

From a more specialized standpoint in materials science and technology, this is made possible by ongoing investigations into and advancements in materials, specifically functional materials. As the name implies, a material's "function" refers to its capacity to carry out a specific task in response to a specific stimulus.

This broad definition includes many materials and various material properties and applications. Nonetheless, materials whose "function" is connected to their optical, magnetic, or electric characteristics are typically categorized as functional materials.

The primary constituents of this class of functional materials are dielectrics, pyroelectrics, piezoelectric, ferroelectrics, ferroelectric relaxor, incipient ferroelectrics, semiconductors, ionic conductors, superconductors, electro-optics, and magnetic materials.

Functional materials are commonly classified into classes with applications such as Information Technology Materials, Electrical Energy Conversion Materials, Biomedical Materials, and Space Technology Materials, among others. Most, if not all, industrial sectors use functional materials for sensing and actuation. Electroactive materials, such as ferroelectrics, pyroelectrics, and piezoelectrics, are examples of such materials.

This covers automotive (solid state piezoelectric fuel injectors), aerospace (accelerometers and micro-positioners), medical diagnostics (ultrasonic imaging), and chemical and process control (which needs use of thermal, strain, and force sensors). Even though some functional materials' properties were discovered in the nineteenth century, some of these materials didn't become useful until the second World War, while others were discovered relatively recently.

Their special qualities, such as magnetoresistance, piezoelectricity, spontaneous polarization, and superconductivity, are reflected in these applications. The manufacturing process, singularities in the crystallographic structure, and chemical composition all directly affect these attributes. It is possible to produce better materials for both known applications and new ones in an economical manner by understanding relationships between composition, structure, processing, and properties. For years, physicists, chemists, and materials scientists have been looking for these connections. Their primary focus in the recent past has been on characterizing the bulk properties of functional materials and describing them through theoretical models. These days, a large body of knowledge can be established for a wide range of functional materials regarding the relationships between composition, microstructure, processing, and macroscopic properties. However, the "materials bulk approach" has recently changed as researchers work to optimize materials to meet the demands of the future. In the near future, devices with nanometric dimensions will be required if the current rate of electronic component miniaturization continues at the same pace as previous decades.

The development of functional materials will continue to be driven by the need for greater functionality and miniaturization, if history is any indication. Through molecular engineering, new functional materials with significantly better performance than current ones are anticipated. Nonetheless, for conventional functional materials like those this chapter describes, future advancements search for:

(1) devices that can be adjusted, meaning that materials with extremely low loss in microwave frequency range should be used
(2) a thorough grasp of the effects of size on functional materials
(3) establishing the connections between the properties of the nanoscale structure and macroscale properties
(4) extremely low processing temperatures for simple integration with various materials
(5) creation of environmentally friendly materials with optimized properties and
(6) creation of heterostructures with maximum properties.

As we need to explore more about ceramic materials, let us understand the classification of ceramics materials in various ways based on different criteria such as composition, structure, processing method, and properties.

Broad classification of ceramic materials is as follows:

1. Traditional Ceramics such as Earthenware, Stoneware and Porcelain
2. Advanced Ceramics as Oxide Ceramics, Non-oxide Ceramics and Composite Ceramics
3. Refractory Ceramics as Refractories
4. Bio-ceramics such as Biocompatible Ceramics
5. Electro-ceramics:

   - Dielectric Ceramics: Ceramics with high dielectric constants and low dielectric losses, used in capacitors, insulators, sensors, and electronic components.

- Piezoelectric Ceramics: Ceramics that exhibit the piezoelectric effect, converting mechanical stress into electrical charge and vice versa. Piezoelectric ceramics are used in actuators, sensors, ultrasonic devices, and energy harvesting applications.

Exploring the properties of materials is crucial for progressing in areas like materials science, electronics, energy tech, communication systems, sensors, actuators, and many other fields. This research helps create materials, devices, and technologies that offer performance, functionality, and dependability.

Further Reading

Emsley, J. (2011). *Nature's building blocks: an AZ guide to the elements.* Oxford University Press, USA.

References

1. Myers, H.P.: Introductory Solid State Physics, 2nd edn. CRC Press, London, U.K. (1998)
2. Neamen, D.: An Introduction to Semiconductor Devices, 1st edn. McGraw-Hill (2006)
3. Kittel, C.: Introduction to Solid State Physics, 6th edn. Wiley, New York (1986)
4. Saxena, B.S., Saxena, P.N., Gupta, R.C., Mandal, J.N.: Fundamentals of Solid State Physics, 32nd edn. Pragati Prakashan, Meerut (2022)
5. Hydrargyram (2011). https://commons.wikimedia.org/wiki/File:Bragg_diffraction_2.svg
6. Sun, C., Zhang, C., Jiang, C., Yang, C., Du, Y., Zhao, Y., Hu, B., Zheng, Z., Christe, K.O.: Synthesis of AgN5 and its extended 3D energetic framework. Nat. Commun. **9**(1), 1269 (2018). https://commons.wikimedia.org/wiki/File:Topological_representation_of_the_3D_framework_in_Ag%28NH3%292%28Ag3%28N5%294%29.jpg
7. Anderson, P.W.: Localized magnetic states in metals. Phys. Rev. **124**(1), 41 (1961)
8. Rowe, D.: Improving workshop design. Train. & Dev. **41**(5), 8–9 (2014). https://commons.wikimedia.org/wiki/File:Types_of_Magnetism.jpg
9. Tzafaras N., Adlhart W., Jagodzinski, H. (2009). https://commons.wikimedia.org/wiki/File:Magnetite_structure.jpg
10. Tem5psu (2014). https://commons.wikimedia.org/wiki/File:Magnetic_hysteresis.png
11. Roseberg, H.M.: The Solid State. Oxford University Press, Oxford (1995)

Dr. Usha Praveena V J completed M.Sc from Osmania University, M.Tech in CSc & Engg. from Acharya Nagarjuna University and M.Phil from Manonmaniam Sundaranar University. She was awarded Ph.D. in the area of Materials Science from Osmania University. Till date she has more than 12 publications in International and National journals and more than 10 publications in conferences. Presently, she is working as Associate Professor in the department of Physics, St. Francis College for Women, Hyderabad.

Dr. Shikha Chander has a Master's Degree from Dr Babasaheb Ambedkar Marathwada University (1994) and a Ph.D. in Chemistry and is currently an Assistant Professor at St. Francis College for Women, India in the Department of Chemistry. She has expertise in Chemistry with emphasis in the field of Nanomaterials, Carbon Nanotubes, Supercapacitors and Reduced graphene Oxide.

Defects in Electroceramics

B. Shivaleela and S. M. Hanagodimaht

Abstract Solid-state physics is the study of rigid matter or solids. The solids are categorized as conductors, semiconductors, or insulators based on how they behave electrically. It should come as no surprise that most texts focus on either ceramics technology or one of the relevant basic solid-state sciences. The term "ceramics" refers to an engineering process that includes ceramic component design and manufacture. Ceramics are susceptible to an excessive number of extrinsic and intrinsic defect types, such as vacancies, interstitials, antisite defects, and so on. In crystalline materials, imperfections are often categorized based on their dimensions, such as Point defects, line defects, surface defects, and volume defects. Electroceramics is developing, at the necessary level, a quantitative wide range of solids' physical characteristics, such as their conductivity, electricity, optical qualities, piezoelectricity, and magneticity. DFT studies of electroceramics with defects have also been carried out. Electroceramics with defects have many applications; they can used as sensors in thermistors to control resistance with respect to temperature, and also in ferroelectric materials.

Keywords Solids · Crystalline · Amorphous · Ceramics · Electroceramics

1 Introduction

The proposed effort aims to further our fundamental knowledge of atomic-level ceramic defects, defect mechanisms, and interfaces. Furthermore, the study will concentrate on how interfaces affect dynamic phase transitions and the transport characteristics of gases, ions, electronic defects, and ions in materials important for energy, environmental, and national security applications. The work will be concentrated on surface/interface features, such as the impact of radiation, the function of hydrogen and surface structure on electronic and ionic transport, and defect-interface interactions and transport. The solids are classified as conductors, semiconductors,

B. Shivaleela (✉) · S. M. Hanagodimaht
Department of Physics, Gulbarga University, Kalaburagi, India
e-mail: shivaleelab17@gmail.com

U. Kumar (ed.), *Defects Engineering in Electroceramics for Energy Applications*, Engineering Materials, https://doi.org/10.1007/978-981-97-9018-0_2

or insulators based on their electrical behavior. It should not be surprising that most of the texts on the subject concentrate on one of the pertinent fundamental solid state sciences or ceramics technology. The word "ceramics" describes an engineering process that includes design and production of ceramic components. A comprehensive variety of quantitative solid physical characteristics, including conductivity, dielectricity, optical properties, piezoelectricity, and magneticity, are being developed by electroceramists at the required level. In general, it is believed that material flaws adversely affect the characteristics and functionality of ceramic materials. Therefore, great attempts have been made to either reduce their concentration or balance off their negative effects.

A new generation of ceramic materials, known as electroceramics, has emerged due to the significant technical breakthroughs and discoveries in which the ceramic industry has played a pivotal role. These materials are ready for new applications after demonstrating their excellence in high-temperature, electro-optic, pyroelectric, and piezoelectric applications. While electroceramic is a broad category of inorganic materials that include oxides, nitrides, niobates, titanates, carbides, and other minerals, conventional ceramic is silicate/clay (Fig. 1).

Section 2 The basic information on imperfection or defect has been given. In Sect. 3, explanation for classifying defects in electroceramic according to dimensions. Section 4 describes defects and their role in materials. Section 5 highlights the various characterization techniques for identifying defects in electroceramics. Section 6 illustrates the applications of defects in different electroceramic materials.

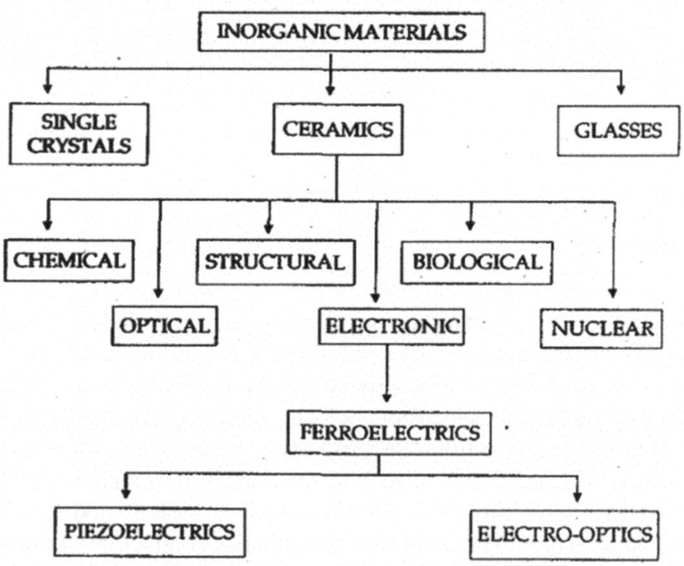

Fig. 1 Classification of electroceramics

The field of ceramics has one important subfield known as electroceramics. Unlike structural ceramics, electroceramics rely on their superior electrical properties to perform a wide range of tasks in automation, communications, electronics, energy conversion and storage, and so on. In the past, advancements in the various electroceramic subclasses—such as dielectrics, which include high dielectric capacitors; ferroelectrics, which include nonvolatile memory and data storage; and piezoelectrics, which include sensors, actuators, and transducers—have increased in tandem with new device technologies. The majority of the aforementioned technologies are gained by controlling certain materials' dielectric, ferroelectric, and piezoelectric characteristics. The question for faults in electroceramics is how such imperfections affect the above-described qualities. The growing number of publications published suggests that research on flaws in electroceramics has received much attention [1].

2 Imperfections or Defects

Imperfections or flaws: In a crystal, everything that deviates from the ideal arrangement of atoms is considered to have imperfections or defects. Since these traits are frequently exploited on purpose to alter a material's mechanical properties, calling them defects is misleading. One method of producing a crystal flaw in a metal is adding alloying elements. Numerous crystal characteristics, including strength, electrical conductivity, and ferromagnet hysteresis loss, are strongly influenced by crystal defects. Therefore, defects and the characteristics of the host crystals both have a significant influence on several significant crystal properties. Minute amounts of chemical impurities solely cause certain semiconductors' conductivity. Flaws and impurities cause many crystals' color and brilliance. Impurities or defects have the potential to increase atomic diffusion greatly. Generally, flaws govern the mechanical and plastic characteristics [2].

3 Imperfections Classification

Public relations is hampered by the persistent belief that flaws are the "bad guys." Defects have typically been written off as not significantly affecting the properties and functionality of ceramic materials and equipment. Consequently, significant efforts have been made to reduce their concentration or offset their negative effects. On the other hand, we are aware that flaws may provide value. For instance, the very desired hue in gem-quality diamonds is caused by imperfections and impurities (see sidebar). The "bad guy" image of flaws is starting to be questioned in light of recent developments in the synthesis, characterization, and theoretical modeling of ceramics.

Ceramics are susceptible to excessive extrinsic and intrinsic defect types, such as vacancies, interstitials, antisite defects, and so on. In crystalline materials, imperfections are often categorized based on their dimensions.

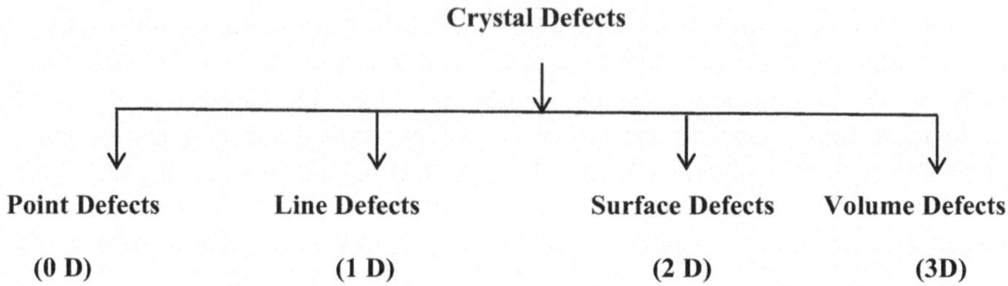

3.1 Point Defects

A point defect is an area inside the lattice structure where one atom is missing or arranged erratically. Point defects include substitutional atoms, vacancies, self-interstitial atoms, and interstitial impurity atoms (Fig. 2). A self-interstitial atom is an extra atom that has pushed its way into an interstitial vacuum in the crystal structure. Self-interstitial atoms are only trace numbers in metals because they significantly stress and modify the closely packed lattice structure. A replacement impurity atom is an atom of a different type from the bulk atoms that has replaced one of the bulk atoms in the lattice. Substitutional impurity atoms usually comprise 15% or less of the bulk atom in size. One example of substitutional impurity atoms is the zinc atoms in brass. Zinc atoms, which have a radius of 0.133 nm, have partially replaced some of the copper atoms in brass, which have a radius of 0.128 nm. Interstitial impurity atoms are substantially smaller than atoms in the bulk matrix. The vacant space in the lattice structure between the bulk atoms is occupied by interstitial impurity atoms [3].

Interstitial impurities include, for example, the carbon atoms added to iron to make steel. The bigger iron atoms (0.124 nm) are neatly spaced apart by the 0.071 nm radius of carbon atoms. Where there should be an atom but isn't is called a vacancy. They are widespread, particularly at high temperatures when atoms move randomly and often, creating voids in the lattice. Diffusion, or mass transfer via atomic motion, is typically only possible in the presence of vacancies. A Schottkey flaw occurs when an atom, liberated from a normal location, migrates through a series of steps before settling at the surface of the crystal. This can be achieved in two ways: either an equivalent charged interstitial forms for each vacancy in an ionic crystal, or there are an equal number of cation and anion vacancies. Combinations of anion cation vacancies (in pairs) result in Schottkey defects. A void and interstitial together result in a Frankel defect (Fig. 3).

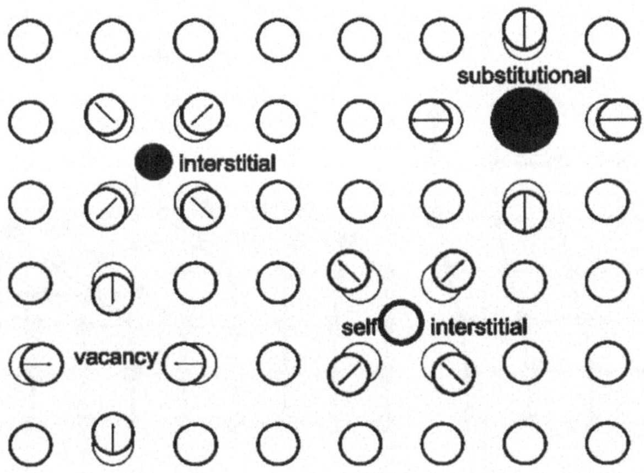

Fig. 2 Various types of point defects

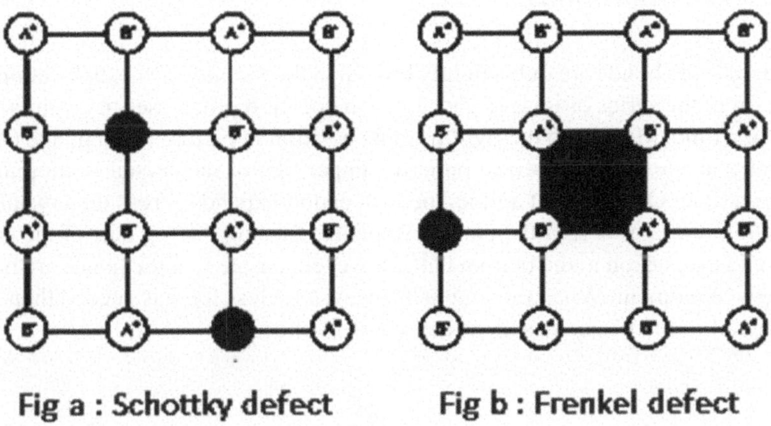

Fig. 3 Images of **a** Schotky and **b** Frankel defect

3.2 Line Imperfections

Atomic groups are positioned irregularly in linear defects. Dislocations are a frequent term for linear flaws. Any deviation from an atomic arrangement along a line that is completely periodic is called a line imperfection. Since the distortion in this instance is limited to a single line, the imperfection may be defined as the separation between two perfectly formed but out-of-register sections of a surface. A dislocation is a line flaw that is located in the slip plane and serves as the border between the slipped and un-slipped area. When stress is applied, dislocations are created and move. Dislocations regulate a metal's ductility and strength. Screw and Edge dislocations are two extreme forms of dislocations that can be distinguished (Fig. 4).

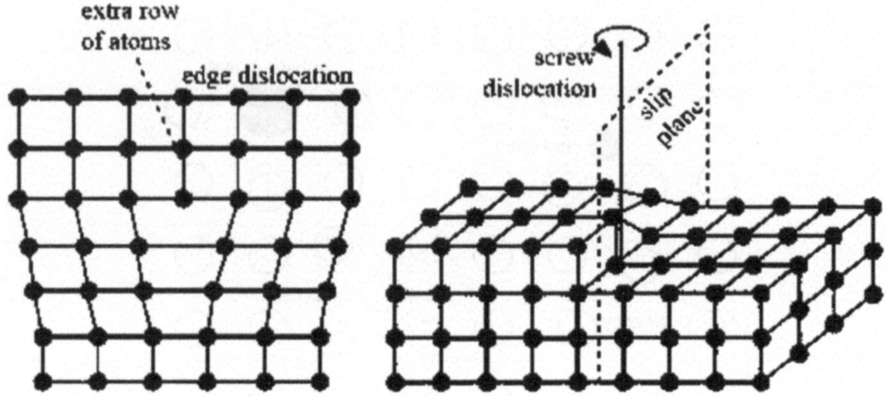

Fig. 4 Illustration of edge and screw dislocation

3.3 Edge Dislocations

The interatomic bonds are only slightly broken in the vicinity of the dislocation line. As shown in the series of images above, the dislocation progresses in a similar way, a little at a time. Moving to the right from its position in picture (a) to image (b) and eventually image (c), the dislocation in the upper half of the crystal is slipping one plane at a time. One plane at a time, the dislocation expands across the crystal. The dislocation migrates across the plane, eventually causing the top half of the crystal to move in proportion to the bottom half. However, very few of the bonds are broken at any given moment. When moving in this way, a lot less force is needed than when breaking all of the ties across the center plane at once [4].

3.4 Screw Dislocations

Visualizing the screw dislocation is a little trickier. Shear force likewise causes a screw to dislocate however, instead of moving parallel to the direction of the stress and atom displacement, this time the defect line moves perpendicularly. To visualize a screw dislocation, consider a block of metal with a shear force applied across one end, causing the metal to begin to rip. The graphic on the upper right shows this. The atom plane just above the rupture is seen in the picture on the lower right. The blue circles indicate the atoms that have not yet shifted from their initial positions. The red circles indicate the locations of the atoms that have restored metallic connections and relocated to new positions within the lattice. It is evident that only a small percentage of the bonds break at any one moment. As with the edge dislocation, moving in this manner requires a great deal less force than breaking all the links across the central plane at once. If the shear force is increased, the atoms will continue to slide to the right. A row of green atoms will return to their proper location in the lattice and turn red, whereas a row of blue atoms will shift and turn green. As a result, screw dislocation in the picture will migrate upward in a direction perpendicular to the stress. Interfaces between homogenous material sections are known as planar faults. Grain boundaries, stacking faults, and exterior surfaces are examples of planar defects.

3.5 Stacking Faults and Twin Boundaries

A rupture in the long-range stacking sequence can give rise to two more common types of crystal defects: a twin region and a stacking fault. A change in the stacking sequence over a small number of atomic spacings causes a stacking defect, while a change over a large number of atomic spacings results in a twin area. An interruption of one or two layers in the atom plane stacking process is called a stacking defect. Although stacking faults may arise in many different types of crystal formations, they are most easily observed in closely packed systems. For instance, it is known from a previous debate that the only way that face centered cubic (fcc) structures and hexagonal close packed (hcp) structures are different is in the order in which they stack. The initial two layers of hcp and fcc structures organize themselves in a similar manner and are referred to as having an AB arrangement. If the third layer is stacked with its atoms directly above those of the first (A) layer, the stacking will be ABA. This is the hcp structure, and it continues ABABABAB. The atoms in the third layer, however, have the ability to align themselves with those in the first layer to form the ABC arrangement of the fcc structure. Therefore, if the hcp structure is functioning as ABABAB and then suddenly switches to ABABABCABAB, there is a stacking error (Fig. 5).

Alternatively, the pattern is ABCABCABC in the fcc arrangement. In a fcc structure, a stacking problem would manifest as the absence of one of the C planes. Stated differently, the pattern would change to ABCABCAB_ABCABC. A stacking error

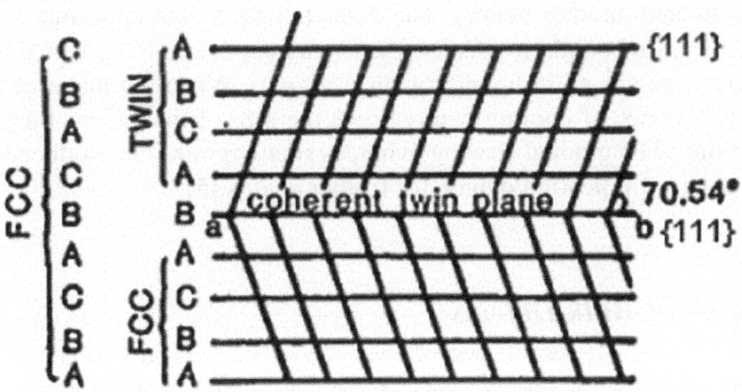

Fig. 5 Stacking faults and twin boundaries

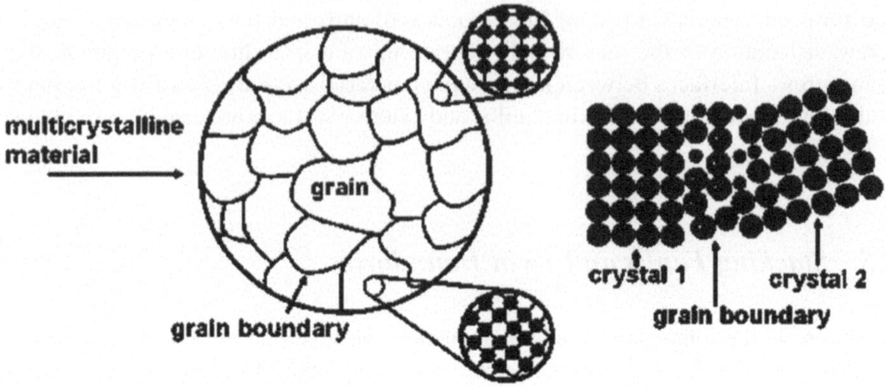

Fig. 6 Grain boundaries in polycrystalline

will create its twin, a second stacking fault, if it does not cure itself right away and persists beyond a certain amount of atomic spacings. Two twin stacking faults are created, for instance, if the stacking pattern is ABABABAB but briefly changes to ABCABCABC before returning to ABABABAB. The twin plane is the red area in the stacking sequence ABCABCACBACBABCABC, and the twin limits are the A planes at each end of the highlighted area.

3.6 Polycrystaline Grain Boundaries

Grain boundary defects are another kind of planer defects. The topic of single crystal flaws has dominated the conversation up to this point. On the other hand, solids are often made up of many grains or crystallites. Grains are typically oriented in a rotational manner with regard to their surrounding grains, and their sizes can vary from nanometers to millimeters wide (Fig. 6) A grain boundary is the point at which one grain ends and another begins. The dislocations' lengths and movements are constrained by grain boundaries. Consequently, a material is strengthened by having smaller grains (greater grain boundary surface area). When the material is cast or heat-treated, the pace of cooling can regulate the size of the grains. Larger grains are often produced by gradual cooling, whereas smallergrains are produced by quick cooling. See the solidification debate for further details [5].

3.7 Volume or Bulk Defects

Compared to other crystal defects covered in this section, bulk faults happen on a far larger scale. Nevertheless, some of the most prevalent bulk faults will be discussed

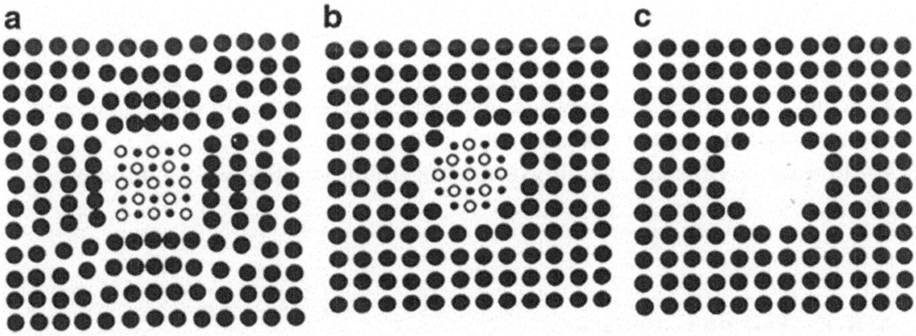

Fig. 7 SEM images of volume defects

for the sake of thoroughness and because they do have an impact on the movement of dislocations. Areas known as voids occur when a significant portion of the lattice's atoms are absent (Fig. 7). A hole in a metal object is shown in the image to the right. An SEM, or scanning electron microscope, was used to get the picture. There are several causes for voids. Porosity is the term used to describe voids that form when a substance freezes as a result of trapped air bubbles. Cavitation is the term for a void created when a substance contracts during the solidification process. When impurity atoms group together to produce tiny patches of a different phase, this is another sort of bulk defect. The region of space filled by a physically homogenous substance is referred to as its "phase." These areas are typically called precipitates or inclusions.

Activity 1 Point Defects

- Figure out how point flaws are taken into consideration when the crystallization process proceeds quickly. The primary cause of these faults is a variation in the way the constituent particles are arranged.
- How the concentrations of point defects react to variations in pressure and temperature, where the distance between sites of repeated growth—which act as sources and sinks for point defects—determines the suitable diffusion length.
- Distingiush between Schotky and Frankel defect, whether they are vacancies or interstetial defects.
- Derive the carrier concentration for Frankel and Shotcky barrier for zero dimensional or point defects.

Activity 2 Line Defect

- What is the difference between line and plane defects, are they same or is there any difference between them?
- How the ductility and ductility of electroceramics is effected by line defects?

Activity 3 Grain Boundaries

- Give the examples of crystal systems with their bravice lattice with examples.
- Ferroelectric domains or grain boundaries explain with neat diagram.

Activity 4 Bulk Defect

- Scanning electron microscope how is it helpful for the defect identification. Write its working principle. Collect SEM images of some electroceramic materials.

4 Defects and Their Role in Materials

There is a growing awareness that flaws are not always negative things. In fact, they have a set of characteristics and affects that, when recognized and managed, may offer a level of utility and control that may be unmatched by any other aspect. The insertion of deliberate flaws with control over their kind, concentration, and placement offers a chance to exploit "good guy" flaws. Defects play a new role in which they may be used as a tool to improve and tune qualities, as well as to enable new functionality. This perspective highlights methodological advancements and scientific instances where faults have been accepted and used to improve material function in order to highlight present work and future prospects for this role redefining.

The function of defects in materials One function of contemporary materials science is to offer a basis on which engineers and scientists from various disciplines might build next-generation technologies to answer the demands of present and future social concerns. Not only is the creation of sophisticated materials with new or improved performance and attributes essential to these advancements, but so is the understanding of how to deterministically synthesis and process these materials to best use their properties. The foundation of materials science is the idea that a material's structure, characteristics, processing, and, ultimately, performance are all closely related. Furthermore, as the subject has developed, materials scientists and engineers have come to understand that failing to take into consideration and deal with the function of material flaws may seriously hinder even our best attempts to manage these principles. Defects are crucial to the evolution of material properties in all areas of materials science, from critical flaws that can drastically reduce a material's strength to the deliberate introduction of desired defects needed to produce modern electronic materials.

From traditional to contemporary: Today's ceramic flaws Chemical alloying is the most well-known and conventional example of using flaws for ceramic property management and improvement. This strategy has typically been employed to enhance qualities by counterbalancing unfavorable impacts of other faults. Chemical alloying, for instance, has long been employed in electroceramics to lessen electrical leakage by balancing charges that are (partially) introduced to lattice by off-stoichiometry or impurities. Direct applications of chemical dopants have also been used to achieve the intended outcomes. For instance, it has been demonstrated that aliovalent substitution of A- and B-site cations in perovskite (ABO3) materials enhances ionic conduction. Although chemical alloying has a long history, it has primarily been directed by chemical intuition and empirical data. Consequently, the resulting "better" performance is often poorly understood or just post-rationalized [6].

Activity 5 Good Guys

- How these defects are play important role in controlling properties of electroceramic materials with examples.

5 Characterization of Defects

Materials science is based on the fundamental concept of characterization, and advancements in this field are revealing hitherto unheard-of details about the behaviors of macroscopic defects, the structures of local and microscopic defects, the kinetics of in situ defects in response to stimuli, and more. The application of both new and ancient techniques created outside of ceramics has resulted in several advancements. Here are some instances of how ceramics are now being affected by cutting-edge methods for flaw characterization.

5.1 *Electrical Characterization*

In addition to the conventional method of fitting current–voltage curves for assessing transport processes in electroceramics, several characterisation approaches based on macroscopic defect response under applied fields have been investigated recently. The following methods are becoming more popular again: Thermally-stimulated depolarization current spectroscopy measures depolarization currents while heating a material at a steady rate to provide information about carrier trap energies, densities, etc., especially for broad bang-gap materials; impedance spectroscopy measures AC impedance as a function of frequency and models it using circuit components to distinguish between grain boundary vs. bulk conduction, ionic versus electronic conduction, and other phenomena. Deep level transient spectroscopy is a transient capacitance method that analyzes carrier trap energies, densities, and captures cross-sections while monitoring the capacitance of a depletion zone under a voltage pulse [7–10].

Activity 6 Electrical Properties

- How the electrical conductivity or resistivity of electro ceramics changes with defects, collect the information from literature.

5.2 *Optical and Magnetic Characterization*

Conventional optical techniques, including as photoluminescence and cathodoluminescence, have been used for years by scientists to gather information on band gap and intra-gap states and to produce light during recombination. Additional contemporary techniques are photoinduced current transient spectroscopy, which, especially for wide band-gap materials, uses light to excite carriers into trap states and measures current transients, thereby providing information about intra-gap traps; and ellipsometry, which measures variations in light polarization as it is reflected from the material, allowing the extraction of the dielectric constant along with information about surface defects (Fig. 8).

Electron magnetic resonances: these use microwaves to move electrons between spin states split by the Zeeman effect under a magnetic field to determine the g factor of a paramagnetic in ion. They are particularly effective at detecting the kind and orientation of defect dipoles (i.e., complexes of charged point defects). Direct imaging and mapping methods now in use provide macroscopic, average fault probes in addition to local, atomic-level characterization [11, 12].

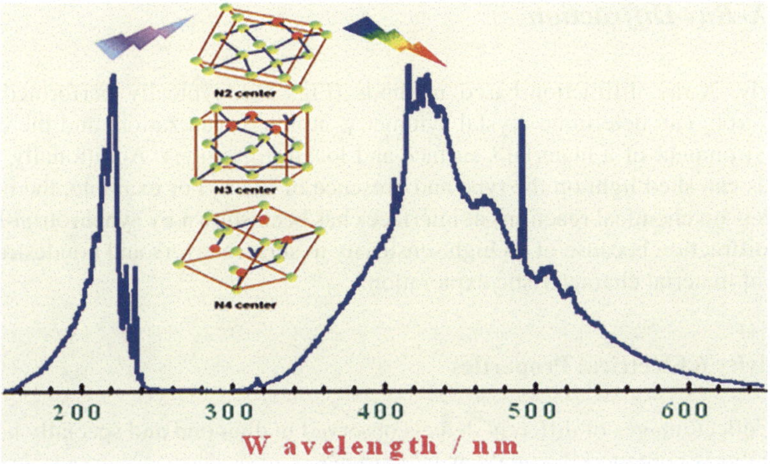

Fig. 8 Absorption spectra of nitrogen defect in diamond

5.3 *Scanning Electron Microscope*

Recent developments in scanning transmission electron microscopy (STEM) and in situ electron energy loss spectroscopy (EELS) have made these methods broadly applicable and very successful for directly examining (and possibly even producing) local defect structures and chemistry. For example, direct mapping of defect migration [13] at the atomic level and surface reconstruction of $SrTiO_3$ single crystals at elevated temperatures have been carried out and shown (Fig. 9).

Fig. 9 SEM images of defect in electroceramic

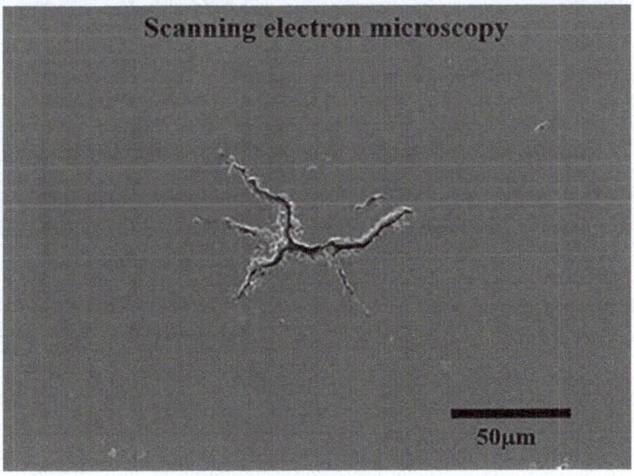

5.4 X-Ray Diffraction

Similarly, X-ray diffraction-based methods (Fig. 10), typically performed in a synchrotron, can determine crystal symmetry, atomic organization, and the chemical environment of a material's surface and localized regions. Additionally, these methods can shed light on the type and presence of flaws. For example, the impact of defects on chemical reactions at interfaces has been shown by synchrotron-based X-ray diffraction because of its high sensitivity to surface layers and nondestructive nature of material characteristic exploration.

Activity 6 Electrical Properties

- Collect images of different defects observed in diamond and specially how color is going to effect defects in diamond.

Defects: Theory, modeling, and computational techniques New insights into functional materials with more complex crystal structures and chemistries have been gained as a result of recent advancements in computer infrastructure, going beyond investigations of conventional metallic compounds or single/binary semiconductor materials. Multiscale and modal approaches may anticipate and comprehend material phenomena like never before, from atomic- to mesoscopic-level structures [14].

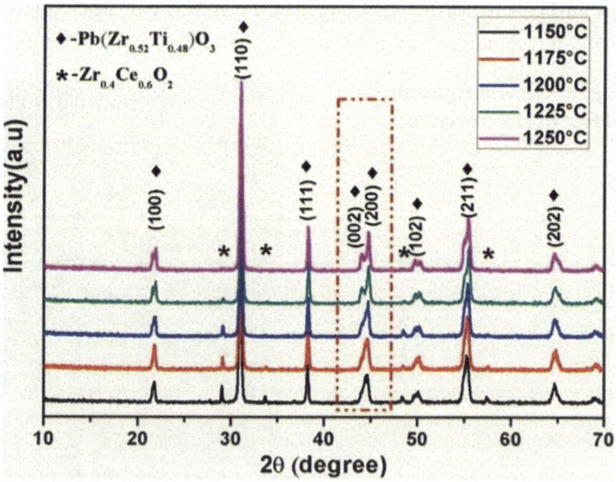

Fig. 10 XRD spectra of lead zirconium titanate for study of defects

Activity 7 XRD Spectra

- XRD images of crystalline and noncystralline electroceramic materials collect them and try to differentiate the peaks or planes with miller indices.

5.5 DFT Studies

Modern techniques and more effective approaches allow us to examine large-scale simulated cells at limited temperatures, which mimic more realistic defect structures, and to perform time-resolved studies of defect evolution under external disturbance. Our understanding of static and dynamic defect patterns, particularly in oxide materials, has increased with the use of density functional theory (DFT), Monte Carlo simulations, phase-field modeling, and molecular dynamics (MD) simulations. Defect-induced electronic/magnetic properties, phase competition, kinetics, stability of defect configurations, etc. can all be obtained at the unit-cell level by DFT simulations. Mesoscopically, pseudopotential developments provide the basis for more and more Monte Carlo and MD simulations, especially when it comes to defect dynamics. In the electrochemical uses of ceramics, for instance, understanding the lattice interactions and ionic migration pathways in materials with various structures (fluorites, perovskites, etc.) has become crucial for developing next-generation solid-oxide fuel cells. Similarly, the creation of exceptionally stable and efficient electrochemical electrode materials has been greatly aided by studies on gas interactions at the solid–vapor interface and the kinetic integration of protons or ions in oxide lattices. MD models also shed light on the dynamic lattice response of ferroelectric materials under electric bias or optical stimulation (Fig. 11). Comparable techniques are being researched to understand defect dynamics [15–20].

Activity 8 DFT Studies by Gaussian

- How Gaussian software is helpful for the DFT studies? Collect some information on Gauss view 6. How to obtain Molecular orbital structure from it and also How it is helpful for spectrosctropic analysis.

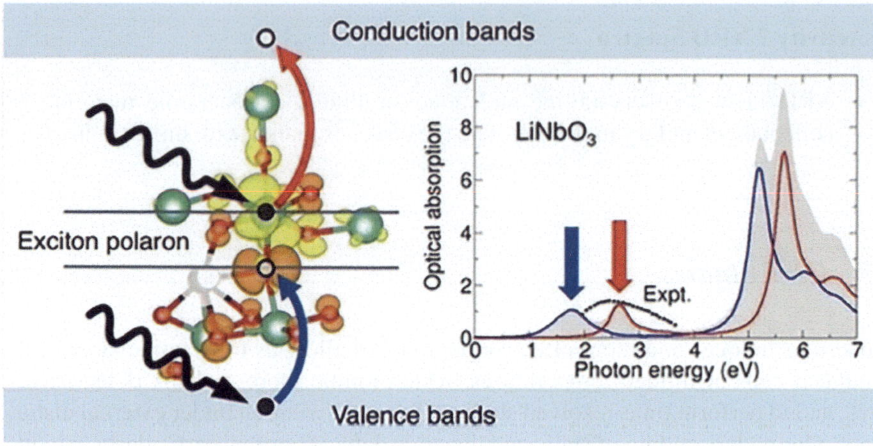

Fig. 11 DFT studies on defects in lithium niobate

6 Defects in Electroceramic Sensors, Thermistors, Ferroelectric Materials

Products in the multibillion dollar electroceramics industry include Mn-Zn ferrites, PZT transducers, SiO_2 optical fibers, $BaTiO_3$, multilayer capacitors, ZnO varistors, and $A1_2O_3$ packages. Roughly speaking, there are six equal market segments (Japan Electronics Almanac 1986). The following technologies are emerging quickly within each market segment: tellurium oxide coatings for video discs, barium hexaferrites for perpendicular recording, buried resistors and capacitors in ceramic packages, PZT piezoelectric motors, and catalytic coatings for chemical sensors [2, 21]. The convergence of several technologies in a synergistic manner is exciting, as is the case with all rapidly expanding fields of science and engineering (Fig. 12).

The most common and least expensive kind of sensors are those with ceramic bodies (Kulwicki 1984). The structure–property correlations in ceramic sensors will be emphasized through the discussion of six different types of sensors: three thermistors and three chemical sensors. The sensors use a variety of electrical phenomena that react to even the smallest changes in chemical composition or temperature.oxygen sensors Superionic conductors are used in automobile engines to measure the mixture of fuel and air (Eddy 1974). One surface of the conductor is in contact with exhaust gas through a porous electrode, and the other surface is in contact with air (Fig. 13). The diffusion of oxygen ions through the solid electrolyte results in a voltage between the electrodes. As the sensor gets hotter, the conductivity level increases. Monitoring the composition of the air–fuel mixture has the dual benefits of lowering air pollution and improving fuel efficiency. Stabilized zirconia is the most often utilized ionic conductor in oxygen sensors. When zirconium is partially replaced with Mg^{2+}, Ca^{2+}, or Y^{3+}, a sizable concentration of oxygen vacancies stabilizes the cubic fluorite structure [22, 23].

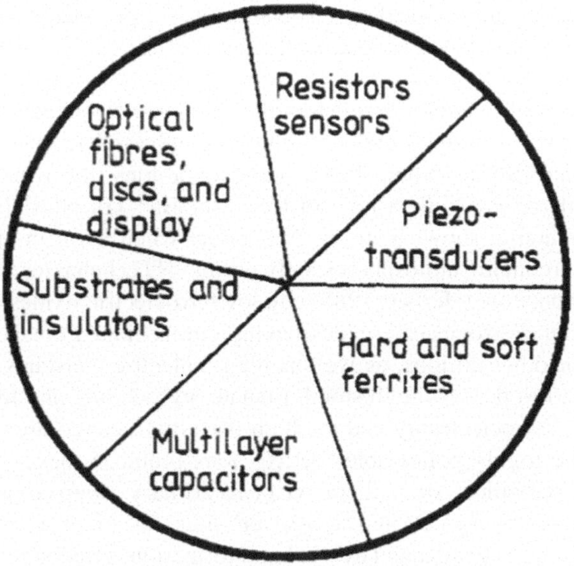

Fig. 12 Electroceramic market and applications over defects

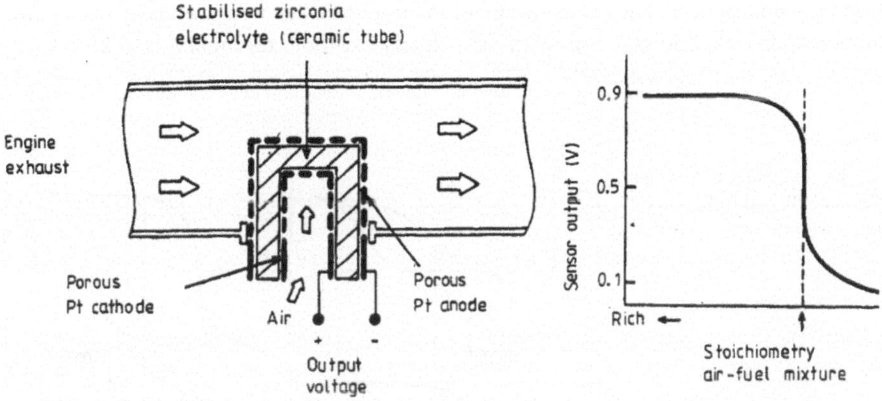

Fig. 13 Instrumentation for oxygen sensor as electro ceramics

According to the structural formula (Zr, -xMgx)(O_{2-x} ~ x), an oxygen vacancy is created for each zirconium that is substituted with magnesium. In the cubic cation sites of the fluorite structure, magnesium takes the place of zirconium, with oxygen vacancies explaining the decreased cation charge. The cations in the fluorite structure form a face-centered cubic lattice, while the anions are organized in a simple cubic arrangement. The anion sites' close proximity (−2.5 A) facilitates oxygens' easy migration to a nearby unoccupied site. This improves ionic conductivity in fluorite-family oxides with fault structure, making them useful as oxygen sensors (McGeehin and Hooper 1977).

The three types of ceramic thermistors that are most frequently used are NTC, PTC, and critical temperature. Ceramics composed of barium titanate are commonly used in PTC thermistors. When doped with donor ions, the resistivity material exhibits a significant PTC effect La^{3+} or Ce^{3+} (for Ba^{2+}) or NbS^{+} (for Ti^{4+}). This low resistivity material shows a strong PTC effect when burnt in air. In ceramics manufactured in reducing atmospheres, only normal NTC behavior is seen [24, 25].

Multilayer capacitors (Herbert 1985) employ ferroelectric oxides (Fig. 14), such as barium titanate. Ferroelectric oxides having corner-linked octahedral networks of Ti^{4+}, Nb^{5+}, or other do ions, as well as high dielectric constants and refractive indices, are found in perovskite, bismuth titanate, pyrochlore, and tungsten bronze layer structures. Ferroelectricity and the high permittivities required for capacitors are driven by the highly polarisable "active" ions found in these transition-metal elements. In the periodic system, there exist two primary groups of active ions, and they are both located near electronic "crossover" sites - locations where the energy of many atomic orbitals is similar and hybrid bond formation is frequent. The first group consists of the three ions Ti $^{4+}$, Nb, and W6', which are octahedrally coordinated to oxygen. A range of molecular orbitals for the (TiO_6)8-complex are formed when the 3d, 4 s, and 4p orbitals of Ti^{4+} electronically crossover with the (Ti and 7 orbitals of its six O_2-neighbors). The bond energy of the complex can be lowered by converting the octahedron to a lower symmetry. This leads to dipole moments, ferroelectricity, and high dielectric constants [26, 27].

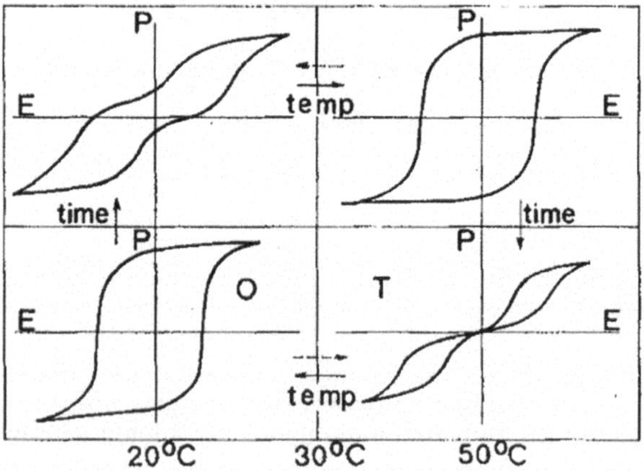

Fig. 14 Ferroelectric oxides with defects and hysteresis curve

Activity 6 Electrical Properties

- What is the significance of Hysteresis loop in ferroelectric materials? Collect the Information of Hysteresis loss in ferroelectric materials due to defects in electroceramics.

7 Conclusions

Every material that we are aware of has some kind of flaw, whether it be composite, metallic, ceramic, or made of polymers. Ceramic flaws, as well as flaws in other materials and their effects We'll learn about some of the most typical ceramic flaws in this blog post, along with the kinds of imperfections that may be found in various ceramic systems. Based on dimensions, there are four different sorts of defects: zero, one, two, and three dimensional, which stand for point, line, surface, and volume defects, in that order. Once more, they are categorized based on the process by which flaws are formed. As a result, the main physical processes were determined after the impacts of defects on the crystal lattice were compiled. Several programs are connected to those issues. Within the fields of condensed matter physics and inorganic chemistry, research on defects has rapidly grown since the beneficial effects of imperfections on electroceramic performance were discovered. This study examined the effects of both internal and external flaws on the properties of materials, with a focus on the dielectric, ferroelectric, and piezoelectric properties. We also covered defect characterization techniques and described the many kinds of faults in electroceramics.

8 Further Reading

For an overview of defects in solid J Kittel book is helpful and for electroceramic material, properties and their applications by A.J. Moulson and J. M. Herbert is devoted entirely to field of electroceramics.

References

1. Saremi, S., Dasgupta, A., Gao, R., Martin, L.W.: New facets for the role of defects in ceramics. Am. Ceram. Soc. Bull. **97** (2018)
2. Nageshwar Rao, G., Umakanthum, K.: Application of electroceramics. Everyman Sci. **33**(3) (1998)

3. Hadi, B.H.: Notes engineering materials lecture 6: defects in solids: point defects and line defects (2016)
4. Nagra, J.S.: Modelling microstructure-property relationships in polycrystalline metals using new fast fourier transform-based crystal plasticity frameworks (2019)
5. Lee, C.-H., et al.: Exploiting dimensionality and defect mitigation to create tunable microwave dielectrics. Nature **502**, 532–536 (2013)
6. Saremi, S., et al.: Enhanced electrical resistivity and properties via ion bombardment of ferroelectric thin films. Adv. Mater. **28**, 10750–10756 (2016)
7. Saremi, S., et al.: Electronic transport and ferroelectric switching in ion-bombarded, defect engineered BiFeO3 thin films. Adv. Mater. Inter. (2017). Accepted
8. Tsur, Y., et al.: Crystal and defect chemistry of rare earth cations in BaTiO3. J. Electroceram. **7**, 25–34 (2001)
9. Erhart, P., Eichel, R.-A., Träskelin, P., Albe, K.: Association of oxygen vacancies with impurity metal ions in lead titanate. Phys. Rev. B: Condens. Matter Mater. Phys. **76**, 174116 (2007)
10. Shimada, T., Ueda, T., Wang, J., Kitamura, T.: Hybrid hartree fock density functional study of charged point defects in ferroelectric PbTiO3. Phys. Rev. B: Condens. Matter Mater. Phys. **87**, 174111 (2013)
11. Yao, Y., Fu, H.: Charged vacancies in ferroelectric PbTiO3: formation energies, optimal fermi region, and influence on local polarization. Phys. Rev. B: Condens. Matter Mater. Phys. **84**, 064112 (2011)
12. Li, F., Zhang, S.J., Yang, T.N., Xu, Z., Zhang, N., Liu, G., Wang, J.J., Wang, J.L., Cheng, Z.X., Ye, Z.G., et al.: The origin of ultrahigh piezoelectricity in relaxor-ferroelectric solid solution crystals. Nat. Commun. **7**, 13807 (2016)
13. Sun, E.W., Cao, W.W.: Relaxor-based ferroelectric single crystals: growth, domain engineering, characterization and applications. Prog. Mater. Sci. **65**, 124–210 (2014)
14. Zheng, T., Wu, J.G., Xiao, D.Q., Zhu, J.G.: Recent development in lead-free perovskite piezoelectric bulk materials. Prog. Mater. Sci. **98**, 552–624 (2018)
15. Gerson, R.: Variation in ferroelectric characteristics of lead zirconate titanate ceramics due to minor chemical modifications. J. Appl. Phys. **31**, 188–194 (1960)
16. Zhu, J.-H., Dai, J.-Q., Xu, J.-W., Li, X.-Y.: Effect of Zn and Ti Co doping on structure and electrical properties of BiFeO3 ceramics. Ceram. Int. **44**, 9215–9922 (2018)
17. Lu, H.C., Lin, M.Y., Chou, S.L., Peng, Y.C., Lo, J.I., Cheng, B.M.: Identification of nitrogen defects in diamond with photoluminescence excited in the 160–240 nm region. Anal. Chem. **84**(21), 9596–9600 (2012)
18. Schmidt, F., Kozub, A.L., Gerstmann, U., Schmidt, W.G., Schindlmayr, A.: A density-functional theory study of hole and defect-bound exciton polarons in lithium niobate. Crystals **12**(11), 1586 (2022)
19. Chen, Y., Xie, S., Wang, Q., Fu, L., Nie, R., Zhu, J.: Correlation between microstructural evolutions and electrical/mechanical behaviors in Nb/Ce co-doped Pb (Zr0. 52Ti0. 48) O3 ceramics at different sintering temperatures. Mater. Res. Bull. **94**, 174–182 (2017)
20. Chouprik, A., Negrov, D., Tsymbal, E.Y., Zenkevich, A.: Defects in ferroelectric HfO 2. Nanoscale **13**(27), 11635–11678 (2021)
21. Chandrasekaran, A., Damjanovic, D., Setter, N., Marzari, N.: Defect ordering and defect−domain-wall interactions in PbTiO3: a first-principles study. Phys. Rev. B: Condens. Matter Mater. Phys. **88**, 214116 (2013)
22. Garipov, R.R., Spaeth, J.M., Keeble, D.J.: Measurement of Cu(II) copper defect dipole in ferroelectric PbTiO3 using electron nuclear double resonance. Phys. Rev. Lett. **101**, 247604 (2008)
23. Ge, F.F., Wu, W.D., Wang, X.M., Wang, H.P., Dai, Y., Wang, H.B., Shen, J.: The first-principle calculation of structures and defect energies in tetragonal PbTiO3. Phys. B **404**, 3814–3818 (2009)
24. Ge, F.-F., Wu, W.-D., Cao, L.-H., Wang, X.-M., Wang, H.-P., Dai, Y., Wang, H.-B., Shen, J.: The structure and defect formation energy in tetragonal PbTiO3: ab initio calculation. Ferroelectrics **401**, 154–160 (2010)

25. Keeble, D.J., Loyo-Menoyo, M., Booq, Z.I.Y., Garipov, R.R., Eremkin, V.V., Smotrakov, V.: Fe3+ defect dipole centers in ferroelectric PbTiO3 studied using electron paramagnetic resonance. Phys. Rev. B: Condens. Matter Mater. Phys. **80**, 014101 (2009)
26. Mackie, R.A., Peláiz-Barranco, A., Keeble, D.J.: Vacancy defects in PbTiO3 and lanthanide-ion-doped PbTiO3: a study of positron lifetimes. Phys. Rev. B: Condens. Matter Mater. Phys. **82**, 024113 (2010)
27. Gong, J.J., Li, C.F., Zhang, Y., Li, Y.Q., Zheng, S.H., Yang, K.L., Huang, R.S., Lin, L., Yan, Z.B., Liu, J.M.: Interactions of charged domain walls and oxygen vacancies in $BaTiO_3$: a first-principles study. Mater. Today Phys. **6**, 9–21 (2018)

Preparation of Ceramics: Different Approaches

Gobinath Velu Kaliyannan, Rajasekar Rathanasamy, Raja Gunasekaran, Santhosh Sivaraj, Suganeswaran Kandasamy, and Subha Krishna Rao

Abstract Ceramics were the most prominent materials, holding higher strength and electrical resistivity even under elevated temperatures. Ceramics, with their unique combination of mechanical, thermal, and electrical properties, find applications across numerous industries, ranging from traditional pottery to cutting-edge electronic devices. Ceramics are in worldwide need due to their versatility and potential use in many other industries, including electronics, medicine, space, sports, etc. Numerous ceramic substances and preparation techniques are available and tailored for certain applications. The properties, including corrosion resistance, remarkable optical and electrical characteristics, hardness, and anti-aging, are influenced by material selection, preparation methods, and processing conditions. The preparation of ceramics for various applications includes selecting raw materials, mixing with binder materials, forming, applying heat, sintering, firing, and final processing. Due to their resistance to oxidation, ceramics were used in moist environments. The strength of ceramics was due to the covalent or ionic bonding between their atoms. This chapter deals with the preparation of ceramic materials using various approaches, encompassing traditional techniques such as slip casting, Tape casting, and sintering, as well as contemporary methods like sol–gel synthesis, micro stereolithography, and 3D printing based on electrical, magnetic, optical, and structural applications.

G. Velu Kaliyannan (✉) · R. Rathanasamy · S. Kandasamy
Department of Mechatronics Engineering, Kongu Engineering College, Erode 638060, Tamil Nadu, India
e-mail: gobinath.v.k@gmail.com

R. Gunasekaran
Department of Mechanical Engineering, Velalar College of Engineering and Technology, Erode 638012, Tamil Nadu, India

S. Sivaraj
Department of Robotics and Automation, Easwari Engineering College, Chennai 600089, India

S. Krishna Rao
Centre for Nanoscience and Nanotechnology, International Research Centre, Satyabama Institute of Science and Technology, Chennai 600119, India

U. Kumar (ed.), *Defects Engineering in Electroceramics for Energy Applications*, Engineering Materials, https://doi.org/10.1007/978-981-97-9018-0_3

Keywords Advanced ceramics · Corrosion resistance · Sol–gel · Sintering · Slip casting · Tape casting · Micro stereolithography · 3D printing

1 Introduction

The word "ceramics" originates from the ancient Greek word "keramicos," which refers to things that have been burned. They are made from naturally existing raw materials and have been known for millennia. C ceramics are inorganic, non-metallic solids typically made from powdered materials with generally high melting points that need high temperatures for production and utilization [1]. Ceramics include various materials, such as metal oxides, borides, carbides, nitrides, and various combinations of these elements [2]. In a broader context, the word "ceramics" is associated with pottery, which signifies a type of substance, brittleness, which is its defining feature, and frequent utilization in the domestic and industrial sectors. In context, the word "ceramics" is significantly more general in its material definition, encompassing a diverse range of predominantly polycrystalline inorganic non-metallic substances. These substances are frequently machined into engineered forms using granules and ambient temperature. Their distinctive physical characteristics are obtained by a high-temperature burning method. Additionally, it comprises inorganic cementitious substances such as lime, cement, and gypsum, which attain their ultimate characteristics without burning [3].

Ceramics represent a diverse material class that has played a significant role in human civilization for millennia. These ceramics may be broadly classified into two main categories: advanced and traditional. Both have distinctive characteristics and functions. With its ancient origins, conventional ceramics includes rich clayware, clay-based pots (earthenware, stoneware), porcelain, glass, cement, and abrasives for decorative and functional uses. The primary constituents of conventional ceramics are inorganic, non-metallic substances such as feldspar, silica, and clay. Whereas silica offers excellent stability at extreme temperatures and enhanced melting points, clay offers stiffness and flexibility. The glass state is formed when ceramics undergo the firing process, with feldspar playing a significant role in this transformation. Advanced ceramics, conversely, are produced from a range of substances that include oxides, nitrate elements, carbides, and non-silicate glasses.

Conventional ceramics can be made from various materials and can occur in either amorphous or solid forms. Ceramic materials frequently exhibit porosity across several size scales, ranging from nano to macro levels, which can significantly impact their inherent characteristics [4]. Traditional ceramics have been utilized since prehistory. It has been utilized for different uses throughout human history, including the manufacturing of floor and ceiling tiles, hygiene products, glass, porcelain, tableware, and concrete.

Conventional ceramics are predominantly made up of silicate and aluminosilicate substances that occur naturally; they consist mainly of oxygen, silicon, and aluminium. The production of conventional ceramics started with the development of ceramic mixtures and concludes with the sintering process. Calcium, magnesium, sodium, and potassium are some of the other elements that could be included in these compounds. The utilization of extremely standardized and uniform raw materials, along with massive automation-enabled production, provides constant efficiency and high-quality products. Constant research and development activities have resulted in advancements within the field of conventional ceramics.

The utilization of computer-aided design and manufacturing (CAD/CAM) has facilitated the advancement of layouts and the development of product variations. Because of these advancements, conventional ceramics are able to compete with non-ceramic substances in a wide range of creative, attractive, and practical uses. Moreover, conventional ceramics are no longer obsolete or rigid. Continuous progress has been made in a wide range of areas, including fundamental components, equipment, processes, combustion systems, and ornamentation, which has enhanced their productivity and viability. The aforementioned developments have resulted in lower energy and labor expenses, higher efficiency, higher standards, and reduced manufacturing costs. Furthermore, these developments have also contributed to a substantial decrease in environmental effect [5].

In contrast, advanced ceramics serve as an embodiment of contemporary material science and engineering, exhibiting a broad variety of specialized characteristics that are precisely suited to the requirements of high-performance applications. Advanced ceramics are a distinct category of ceramic materials that are produced by synthetic substances of exceptional purity. In terms of chemical composition, these ceramics consist of oxides, nitrides, and carbides; diamond and graphite can be included within the same group. In order to facilitate the forming process, organic binders are frequently used.

Advanced ceramics are often divided into oxide and non-oxide categories according to their constituents. Binary oxides, aluminates, ferrites, zirconates, and more are included in the category of oxide ceramics. Carbon-based materials, carbides, nitrides, and borides are examples of non-oxide ceramics. This broad category also includes ceramic composites composed of both oxides and non-oxides.

Advanced ceramics initially emerged to maximize an individual fundamental feature for a particular use. Technological advancements, however, have produced modern ceramics with multipurpose characteristics that may combine features that were previously thought to be incompatible. Traditionally, obtaining versatility has required limitations, either by reducing a particular feature to improve another or by creating composite arrangements. The development of ceramic materials with unique features continues for a range of functions, including structural, functional, and biologic uses, as ceramic parts continue to decrease in size. In research, these ceramics are frequently denoted as "fine," "unique," "technical," or "outstanding performance."

On the basis of their primary usage, advanced ceramics may be classified as functional ceramics, structural or engineering ceramics, or nuclear ceramics.

Structural Ceramics: These ceramics are well known for having exceptional mechanical characteristics, including resilience to wear, hardness, and toughness, particularly in abrasive conditions and at elevated temperatures. They found utility in a variety of sectors, especially dental restoration, and are frequently substituted for conventional substances like metals.

Functional Ceramics: Functional ceramics are distinguished by their distinctive functionalities, which are attained by deliberate manipulation of the materials and characteristics. A wide variety of features may be seen in these materials, encompassing electrical insulation, semiconductor behavior, superconductivity, as well as ceramic characteristics such as a dielectric, ferroelectric, piezoelectric, thermoelectric, and multiferroic capabilities. Although the current application of dental implants in dental restoration is restricted, there is potential for their future utilization in the field of biting force monitoring.

Nuclear Ceramics: Uranium oxide and nitride are utilized as nuclear energy sources, constituting a specific category within the realm of advanced ceramics sometimes referred to as nuclear ceramics [3].

Advanced ceramics are crystalline compounds whose constituents are thoroughly regulated. When manufacturing this ceramics, precise standards are followed, and the raw materials used are thoroughly purified and categorized in order to attain particular properties. These substances may consist of borides, oxides, carbides, nitrides, silicides, and even more [1]. This type of material encompasses several kinds of ceramics, such as structural, electro and bio-ceramics. These ceramics play a crucial role in driving innovation in areas such as aerospace, electronics, healthcare, and other specialized industries. Electrical and magnetic ceramics are of greatest significance in an extensive range of electronic and magnetic applications, encompassing semiconductors as well as numerous carbon compounds. These materials are renowned for their remarkable characteristics, including outstanding durability, resistance to heat, conductive properties, and compatibility with biological systems, rendering them important in numerous revolutionary sectors.

Ceramics, encompassing both conventional and advanced forms, are widely recognized for their diverse characteristics offering them important material across several sectors. Several important characteristics of these materials include their great mechanical durability and hardness, exceptional thermal and chemical resistant properties, as well as notable thermal, optical, electrical, and magnetic functions. Despite possessing great durability and hardness, ceramics have a noticeable brittleness and are susceptible to experiencing major damage even under little flexural stress. Ceramics exhibit contrasting mechanical properties when compared to metals. While metals possess non-brittle and ductile characteristics as a result of their metallic connections, ceramics demonstrate strength under compression but weakness under tension. The brittle behaviour exhibited by ceramics can be attributed to the structure of interatomic bonding, which represents in direct contrast to the elastic and flexible behaviour observed in materials. The degree of transparency exhibited by ceramics can exhibit considerable variation, spanning from high translucency to high opaque. The visibility of ceramics is often influenced by the microstructure of the material. These structures with a higher degree of glassiness often exhibit more translucency, whereas microstructures with a higher level of crystallinity typically display increased transparency. The total translucency of ceramic substances is influenced by several parameters, such as size, density, refractive index, and porosity of the particle [6].

Ceramics are distinct from other materials due to their many intriguing characteristics. These consist of minimal temperature and electrical conductivity, resistance, chemical inertness, and economics [7]. Ceramics also provide a large specific surface area, low weight, superior shock and heat resistance, and thermal insulation. Long operating cycles usually do not cause them to distort, which lowers the cost of maintenance [8]. Furthermore, ceramics demonstrate exceptional stability when exposed to a variety of solvent-based substances and, owing to their adaptable porous structure and chemical inertness, do not contribute adversely to the marine environment. Ceramics are utilized in numerous fields, including health care, chemical processing, equipment, electronic devices, and aviation, due to their numerous exceptional properties.

Ceramics are produced by a range of techniques, such as sintering, firing, and extreme temperature processing. These procedures frequently include subjecting the raw materials to elevated temperatures, so this can facilitate their fusion and subsequent solidification that transforms them into an inorganic substance. The selection of a methodology is determined by the particular usage and intended characteristics of the ceramic substance. Ceramic parts are typically fabricated by shaping a combination of powders, either with or without adhesives and other substances, by standard techniques such as die pressing, injection moulding, gel and tape casting, in order to get the required structures [9].

Section 1 provides an introduction to ceramics, offering an overview of ceramics, including their classification, properties, applications and preparations showcasing their significance in numerous industries. Subsequently, Sect. 2 explores the preparation of ceramics through various methods, including Roll forming and Tape casting, micro-stereolithography technology, cold sintering, spark plasma sintering, digital light processing (DLP) 3D printing technology, low-pressure sintering, freeze drying, and low-temperature sintering. Finally, the chapter ends with a conclusion (Sect. 3).

2 Preparation of Ceramics

2.1 Preparation of Laminated B4C–TiB_2/BN Ceramics by Roll Forming and Tape Casting

Boron carbide (B4C) has emerged as a prominent challenger for several technical uses [10]. The substance exhibits exceptional physical and chemical characteristics, including an exceptionally low density of 2.52 g/cm^3, an outstanding hardness, and a high melting point of 2427 °C [11]. Due to its exceptional characteristics, B4C ceramics exhibit significant promise for a wide range of structural usage under ambient conditions. Boron carbide (B4C) has numerous applications, serving as an armor material, a neutron absorbent, a finishing medium, and as a constituent for wear-resistant equipment [12–16]. Various techniques are employed to acquire separate microlayers, such as electrophoretic deposition, roll formation, tape casting, and slip casting.

Tape casting permits the design of materials' microstructures and macrostructures. An additional benefit is that a cast tape's thickness may range considerably from 1 to 3000 μm [17]. Hexagonal boron nitride (h-BN) is sometimes referred to as white graphite due to its structural resemblance to graphite because of its layered atomic arrangement [18]. This approach involves a unique combination of two molding techniques to overcome the limitations of each individual process, resulting in a significant improvement of the interfacial bonding. The preparation process involved the development of B4C–TiB_2 and h-BN by tape casting and roll forming methods, press sintering to produce laminated B4C–TiB_2/BN ceramics. The fracture toughness of laminated B4C-TiB_2/BN ceramics was found to be considerably greater compared to composite B4C, demonstrating the presence of a laminar toughening effect.

Figure 1 depicts the method of producing laminated ceramics. The procedure involved the ball-milling of a powder combination consisting of 75 wt% B4C, 15 wt% TiB2, and 10 wt% $Mg_2Al_4Si_5O_{18}$ (MAS) in an ethanol solution for a duration of 24 h. The mesoporous aluminum silicate (MAS) was synthesized by sol–gel technique. In order to achieve a homogeneous dough, a binder, namely polyvinyl alcohol, and a plasticizing substance specifically glycerin, were included into the combined powder. Subsequently, a rolling compaction equipment was employed in order to generate green sheets. The residual water was removed by the process of oven drying in order to acquire green sheets composed of B_4C-TiB_2.

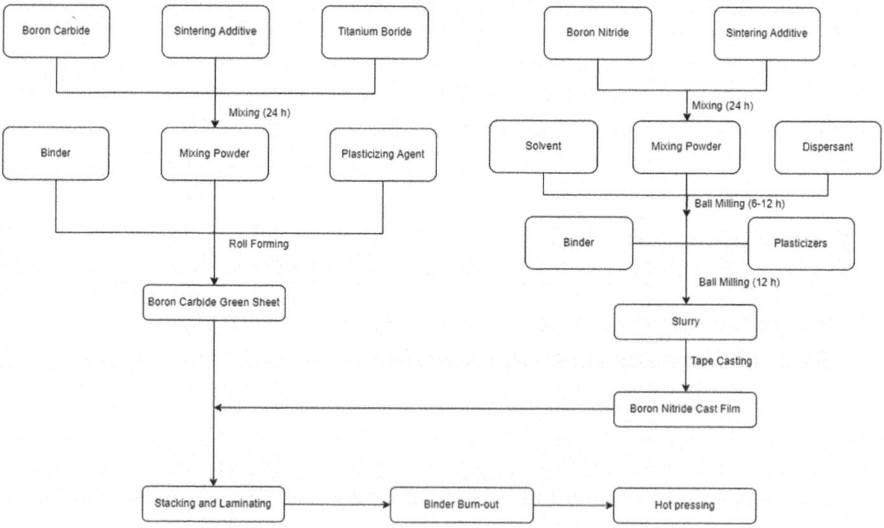

Fig. 1 Flow chart of the fabrication process of the laminated B4C–TiB2/BN ceramics [19]

Table 1 The composition of the fundamental elements in the cast slurry [19]

Role	Additive	Content/%
Solvent	Anhydrous ethanol	32
Dispersant	Triethyl phosphate	30
Binder	Polyvinyl butyral	4
Dispersant	Dibutyl phthalate	4

The powder mixture comprising of hexagonal boron nitride (h-BN) and magnesium aluminium silicate (MAS) was subjected to ball milling in an ethanol medium for a duration of 24 h. The h-BN powder, which had been sieved, was subjected to ball milling using a planetary ball mill for a duration of 6–12 h. The ball milling process was carried out using anhydrous ethanol solvent and a dispersing agent, with low energy levels. Following homogeneous dispersion, sufficient volumes of adhesive and plasticizer were included and the resulting mixture was ball-milled at low energy using a planetary ball mill for the duration of 12 h.

(Table 1) illustrates the additive quantities. Subsequently, the process of tape casting was carried out, whereby the cast scraper was adjusted to a height of 180 μm, resulting in the production of h-BN green sheets. These sheets undergo lamination and pressing at 100 MPa. After that, the binder burn-out procedure was carried out in an air-filled muffle furnace, following the procedure represented in Fig. 2. Ultimately, the provided sample underwent the process of sintering by hot pressing at a temperature of 1850 °C for a duration of 1.5 h, while being subjected to a pressure of 35 MPa within an environment consisting of argon gas.

The overall density is decreased from 2.68 to 2.56 g/cm^3. The drop in bulk density is because of the lack of strong bonding at the point of contact between B_4C–TiB_2 and h-BN, which becomes more prominent with an increase in the number of interfaces. Furthermore, B_4C–TiB_2 has a greater theoretical density of 2.71 g/cm^3 than

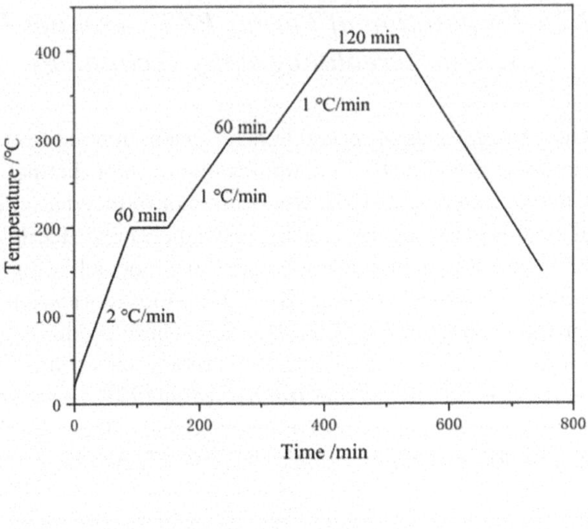

Fig. 2 Pyrolysis process curve [19]

h-BN, which has a density of 2.28 g/cm^3. In general, the laminated B_4C–TiB_2/BN composite exhibited a significantly elevated bulk density, due to the incorporation of MAS sintering aids and TiB_2, which facilitated the densification process during sintering. The laminated B_4C–TiB_2/BN exhibited enhanced durability and hardness as the density increased. Cracks were more easily deflected by the multilayer structure. Nevertheless, the h-BN ceramic exhibited mechanical characteristics that were comparatively lower than B_4C–TiB_2 ceramic. Additionally, the incorporation of the h-BN layer resulted in the generation of additional flaws within the ceramic materials, thereby causing a decrease in its flexibility.

The potential for use of B_4C ceramics is currently limited owing to their extreme brittleness. However, the fracture toughness of laminated B_4C–TiB_2/BN ceramics, specifically those with 7 h-BN layers, exhibited a notable increase, reaching a value of 9.01 ± 0.61 MPa m$^{1/2}$. This represents a considerable enhancement in fracture toughness over B_4C ceramic. The substantial increase in durability can be primarily caused by two aspects. TiB_2 enhanced the toughness of the B_4C matrix layer, while the fracture toughness of the ceramic substance was compounded by the distinctive laminated arrangement. Both boron carbide (B_4C) and titanium diboride (TiB_2) are examples of covalently bonded solids that exhibit exceptional strength and hardness properties. Even so, it has been shown that Titanium Diboride (TiB_2) possesses the ability to function as a suppressor of grain development. This suppression occurs by impeding the formation of Boron Carbide (B_4C) grains through a process known as pegging. Consequently, this leads to enhanced densities in B_4C ceramics, ultimately resulting in improved mechanical characteristics. The multilayered structure of the laminated B_4C–TiB_2/BN ceramic was primarily responsible for the remarkable toughness exhibited by the material. The significant enhancement in fracture toughness can be due to several toughening processes, including as crack deviation, crack bifurcation, and delamination.

2.2 Preparation of Porous PZT Ceramics Using Micro-Stereolithography Technology

The current focus of researchers has been directed towards low-density Porous Lead zirconate titanate (PZT) ceramics due to their distinct benefits. One example is the utilization of porous PZT, where air is introduced as a secondary phase. The addition of air aids to reduce the acoustic resistance, hence facilitating the acoustic matching of PZT with low-acoustic-impedance substances like lightweight metallic substances, biological tissues, or water [20–22]. Various fabrication techniques are now employed for the production of PZT ceramics. These methods encompass polymeric sponge approaches [23], gel-casting [24], replamine-form [25] and solid free-form (SFF) fabrication [26], as well as freeze casting [27]. The burnable plastic sphere (BURPS) approach has emerged as a unique and extensively employed method for the preparation of porous ceramics due to its cost-effectiveness and straightforward application

[28–30]. Pore-forming agents (PFAs) are added into ceramic particles which then undergo sintering to cause the formation of pores inside the ceramic matrix. The above methods often required the usage of molds or relied heavily on inorganic materials (gel-casting methods), which made the process more complicated and restricted the capability of the technology to only produce objects with basic shapes. Additionally, due to the brittle nature of ceramics, it becomes harder to produce complicated structures using conventional machining processes. This is because ceramics tend to crack easily.

On the other hand, additive manufacturing (AM) enables the direct production of complicated physical objects of different forms and sizes by employing 3D printing, CAD and layer-by-layer stacking methods [31, 32]. Micro-stereolithography (μSL) is regarded as an effective additive manufacturing (AM) technique for the production of ceramic parts. This method of three-dimensional printing(3D) employs a photocurable resin instead of high-energy laser beams. The objects are acquired with an exceptional degree of precision and an appropriate surface finish, while also exhibiting a high manufacturing speed [33–35]. Micro-stereolithography (μSL) is an easy method utilized for the production of complicated three-dimensional (3D) components suitable for implementation in piezoelectric usages. This process is primarily used for producing dense PZT ceramics with complicated objects [34–37]. However, the combination of Micro-stereolithography (μSL) with the BURPS process demonstrates possibility in the production of complicated and diverse porous PZT ceramics. This method is particularly suitable for applications requiring low-acoustic-impedance materials. In this experimental procedure, Micro-stereolithography (μSL) and BURPS technique was utilized, in order to produce porous PZT ceramics. Polymethyl methacrylate was employed for the production of spherical pores.

2.2.1 The Process of Preparing Ceramic Suspensions

The raw material utilized in this process was PZT-5H, which was in the form of powder. The powder had an average particle size of 1.5 μm and a density of 7.6 g/cm^3. The photosensitive resin employed in this process was trimethylpropane triacrylate (TMPTA), which comprised 50% by weight of the PZT powder. To attain a homogeneous solution, Triton X-100 is introduced, at a concentration of 1 weight percent equivalent to the PZT powder, and it was employed as a dispersing agent to decrease the viscosity. Diphenyl (2, 4, 6-trimethyl benzoyl) phosphine oxide was utilized as the photo-initiator due to its known characteristics that it possesses a high level of photoactivity. The mass fractions of Polymethyl methacrylate in PZT ceramics for the mixed particles were as follows: 5, 10, 15, 20, 30, and 35% by weight (wt%). Zirconia balls were used to grind all of the powders in TMPTA for the duration of two hours. Following the milling process, the suspensions undergo evacuation in order to take out any interior bubbles.

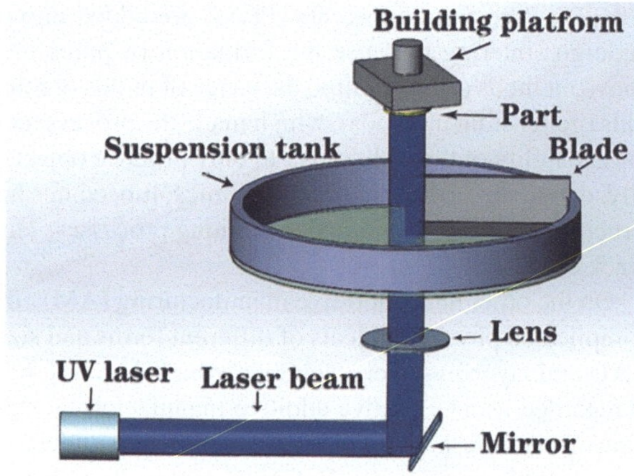

Fig. 3 Schematic of the μSL technique [38]

2.2.2 The Production of Green Parts

Computer-aided design (CAD) software is utilized for the construction of three-dimensional models of the green body. Figure 3 displays a diagrammatic representation of the μSL technology. The suspensions of this ceramic were evenly distributed over the resin container using a doctor blade in order to get a consistent and thin coating. The software of micro-scale stereolithography (μSL) equipment automatically slices the three-dimensional model into several single-layer slices with a thickness of 20 μm. This slicing process is designed to enhance the bonding between the cured layers. By emitting ultraviolet (UV) light from the translucent resin container's base, the ultraviolet (UV) lamp delineated the model's structure. The suspensions were subjected to Ultra Violet radiation, resulting in the cross-linking polymerization process. This led to the formation of a solid layer of the ceramics. As a result, disk-shaped green bodies of 15 mm in diameter and 1.5 mm in thickness were acquired.

2.2.3 The Process of Debinding of PZT Green Components

Following the printing of the PZT green bodies, organic matter is extracted via low-temperature thermal debinding procedure and the PZT ceramics are densified via high-temperature sintering. To prevent the crack occurrence in this ceramics, thermal debinding method was implemented in a two-step approach. The initial stage of this process was conducted under an argon environment to reduce the formation of flaws [36]. Each minute, the temperature was increased to 1 degree Celsius. A series of temperatures, that is, 100, 200, 300, 350, 400, 500, and 600 °C, were selected in order to facilitate the complete thermal degradation and evaporation of organic substances. These temperatures were accompanied by a dwell duration of 2 h. The majority of organic additives undergo conversion into carbon during the first stage of this process.

In order to eliminate any excess carbon and binders, these bodies were exposed to air at 800 °C for three hours during the second debinding stage. Following the debinding process, the PZT ceramics underwent sintering at a temperature of 1200 °C for a duration of 3 h in an ambient air atmosphere. To prevent any loss of lead in sintering, the samples are immersed in PZT powder and enclosed within an alumina crucible.

The hysteresis loops of the ceramics had a substantially "squared" form, indicating that these ceramics possess a high level of ferroelectricity that is equivalent to that found in some dense piezoelectric ceramics. The PZT ceramics showed a dielectric constant of around 2417 at a frequency of 1 kHz, demonstrating their ability to store electrical energy. Additionally, the dielectric loss of about 0.025 indicates low energy dissipation, causing these ceramics suitable for piezoelectric usage in acoustic equipment. At 1 MHz, the dielectric constant and dielectric loss exhibited a corresponding modification to 2135 and 0.04 as the frequency ascended. The degradation of dielectric characteristics observed at extreme frequencies has been suggested to the pore formation. The conclusion suggests that the porous PZT ceramics produced by μSL had a significant dielectric constant and minimal loss throughout a broad spectrum of frequencies. The dielectric constant generally exhibits a drop as porosity increases.

2.3 *Preparation of High-Density Bi_2O_3 Ceramics by Cold Sintering*

The cold sintering process (CSP) is a novel technique for sintering ceramics, which has the capability to produce ceramics with high density even at processing temperatures below 300 °C [39–41]. The primary use of CSP involves the utilization of a co-solvent solution to facilitate the sintering process of ceramics. "Cold sintering" (CSP) was officially named in 2016 and was first used to make Li_2MoO_4 ceramics in 2014. Cold sintering process (CSP) has received considerable interest due to its shown benefits, including its cost-effectiveness, minimal consumption of energy, and its versatility to deal with various materials like polymer. The application of CSP technology has demonstrated outstanding results in a wide range of materials, encompassing conventional oxide ceramics [42], ceramic/ceramic composites [43], ceramic with microwave dielectric property [44], ceramic with ferroelectric and piezoelectric property [45], solid electrolytes [46], thermoelectric materials [47], and nanomaterial composites [48].

Ceramics composed of Bi_2O_3, which exhibit desirable dielectric characteristics, hold considerable significance within the context of wireless communication. Bi_2O_3 has low solubility in water and alkali solutions, making it insoluble in these solvents. However, it exhibits solubility in acetic acid, making it a suitable choice as a co-solvent for the cold sintering of Bi_2O_3 ceramics. During the preparation process, Bi_2O_3 ceramics with a high density of around 99.13% were subjected to sintering at

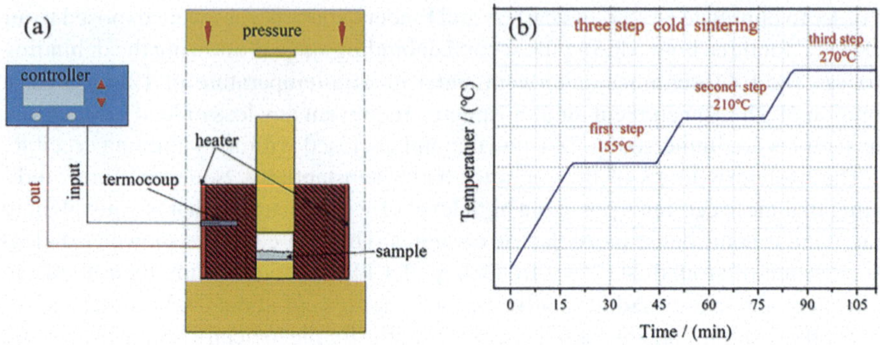

Fig. 4 **a** An illustration of the CSP mold in schematic form, **b** The Bi_2O_3 ceramics' TS-CSP sintering curve [49]

a temperature of 270 °C and a pressure of 300 MPa using a three-step CSP (TS-CSP) method.

2.3.1 Experimental Methodology

In order to conduct the ceramic CSP experiment, a partially sealed mold and a heating jacketed band heater is required. The mold possesses the capacity to apply uniaxial pressure, while the heater enables the mold to heated up to a temperature of 450 °C. In Fig. 4a, a thermocouple was placed in a direct alignment between the heater and the sample in order to accurately gauge the temperature. Furthermore, Fig. 4b illustrates the TS-CSP sintering temperature curve.

2.3.2 Preparation of Ceramic

In the experimental procedure, a quantity of 7.0 g of Bi_2O_3 powder was introduced into a mortar, followed by the addition of an acetic acid solution with a concentration of 3 mol/L, which was prepared to have a weight percentage of 15%. Subsequently, the mixed sample was homogeneously ground using an agate mortar, and the resulting slurry substance was transferred into the mold. At last, the samples are produced in accordance with three distinct sets of experimental conditions: For the initial group, the mold was heated to a temperature of 300 °C at a rate of 6 °C/min under uniaxial pressure ranging from 150 to 350 MPa. In the next experimental group, the mold had a pressure ranging from 350 to 150 MPa, with intervals of 50 MPa. Simultaneously, the mold was subjected to a gradual heating process at rates of 6 °C/min, reaching temperatures of 142–157, 145–160, 160–175, 190–205, and 230–245 °C, respectively. Each temperature level was maintained isothermally for a duration of 25 min. Finally, 300 MPa uniaxial pressure was applied to the ceramic mold. After that, the

TS-CSP was used to sinter the ceramic samples. The mold is heated to 155 °C at a pressure of 300 MPa as the initial step in this procedure. The system was maintained at a constant temperature from 0 to 45 min. Subsequently, the mold was raised to a temperature of 210 °C at a pressure of 300 MPa and maintained at this level for a duration of 25 min. Following this, the temperature was further increased to 270 °C at 300 MPa and maintained at this level for an additional 25 min to enhance the density of the product. The ceramic object had a diameter of 10 mm and a thickness of 5 mm after it had been left to settle down to the room temperature.

2.4 Preparation of Lightweight Porous Ceramics Using Metallurgical Steel Slag

Porous ceramics are a type of ceramics that have a network-like structure in three dimensions [50–52]. The material exhibits significant properties such as elevated porosity, decreased thermal conductivity, reduced density, exceptional heat resistance, enhanced resistance to corrosion, and excellent mechanical characteristics [53, 54]. At present, an extensive array of approaches exists for the production of porous ceramics. These include the direct foaming method, the freezing gel method, the freeze-drying method, and the sol–gel method. The organic foam impregnation method is widely utilized in many applications owing to its simplicity and cost-effectiveness [55–57]. This process involves the application of a slurry onto a polyurethane sponge, followed by the removal of leftover slurry. Subsequently, the structure is subjected to a drying and sintering process [58–60]. The raw materials used are metallurgical steel slag and kaolin. Finally, the ceramics were prepared with high porosity and low bulk density.

2.4.1 Raw Materials

Steel slag is utilized as raw materials. The chemical composition of the steel slag and kaolin is presented in Table 2. To identify the crystalline phases of the samples, X-ray diffraction (XRD) was used. The XRD analysis of steel slag and kaolin is illustrated in Fig. 5. Table 2 illustrates that the composition of steel slag primarily comprises SiO_2 and Al_2O_3, with a few particles of CaO, MgO, Fe_2O_3, K_2O, Na_2O, SO_3, and TiO_2. The graphical representation presented in Fig. 4 illustrates the main crystalline phases observed in steel slag, which include calcium silicate, calcite, andradite, akermanite, monticellite, merwinite, and enstatite. The graphical representation presented in Fig. 5 illustrates the primary crystalline phases observed in steel slag, which include calcium silicate, calcite, andradite, akermanite, monticellite, merwinite, and enstatite. The steel slag undergoes grinding processes for durations of 30, 45, 60, 90, and 120 min, correspondingly. Their average particle size is shown in Table 2, and their particle size dispersion is shown in Fig. 6 (Table 3).

Table 2 The chemical composition of raw materials, expressed as mass fractions in percentage [61]

Oxides	SiO_2	CaO	Al_2O_3	MgO	Fe_2O_3	K_2O	Na_2O	SO_3	TiO_2	TF	L.O.I
Steel Slag	31.10	38.32	2.86	6.01	9.72	0.18	0.07	0.21	0.52	10.48	0.53
Kaolin	54.34	0.05	41.85	0.07	0.63	0.43	0.30	0.12	0.31	-	1.42

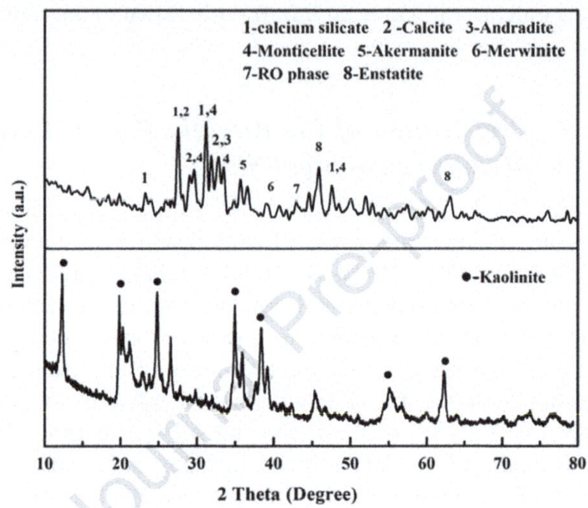

Fig. 5 XRD pattens of Steel slag and Kaolnite [61]

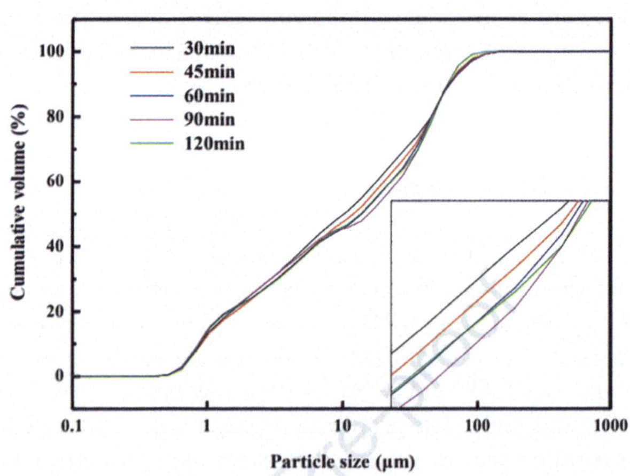

Fig. 6 Particle size distribution of steel slag [61]

Table 3 Average particle size of steel slag [61]

Grinding time/min	30	45	60	90	120
Average particle size/μm	35.321	34.719	34.633	32.589	33.725

2.4.2 Preparation of Sample

Steel slag was ground using a planetary ball mill for the following durations: 30, 45, 60, 90 and 120 min, respectively. Using a sieve with a mesh size of 100, the ground steel slag was prepared for sieving. The flow chart for the preparation of porous ceramics is depicted in Fig. 7. There are two main methods that were used in the experiment. The preparation of porous ceramics using a polyurethane template is first approach. Initially, ball milling is used to ground the steel slag. Then it underwent the process of sieving. Slurry was prepared by the uniform mixture of raw materials and additives. he kaolin, boric acid, and steel slag mixture was ground into an even powder form using a Simoloyer grinder for various durations. Finally, the powder was combined with water and thoroughly mixed to produce a homogeneous slurry. To hang the slurry, the pre-treated polyurethane foam was subsequently submerged in it. Multiple repetitions of immersion and squeezing were performed and residual slurry was squeezed out. The polyurethane material, which had been infused with the ceramic slurry, was subjected to a drying process in an oven set at a temperature of 60 °C for a duration of 24 h. This procedure was carried out in order to produce the porous ceramic structure. At last, the dried sample was introduced into a muffle furnace heating at elevated temperatures to undergo the process of sintering.

Next approach involves incorporating a surfactant into the slurry preparation process, based upon the first procedure. A rich foam slurry was prepared by stirring the slurry mixture at high speed. The polyurethane template was submerged in the slurry. Then it was subjected to a drying process at a temperature of 80 °C for a duration of 12 h. Following this, the template was sintered to produce porous ceramics. The process employed in the preparation of the porous ceramic results in the formation of a 3D network structure. This structure exhibits fine pores distributed throughout its skeleton and have a low density.

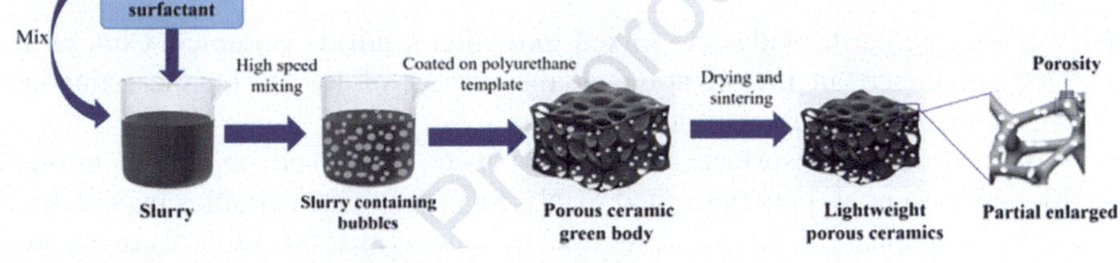

Fig. 7 Experimental flowchart [61]

The strength of ceramics produced from steel slag ranges from 20 to 50 MPa. The primary reason is due to the presence of calcium carbonate in steel slag. At high temperatures, calcium carbonate breaks down easily into CO_2gas. The liberation of gas leads the ceramics to an uneven structure, resulting in an increased presence of internal pores and a less uniform internal structure. This causes a reduction in the mechanical strength of ceramic. It has been observed that the compressive strength of the sample rises in proportion to the amount of ZrO_2. The addition of 3% by weight in ZrO_2 results in a corresponding increase of 16.98% in the compressive strength of the sample. Zirconia exhibits an exceptional level of hardness when incorporated into a sample, and its presence may influence stress-induced phase transformation and strengthening.

2.5 Preparation of Si_3N_4/O'-SiAlON Composite Ceramics

Silicon nitride ceramics has been extensively utilized in various applications such as bearings, high-speed cutting tools and engine blades owing to its significant characteristics including increased strength, enhanced hardness, and exceptional chemical stability [62–64]. Nevertheless, it should be noted that silicon nitride ceramics, despite being referred to as nonoxidizing ceramics, exhibit susceptibility to oxidation when exposed to high temperatures and oxygen. This limitation limits their prolonged use and has an adverse effect on their inherent characteristics [65]. The oxidation resistance characteristics of this ceramics can be enhanced by following methods:

i. The development of an antioxidant layer on the surface of the sample by coating or in-situ production. In order to enhance the antioxidation characteristics of this ceramic, Lee et al. [66] employed the hot spray technique to produce SiO_2 layers on the material surface.
ii. Ceramics made of silicon nitride composites are produced by using second phase addition. In order to enhance the antioxidation characteristic of silicon nitride ceramics, $MoSi_2$, TiC, and SiC were included [67, 68].
iii. Preparation of Si_2N_2O Ceramic. Lin et al. [69] produced Si_2N_2O ceramics and found that the mass remained mostly the same at temperatures between 1200 and 1300 °C.
iv. Various rare earth oxides are mixed into silicon nitride ceramics. Choi et al. [70] conducted an investigation on the impact of Re_2O_3 on the oxidation characteristics of silicon nitride ceramics.
v. Post-treatment and surface modification were performed on silicon nitride. Kovalcikova et al. [71] fabricated Si_3N_4–SiC composite ceramics by sintering in the liquid phase, followed by thermal treatments at 1680–1800 °C. In conclusion, thermal treatment resulted in an improvement of the sample's resistance to oxidation.

Table 4 Chemical compositions of the samples [77]

Sample	α-Si_3N_4/wt. %	β-Si_3N_4/wt. %	SiO_2/wt. %	Al_2O_3/wt. %	Y_2O_3/wt. %
S0	70	20	0	4	6
S1	67.5	20	2.5	4	6
S2	65	20	5	4	6
S3	62.5	20	7.5	4	6
S4	60	20	10	4	6
S5	57.5	20	12.5	4	6

Sialon ceramics are prepared by the combination of Si_3N_4–Al_2O_3–AlN–SiO_2, and may be classified into four distinct groups: O′ -sialon, α-sialon, β-sialon, and X-sialon [72]. One of these compounds, O′-sialon, is a solid solution that results from the substitution of Al–O bonds for the Si–N bonds in Si3N4 [73]. In addition, the O′-sialon ceramic exhibits exceptional resistance to oxidation and has the maximum amount of oxygen. In this process, Si_3N_4/O′-SiAlON composite ceramics were produced by the addition of varying amounts of SiO_2.

In this method, α-Si_3N_4, β-Si_3N_4, and SiO_2 were utilized as the primary constituents, with an average particle size of 5 μm. Additionally, Y_2O_3 and Al_2O_3 were used as sintering agents, with respective weight percentages of 6 and 4% [74–76]. The mixes were precisely weighed in accordance with the experimental system described in Table 4. Following that, these compounds were in an ethanol-filled ball grinding tank and rest for 14 h. Following the milling process, a rotary evaporator is used for drying the slurry. Following that, the dried powders underwent the process of sieving using 80-mesh sieve screens, they were subsequently formed into rectangular bars with dimensions of 5 mm × 4 mm × 56 mm using uniaxial pressure at 80 MPa. Throughout the sintering procedure, the samples were submerged in a powder bed consisting of Si_3N_4 and BN containing graphite crucible. The samples were sintered in a furnace operating at 1,750 °C for a duration of 2 h, at a rate of 5 °C per minute, under a N_2 gas pressure of 3 MPa.

Oxide layers were considered porous at the initial phase of oxidation, and interface reactions typically impeded the rate of oxidation. As an additive cation, Al3+ underwent diffusion into the oxide layer. In contrast to Y3+ ions, the diffusion rate of Al3+ ions were determined to be negligible due to the minor Al deficiency under the oxide layer. The oxidation process resulted in the formation of an oxide layer comprising $Y_2Si_2O_7$, SiO_2, and Y–Si–Al–O glassy phases.

2.6 *Preparation of xY_2O_3-(1-x) $LaCrO_3$ Thermosensitive Ceramics by Spark Plasma Sintering*

The Negative Temperature Coefficient (NTC) thermistor is a type of thermistor that exhibits a reduction in resistance as temperature increases. NTC possesses several benefits, including enhanced flexibility, rapid responsiveness, compact size, and cost-effectiveness. Therefore, it is extensively utilized in many conditions, including the reduction of in-rush current, monitoring of temperature, and temperature compensation [78–80]. The primary classifications of NTC ceramics are spinel, perovskite and composite type. Thermal sensitive ceramics of the spinel type commonly consist of elements such as Cu, Co, Mn, Ni, Fe, and others. However, their temperature measurement capacities are generally limited to values below 300 °C [81]. Magnesia alumina spinel ceramics are an advanced class of spinel-type ceramics that are capable of withstanding temperatures over 300 °C. Mg (Al1-xCrx) $2O_4$ were subjected to a sintering process at a temperature of 1650 °C in an ambient air atmosphere. This particular type of thermistor is suitable for temperature measurements within the range of 500–1000 °C. The electrical properties of the thermistor may be changed by altering the concentration of Cr [82]. Ceramics of the perovskite type have the capability to reliably measure temperatures ranging from ambient temperature to 1000 °C. Zhang et al. [83] conducted the sintering process of composite ceramics consisting of $0.6Y_2O_3$-$0.4YCr_{0.5}Mn_{0.5}O_3$, with the addition of La_2O_3 as a dopant. The sintering was carried out under atmospheric conditions at temperatures of 1550 and 1600 °C. This indicate that by altering the concentration of La_2O_3, it is possible to change the electrical characteristics of the material. Furthermore, these ceramic materials exhibit the capability to accurately measure temperatures ranging from room temperature to 1100 °C. Owing to its excellent electrical characteristics, $LaCrO_3$ has significant attention in academic research. The study conducted by Chen et al. [84] focused on the investigation of $La1\text{-}xBaxCrO_3$ NTC ceramics, whereby the researchers determined that, at −75 to 50 °C, this particular material system exhibited desirable electrical characteristics.

Spark plasma sintering (SPS) is a sintering technique that offers several benefits over traditional sintering methods. These advantages include reduced sintering cycle duration and enhanced density of the resulting samples [85]. SPS is widely employed for structural applications [86, 87]. Currently, the utilization of spark plasma sintering (SPS) for the sintering of NTC composite ceramics is not often [88]. SPS method is employed in the preparation of xY_2O_3-$(1\text{-}x)LaCrO_3$ composite ceramics.

2.6.1 Preparation of Material

The raw materials utilized in the experiment consisted of La_2O_3, Cr_2O_3, and Y_2O_3 with stoichiometric quantities. These materials have 99.99% purity and 500 nm in size. The powders of La_2O_3 and Cr_2O_3 were thoroughly mixed and subjected to ball milling which contains ethanol for a duration of 8 h. Agate ball is used as the milling media. Subsequently, the mixture underwent vacuum drying within a drying cabinet set at a temperature of 120 °C. It was then subjected to sieving using a 100-mesh sieve, followed by calcination in an extreme temperature resistance furnace at 1100 °C for a duration of 5 h. $LaCrO_3$ powder is obtained by sieving the calcined powder with 100-mesh sieve. The $LaCrO_3$ and Y_2O_3 powders are measured and combined.

The mixes were subjected to ball milling, vacuum drying, and sieving procedures, as previously mentioned, in order to acquire the powders of the sample.

The combined powders are introduced into the cylindrical graphite container. The particles were segregated using graphite paper, which had 0.2 mm of thickness. After that, the solution underwent sintering within a spark plasma sintering furnace under a uniaxial pressure of 30 MPa in a vacuum environment. The combined powders with varying compositions were sintered at temperatures of 1300, 1400, 1500, and 1600 °C, respectively, with a heating rate of 100 °C/min. Once the sintered samples had attained the final sintering temperature, they were hold for a duration of 5 min. After the holding period, the pressure was released.

The rate of cooling was 50 °C per minute. Once the temperature of the chamber decreased from 60 °C, the cylinder was extracted. A mechanical device known as a press was utilized to extract the sample from the cylinder. After that, sandpaper was used to evacuate the graphite layer from the sample. The experimental method for the preparation of composite ceramics is depicted in Fig. 8.

As the temperature reached a value of 1400 °C, there was a progressive rise in grain size and a corresponding increase in microstructure density. Upon reaching a temperature of 1500 °C, the sintered samples exhibited a high level of density and a uniform grain size, suggesting the completion of the sintering process. Upon reaching a sintering temperature of 1500 °C, the grains exhibited a substantial increase in size, indicating an excessive sintering temperature. As a result, it was determined that the optimal temperature for the xY_2O_3-(1-x)$LaCrO_3$ ceramics, fabricated using Spark Plasma Sintering (SPS) technique, was 1500 °C.

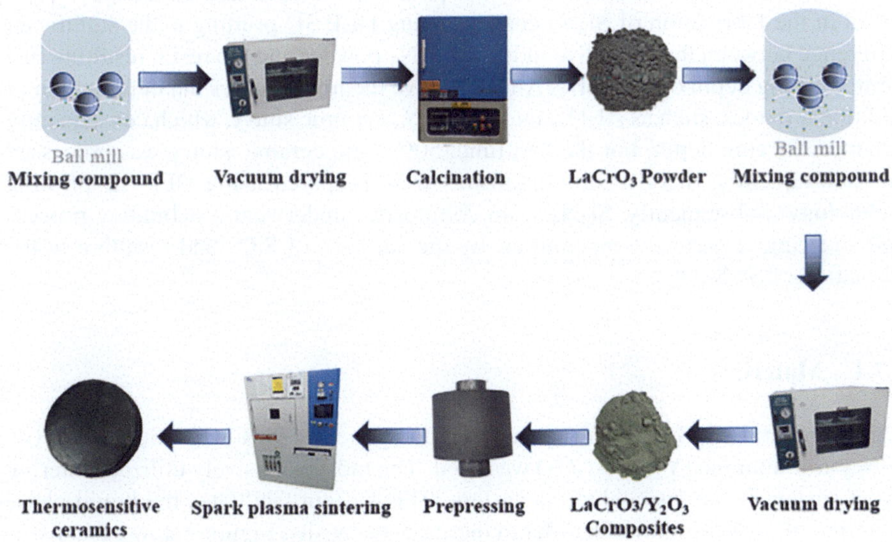

Fig. 8 The experimental procedures of composite ceramics [89]

2.7 *Preparation of Broadband Transparent Si_3N_4-SiO_2 Ceramics by Digital Light Processing (DLP) 3D Printing Technology*

Wave-transmitting materials are of significant importance in the construction of missiles, aircraft radomes, and antenna window panels. Nitride ceramics, such as Si_3N_4, BN, SiBN, and others, have exceptional thermal stability, resilience to thermal shock, low dielectric constants, and low dielectric loss. The thermophysical characteristics of the material remain optimal, exhibiting stability in dielectric properties throughout an entire range of temperatures. The mechanical strength of Si_3N_4 is very high, and also it has exceptional resistance to corrosion caused by rain [72, 90, 91]. Complex molds are needed for the conventional ceramic molding method. The manufacturing of complex parts poses significant challenges, resulting in considerable increases in both fabrication cost and production durations [92]. Ceramic objects may be produced more quickly and efficiently with additive manufacturing (AM) technology [93].

Digital light processing (DLP) 3D printing is an AM approach that has been around for several years. while printing, this technology can project and combine the whole layer with high printing speed and accuracy. In the production of ceramic objects, the DLP technique employs ceramic slurry as its primary constituent. The preparation of the ceramic slurry involves the combination of ceramic powder, photosensitive resin, initiator, and other relevant materials. After being exposed to a certain wavelength of UV light for a short period of time, the ceramic particles bond together to form complex shaped ceramic objects [94, 95]. The primary challenge experienced in the fabrication of Si_3N_4 ceramic using DLP 3D printing is the significant difference between the refractive index of Si_3N_4 powder and the resin, resulting in a limited curing depth of the slurry. An alternative method involves the introduction of additional oxides, such as Al_2O_3, into the Si_3N_4 ceramic slurry, which consequently enhances its cure depth. For the first time, 50% solid ceramic slurry was processed and subsequently, Si_3N_4-resin composites were prepared using DLP 3D printing technology. Subsequently, Si_3N_4-resin composites underwent a debinding process and sintering at various temperatures, by the addition of SiO_2 and resulting in the formation of Si_2N_2O.

2.7.1 Materials

In this preparation, Silicon nitride ceramic comprised of 90% wt% Si_3N_4 and 10% wt% yttria-alumina (Y_2O_3-Al_2O_3) was used. The most extensively utilized sintering assist system is the yttria-alumina system. This system facilitates the liquid phase sintering of Si_3N_4 ceramics in order to increase its density. Figure 9 shows the distribution of the size of Si_3N_4 powder. The average particle size, denoted as D50, of the raw Si_3N_4 powder is measured to be 3.452 μm. In this process, copolymer containing acidic functional group was utilized as the dispersant. Trimethylolpropane triacrylate

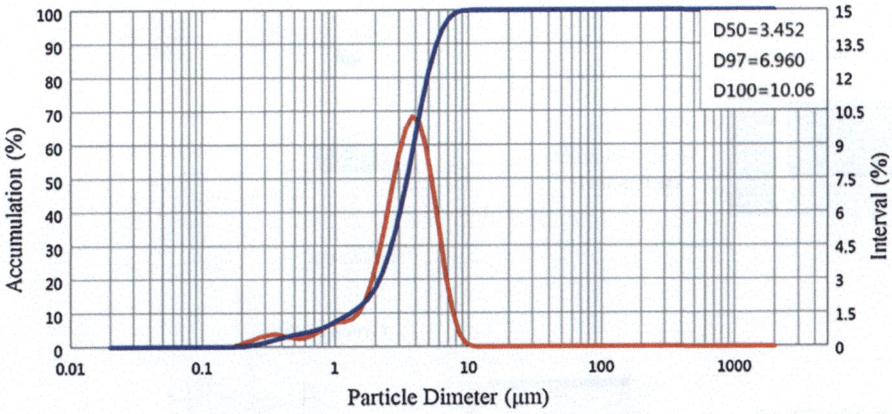

Fig. 9 Particle size distribution of raw Si_3N_4 powder

(TMPTA) is an acrylate resin that exhibits exceptional photo-polymerization activity and printing precision, making it highly compatible with digital light processing (DLP) printing techniques.

2.7.2 Preparation

In order to start the digital light processing 3D printing, it is important to acquire a Si_3N_4 slurry that possesses a high solid loading and high dispersion. Micron-sized silicon nitride (Si_3N_4) particles, trimethylolpropane triacrylate (TMPTA) resin monomers, yttrium oxide-aluminum oxide (Y_2O_3-Al_2O_3) additives, and dispersants were accurately measured and subjected to ball milling process for the duration of a 5 h in a planetary mill. A photoinitiator was introduced into the slurry at a concentration of 1 wt% relative to the total weight of the resin. The mixture was then subjected to ball milling for a duration of 1 h. Thus, a uniform distribution of Si_3N_4 slurry which comprised 50 vol % of solid concentration was successfully acquired. The ceramic objects made up of Si_3N_4 were prepared using the aforementioned processing techniques. The fabrication process is illustrated in Fig. 10.

I. Initially, a computer-aided design (CAD) model was created and subsequently transferred to the DLP 3D printer. Following that, the exposure period and light intensity were modified based on the light-curing capacity of the paste. By cross-linking the photosensitive groups, a polymer network is formed when exposed to ultraviolet light. The development of three-dimensional components was achieved by a layer-by-layer process.
II. Following the printing process, the ceramic body was exposed to sintering in a muffle furnace in order to eliminate the photo-sensitive resin. Figure 11a displays the debinding process of green bodies. The printed green bodies were initially raised from room temperature to 600 °C at a constant rate of 0.5 °C/

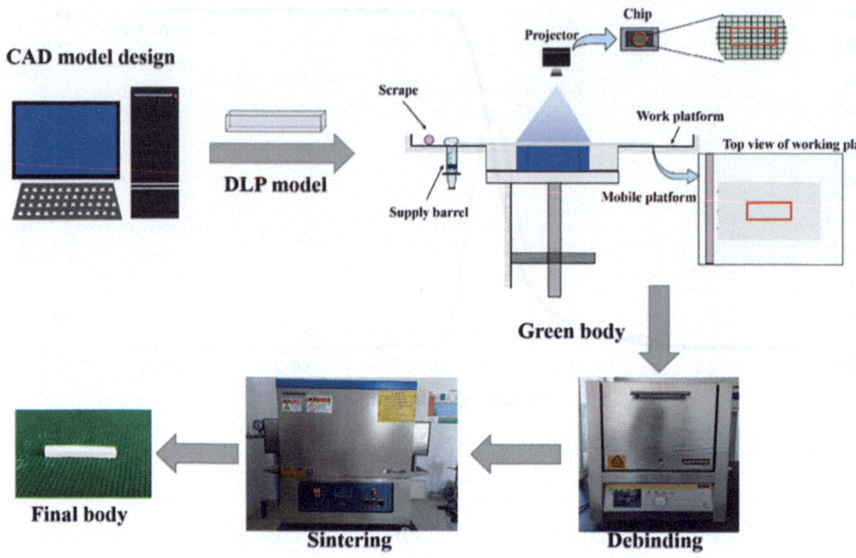

Fig. 10 Flow diagram of Si_3N_4-SiO_2 ceramic

min, then increased to 800 °C at a rate of 1.5 °C/min, and finally remained at 800 °C for the duration of one hour.

III. Lastly, the Si_3N_4-SiO_2 ceramic product was subjected to sintering in an oxygen (O2) environment, utilizing a tube-furnace. Sintering process was illustrated in Fig. 11b. Oxygen filled tube furnace is employed for the polymer to ceramic conversion process.

The SiO_2 concentration exhibits an initial increase followed by a subsequent drop when the sintering temperature is raised, ultimately reaching a value of 61.3% at a temperature of 1350 °C. Between the temperature range of 1250–1350 °C, there was a significant rise in the density of ceramics from 1.60 to 1.76 g/cm^3. Within

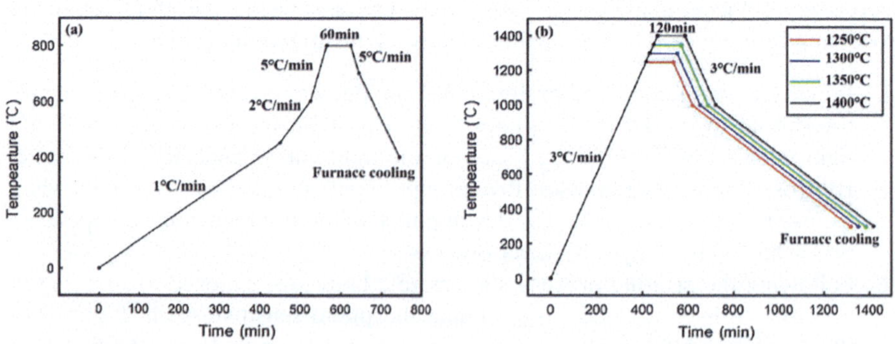

Fig. 11 Debinding and sintering of green bodies: **a** debinding, **b** sintering

the frequency range of 8.2–18 GHz in the electromagnetic spectrum, the materials exhibited notable efficiency in absorbing electromagnetic waves. Broadband wave-transmitting ceramics were prepared by combining 3D printing with heat treatment.

2.8 Preparation of h-BN Based Textured Ceramics by DLP Printing and Low-Pressure Sintering

Hexagonal boron nitride (h-BN) is a significant engineering ceramic material that exhibits numerous distinctive characteristics [96, 97]. Hexagonal boron nitride (h-BN) shows a lamellar structure which is similar to graphite, so it is commonly referred to as "white graphite." Due to its lamellar structure, certain characteristics of the material display anisotropic properties. But the densification of h-BN creates significant challenges due to its unique lamellar crystal structure.

Hot pressing (HP) and spark plasma sintering (SPS) are widely accepted as the primary methods of producing a high-density h-BN ceramic [98]. In addition, sintering additives like as B_2O_3, CaO, Al_2O_3, Y_2O_3, SiO_2, and others, which produce a liquid phase at elevated temperatures, are commonly employed to enhance the sintering capability of h-BN ceramic materials [99, 100]. As a result, h-BN ceramics are typically produced under conditions of elevated temperature, high pressures, or extended period of time. Therefore, the development of this type of ceramics that exhibit a high degree of orientation at relatively low temperatures and pressures has emerged as a novel area of research. The preparation of this ceramic was carried out by the utilization of a DLP-based printing procedure and HP sintering. The utilization of Al_2O_3-Y_2O_3 as a liquid phase sintering additive was identified in the process of selecting sintering additives for Si_3N_4 and SiAlON ceramics [101].

In this method, hexagonal boron nitride (h-BN), yttrium oxide (Y_2O_3) powders and aluminum oxide (Al_2O_3) are utilized as the raw materials which have the diameters of 5.3 lm, 50 nm and 1.5 lm respectively and purity level exceeding 99.5%. The volumetric ratio between h-BN and YAG in the reaction involving Y_2O_3 and Al_2O_3 was 8:2. The powders of the ceramic was introduced into the photosensitive resin and subsequently combined to produce a printing paste by a milling process utilizing Al_2O_3 balls for a duration of 12 h. The printing slurry consisted of ceramic powder, Hexanediol Diacrylate (HDDA), Capryl alcohol, and Polyurethane acrylate (PUA). Finally, polyethylene glycol 300 (PEG-300) was used as a photo initiator, constituting one volume percent of the slurry, and thoroughly combined to produce the suspension of digital light processing. The power of the UV source was measured to be 80 mW, while the duration of the curing process was 20 s (s). The green body was prepared using a stereolithographic-based 3D printing technique, wherein successive layers were added until the desired bulk ceramic structure was achieved. The layer had a thickness of 5 lm. Following the degreasing process, the green body underwent hot press sintering at a temperature of 1800 °C for a duration of 1 h and at a pressure of 20 MPa in a nitrogen atmosphere at atmospheric pressure.

The process employed in the preparation of this ceramic material results in a greater thermal conductivity and exhibits anisotropic characteristics. Additionally, the thermal conductivity has a similarity to that of ceramic materials produced by the use of elevated temperature and pressure [98]. The development and rearrangement of h-BN grains, which in turn affect the thermal conductivity of this ceramic, are significantly influenced by the crystallinity of the h-BN powder, holding time, size of the particle, pressure and sintering temperature.

2.9 Preparation of Porous Silicon Nitride Ceramics by Freeze Drying

Silicon nitride ceramics provide exceptional mechanical characteristics, excellent thermal stability, decreased dielectric constant, and reduced dielectric loss. It is an optimal material for transmitting electromagnetic waves on high-speed aircraft. This type of ceramics includes a combination of advantageous characteristics, including exceptional strength, tolerance to high temperatures, and the lightweight nature specific to silicon nitride. The material's dielectric constant may be changed within the range from 2 to 8. This type of ceramics demonstrated exceptional benefits as radome materials with high Mach numbers, large frequency ranges, and reduced targeting errors. Porous silicon nitride ceramics with low dielectric properties were fabricated using the freeze-drying method. By using the freeze-drying process, porous silicon nitride ceramics with pore distribution were produced, which helped to further improve their porous structure [102].

2.9.1 Raw Materials

Si_3N_4 powder contains more than 93% α-Si_3N_4, with an average particle size of 0.8 μm. It has particles that are about 0.5 μm in size and is made with high purity Y_2O_3 powder and Al_2O_3 powder as sintering aids. During freeze drying process, PVB solution was employed as the binder, polyacrylamide was utilized as the dispersant, and deionized water was utilized as the solvent.

2.9.2 Preparation

1. **Preparation of Slurry**

Silicon nitride slurries with varying solid content were fabricated through the process of planetary ball milling. The raw materials employed in this process included Si_3N_4 powder as the primary constituent, high purity Al_2O_3 and Y_2O_3 powder, polyacrylamide, PVB solution, deionized water, and silicon nitride balls as the milling medium. The slurry exhibits an exceptional degree of suspension and fluidity.

2. **The Process of Vacuum Defoaming**

The slurry that was produced in the first stage is subjected to a vacuum defoamer in order to eliminate the bubbles that were added during the ball milling process.

3. **Freeze Forming**

The slurry containing the vacuum defoaming particles was subjected to freeze drying. Initially, the freeze dryer's temperature/pressure setting was adjusted to −40 °C/13.33 Pa, leading to the freezing of the slurry. Subsequently, the pressure is decreased to a level below 1.33 Pa at a temperature of −60 °C. At this point, the heating system is activated to consistently provide the necessary heat for the sublimation of ice within the slurry. This process effectively removes the moisture present in the slurry, resulting in the production of porous silicon nitride blanks.

4. **Sintering Pressure**

The process for producing this type of ceramics involved sintering a freeze-dried porous silicon nitride body within a gas pressure furnace.

The mechanical characteristics of materials exhibit a rising trend as the solid concentration of slurry increases. The mechanical characteristics of the material increases significantly, particularly as the volume fraction of solid content of the slurry exceeds 25 vol.%. Thus, the mechanical characteristics of materials correlate with the pore structure. The bonding effect observed in porous silicon nitride is caused by the development of nano-pores resulting from the interconnecting and stacking of silicon nitride grains. This process leads to an improvement in the flexural strength of the material and it was found to be 94.7 MPa, while the compressive strength is measured to be 314 MPa.

2.10 *Preparation of Porous Ceramics from Spodumene Flotation Tailings by Low-Temperature Sintering*

Due to its exceptional chemical and physical characteristics, lithium, which has the lowest density among elements, finds extensive use in optical industries, energy storage, and aeronautics, particularly within the realm of lithium battery technology [103–105]. Spodumene, a prominent lithium source, is commonly extracted by the utilization of flotation as a predominant technique for effectively separating spodumene from gangue minerals [106, 107]. The primary constituents of spodumene flotation tailings (SFT) include albite, quartz, and microcline. These components align with the eco-friendly characteristics of ceramic materials [108]. In this method, kaolin was used as the adsorbent material. LPG was utilized to decrease the sintering temperature of ceramic materials. The application of kaolin as a loading adsorption material was used in order to produce adsorption characteristics in porous ceramics.

Table 5 Chemical compositions of main raw materials (wt.%) [109]

Raw material	SiO_2	Al_2O_3	K_2O	Na_2O	CaO	Fe_2O_3	P_2O_5	Others
SFT	79.75	12.04	5.39	0.96	0.25	0.81	0.23	0.57
Kaolin	54.30	42.43	0.13	0.00	0.42	0.67	0.30	1.75
LPG	2.79	24.77	14.86	17.81	0.05	0	35.5	4.2

2.10.1 Raw Materials

Table 5 demonstrates the chemical composition of raw materials. Sodium cardiomyopathy cellulose (CMC-Na) was selected as the reinforcing agent and it's formula is $[C_6H_7O_2(OH)_2OCH_2COONa]n$. The average median particle diameters (d50) of SFT, kaolin, and LPG were 61.62, 16.47, and 14.23 μm.

2.10.2 Preparation

The preparation procedures of porous ceramics are depicted in Fig. 12. The NJ160 cement paste mixer was used to combine the raw materials with water for a duration of four minutes. After that, hydrogen peroxide (H_2O_2) was introduced into the mixture, and the process of stirring was repeated for a duration of one minute. Ceramic materials were formed into rectangular prism alloy molds of 120 mm × 20 mm × 20 mm, as well as cylindrical molds with a diameter of 20 mm and a height of 20 mm. Following the demoulding process, they were subjected to a drying process at 120 °C for a duration of 24 h. After that, a saw was employed to remove any excess material that had grown beyond the borders of the mould during the foaming process. XZWL−14−12Y test furnace was used to sinter the samples at medium temperature. At the optimum temperature, 8 °C/min was the rate of heating of the mixtures and the holding duration was 2 h. Upon completion of the sintering process, the samples underwent a controlled cooling procedure at a rate of 8 °C per minute until reaching a temperature of 50 °C. When the sintering temperature of the porous ceramic increased, the flexural strength and compressive strength were also increased and absorption characteristic was decreased.

The mechanism of this process may be categorized into three distinct processes, namely the foaming, sintering and adsorption process. The introduction of hydrogen peroxide (H_2O_2) in the foaming process resulted in the development of a significant number of pores. Additionally, the presence of mineral particles led to the formation of additional pores through accumulation. Schematic diagram of mechanism was demonstrated in Fig. 13.

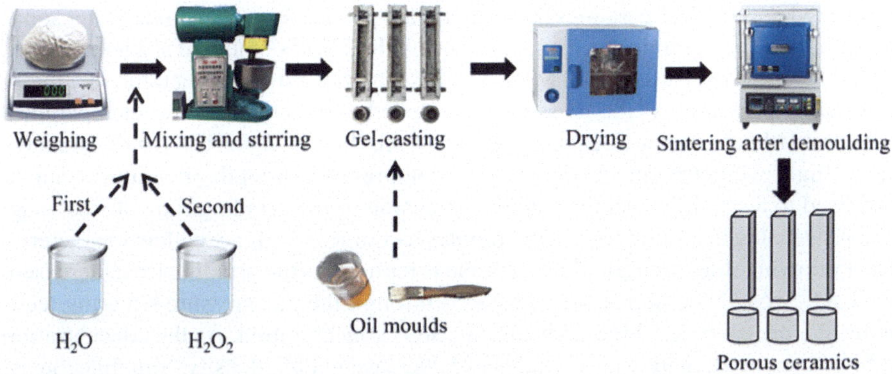

Fig. 12 Preparation process of porous ceramics [109]

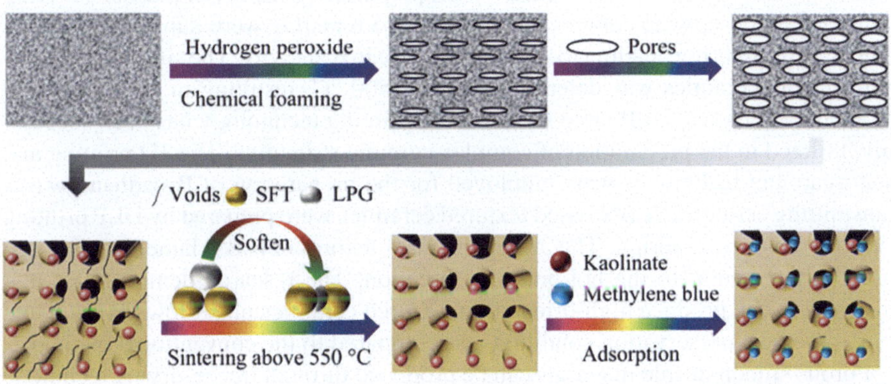

Fig. 13 Schematic diagram of mechanism [109]

3 Conclusion

In summary, the materials titanium, alumina, silicon, boron, zirconia and their nitride derivatives have enormous potential for the production of ceramic components. Various synthetic techniques for the preparation of ceramics were described. Hot-press sintering was utilized to produce laminated B_4C–TiB_2/BN ceramics subsequent to sequential lamination and high-temperature pyrolysis procedures. This ceramic showed anisotropic mechanical characteristics. The bending strength of the ceramic exhibited a progressive drop in both the perpendicular and parallel directions as the number of stacked layers increased. Porous ceramics made of Lead zirconate titanate (PZT) were produced by the combination of micro-scale stereolithography (μSL) with the BURPS technology. The porous PZT ceramic prepared by μSL technology showed excellent performance under ambient conditions. The acoustic impedance of the sample was reduced, while the porosity was increased. These changes show that the printed porous PZT ceramics have significant potential usage in underwater sonar detectors. Bi_2O_3 Ceramics were prepared using the TS-CSP method at a temperature of 270 °C and a pressure of 300 MPa. The first crystallization temperature is greatly affected by pressure. As the pressure increases, the first crystallization temperature decreases. Metallurgical steel slag and kaolin are the raw materials for the preparation of light weight porous ceramics. The compressive strength of ceramics can be increased to a specific level through the extension of the grinding time of steel slag. When steel slag is ground for ninety minutes, ceramics exhibit excellent characteristics. Following this, strength of the ceramic is reduced as the grinding time increased. Si_3N_4/O′-SiAlON ceramics were prepared through the gas pressure sintering technique at a pressure of 3 MPa with varying SiO_2 concentration. As the concentration of SiO_2 increased, there was a progressive decrease in bulk density. Simultaneously, the apparent porosity showed a similar rise from 7.07 to 12.66%. The chemical formula xY_2O_3-(1-x) represents a solid compound consisting of Composite ceramics with varying compositions of $LaCrO_3$ were prepared by the spark plasma sintering method. The composite ceramics of xY_2O_3-(1-x)$LaCrO_3$ were subjected to high-temperature sintering, leading to a substantial rise in resistivity. The sintering temperature of this ceramics was determined to be 1500 °C, resulting in excellent electrical characteristics. DLP three-dimensional printing technology has been successfully utilized in the preparation of complex ceramic structures. The 3D printing and heat treatment techniques were employed for the preparation of Broadband wave-transmitting ceramics. h-BN based textured ceramics were prepared by DLP printing and low-pressure sintering. The arrangement of texture in h-BN lamellae was seen to be perpendicular to the hot-pressing direction. The research demonstrates that this process can produce high-orientation textured ceramics under low-pressure and short-time heat preservation conditions, as compared to the conventional method.

Porous silicon nitride materials can be produced through freeze-drying technique and by manipulating the solid concentration of the slurry to achieve variable pore structure and directional growth. The weight growth rate of nitriding exhibits a progressive increase when the nitriding temperature is raised. The volumetric composition of slurry exerts a substantial influence on the characteristics of the material. Porous ceramics were effectively synthesized using the H_2O_2 foaming process, and the introduction of LPG resulted in a reduction in the sintering temperature. The tailings exhibited a notable usage rate, displaying a significant apparent porosity and water absorption capacity, with desirable mechanical characteristics, with average flexural and compressive strength. Various fabrication processes, including slip casting, powder injection molding, tape casting, freeze casting, and cold isostatic pressing, can be employed to manufacture ceramics with diverse forms and dimensions. During the preparation, sintering temperatures influences the primary factors such as densification, porosity, and transition of crystal phases of the ceramics. The aforementioned characteristics impart attributes such as high hardness, resistance to corrosion, excellent conductivity, and effective microwave absorption onto ceramics. The process of machining is employed to transform ceramics into the appropriate form, surface quality, mechanical properties, and dimensions. Nevertheless, traditional machining accounts for approximately 60–80% of the total cost of producing ceramics. In particular, 3-D printing is becoming more and more popular due to its increased potential for widespread commercialization. Moreover, the extensive range of applications of ceramics in the fields of biomedicine, environmental science, and electrical engineering makes it an extremely desirable material. The utilization of ceramics in power electronics is an increasing field of study, showcasing promising prospects for application in high temperature cycling conditions. This is mostly attributed to its advantageous attributes, such as reduced switching loss and enhanced voltage breakdown capabilities.

References

1. Otitoju, T.A., Okoye, P.U., Chen, G., Li, Y., Okoye, M.O., Li, S.: Advanced ceramic components: materials, fabrication, and applications. J. Ind. Eng. Chem. **85**, 34–65 (2020)
2. Kingery, W.D., Bowen, H.K., Uhlmann, D.R.: Introduction to Ceramics, vol. 17. Wiley (1976)
3. Salamon, D.: Advanced ceramics. In: Advanced Ceramics for Dentistry, pp. 103–122. Elsevier, Amsterdam (2014)
4. Hennicke, H.W., Hesse, A.: Traditional ceramics. In: Concise Encyclopedia of Advanced Ceramic Materials, 488–494. Elsevier (1991)
5. Maiti, K.N.: Technological innovations in traditional ceramics with emphasis on production of sanitarywares in the international and Indian scenario. Trans. Indian Ceram. Soc. **62**(4), 229–237 (2003)
6. McLaren, E., Giordano, R.: Ceramics overview: classification by microstructure and processing methods. Compend. Contin. Educ. Dent. **31**, 682–684 (2010)
7. Davidge, R.W.: Mechanical properties of ceramic materials. Contemp. Phys. **10**(2), 105–124 (1969)
8. Lv, Y., Liu, H., Wang, Z., Liu, S., Hao, L., Sang, Y., Liu, D., Wang, J., Boughton, R.I.: Silver nanoparticle-decorated porous ceramic composite for water treatment. J. Membr. Sci. **331**(1–2), 50–56 (2009)
9. Chen, Z., Li, Z., Li, J., Liu, C., Lao, C., Fu, Y., Liu, C., Li, Y., Wang, P., He, Y.: 3D printing of ceramics: a review. J. Eur. Ceram. Soc. **39**(4), 661–687 (2019)
10. Thevenot, F.: Boron carbide—a comprehensive review. J. Eur. Ceram. Soc. **6**(4), 205–225 (1990)
11. Thevenot, F.: A review on boron carbide. Key Eng. Mater. **56**, 59–88 (1991)
12. Halverson, D.C., Pyzik, A.J., Aksay, I.A., Snowden, W.E.: Processing of boron carbide-aluminum composites. J. Am. Ceram. Soc. **72**(5), 775–780 (1989)
13. Kanno, Y., Kawase, K., Nakano, K.: Additive effect on sintering of boron carbide. Yogyo-Kyokai-Shi **95**(11), 1137–1140 (1987)
14. Mashhadi, M., Saraf, M.R., Baghshahi, S.: Synthesis of boron carbide power via carbothermic reaction of boron oxide (2005)
15. Mashhadi, M., Taheri-Nassaj, E., Sglavo, V.M., Sarpoolaky, H., Ehsani, N.: Effect of Al addition on pressureless sintering of B4C. Ceram. Int. **35**(2), 831–837 (2009)
16. Thevenot, F.: Sintering of boron carbide and boron carbide-silicon carbide two-phase materials and their properties. J. Nucl. Mater. **152**(2–3), 154–162 (1988)
17. Jabbari, M., Bulatova, R., Tok, A.I.Y., Bahl, C.R.H., Mitsoulis, E., Hattel, J.H.: Ceramic tape casting: a review of current methods and trends with emphasis on rheological behaviour and flow analysis. Mater. Sci. Eng. B **212**, 39–61 (2016)
18. McColm, I.J.: Ceramic Science for Materials Technologists (1983)
19. Wang, B., Cai, D., Xue, C., Niu, B., Yang, Z., Duan, X., Li, D., Jia, D., Zhou, Y.: Preparation and mechanical properties of laminated B4C–TiB2/BN ceramics. Ceram. Int. **47**(22), 31214–31221 (2021)

20. White, E.W., Weber, J.N., Roy, D.M., Owen, E.L., Chiroff, R.T., White, R.A.: Replamineform porous biomaterials for hard tissue implant applications. J. Biomed. Mater. Res. **9**(4), 23–27 (1975)
21. Arai, T., Ayusawa, K., Sato, H., Miyata, T., Kobayashi, K.K.: Properties of hydrophone with porous piezoelectric ceramics. Jpn. J. Appl. Phys. **30**(9S), 2253 (1991)
22. Boumchedda, K., Hamadi, M., Fantozzi, G.: Properties of a hydrophone produced with porous PZT ceramic. J. Eur. Ceram. Soc. **27**(13–15), 4169–4171 (2007)
23. Creedon, M.J., Gopalakrishnan, S., Schulze, W.A.: 3-3 composite hydrophones from distorted reticulated ceramics. In: Proceedings of 1994 IEEE International Symposium on Applications of Ferroelectrics. IEEE (1994)
24. Yang, A.K., Wang, C.-A., Guo, R., Huang, Y., Nan, C.-W.: Porous PZT ceramics with high hydrostatic figure of merit and low acoustic impedance by TBA-based gel-casting process. J. Am. Ceram. Soc. **93**(5), 1427–1431 (2010)
25. White, R.A., Weber, J.N., White, E.W.: Replamineform: a new process for preparing porous ceramic, metal, and polymer prosthetic materials. Science **176**(4037), 922–924 (1972)
26. Smay, J.E., Cesarano III, J., Tuttle, B.A., Lewis, J.A.: Piezoelectric properties of 3-X periodic Pb (Zr x Ti 1− x) O 3–polymer composites. J. Appl. Phys. **92**(10), 6119–6127 (2002)
27. Zhang, Y., Roscow, J., Xie, M., Bowen, C.: High piezoelectric sensitivity and hydrostatic figures of merit in unidirectional porous ferroelectric ceramics fabricated by freeze casting. J. Eur. Ceram. Soc. **38**(12), 4203–4211 (2018)
28. Bowen, C.R., Perry, A., Lewis, A.C.F., Kara, H.: Processing and properties of porous piezoelectric materials with high hydrostatic figures of merit. J. Eur. Ceram. Soc. **24**(2), 541–545 (2004)
29. Zhang, H.L., Li, J.-F., Zhang, B.-P.: Microstructure and electrical properties of porous PZT ceramics derived from different pore-forming agents. Acta Mater. **55**(1), 171–181 (2007)
30. Zeng, T., Dong, X.L., Chen, S.T., Yang, H.: Processing and piezoelectric properties of porous PZT ceramics. Ceram. Int. **33**(3), 395–399 (2007)
31. Levy, G.N., Schindel, R., Kruth, J.-P.: Rapid manufacturing and rapid tooling with layer manufacturing (LM) technologies, state of the art and future perspectives. CIRP Ann. **52**(2), 589–609 (2003)
32. Travitzky, N., Bonet, A., Dermeik, B., Fey, T., Filbert-Demut, I., Schlier, L., Schlordt, T., Greil, P.: Additive manufacturing of ceramic-based materials. Adv. Eng. Mater. **16**(6), 729–754 (2014)
33. Yang, Y., Chen, Z., Song, X., Zhu, B., Hsiai, T., Wu, P.I., Xiong, R., Shi, J., Chen, Y., Zhou, Q.: Three dimensional printing of high dielectric capacitor using projection based stereolithography method. Nano Energy **22**, 414–421 (2016)
34. Huang, R.-J., Jiang, Q.-G., Wu, H.-D., Li, Y.-H., Liu, W.-Y., Lu, X.-X., Wu, S.-H.: Fabrication of complex shaped ceramic parts with surface-oxidized Si3N4 powder via digital light processing based stereolithography method. Ceram. Int. **45**(4), 5158–5162 (2019)
35. Song, X., Chen, Z., Lei, L., Shung, K., Zhou, Q., Chen, Y.: Piezoelectric component fabrication using projection-based stereolithography of barium titanate ceramic suspensions. Rapid Prototyp. J. **23**(1), 44–53 (2017)
36. Chen, W., Wang, F., Yan, K., Zhang, Y., Wu, D.: Micro-stereolithography of KNN-based lead-free piezoceramics. Ceram. Int. **45**(4), 4880–4885 (2019)
37. Wang, W., Sun, J., Guo, B., Chen, X., Prem Ananth, K., Bai, J.: Fabrication of piezoelectric nano-ceramics via stereolithography of low viscous and non-aqueous suspensions. J. Eur. Ceram. Soc. **40**(3), 682–688 (2020)
38. Hu, X., Li, X., Yan, K., Qi, X., Chen, W., Wu, D.: Fabrication of porous PZT ceramics using micro-stereolithography technology. Ceram. Int. **47**(22), 32376–32381 (2021)
39. Guo, H., Baker, A., Guo, J., Randall, C.A.: Cold sintering process: a novel technique for low-temperature ceramic processing of ferroelectrics. J. Am. Ceram. Soc. **99**(11), 3489–3507 (2016)
40. Guo, J., Guo, H., Baker, A.L., Lanagan, M.T., Kupp, E.R., Messing, G.L., Randall, C.A.: Cold sintering: a paradigm shift for processing and integration of ceramics. Angew. Chem. **128**(38), 11629–11633 (2016)

41. Maria, J.-P., Kang, X., Floyd, R.D., Dickey, E.C., Guo, H., Guo, J., Baker, A., Funihashi, S., Randall, C.A.: Cold sintering: current status and prospects. J. Mater. Res. **32**(17), 3205–3218 (2017)
42. Funahashi, S., Guo, J., Guo, H., Wang, Ke., Baker, A.L., Shiratsuyu, K., Randall, C.A.: Demonstration of the cold sintering process study for the densification and grain growth of ZnO ceramics. J. Am. Ceram. Soc. **100**(2), 546–553 (2017)
43. Guo, J., Berbano, S.S., Guo, H., Baker, A.L., Lanagan, M.T., Randall, C.A.: Cold sintering process of composites: bridging the processing temperature gap of ceramic and polymer materials. Adv. Func. Mater. **26**(39), 7115–7121 (2016)
44. Wang, D., Zhou, D., Zhang, S., Vardaxoglou, Y., Whittow, W.G., Cadman, D., Reaney, I.M.: Cold-sintered temperature stable Na0. 5Bi0. 5MoO4–Li2MoO4 microwave composite ceramics. ACS Sustain. Chem. & Eng. **6**(2), 2438–2444 (2018)
45. Wang, D., Guo, H., Morandi, C.S., Randall, C.A., Trolier-McKinstry, S.: Cold sintering and electrical characterization of lead zirconate titanate piezoelectric ceramics. APL Mater. **6**(1) (2018)
46. Berbano, S.S., Guo, J., Guo, H., Lanagan, M.T., Randall, C.A.: Cold sintering process of Li1. 5Al0. 5Ge1. 5 (PO4) 3 solid electrolyte. J. Am. Ceram. Soc. **100**(5), 2123–2135 (2017)
47. Funahashi, S., Guo, H., Guo, J., Baker, A.L., Wang, K., Shiratsuyu, K., Randall, C.A.: Cold sintering and co-firing of a multilayer device with thermoelectric materials. J. Am. Ceram. Soc. **100**(8), 3488–3496 (2017)
48. Guo, J., Legum, B., Anasori, B., Wang, K., Lelyukh, P., Gogotsi, Y., Randall, C.A.: Cold sintered ceramic nanocomposites of 2D MXene and zinc oxide. Adv. Mater. **30**(32), 1801846 (2018)
49. Song, J., Zhu, G., Xu, H., Fu, W., Xu, J., Zhang, J., Huang, S., Yu, A.: Preparation and properties of high-density Bi2O3 ceramics by cold sintering. Ceram. Int. **46**(9), 13848–13853 (2020)
50. Ding, Y., Zhang, X., Wu, B., Liu, Bo., Zhang, S.: Highly porous ceramics production using slags from smelting of spent automotive catalysts. Resour. Conserv. Recycl. **166**, 105373 (2021)
51. Wang, H., Sun, Y., Liu, L., Ji, R., Wang, X.: Integrated utilization of fly ash and waste glass for synthesis of foam/dense bi-layered insulation ceramic tile. Energy Build. **168**, 67–75 (2018)
52. Xi, C., Zheng, F., Jiahe, Xu., Yang, W., Peng, Y., Li, Y., Li, P., Zhen, Q., Bashir, S.: Jingbo Louise Liu, Preparation of glass-ceramic foams using extracted titanium tailing and glass waste as raw materials. Constr. Build. Mater. **190**, 896–909 (2018)
53. Wan, Li., Guo, X., Li, K., Yan, J.: Preparation of porous SiC-Al2O3 ceramics with spherical shell structures of large surface area and high strength. Ceram. Int. **46**(8), 10325–10331 (2020)
54. Ren, X., Ma, B., Su, C., Qian, F., Yang, W., Yuan, L., Yu, J., Liu, G., Li, H.: In-situ synthesis of FexSiy phases and their effects on the properties of SiC porous ceramics. J. Alloy. Compd. **784**, 1113–1122 (2019)
55. Zhang, Bo., Huang, H., Lu, X.: Fabrication and properties of C/SiC porous ceramics by grinding-mould pressing-sintering process. J. Eur. Ceram. Soc. **39**(5), 1775–1780 (2019)
56. Tian, C., Huang, X., Guo, W., Gao, P., Xiao, H.: Preparation of SiC porous ceramics by a novel gelcasting method assisted with surface modification. Ceram. Int. **46**(10), 16047–16055 (2020)
57. Tange, M., Manonukul, A., Srikudvien, P.: The effects of organic template and thickening agent on structure and mechanical properties of titanium foam fabricated by replica impregnation method. Mater. Sci. Eng. A **641**, 54–61 (2015)
58. Saraji, M., Ghani, M., Rezaei, B., Mokhtarianpour, M.: Highly porous nanostructured copper foam fiber impregnated with an organic solvent for headspace liquid-phase microextraction. J. Chromatogr. A **1469**, 25–34 (2016)
59. Yuan, L., Jin, E., Li, C., Liu, Z., Tian, C., Ma, B., Yu, J.: Preparation of calcium hexaluminate porous ceramics by novel pectin based gelcasting freeze-drying method. Ceram. Int. **47**(7), 9017–9023 (2021)

60. Gu, T., Chen, F., Shen, Q., Zhang, L.: Low-temperature preparation of porous SiC ceramics using phosphoric acid as a pore-forming agent and a binder. Ceram. Int. **45**(13), 16470–16475 (2019)
61. Wu, Q., Huang, Z.: Preparation and performance of lightweight porous ceramics using metallurgical steel slag. Ceram. Int. **47**(18), 25169–25176 (2021)
62. Bocanegra-Bernal, M.H., Matovic, B.: Dense and near-net-shape fabrication of Si3N4 ceramics. Mater. Sci. Eng. A **500**(1–2), 130–149 (2009)
63. Zou, C., Zhang, C., Li, B., Wang, S., Cao, F.: Microstructure and properties of porous silicon nitride ceramics prepared by gel-casting and gas pressure sintering. Mater. Des. **44**, 114–118 (2013)
64. Li, D., Li, B., Yang, X., Gao, S., Zheng, Y.: Fabrication and properties of in situ silicon nitride nanowires reinforced porous silicon nitride (SNNWs/SN) composites. J. Eur. Ceram. Soc. **38**(6), 2671–2675 (2018)
65. Klemm, H.: Silicon nitride for high-temperature applications. J. Am. Ceram. Soc. **93**(6), 1501–1522 (2010)
66. Lee, S.-H., Rixecker, G., Aldinger, F., Choi, S.-C., Auh, K.-H.: Effect of spray coated SiO2 layers on the low temperature oxidation of Si3N4. J. Eur. Ceram. Soc. **23**(8), 1199–1206 (2003)
67. Koh, Y.-H., Kim, H.-W., Kim, H.-E.: Mechanical properties and oxidation resistance of Si3N4-SiC nanocomposites. Scripta Mater. **44**(8–9), 2069–2073 (2001)
68. Deschaux-Beaume, F., Fréty, N., Cutard, T., Colin, C.: Oxidation modeling of a Si3N4–TiN composite: comparison between experiment and kinetic models. Ceram. Int. **35**(5), 1709–1718 (2009)
69. Lin, S., Ye, F., Ding, J., Yang, C., Ma, J., Dong, S., Liu, Q.: Effects of pore diameters on phase, oxidation resistance, and thermal shock resistance of the porous Si2N2O ceramics. J. Am. Ceram. Soc. **100**(5), 2190–2198 (2017)
70. Choi, H.-J., Lee, J.-G., Kim, Y.-W.: Oxidation behavior of hot-pressed Si3N4 with Re2O3 (Re= Y, Yb, Er, La). J. Eur. Ceram. Soc. **19**(16), 2757–2762 (1999)
71. Kovalčíková, A., Dusza, J., Šajgalík, P.: Influence of the heat treatment on mechanical properties and oxidation resistance of SiC–Si3N4 composites. Ceram. Int. **39**(7), 7951–7957 (2013)
72. Riley, F.L.: Silicon nitride and related materials. J. Am. Ceram. Soc. **83**(2), 245–265 (2000)
73. Kurama, S., Schulz, I., Herrmann, M.: Wear behaviour of α-and α/β-SiAlON ceramics stabilized with Nd2O3 and Y2O3. J. Eur. Ceram. Soc. **29**(1), 155–162 (2009)
74. Qadir, A., Fogarassy, Z., Horváth, Z.E., Balazsi, K., Balazsi, C.: Effect of the oxidization of Si3N4 powder on the microstructural and mechanical properties of hot isostatic pressed silicon nitride. Ceram. Int. **44**(12), 14601–14609 (2018)
75. Kovalčíková, A., Balázsi, C., Dusza, J., Tapasztó, O.: Mechanical properties and electrical conductivity in a carbon nanotube reinforced silicon nitride composite. Ceram. Int. **38**(1), 527–533 (2012)
76. Pasupuleti, S., Peddetti, R., Santhanam, S., Jen, K.-P., Wing, Z.N., Hecht, M., Halloran, J.P.: Toughening behavior in a carbon nanotube reinforced silicon nitride composite. Mater. Sci. Eng. A **491**(1–2), 224–229 (2008)
77. Ma, H., Bao, C.: Preparation, oxidation property and mechanism of Si3N4/O′-SiAlON composite ceramics. Ceram. Int. **47**(11), 15383–15391 (2021)
78. Park, K., Lee, J.K.: The effect of ZnO content and sintering temperature on the electrical properties of Cu-containing Mn1. 95− xNi0. 45Co0. 15Cu0. 45ZnxO4 ($0 \leq x \leq 0.3$) NTC thermistors. J. Alloys Compd. **475**(1–2), 513–517 (2009)
79. Luo, Y., Liu, X.: High temperature NTC BaTiO3-based ceramic resistors. Mater. Lett. **59**(29–30), 3881–3884 (2005)
80. Park, K., Yun, S.J.: Influence of the composition on the electrical properties of (Mn 2.1− x Ni 0.9 Si x) O 4 negative temperature coefficient thermistors. J. Mater. Sci.: Mater. Electron. **15**, 359–362 (2004)

81. Houivet, D., Bernard, J., Haussonne, J.-M.: High temperature NTC ceramic resistors (ambient–1000° C). J. Eur. Ceram. Soc. **24**(6), 1237–1241 (2004)
82. Yang, T., Zhang, Bo., Zhao, Q., Luo, P., Chang, A.: New high temperature NTC thermistors based on the Mg (Al1− xCrx) 2O4 ceramics. J. Alloy. Compd. **685**, 287–293 (2016)
83. Zhang, B., Zhao, Q., Chang, A., Huang, X., Hou, J., Zhao, P., Ji, G.: La2O3-doped 0.6 Y2O3–0.4 YCr0. 5Mn0. 5O3 composite NTC ceramics for wide range of temperature sensing. J. Alloys Compd. **581**, 573–578 (2013)
84. Chen, M., Zhang, H., Liu, T., Jiang, H., Chang, A.: Preparation, structure and electrical properties of La 1–x Ba x CrO 3 NTC ceramics. J. Mater. Sci. Mater. Electron. **28**, 18873–18878 (2017)
85. Frage, N., Cohen, S., Meir, S., Kalabukhov, S., Dariel, M.P.: Spark plasma sintering (SPS) of transparent magnesium-aluminate spinel. J. Mater. Sci. **42**, 3273–3275 (2007)
86. Wang, S., Yuan, J., Han, W., Yin, Z.: Microstructure and mechanical properties of B4C-TiB2 composite ceramic fabricated by reactive spark plasma sintering. Int. J. Refract Metal Hard Mater. **92**, 105307 (2020)
87. Yin, Z., Yuan, J., Xu, W., Ye, J., Liu, K., Yan, S.: Improvement in microstructure and mechanical properties of Ti (C, N) cermet prepared by two-step spark plasma sintering. Ceram. Int. **45**(1), 752–758 (2019)
88. Zhang, B., Zhao, Q., Chang, A., Wu, Y., Li, H.: Spark plasma sintering of MgAl2O4–LaCr0. 5Mn0. 5O3 composite thermistor ceramics and a comparison investigation with conventional sintering. J. Alloy. Compd. **675**, 381–386 (2016)
89. Chen, H., Yin, Z., Hao, X., Yuan, J.: Microstructure and electrical properties of xY2O3-(1–x) LaCrO3 thermosensitive ceramics prepared by spark plasma sintering. Ceram. Int. **47**(18), 26447–26454 (2021)
90. Zhang, J., Chen, J.W., Wang, H.J.: High performance porous wave-transmitting silicon nitride ceramics by gel-casting technique. Key Eng. Mater. **512**, 869–872 (2012)
91. Cheng, C.B., Fan, R.H., Yan, K.L., Sun, K., Chen, M., Hou, Q., Li, Q.Q.: Fabrication and property of porous silicon nitride ceramics with Y2O3-Yb2O3 additions. Key Eng. Mater. **655**, 6–10 (2015)
92. Pavese, M., Valle, M., Badini, C.: Effect of porosity of cordierite preforms on microstructure and mechanical strength of co-continuous ceramic composites. J. Eur. Ceram. Soc. **27**(1), 131–141 (2007)
93. Shao, H., Zhao, D., Lin, T., He, J., Wu, J.: 3D gel-printing of zirconia ceramic parts. Ceram. Int. **43**(16), 13938–13942 (2017)
94. Hu, K., Wei, Y., Lu, Z., Wan, L., Li, P.: Design of a shaping system for stereolithography with high solid loading ceramic suspensions. 3D Print. Addit. Manuf. **5**(4), 311–318 (2018)
95. Yang, Y., Song, X., Li, X., Chen, Z., Zhou, C., Zhou, Q., Chen, Y.: Recent progress in biomimetic additive manufacturing technology: from materials to functional structures. Adv. Mater. **30**(36), 1706539 (2018)
96. Ni, D.-W., Zhang, G.-J., Kan, Y.-M., Sakka, Y.: Textured h-BN ceramics prepared by slip casting. J. Am. Ceram. Soc. **94**(5), 1397–1404 (2011)
97. Duan, X., Yang, Z., Chen, L., Tian, Z., Cai, D., Wang, Y., Jia, D., Zhou, Yu.: Review on the properties of hexagonal boron nitride matrix composite ceramics. J. Eur. Ceram. Soc. **36**(15), 3725–3737 (2016)
98. Kusunose, T., Sekino, T.: Thermal conductivity of hot-pressed hexagonal boron nitride. Scripta Mater. **124**, 138–141 (2016)
99. Zhang, Xi., Chen, J., Zhang, J., Wan, D., Zhou, Y.: High-temperature mechanical and thermal properties of h-BN/30 vol% Y2SiO5 composite. Ceram. Int. **41**(9), 10891–10896 (2015)
100. Tian, Z., Duan, X., Yang, Z., Ye, S., Jia, D., Zhou, Yu.: Ablation mechanism and properties of in-situ SiAlON reinforced BN–SiO2 ceramic composite under an oxyacetylene torch environment. Ceram. Int. **40**(7), 11149–11155 (2014)
101. Motealleh, A., Ortiz, A.L., Borrero-López, O., Guiberteau, F.: Effect of hexagonal-BN additions on the sliding-wear resistance of fine-grained α-SiC densified with Y3Al5O12 liquid phase by spark-plasma sintering. J. Eur. Ceram. Soc. **34**(3), 565–574 (2014)

102. Xiao, C.-F., Han, B.: Preparation of porous silicon nitride ceramics by freeze drying. J. Market. Res. **8**(6), 6202–6208 (2019)
103. Costa, C.M., Lee, Y.-H., Kim, J.-H., Lee, S.-Y., Lanceros-Méndez, S.: Recent advances on separator membranes for lithium-ion battery applications: from porous membranes to solid electrolytes. Energy Stor. Mater. **22**, 346–375 (2019)
104. Hannan, M.A., Hossain Lipu, M.S., Hussain, A., Mohamed, A.: A review of lithium-ion battery state of charge estimation and management system in electric vehicle applications: challenges and recommendations. Renew. Sustain. Energy Rev. **78**, 834–854 (2017)
105. Xiong, R., Pan, Y., Shen, W., Li, H., Sun, F.: Lithium-ion battery aging mechanisms and diagnosis method for automotive applications: Recent advances and perspectives. Renew. Sustain. Energy Rev. **131**, 110048 (2020)
106. Tian, J., Xu, L., Deng, W., Jiang, H., Gao, Z., Hu, Y.: Adsorption mechanism of new mixed anionic/cationic collectors in a spodumene-feldspar flotation system. Chem. Eng. Sci. **164**, 99–107 (2017)
107. Wu, H., Tian, J., Xu, L., Fang, S., Zhang, Z., Chi, R.: Flotation and adsorption of a new mixed anionic/cationic collector in the spodumene-feldspar system. Miner. Eng. **127**, 42–47 (2018)
108. Lemougna, P.N., Yliniemi, J., Adesanya, E., Tanskanen, P., Kinnunen, P., Roning, J., Illikainen, M.: Reuse of copper slag in high-strength building ceramics containing spodumene tailings as fluxing agent. Miner. Eng. **155**, 106448 (2020)
109. Yang, J., Xu, L.-H., Wu, H.-Q., Jin, J.: Preparation and properties of porous ceramics from spodumene flotation tailings by low-temperature sintering. Trans. Nonferrous Metals Soc. China **31**(9), 2797–2811 (2021)

Emerging Strategies for Electroceramic Preparation: Contemporary Methods and Novel Techniques

Nisha Shankhwar, Pinki Singh, Jewel Thomas, Maya Rajnarayan Ray, and Satyendra Singh

Abstract Electroceramics belong to a functional category of ceramics where the materials are specifically engineered to showcase diverse electrical properties; piezoelectric, ferroelectric, and dielectric behaviours. Electroceramics have become increasingly important in a wide range of crucial technologies, including communications, energy conversion and storage, electronics, automation, high-capacity dielectric devices, non-volatile memories, dynamic random access memory (DRAMs), as well as materials for energy storage and conversion. Furthermore, they find application in advanced high-density and high-speed computer memories. As a result, the processing techniques employed for electroceramics must be meticulously designed to produce materials of exceptional quality. There exists a diverse array of techniques for processing electroceramics, which can be applied to different forms such as bulk ceramic, nano ceramics, film, and single crystal. The objective of this chapter is to present the latest advancements in methods and technologies utilized in the processing of electroceramics. The chapter focuses on discussing the Modern approaches like solid-state synthesis, sol gel, laser ablation, spin casting, as well as advanced technologies like molecular beam epitaxy (MBE), RF magnetron sputtering, Liquid Phase Epitaxy, high temperature solution growth (HTSG) or flux growth method, Czochralski Growth, Top Seeded Solution Growth (TSSG), and Hydrothermal Growth. Dedicated sections of this chapter provide an outline of each processing technique, with a particular emphasis on novel research directions and emerging areas of applicability.

Keywords Electroceramics · Sol–gel · Laser ablation · Sputtering · Molecular beam epitaxy · Energy storage · Liquid phase epitaxy · High energy density · Contemporary techniques · Thin film deposition · Sol–gel processing · Chemical

N. Shankhwar (✉) · P. Singh · S. Singh
Special Centre for Nanoscience, Jawaharlal Nehru University, New Delhi, India
e-mail: nisha2023@jnu.ac.in; shankhwarnisha3july@gmail.com

J. Thomas
School of Nanoscience and Nanotechnology, Mahatma Gandhi University, Kerala, India

M. R. Ray
Department of Civil Engineering, BIT Sindri, Dhanbad, Jharkhand, India

U. Kumar (ed.), *Defects Engineering in Electroceramics for Energy Applications*, Engineering Materials, https://doi.org/10.1007/978-981-97-9018-0_4

vapor deposition · Top-seeded solution growth · Czochralski growth method · Templated growth · Green synthesis

1 Introduction

In the realm of electroceramics, situated at the confluence of application-driven and technology-centric domains, the discipline draws from the fundamentals of ceramic processing, chemistry, and solid-state physics. Its adaptable products find utility across a spectrum of fields, including industrial process control, healthcare, consumer electronics, and environmental monitoring. Electroceramics utilize an extensive toolkit that spans the majority of the periodic table, encompassing inorganic and organo-metallic chemistry, solid-state chemistry, crystallography, and various tools for structural and chemical characterization [1]. Moreover, they apply experimental physics, modeling techniques, and device engineering. This field inherently embraces a transdisciplinary nature, evident in both its foundational principles and its multi-faceted applications. This chapter highlights the importance of processing techniques and their influence on various properties of electroceramics.

The processing of electroceramics is a critical and indispensable aspect of their use in diverse technological applications. Unlike traditional ceramics, electroceramics often necessitate intricate fabrication techniques for specific micro/nano structures and desired properties crucial for functionality [2]. Precise control over composition, crystal structure, and grain size/boundaries during processing is vital for tailoring electroceramics for various applications. The chosen processing methods significantly influence the electrical, mechanical, and thermal characteristics, directly impacting performance and efficiency in electronic/energy storage devices [1, 2]. Furthermore, meticulous processing allows for the synthesis of thin films, multilayer structures, and complex shapes, expanding the range of applications and enhancing the integration of electroceramics into miniaturized and multifunctional devices.

The processing of electroceramics is far more than a routine technical procedure; it is a fundamental and pivotal stage in unleashing the complete capabilities of these materials. Rather than being a mere formality, the processing of electroceramics serves as a critical gateway that allows these materials to reach their maximum potential. This transformative process is what empowers electroceramics to contribute to the realization of advanced technologies that have become indispensable in shaping our modern way of life. Essentially, this chapter recognizes the processing stage as a foundational step, emphasizing its transformative role in converting raw electroceramic materials into customized components.

2 Processing of Electroceramics

Harnessing the reproducible physical and structural properties of electroceramics for diverse applications requires meticulous processing [2, 3]. This section explores the foundational considerations in various electroceramic categories, including bulk ceramic, thin and thick film, single crystal, and nanostructured ceramics.

2.1 Approaches for Bulk Electroceramics Processing

Bulk electroceramic preparation methods are essential for unlocking the full potential of advanced materials across diverse technological applications. These methods involve precise techniques, including solid-state synthesis, sol–gel processes, and various deposition techniques. The choice of method depends on specific material requirements, showcasing the interdisciplinary nature of bulk electroceramic preparation. These methods empower researchers and engineers to finely tune electroceramic structures, driving innovations in electronics, energy storage, and diverse technological domains. This section will examine these methods in relation to their impact on the structural and other properties of electroceramics.

2.1.1 Solid-State Synthesis

Solid-state synthesis stands as a fundamental method in the preparation of electroceramics, playing a pivotal role in tailoring their structural and functional characteristics. This technique involves the intimate mixing of precursor materials, typically in powdered form, followed by a high-temperature sintering process. During sintering, the powders react and fuse together, creating a homogeneous and crystalline structure. The temperature, duration, and composition of the precursor materials are carefully controlled to achieve the desired electroceramic properties. Solid-state synthesis is known for its versatility in producing a wide range of materials, and its application extends to various electroceramic families, including ferroelectrics, piezoelectrics, and dielectrics. This method's significance lies not only in its efficiency in creating bulk structures but also in its ability to influence the phase composition and microstructure, thereby influencing the overall performance of electroceramic materials in technological applications.

The solid-state route comprises several crucial steps, each with distinct roles that significantly impact the resultant properties. Fig. 1 depicts the standard phases of the solid-state pathway.

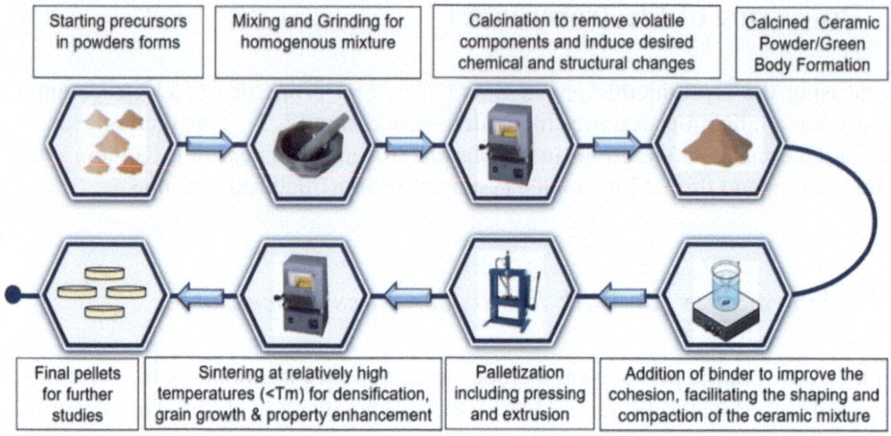

Fig. 1 A block diagram outlining steps involved in solid state synthesis

Machining and Polishing of Ceramics (Powder Compaction)

Ceramics undergo routine machining and polishing to meet specific application requirements. Room Temperature Uniaxial Pressing (RTUP) (Fig. 2), a well-established method, is highlighted for its simplicity and efficiency, making it ideal for fundamental studies and device fabrication. There are alternative methods like Hot Isostatic Pressing (HIP), Cold Isostatic Pressing (CIP), and Low Temperature Sintering (LTS). Each method has distinct features and complements Room Temperature Uniaxial Pressing (RTUP), contributing to advancements in ceramic processing [2]. Ceramic powder processing involves several stages beginning with the production of powder through milling or grinding. This is followed by the creation of a green product, which undergoes consolidation to achieve the final product. The powder itself comprises fine particles, and its synthesis includes preparatory steps like crushing, grinding, impurity removal, powder blending, and drying to form soft agglomerates suitable for shaping. Various methods such as compaction, tape casting, slip casting, injection molding, and extrusion are then employed to shape the processed powders into what is termed as green ceramic. Subsequently, the green ceramic undergoes further consolidation through a high-temperature process known as sintering or firing. In the initial stages, raw materials are subjected to milling or grinding to reduce particle size and separate desired minerals from unwanted gangue material. Wet milling is particularly common in ceramic processing compared to metal processing. The mixing of dry powders with a dispersant, typically water, results in a slurry. Ball and vibratory milling are utilized to further refine mineral size and blend various powders. The shaping of ceramic powders encompasses techniques such as casting, compaction, extrusion, hydro-plastic forming, and injection molding.

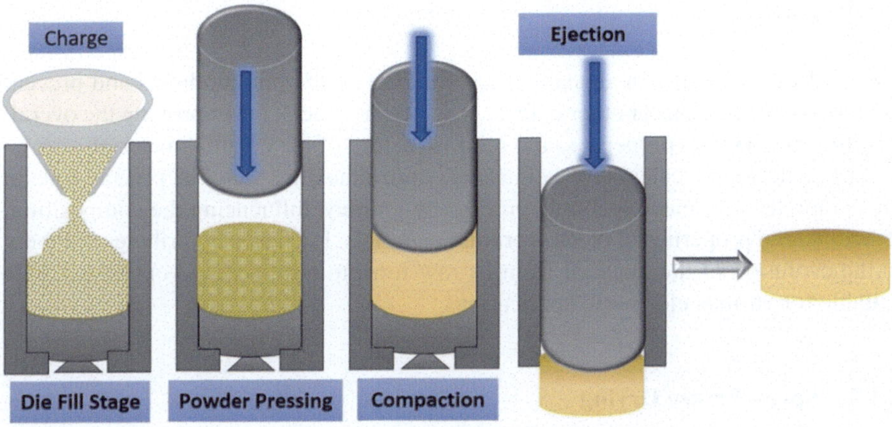

Fig. 2 Steps of the experimental procedure for Room Temperature Uniaxial Pressing (RTUP)

Heat Treatment in Ceramic Processing

This section focuses on the pivotal role of the three primary heat treatment methods—calcination, sintering, and annealing—in the ceramic processing journey. It explores how these processes initiate the formation of green ceramics, achieve densification, and enhance density while preventing cracking. Let's elaborate on each of the three fundamental heat treatment methods:

Calcination: Calcination is an initial heat treatment process that involves the removal of volatile components, such as water and organic matter, from a precursor material. This step serves to transform the raw materials into a reactive intermediate, often referred to as a "green body" or "calcined powder". During calcination, chemical reactions occur that result in physical and chemical changes, setting the stage for subsequent processing steps. The temperature and duration of calcination are crucial factors influencing the properties of the final ceramic material.

Sintering: Sintering is a critical step aimed at achieving densification of the calcined powder. It involves heating the material to high temperatures, typically below its melting point, causing the particles to bond and grow together. This process enhances the mechanical strength and reduces porosity, leading to the formation of a solid and durable ceramic structure. Particle rearrangement, neck formation, and grain growth are prominent phenomena during sintering. The control of temperature, time, and pressure is vital to achieving the desired density and structure in the final ceramic product.

Annealing: Annealing is a post-sintering heat treatment process that follows the attainment of the desired density. It involves controlled cooling of the sintered material to relieve internal stresses and enhance certain material properties, such as crystallinity and grain size.

Key Processes

The controlled cooling in annealing helps minimize thermal gradients and prevent the formation of cracks or other defects. This step is crucial for improving the overall stability, mechanical properties, and microstructure of the ceramic material.

The synergy of calcination, sintering, and annealing plays a pivotal role in the comprehensive electroceramic processing journey, influencing the composition, structure, and properties of electroceramic materials. Each step contributes uniquely to the evolution of the material from its raw form to a refined and optimized state suitable for various electronic applications.

2.1.2 Spray/Freeze Drying

Spray drying and freeze drying are two widely employed methods in the fabrication of electroceramics, offering unique advantages in terms of powder morphology and material properties. Spray drying involves the atomization of a liquid slurry containing electroceramic precursors into fine droplets, which are then rapidly dried in a hot air stream, yielding spherical or irregularly shaped particles. This method is favored for its scalability and cost-effectiveness, providing a high surface area for subsequent sintering processes. On the other hand, freeze drying, or lyophilization, relies on the controlled removal of solvent from a frozen electroceramic suspension. This technique results in porous, agglomerated powders with enhanced reactivity due to the preservation of the porous structure. Both methods play pivotal roles in tailoring the microstructure and properties of electroceramic materials, influencing their performance in applications such as capacitors, sensors, and electronic devices. Researchers and manufacturers continue to explore and optimize these drying techniques to achieve the desired characteristics in electroceramic powders for advanced technological applications.

2.1.3 Spray Pyrolysis

Spray pyrolysis is a versatile and effective method employed in the fabrication of electroceramics, offering precise control over film thickness and composition. In ceramic preparation through spray pyrolysis, a precursor solution is meticulously formulated by dissolving ceramic precursors in a suitable solvent. The chosen precursors align with the composition and properties required for the ceramic thin film. Subsequently, the substrate, often glass or silicon, undergoes thorough cleaning to ensure optimal adhesion. The precursor solution is then atomized into fine droplets using an atomizer and sprayed onto a heated substrate. The evaporation of the solvent occurs, leaving behind a thin layer of the ceramic precursor material. Concurrently, pyrolysis transpires, involving the thermal decomposition of the precursor, resulting in the

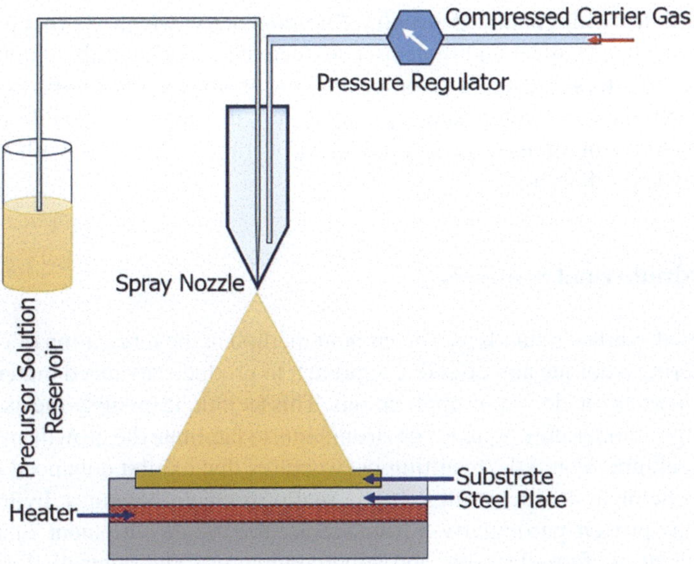

Fig. 3 The scheme of the spray pyrolysis setup

formation of a solid ceramic film on the substrate. Control over the substrate temperature influences the film's characteristics. The process can be iterated for additional thickness, and after deposition, the substrate gradually cools (Fig. 3).

In the context of electroceramics, such as ferroelectric or piezoelectric materials, spray pyrolysis provides distinct advantages. The fine control over deposition parameters, such as spray rate, precursor concentration, and substrate temperature, enables the synthesis of thin films with tailored properties, essential for applications in sensors, actuators, and electronic devices. Additionally, the scalability and cost-effectiveness of the spray pyrolysis process make it a promising method for large-scale production of electroceramic thin films, further enhancing its relevance in the field of advanced materials for electronic applications.

2.1.4 Emulsion Synthesis

Emulsion synthesis methods have emerged as a promising approach in the realm of electroceramics, offering a versatile and controlled means of fabricating advanced materials with tailored properties. In this process, a stable emulsion, typically comprising an aqueous phase containing metal precursors and an organic phase containing surfactants and solvents, serves as the template for the formation of ceramic nanoparticles. The electroceramic nanoparticles, such as ferroelectrics or piezoelectrics, are intricately crafted within the emulsion, allowing for precise control over size, morphology, and composition. This method enables the synthesis of uniform particles with enhanced structural homogeneity, which is crucial for optimizing the electrochemical and dielectric performance of materials in applications like sensors, actuators, and energy storage devices. Moreover, the emulsion synthesis approach facilitates scalability and reproducibility, making it a valuable technique for the development of next-generation electroceramic materials with improved functionality and reliability.

2.1.5 Hydrothermal Synthesis

Hydrothermal synthesis stands as a prominent method in the fabrication of electroceramics, offering a unique and effective approach to produce advanced materials with tailored properties for electronic applications. This technique involves the use of high-pressure, high-temperature aqueous environments to facilitate the growth and crystallization of ceramic materials, resulting in structures that exhibit enhanced electrical and electrochemical characteristics. In the realm of electroceramics, hydrothermal synthesis has proven particularly advantageous for the development of materials like piezoelectrics, ferroelectrics, and superconductors. The controlled conditions of hydrothermal processing enable the precise manipulation of crystal structures, grain sizes, and defect concentrations, ultimately influencing the functional properties of electroceramic materials. This method not only offers a versatile means of synthesizing a variety of electroactive ceramics but also allows for the exploration of novel compositions and phases that can push the boundaries of electronic device performance. As research in hydrothermal synthesis for electroceramics continues to advance, it holds the promise of unlocking new possibilities in the development of next-generation electronic components with improved efficiency and reliability (Table 1).

2.2 Approaches for Nanostructured Electroceramics Processing

In recent times, the trend toward miniaturization has significantly influenced the field, with current research and development efforts concentrated on electroceramic thin films. While thin films are a focal point of interest, the evolution extends beyond them to encompass thick films. Notably, there is a shift in industrial interest from discrete components to the development of electroceramic systems and multifunctional electroceramics. The impact of size effects on electroceramics has been a longstanding focus of thorough investigations. However, the current surge of interest in nanotechnology has been primarily sparked by the downsizing of devices and components reliant on electroceramics. The growing availability of advanced experimental tools for directly observing nanoscale features, coupled with the broader contemporary

Table 1 The table presents a comparison of various synthesis methods based on compositional control, morphology control, powder reactivity, particle size, purity, agglomeration, calcination step, milling step, and costs [4, 5]

Synthesis method	Calcination step	Milling step	Compositional control	Morphology control	Powder reactivity	Particle size (nm)	Purity (%)	Agglomeration	Costs
Solid-state reaction	Yes	Yes	Poor	Poor	Poor	>1000	<99.5	Moderate	Low-moderate
Coprecipitation	Yes	Yes	Good	Moderate	Good	>10	>99.5	High	Moderate
Sol–gel	Yes	Yes	Excellent	Moderate	Good	>10	>99.9	Moderate	Moderate-high
Spray/freeze drying	Yes	Yes	Excellent	Moderate	Good	>10	>99.9	Low	Moderate-high
Spray pyrolysis	No	No	Excellent	Excellent	Good	>10	>99.9	Low	High
Emulsion synthesis	Yes	Yes	Excellent	Excellent	Good	>100	>99.9	Low	Moderate
Hydrothermal synthesis	No	No	Excellent	Good	Good	>100	>99.5	Low	Moderate

fascination with nanoscience, serves as additional significant factors contributing to this heightened interest.

2.2.1 Co-precipitation Method

Coprecipitation methods represent a notable approach in the synthesis of electroceramics, offering a versatile and effective means of tailoring material properties. In this technique, precursor ions are simultaneously precipitated from a solution, resulting in the formation of a gel-like solid. The subsequent thermal treatment of this gel yields the desired electroceramic material. Coprecipitation starts with the meticulous preparation of precursor solutions, each containing ions of the elements intended for co-precipitation.

These precursor solutions typically comprise soluble salts or complexes dissolved in an aqueous medium. Success depends on achieving absolute homogeneity in the mixed precursor solution, ensuring an even distribution of ions from various elements. Following the amalgamation of precursor solutions, a specialized precipitating agent is introduced into the precursor solution. This agent is a chemical compound that orchestrates the simultaneous precipitation of ions from different elements. Commonly employed precipitating agents include hydroxides, carbonates, or sulfides. The introduction of the precipitating agent instigates the formation of a solid co-precipitate, embodying both cations and anions essential for electroceramic properties.

Once the co-precipitate materializes, the subsequent step involves the meticulous separation of the solid from the remaining liquid. Techniques such as filtration or centrifugation are typically employed, facilitating the isolation of the co-precipitated ceramic material. To ensure the product's purity, the separated co-precipitate undergoes a washing step, meticulously designed to remove any impurities or soluble by-products. Following this, co-precipitate is subjected to a drying process, systematically eliminating any lingering liquid and culminating in the attainment of the final solid co-precipitated product. Optionally, and particularly relevant for achieving desired electroceramic properties, the co-precipitate may undergo additional heat treatment, known as calcination or annealing. Figure 4 explaining these steps involved. The versatility of coprecipitation lies in its ability to control particle size, composition, and homogeneity by adjusting parameters such as pH, temperature, and the choice of precipitating agents. The method is particularly advantageous for producing fine powders with uniform characteristics, crucial for enhancing the performance of electroceramics. Additionally, coprecipitation is well-suited for the synthesis of multi-component materials, allowing for the incorporation of various elements in precise ratios. This method's adaptability and efficiency make it a valuable tool in the quest to engineer electroceramics with tailored properties for diverse technological applications, including sensors, actuators, and energy storage devices.

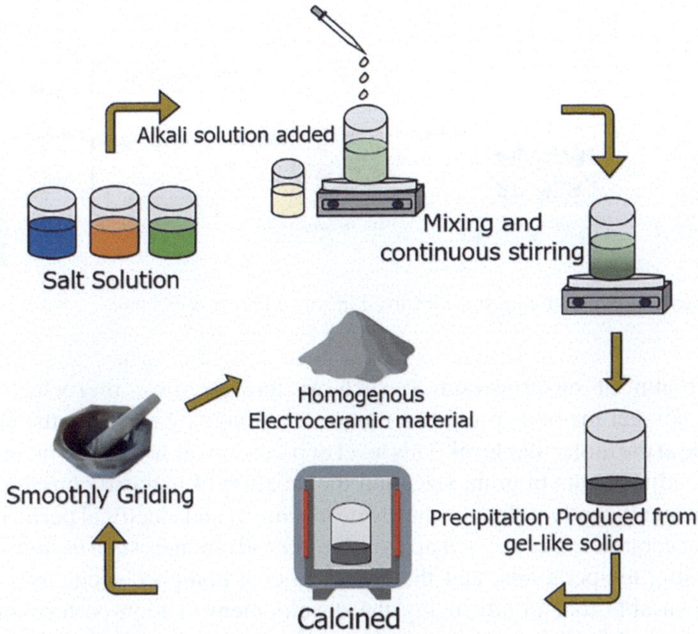

Fig. 4 Steps of the experimental procedure for coprecipitation methods

2.2.2 Sol–Gel Method

Sol–gel methods play a pivotal role in the fabrication of electroceramics, offering a versatile and precise approach to synthesizing materials with tailored properties for various electronic applications. This innovative technique involves the transformation of a sol, a colloidal suspension of nanoparticles, into a gel and subsequent solidification to form a ceramic material.

The sol–gel process begins with the selection of precursor compounds, typically metal alkoxides or metal salts, dissolved in a solvent made to the desired characteristics of the final ceramic product (Fig. 5). The introduction of water or a hydrolysing agent initiates hydrolysis, breaking chemical bonds in the precursors and forming hydroxyl groups, laying the foundation for metal oxide or metal hydroxide species. Subsequent condensation reactions among the hydrolysed species create a three-dimensional network, transforming the sol into a gel, a crucial intermediate structure influencing porosity and mechanical properties. Following maturation, crucial for enhancing mechanical strength and influencing properties like density and hardness, the gel undergoes controlled drying to form a porous xerogel. The drying method shapes the microstructure, porosity and electrical properties of the ceramic. Optionally, the dried xerogel may undergo calcination at elevated temperatures, eliminating residual organics, consolidating the material, and inducing crystallization.

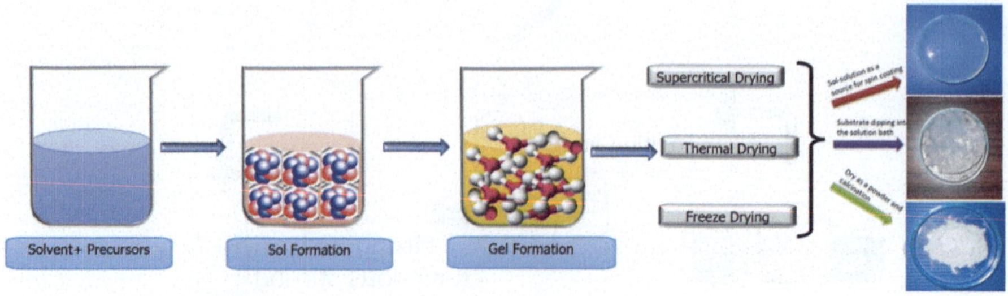

Fig. 5 A block diagram outlining steps involved in sol–gel process of films

In the realm of electroceramics, such as ferroelectrics, piezoelectrics, and dielectrics, sol–gel methods provide a unique advantage by enabling the control of composition at the molecular level. This level of precision allows for the incorporation of dopants, adjustments in grain size, and the creation of nanostructured materials, all of which significantly influence the electrochemical and electrical performance of the resulting ceramics. The sol–gel approach offers advantages such as homogeneity, low processing temperatures, and the ability to coat complex geometries, making it an indispensable tool in advancing the development of high-performance electronic devices and components. The ability to fine-tune material properties through sol–gel methods opens new avenues for designing electroceramics with enhanced functionality and efficiency in a wide range of technological applications.

2.3 Approaches for Thin Film Electroceramics Processing

In the pursuit of advancing thin film electroceramics processing, researchers and engineers explore innovative approaches to overcome persistent challenges and enhance the efficiency of fabrication techniques. This field, at the intersection of materials science and electrical engineering, focuses on developing methodologies for depositing thin films of electroceramic materials with precision and reliability. From addressing issues related to phase stability and stoichiometry control to optimizing substrate selection and deposition conditions, various strategies are employed to propel the development of thin film electroceramics. This section provides the details of the diverse approaches that contribute to the evolution of thin film processing in the realm of electroceramics.

2.3.1 Pulsed Laser Deposition (PLD) Method

Pulsed Laser Deposition (PLD) is an advanced technique utilized for growing simple and complex oxides. Despite initial obstacles, the importance of PLD increased in the late 1980s following the breakthrough in high-temperature superconductivity.

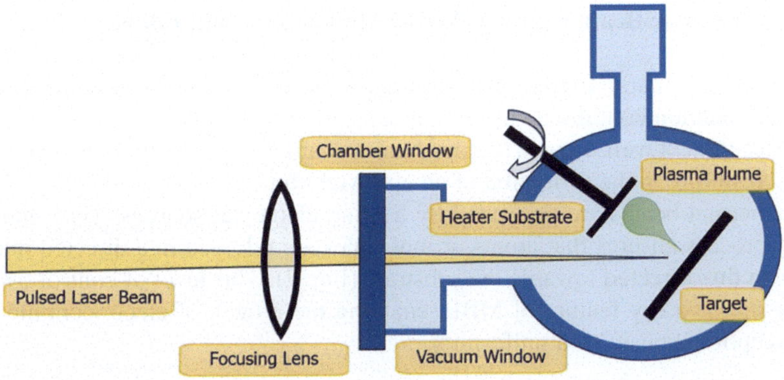

Fig. 6 A block diagram outlining pulsed laser deposition method

In this technique, a high-energy laser pulse is precisely directed onto a target material, typically a ceramic oxide chosen for its electroceramic properties. The laser's intense pulse rapidly heats the target, leading to ablation and the formation of a plume or plasmatic cloud containing vaporized atoms and ions. A carefully selected substrate, chosen to match the desired composition and electrical properties, serves as the growth surface. As the plume condenses on the substrate, it results in the formation of a thin film or layer of the desired ceramic material (Fig. 6).

Control over laser parameters such as energy, duration, and fluence, along with meticulous selection of the target material and deposition conditions, enables the precise fabrication of high-quality electroceramic films with tailored properties. One significant advantage of PLD is its ability to maintain the stoichiometry, ensuring an accurate reproduction of the composition and crystal structure of the desired material. The pulsed nature of the laser facilitates precise control over the deposition process, influencing factors like thickness and uniformity of the deposited film. The discovery of high-temperature superconductivity significantly boosted the relevance of PLD in electroceramic synthesis. Researchers recognized PLD as a reliable and effective method for growing thin films of high-temperature superconducting materials. This application has furthered the investigation and understanding of electroceramic properties, expanding the potential applications of these materials [6]. Continuous refinement and optimization of PLD have led to advancements in deposition techniques, laser technologies, and target material choices, making it a versatile tool in electroceramic synthesis. Its widespread application spans across various fields, including electronics, photonics, energy storage, and advanced materials research. The precise control offered by PLD has opened up new avenues for developing functional electroceramic devices such as sensors, actuators, thin-film transistors, and integrated circuits. Additionally, it facilitates the exploration of novel physical phenomena and emergent properties in electroceramic systems.

2.3.2 Molecular Beam Epitaxy (MBE) Method for Thin Films

Developed in the early 1970s, MBE ensures high-quality single crystal films under ultra-high vacuum conditions. MBE is a deposition technique characterized by the precise reaction of well-defined thermal beams of atoms or molecules on a crystalline surface, resulting in the formation of an epitaxial film.

The thermal beams are generated by heating elemental sources. The controlled heating process ensures that atoms or molecules are released in a directed manner, forming a flux directed towards the substrate (Fig. 7). This level of control over the particle flux is a key feature of MBE, enabling the growth of electroceramic films with exceptional purity and uniformity.

> The substrate, positioned within the growth chamber, acts as the receiving surface for the deposited particles. One distinctive feature of MBE is its precision in controlling the thickness of each atomic layer during growth. This level of control is achieved through real-time monitoring and adjustment of particle flux using sophisticated feedback mechanisms. This capability empowers the creation of intricate heterostructures, where different materials with distinct ceramic properties are precisely layered.

Conducted under ultra-high vacuum (UHV) conditions, MBE ensures minimal impurities, safeguarding the purity and quality of the deposited electroceramic material. The absence of highly energetic species in MBE prevents damage to the growing film, resulting in films with outstanding structural and electrical properties. Moreover, MBE allows in situ characterization and monitoring during growth. Techniques such as reflection high-energy electron diffraction (RHEED) and scanning tunneling microscopy (STM) enable real-time assessment of the electroceramic growth process, facilitating immediate feedback for optimizing film quality. A variant

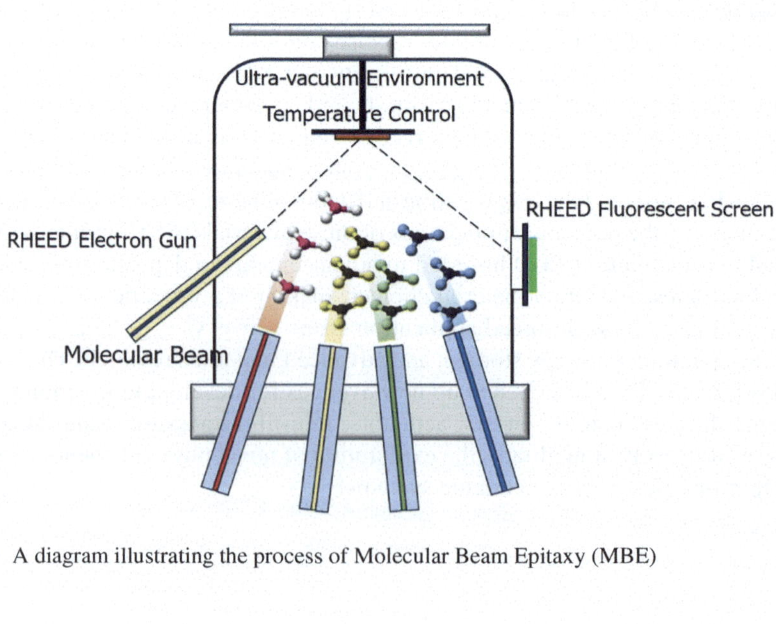

Fig. 7 A diagram illustrating the process of Molecular Beam Epitaxy (MBE)

known as "reactive MBE" enhances MBE's capabilities by introducing molecular beams of gases alongside elemental sources, allowing controlled formation of oxides or nitrides during growth. Reactive MBE broadens the scope of materials that can be synthesized, enabling the fabrication of complex electroceramic compounds and solid solutions. Initially designed for GaAs and (Al, Ga) As growth, MBE's unique capability to control layering at the monolayer level, along with its compatibility with surface-science techniques for real-time growth monitoring, has led to its widespread application in various semiconductors, metals, and insulators.

2.3.3 Chemical Solution Deposition (CSD)

CSD is a method involving the application of a thin film derived from a solution containing essential elements in the desired stoichiometry. Initially, a homogeneous solution is created by combining stoichiometric quantities of precursor substances (typically nitrates, acetates, citrates, and metal alkoxides) in solvents (typically glacial acetic acid, citric acid, 2-methoxethanol). Figure 8 illustrates the procedural steps for CSD methods.

The mixture may need to occur at room temperature or higher, depending on the dissolution characteristics of the precursors in the respective solvents. It's also possible that each precursor needs to be initially dissolved in separate solvents, which are later mixed together. After preparing the solution, viscosity and pH adjustments are made by adding appropriate solutions. Stabilizing agents like ethanolamine may be necessary to prevent ingredient precipitation into a complex. Finally, the prepared solution is examined for turbidity and precipitation over a few days to assess solution stability. In the second stage, the solution is spin-coated onto suitable substrates. Pre-treatment of substrate surfaces, such as cleaning with specially formulated solutions followed by ozonization, is often necessary to make the substrate surface hydrophilic to the solution and remove any dust and organic residues. For the first coating, a few drops of the solution are placed on the substrate on a vacuum-enabled chuck of a spin coater, ensuring complete coverage. The substrate is then spun for a specified duration to remove excess solution, leaving a thin layer on top. Subsequently, the substrate is either placed on a hot plate, oven, or furnace at a preset temperature to dry the coating and eliminate volatile chemicals without crystallizing it. These temperatures are determined through thermal analysis of the solution. After drying

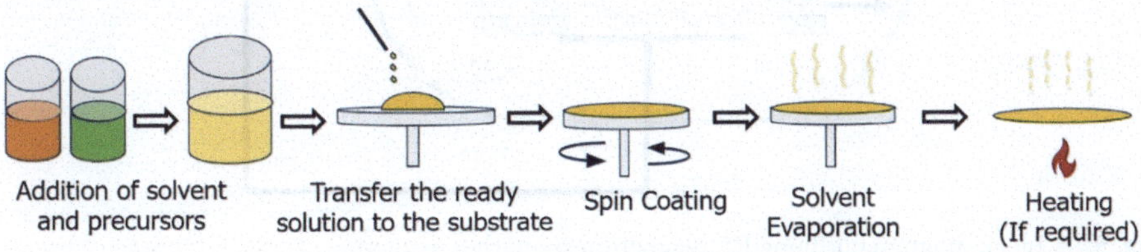

Fig. 8 A schematic outlining steps involved chemical solution deposition of electroceramic films

for a specific duration, the coating and drying procedure is repeated to achieve the desired thickness. Once the desired thickness is attained, the final film undergoes annealing in a selected environment (typically air, nitrogen, or oxygen) for a fixed period to achieve film crystallization. Optimizing various processing parameters is crucial for obtaining high-quality films without defects and with the desired phase.

2.3.4 RF Magnetron Sputtering Method (RFMS)

Magnetron Sputtering is a phenomenon in glow discharge processes where a surface is bombarded by energetic particles, leading to the ejection of target atoms. In this process, target atoms are ejected by the bombardment of energetic species, usually inert gas ions. These sputtered atoms travel through the plasma, undergoing collisions with plasma species and subsequently deposit on a strategically placed substrate. The substrate may be heated depending on the deposition's nature, and sputtering can employ either DC or RF power depending on the target's electrical conductivity [7, 8]. Reactive sputtering can be used to prepare alloys, compounds, solid solutions, or mixtures, and it has been essential for depositing oxides since the 1970s. However, the issue of resputtering during oxide sputtering poses challenges, leading to morphological changes on film surfaces (Fig. 9).

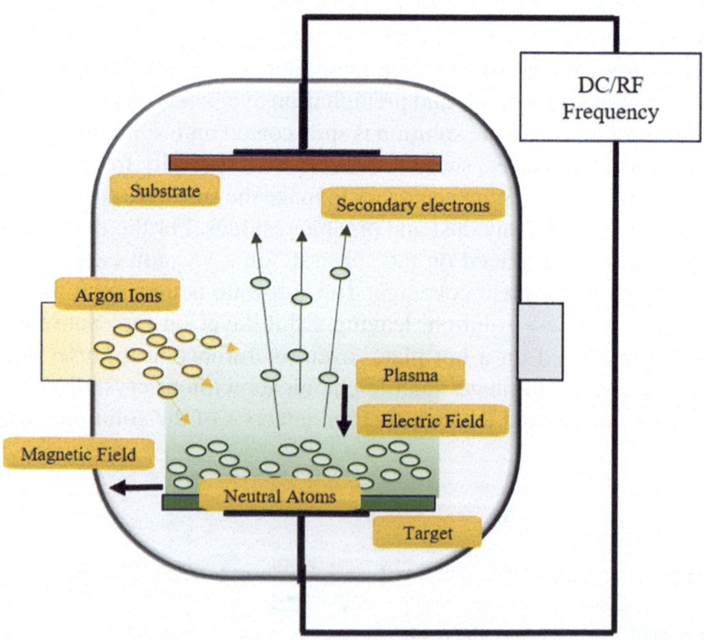

Fig. 9 A block diagram outlining RF magnetron sputtering method

The sputtering method is a physical vapor deposition (PVD) technique for thin film growth across various materials. It involves bombarding a target with high-energy ions or atoms, generating a plasma plume that reaches the substrate, where thin film nucleation and growth occur. RF magnetron sputtering is one variant of this method, utilizing radio-frequency power and magnetic fields to enhance plasma efficiency and sputter efficiency. The interdependence of parameters makes the process complex, and addressing issues like stoichiometry control, epitaxy, substrate nature, and oxygen activity is crucial. Despite the slow deposition rate, sputtering methods, including RF magnetron sputtering, can produce films with good structural quality and composition control, making them suitable for various applications.

2.3.5 Liquid Phase Epitaxy (LPE) Method

Liquid Phase Epitaxy (LPE) is a method used for growing films on a single crystal substrate by utilizing liquid solutions under near-equilibrium processing conditions. This technique involves the precipitation of a crystalline phase from a supersaturated melt onto a substrate, which acts as both the template for epitaxy and the physical support for the film. As the liquid cools, constituents within the solution precipitate out and grow epitaxially on the substrate. The growth rates in LPE generally range from 1 to 10 mm/min (Fig. 10).

LPE offers several advantages that make it an attractive option for film growth. One of the key benefits is its low-cost operation, which makes it accessible for various applications. Additionally, LPE provides precise control over the composition and thickness of the grown films, allowing for fine-tuning of desired properties. Another advantage is the relatively faster deposition rates compared to other epitaxial growth methods. However, it is important to note that LPE may have certain drawbacks. One limitation is the potential for poor surface finish in the resulting films. Achieving a smooth surface can be challenging in LPE, which may affect the film's quality and performance. Furthermore, the chemistry of complex systems can present challenges in LPE, as maintaining the desired composition can be more intricate in such cases. Additionally, the presence of volatile elements in the system can also pose challenges during the growth process. LPE finds practical use in various applications. One notable application is in the production of coated conductors based on YBCO (yttrium-barium-copper oxide) films. LPE offers a viable technique for the deposition of YBCO films required for high-temperature superconducting applications in coated conductors.

2.3.6 Pulsed Laser Ablation

Pulsed Laser Deposition (PLD) and Pulsed Laser Ablation (PLA) are related techniques that involve the use of lasers, but they differ in their primary objectives and applications. PLD focuses on controlled laser ablation of a target material, depositing

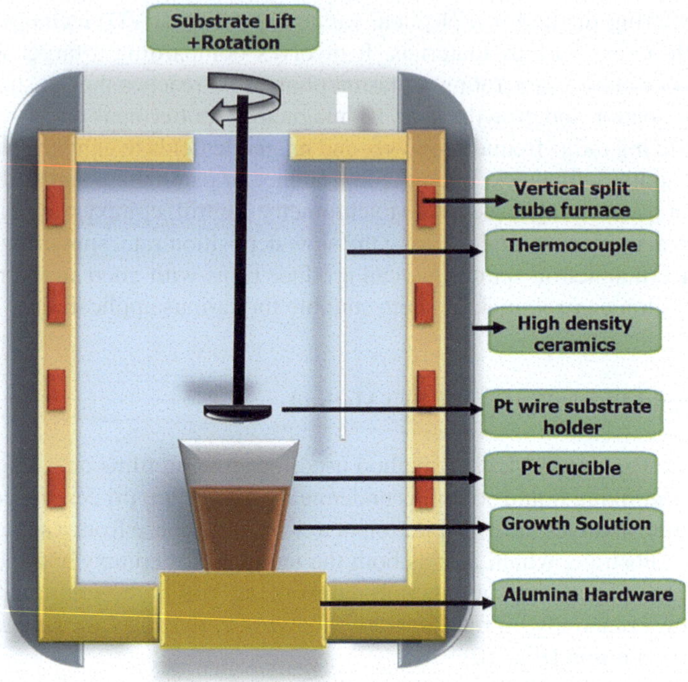

Fig. 10 Schematic outlining steps of Liquid Phase Epitaxy method (LPE)

the ablated material onto a substrate for thin film fabrication, commonly used in electronic devices and functional coatings. In contrast, Pulsed Laser Ablation (PLA) is a broader technique that involves using a pulsed laser to remove material from a target, serving various applications such as material removal, surface modification, and nanoparticle generation, without a specific emphasis on substrate deposition. PLD specializes in thin film creation, while PLA has a wider range of applications beyond deposition onto a substrate. PLA involves focusing a pulsed laser beam onto a target surface at an oblique angle, allowing the ablated material to deposit onto a substrate. Typically, a mixture of Ar and O_2 or pure O_2 gas is used for oxide deposition at pressures ranging from 20 to 50 Pa. UV excimer lasers like ArF (193 nm) and KrF (248 nm) are commonly used. The process starts with the meticulous selection of an appropriate target material, encompassing a spectrum from metals to semiconductors or insulators, contingent upon the specific requirements of electroceramic applications. A UV excimer laser is employed to produce brief pulses of intense light. This laser beam is precision-focused onto the surface of the chosen target material, instigating a phenomenon known as photoablation or laser ablation (Fig. 11).

This intricate interaction induces a swift elevation in the target's temperature, prompting the expulsion of atoms or molecules from its surface and giving rise to the formation of a plume or plasmatic cloud comprised of vaporized target material

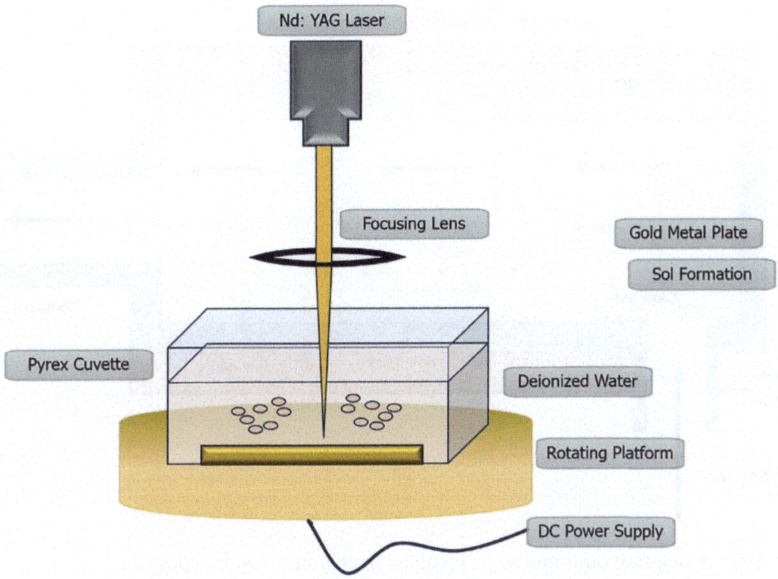

Fig. 11 A diagram outlining steps involved in pulsed laser ablation

[9]. This resulting plume, enriched with vaporized atoms and ions, expands away from the target surface under the influence of the high pressures generated during ablation. Strategically positioned in proximity, a substrate, critically tailored to align with the desired electroceramic properties, diligently collects the ablated material from the plume. The substrate, acting as the growth surface in this electroceramic synthesis process, plays a pivotal role in the subsequent formation of a thin film. As the vaporized atoms and ions precipitate onto the substrate, guided by the specific characteristics dictated by the composition and properties of the target material, the foundation for an electroceramic thin film is established. This phase of condensation ensures that the thin film takes on the electroceramic properties essential for its intended application. Laser ablation has been successful in depositing high Tc superconductors, ferroelectric thin films, and other oxide films. It offers advantages such as precise control over stoichiometry, efficient target use, compatibility with *in-situ* growth monitoring methods, operation at high pressures, potential for multilayer deposition, higher deposition rates, and minimal bombardment effects on the growing film. However, challenges like the creation of small droplets and deposition on larger area substrates still need to be addressed.

2.3.7 Chemical Vapor Deposition (CVD) Processes

In Chemical Vapor Deposition (CVD), precursor compounds are heated to vaporize them, and the resulting vapors are introduced into a heated reaction chamber

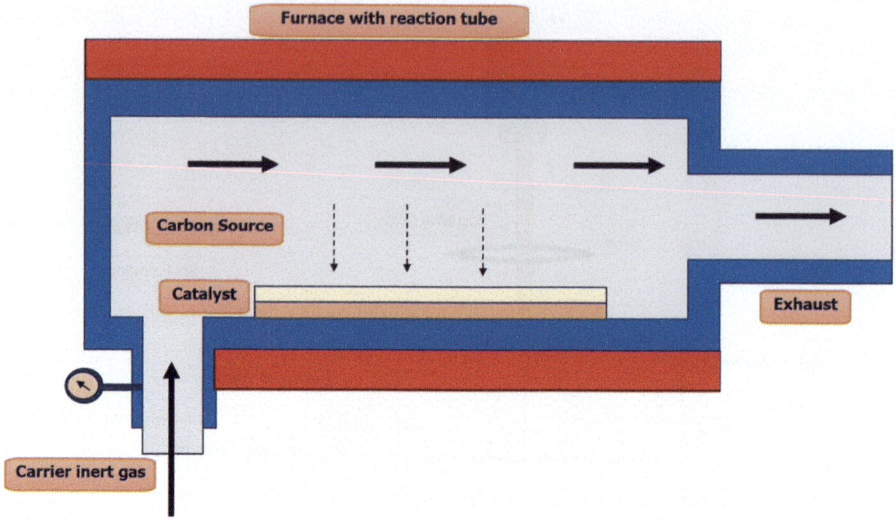

Fig. 12 A block diagram outlining steps involved in chemical vapor deposition

containing O2. The reaction vapors flow over a substrate, producing film species that deposit on the substrate. Precursor materials, often metalorganic compounds, are delivered to a reaction chamber in gaseous form using a carrier gas. The substrate, crucial for electrical properties, is meticulously prepared and placed within the chamber. Through chemical reactions, the precursors generate reactive species that adsorb onto the substrate surface. Surface reactions lead to the formation of a solid ceramic film, and desorption efficiently removes by-products. Control over temperature and pressure within the reaction chamber is pivotal for influencing growth rate, film quality, and electrical conductivity. The thin film continues to grow in a cyclic manner as additional precursor molecules contribute to adsorption, surface reactions, and desorption (Fig. 12).

The final deposited thin film, tailored for electrical applications, undergoes characterization to assess its thickness, composition, and electrical properties [10]. CVD is particularly valuable in producing ceramics for electrical applications, offering precise control over film properties and enabling the fabrication of materials crucial in the development of electronic devices, sensors, and other electrical components. CVD processes, including metallo-organic chemical vapour deposition (MOCVD) and plasma-enhanced MOCVD (PE-MOCVD), offer advantages such as lower deposition temperatures compared to physical vapor deposition (PVD) processes. However, they are complex due to numerous process parameters, and challenges include control of composition, accurate knowledge of phase diagrams, and the expense of MOCVD precursors. Despite these complexities, CVD processes can deposit films over large area substrates.

2.4 *Approaches for Single Crystal Electroceramics Processing*

The growth of bulk single crystals, especially those relevant to electroceramics, is a distinctive processing method that cannot be substituted by other techniques. Crystal growth, both a science and an art, requires patience, diligence, intuition, and sometimes luck. This ancient field has seen the evolution of numerous crystal growth methods. This chapter focuses on oxides, the lead members of the electroceramic group. Crystal growth for oxides falls into three categories: growth from the vapor, growth from the melt, and growth from fluxes. The selection of a method depends on factors such as the melting points of solids and their chemistry at high temperatures. Melting points can be congruent or incongruent, with incongruent melting being more common among oxides. The subsequent sections delve into detail about the three most widely used methods for crystal growth of oxides.

2.4.1 High Temperature Solution Growth (HTSG) or Flux Growth Method

This technique is employed to cultivate crystals of materials that undergo incongruent melting at extremely high temperatures. To achieve single-crystal synthesis, the materials are initially dissolved in high-temperature solvents, also known as fluxes, and then slowly cooled for crystal nucleation. In Fig. 13 we present a procedure for crystal growth using the HTSG or flux method. The crystal growth charge, mixed with a flux, is enclosed in a typically platinum crucible. The growth occurs in the presence of air using resistive heaters. Platinum crucibles are chosen for their chemical inertness, resisting reactions with aggressive melts containing flux and the charge. Although some fluxes may attack platinum at elevated temperatures, the damage is minimal and does not render the crucibles unusable. The crucible, with the charge and flux, must be kept exceptionally clean and is thus enclosed in a covered tall alumina crucible during the growth process.

Various fluxes are used, with the most prominent ones including mixtures of lead oxide (PbO) and vanadium oxide (V_2O_5), PbO and lead fluoride ($PbO–PbF_2$), PbO and boron oxide (B_2O_3), and barium borate ($BaO–B_2O_3$). PbO tends to lose oxygen at high temperatures, leading to the reduction of PbO to metallic Pb in a highly aggressive molten state. To counteract this, PbO_2 is often added. As outlined in Fig. 13, the charge undergoes rapid heating to the maximum temperature (T_m), followed by a soaking period. After the initial cycle, a slight temperature reduction initiates supersaturation of the dissolved charge in the molten flux. Subsequently, a second soaking period commences to achieve a homogenous supersaturated melt. The final stage involves very slow cooling, typically at a rate of 0.5–1 $^{\circ}Ch^{-1}$, creating a thermal range favorable for the nucleation of tiny crystallites. The slow cooling cycle may extend over days or weeks. After completion, the furnace is shut off, allowing it to rapidly reach room temperature. Knowledge of the phase diagram is

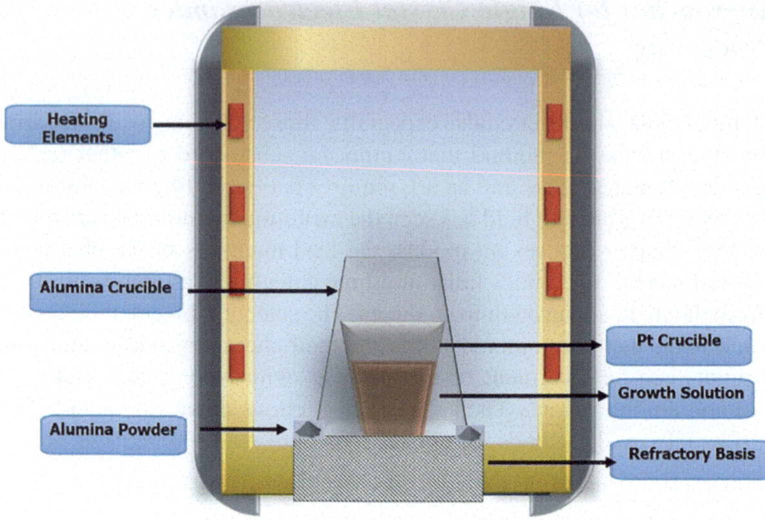

Fig. 13 A block diagram outlining steps involved in high temperature solution growth

beneficial for determining temperatures in different thermal cycles. Without such a diagram, intuitive temperature guessing and trial-and-error methods are employed to find the most suitable temperature ranges. Harvesting the crystals, ranging in size from 1 to 12 mm, involves dissolving the solid mass in an acid bath and filtering out the crystals. This process, albeit time-consuming, is chosen when HTSG is the sole viable method for crystal production. Despite potential impurities from the flux and possible defects, the HTSG method is extensively used for growing oxides when no other method is applicable. Noteworthy oxide crystals, including magnetic garnets, YIG ($Y_3Fe_5O_{12}$), piezoelectric PZN–PT (lead-zirconate-titanate and lead titanate solid solution), antiferroelectric lead–titanate ($PbTiO_3$), ferroelectric barium titanate ($BaTiO_3$), and semiconductor pseudobrookite (Fe_2TiO_5), have been successfully grown using the HTSG method. Crystals' size enhancement is achievable by introducing crucible rotation, melt stirring, a seed crystal, or nutrients (tiny crystallites from previous growth runs) during the growth process. These methods facilitate the adherence and growth of crystallites, resulting in larger crystal sizes.

2.4.2 Czochralski Growth Method

The Czochralski Growth Method, named after Jan Czochralski, who accidentally discovered it in 1916 while experimenting with molten tin, has become pivotal for crystal growth. The accidental discovery gained importance when Gordon Teale and colleagues at Bell Laboratories refined the technique for growing germanium crystals in 1948, leading to the invention of the transistor. This method involves growing

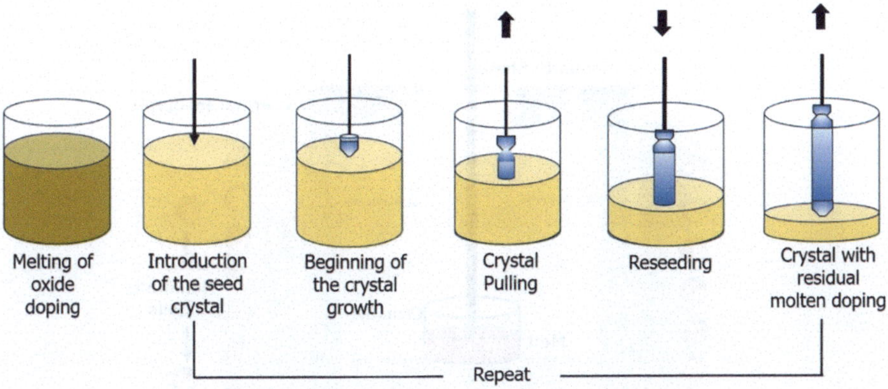

Fig. 14 A block diagram outlining Czochralski growth method

crystals from the melt of congruent materials, making it suitable for materials with congruent melting points. Nucleation begins on a seed by bringing it into contact with the melt, and once established, the crystal continues growing until the melt is exhausted. Successful Cz-growth experiments can produce large crystal boules with the same crystal orientation as the seed. For instance, silicon boules with diameters ranging from 2 to 6 inches and lengths of many feet are routinely grown on an industrial scale.

The Cz-growth system, illustrated in Fig. 14, involves rotating the seed and crucible pedestal in opposite directions to maintain a well-mixed melt. Computer-controlled processes ensure boule quality and diameter control. Pure argon is introduced during growth to expel SiO_2, which can form from the chemical reaction of the melt with the quartz crucible. The Cz-method is extensively used for growing oxide materials, with platinum or iridium crucibles employed due to their inertness to high temperatures and chemically aggressive melts.

2.4.3 Top Seeded Solution Growth (TSSG) Method

The Top Seeded Solution Growth (TSSG) Method is applicable to incongruently melting materials, combining elements of the Cz-growth method and the HTSG method. Understanding the phase diagram is crucial for developing a working protocol for TSSG (Fig. 15). This method is commonly used for growing large crystals of ferroelectric materials like barium titanate (BT) and potassium-tantalate-niobate (KTN), as well as various relaxor ferroelectrics [11].

The TSSG process involves melting the ceramic charge, introducing a seed crystal below the melt, rotating both the seed and crucible, and initiating a slow cooling process. The growth temperature is monitored using an optical pyrometer, while a pre-programmed growth protocol is executed by a computer. The growth can continue for days until a substantial crystal is obtained. A reliable phase diagram guides the

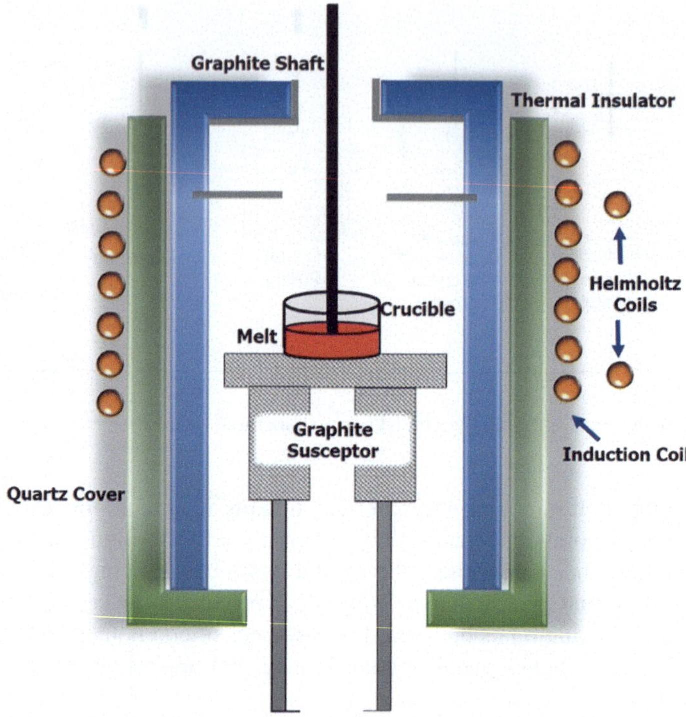

Fig. 15 A block diagram outlining Top Seeded Solution Growth (TSSG) method

cooling range and rate; for example, in the case of KTN, the phase diagram of potassium niobate ($KNbO_3$) and potassium tantalate ($KTaO_3$) is consulted. Slow cooling, often at a rate of 1 Ch^{-1} or less, allows for the growth of crystals with desired compositions. After growing a good-sized crystal, it is carefully pulled out from the melt, rotated to remove adhering liquid, and gradually brought to room temperature to prevent thermal stress-induced cracking. The TSSG method offers a versatile approach for growing crystals of varying compositions and sizes, making it valuable in the fabrication of optical quality large crystals.

3 Challenges

The preparation of electroceramic materials confronts numerous challenges and holds promising future prospects, particularly when considering diverse methods employed in their synthesis. Attaining precise control over composition, microstructure, and properties remains a critical challenge, as subtle variations can significantly impact electrical, magnetic, and thermal characteristics. The intricate processes, often involving high temperatures and intricate chemical reactions, demand meticulous attention to prevent impurities and defects. Scalability issues further complicate large-scale production. Addressing these challenges involves achieving stable phases with precise composition and favorable structure in electroceramics. The selection of device-compatible and cost-effective substrates, along with the ability to perform depositions over larger areas, remains a bottleneck for widespread commercial success in deposition methods. The interplay of parameters such as processing temperature and epitaxy adds complexity to identifying optimal conditions for electroceramic materials. Phase stability and stoichiometry control present additional hurdles due to the non-equilibrium nature of growth processes in electroceramics. Understanding phase evolution under specific process parameters, including deposition temperature and oxygen activity, is crucial, especially with perovskite oxides. Precise control of processing conditions is essential for achieving the correct phase, and stoichiometry control becomes complicated in techniques like sputtering. Challenges arise due to moisture sensitivity, high dielectric constant, and cost.

However, looking ahead, the future of electroceramic preparation methods offers exciting opportunities. Innovations in nanotechnology and manufacturing techniques are poised to transform the field, enabling the creation of materials with tailored properties at the atomic and molecular levels. The integration of electroceramics into emerging technologies, such as flexible electronics and the Internet of Things (IoT), will drive research efforts to address current limitations and explore new frontiers in materials science and technology.

4 Summary

The chapter provides an overview of the preparation of electroceramics, highlighting the adaptability to either bulk or thin film forms depending on the intended application. The bulk processing involves sequential steps such as material selection, weighing, mixing, calcinations, and sintering. In contrast, thin film processes display variability depending on the chosen method. Physical vapor deposition processes transform the material into a plasma and deposit it onto a substrate, while chemical processes involve the reaction and cracking of precursors over a substrate at moderately high temperatures. Although thin film processes offer advantages like achieving desired phase formation and structural perfection at lower temperatures, they come with the drawback of necessitating capital-intensive equipment. Chemical methods like sol-gel and Chemical Vapor Deposition (CVD), utilizing precursor reactions at moderate temperatures. Solution-based methods like Liquid Phase Epitaxy (LPE) and Hydrothermal Growth, which enable single crystal formation. Each technique offers unique advantages: physical methods achieve high structural quality, while chemical methods allow fine control at lower temperatures. However, many of these methods are capital-intensive. In conclusion, selecting appropriate processing techniques is critical for advancing applications in energy storage, computing, and communications. Emerging methods continue to enhance the versatility of electroceramics in modern technology.

Acknowledgements The authors express sincere appreciation to the Department of Biotechnology (DBT), Government of India, for their generous financial backing under project No. DBT/2022-23/July/N/2240 as part of the DBT-RA Program. The authors would like to express their sincere gratitude to Prof. A. Srinivasan and Prof. Aynur Unal for their encouragement, as well as to Dr. Mohit Kumar for his invaluable support.

References

1. Pandey, R.K.: Fundamentals of Electroceramics: Materials, Devices, and Applications. Wiley (2019)
2. Pandey, R.K.: Processing of Electroceramics, vol. 2. Basic Concepts of Equilibrium. Wiley (2018)
3. Moulson, J.M.H.A.J.: Electroceramics (2003). https://doi.org/10.1002/0470867965.fmatter
4. Jmar Al-Amani, A., Sreekantan, S., Ahmad Fauzi, M.N., Ruzuk, K.A.: Current Progress in synthesis and properties of doped bismuth titanate for advanced electronic application. In: Bismuth: Characteristics, Production and Applications (2012)
5. Kong, L.B. et al.: Ceramic powder synthesis. In: Transparent Ceramics, Topics in Mining, Metallurgy and Materials Engineering. Springer, Cham, (2015).https://doi.org/10.1007/978-3-319-18956-7_3
6. York, T.H.: Pulsed Laser Deposition of Electronic Ceramics and Analysis of the Ablation Plume (1996)
7. Safavi, M.S., Surmeneva, M.A., Surmenev, R.A., Khalil-Allafi, J.: RF-magnetron sputter deposited hydroxyapatite-based composite & multilayer coatings: a systematic review from mechanical, corrosion, and biological points of view. Ceram. Int. **47**, 3031 (2021). https://doi.org/10.1016/j.ceramint.2020.09.274
8. Sumets, M., Kostyuchenko, A., Ievlev, V., Dybov, V.: Electrical properties of phase formation in $LiNbO_3$ films grown by radio-frequency magnetron sputtering method. J. Mater. Sci. Mater. Electron. **27**, 7979 (2016). https://doi.org/10.1007/s10854-016-4792-y
9. Mertin, M., Offenberg, D., An, C.W., Wesner, D.A., Kreutz, E.W.: Pulsed laser deposition of electroceramic thin films. Appl. Surf. Sci. **96–98**, 842–848 (1996). https://doi.org/10.1016/0169-4332(95)00569-2
10. El Amrani, A., Bekhtari, A., Mahmoudi, B., Lefgoum, A., Menari, H.: Experimental study of the effect of process parameters on plasma-enhanced chemical vapour deposition of silicon nitride film. Vacuum (2011). https://doi.org/10.1016/j.vacuum.2011.08.003
11. Long, X., Ye, Z.G.: Top-seeded solution growth and characterization of rhombohedral PMN-30PT piezoelectric single crystals. Acta Mater. (2007). https://doi.org/10.1016/j.actamat.2007.08.009

Foundations of Ceramic Synthesis: Processes, Principles, and Potential Biomedical Prospects

Shagun Varshney, Abhishek Nigam, and Nidhi Mishra

Abstract Ceramics have long been recognized for their exceptional properties, ranging from mechanical stability to biocompatibility, making them highly promising candidates for various medical and healthcare innovations. The chapter elucidates the essential processes involved in ceramic synthesis, from precursor selection to material shaping and different heat treatments. It delves into diverse synthesis methods, including traditional solid-state reactions and advanced techniques like sol–gel and hydrothermal routes. The discussion extends to the crucial principles underlying ceramic synthesis, including crystallography, phase transformations, and microstructural development. These principles serve as the cornerstone for designing ceramics with optimal properties for biomedical applications. The chapter explores how variations in synthesis parameters influence material characteristics, leading to tailored properties such as mechanical strength, surface reactivity, and porosity for their applications in tissue engineering, drug delivery, and medical diagnostics. By synergistically addressing the synthesis intricacies and the biomedical landscape, this chapter provides a comprehensive resource for harnessing the potential of ceramics to revolutionize healthcare technologies. Ultimately, it underscores how a deep understanding of synthesis processes and principles lays the groundwork for unlocking the transformative capabilities of ceramics in the dynamic field of biomedicine as well.

S. Varshney
Department of Biomedical Engineering, UV Patel College of Engineering, Ganpat University, Mehsana 384012, Gujarat, India

A. Nigam
Department of Electronics and Communication Engineering, Institute of Engineering and Technology Lucknow, Lucknow 226021, Uttar Pradesh, India

N. Mishra (✉)
Department of Applied Sciences, Indian Institute of Information Technology Allahabad, Prayagraj 211012, Uttar Pradesh, India
e-mail: nidhimishra@iiita.ac.in

U. Kumar (ed.), *Defects Engineering in Electroceramics for Energy Applications*, Engineering Materials, https://doi.org/10.1007/978-981-97-9018-0_5

1 Introduction

Material science has been recognized as a field of study because of global research on metals, polymers, and ceramics over this century. These studies are multidisciplinary, which is unique to the field of ceramics. At the moment, chemistry is contributing more and more significantly to the study, creation, and manufacturing of ceramic materials [80]. The term "ceramics" has historically been used to refer to materials made of clay, including bricks, tiles, pottery, tableware, and glass. These materials are produced from naturally existing minerals including sand, quartz, bauxite, feldspar, etc. Improved mechanical strength, fracture toughness, wear resistance, refractoriness, dielectric, magnetic, and optical properties distinguish advanced ceramics from conventionally synthesized ceramics [55]. Advanced ceramics are inorganic substances with significant characteristics synthetically obtained from highly pure powders to regulate their morphology, structure, and other biological characteristics [65]. Advanced ceramic powders with size lying in the submicron range, thin size delivery, and absenteeism of agglomeration are often prepared using wet-chemical methods, such as Hydrothermal, sol–gel route, freeze-dry route, etc. [48]. These synthetic routes require expensive reagents and chemicals, a long processing period, and expert equipment.

Nanotechnology has developed a focus for many researchers in recent times. Researchers have been interested in the synthesis of nanostructured ceramics owing to their numerous applications and characteristics, such as high surface to volume ratio [83]. With the advent of nanotechnology, there has been considerable interest in the development of ceramic-based nanoparticles as well [11, 22, 60], with improved characteristics and their use in a variety of disciplines, including surface chemistry and chemical synthesis. Silica (SiO_2) [93], Alumina (Al_2O_3), Titania (TiO_2), Silicon Nitride (Si_3N_4), Zirconia (ZrO_2), and Silicon Carbide (SiC), etc. are of the ceramic-based NPs that have been synthesized via different techniques. The bioanalytical dimensions in ceramics have drastically changed as a result of the adaption of these nanostructures for biological, biotechnological applications [81]. Due to their distinct physio-chemical and mechanical characteristics, these ceramic nanoparticles have been used in the fabrication of composites being widely used in science, electronics, catalysis, photonics, etc. [82, 90]. Ceramic-based nanocomposites have been synthesized using a variety of techniques, counting, twin screw extrusion, solvent casting, melt compounding, extrusion blow moulding, etc. Their enhanced mechanical, biological, chemical, and antibacterial abilities have been exploited for the development of materials with high demand in different industries. In the realm of biomedical engineering, these nanocomposites are of great attention because of their enhanced antimicrobial properties. Due to several environmental concerns, green or biogenic synthesis of these ceramic-based nanocomposites has been preferred over other conventionally employed synthetic routes. The different routes for synthesis of ceramic particles are shown in Fig. 1.

The research based on the development and applications of ceramics has attracted imperative consideration. Several factors show a significant role in interesting area

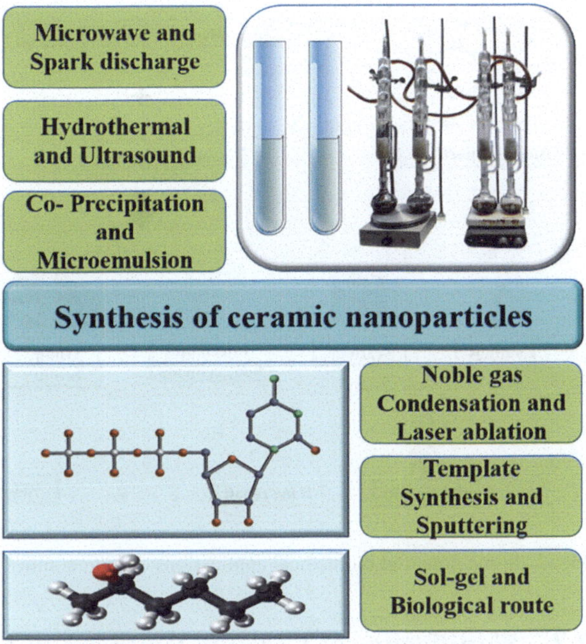

Fig. 1 Different routes for synthesis of ceramic nanoparticles [70]

of ceramics-based research. One of them may, be the potential biomedical applications including antimicrobial activity, drug delivery systems, tissue engineering, etc. [76, [92]. Calcium-based substances, such as carbonates and phosphates, are some of the emerging ceramic-based biomaterials in relations of the rising quantity of publications, and patents being published each year. These ceramic-based biomaterials are commonly termed bio-ceramics. These can be either synthetic or naturally occurring in origin and are meant to integrate with bones and can also be used to replace conventionally used metallic implants [67]. Some potential applications of bio-ceramics are shown in Fig. 2.

These bio-ceramics are significant in the biomedical area due to their occurrence in the human body which can be coordinated by their favorable and harmonious physio-chemical characteristics [1]. For instance, the dental crown treatment in the 18th century employed porcelain which is well-known bio-ceramic. Plaster of Paris was then used in dentistry in the 19th century [17]. Due to improvements in processing technology, ceramic applications in the medical profession expanded over the 20th century [53]. Bio-ceramics outperform metal-based biomaterials in terms of mechanical strength, low malleability, poor degradation, high melting point, and non-corrosive nature. As compared to bone, bio-ceramics have a lower value of Young's modulus, weak fracture stiffness, and are both rigid and brittle [37, 39, 45, 72].

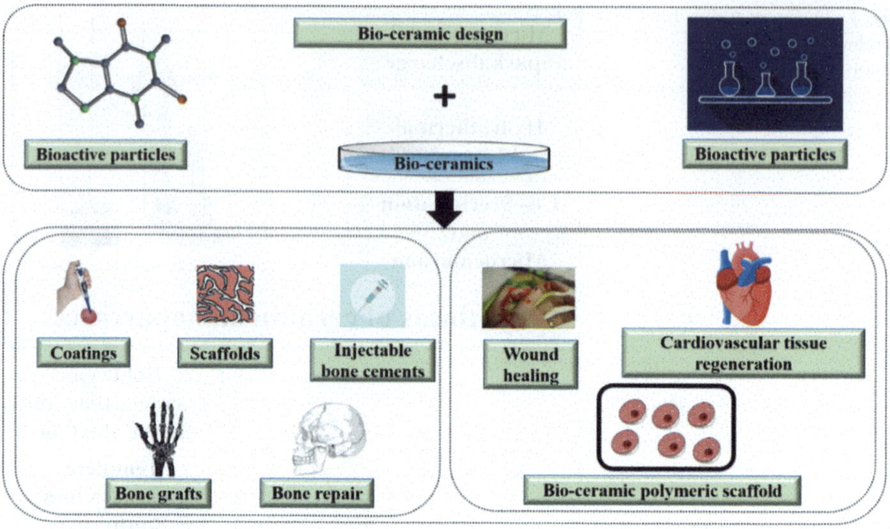

Fig. 2 Some potential biomedical applications of bio-ceramics [64]

1.1 Significance of Ceramic Synthesis in Science and Medicine

Ceramics have been essential in many commercial and scientific applications for centuries, but in recent years, their importance in the fields of biomedicine and healthcare has increased dramatically. Bio-ceramics are ideal candidates for a variety of biomedical applications due to the amalgamation of their characteristics including biocompatibility, high compressive strength, resistance to wear, stability at high temperatures, etc. Some of the areas where these bio-ceramics are employed are listed below:

- **Dental implants**

Dental implants have transformed contemporary dentistry by offering a durable and dependable method of lost teeth restoration. Dental implant components like crowns are often made of ceramic materials, especially Yttria-stabilized Zirconia and Alumina. These are perfect for dental restorations owing to their high power-driven strength, cosmetic aesthetics, and enhanced biocompatibility. Patient safety is also ensured by the corrosion resistance and no allergic response of these ceramics. Some commercially available ceramic-based products are shown in Fig. 3.

- **Tissue engineering**

Using biocompatible materials, bone tissue engineering objectives to restore the damaged tissue and organ. Bio-ceramics including hydroxyapatite and bioactive glasses, have been used for fabricating scaffolds in tissue engineering. The ceramic

Fig. 3 Bio ceramics based products in commercial medical application [84]

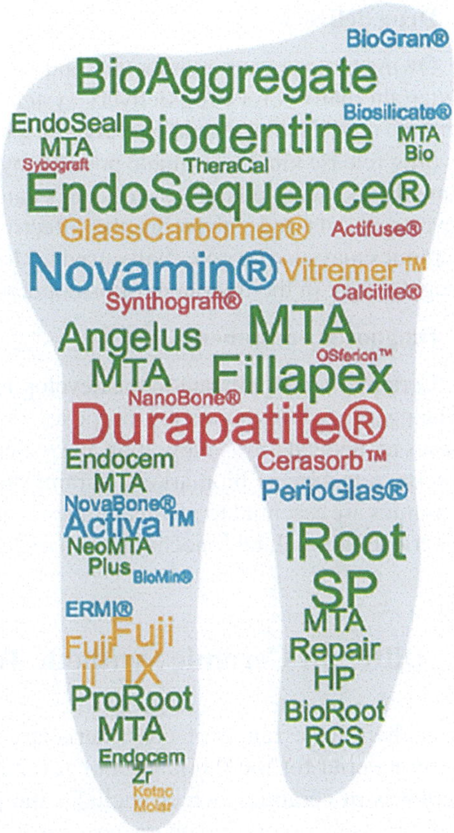

hydroxyapatite, which resembles the bone in its composition, can be effectively employed for the development and osteo-integration with osteoblasts. The material properties, for example porosity, surface characteristics, and mechanical strength, may be combined with other materials using a variety of synthetic techniques to encounter the requirements of bone tissue engineering.

- **Bone repair**

Ceramics are essential in orthopaedic applications, particularly for bone regeneration and repair. Ceramics for example hydroxyapatite and tricalcium phosphate, are frequently utilized in bone transplants and as fillers for filling bone voids. These substances serve as a framework for the development of new bones; get resorbed as natural bone regenerates, and also reduce the possibility of infection or rejection by the human body. Porous ceramic materials synthesized with a specific composition can be produced using customized synthetic techniques to support bone repair.

- **Drug delivery**

Owing to its porous architecture and continuous release nature, ceramics provide better alternatives for drug delivery systems. Mesoporous silica is a type of porous ceramic material that can effectively transport drug molecules. The precise control of its drug release kinetics is made possible by the large surface area and tuneable pore dimensions. Targeted delivery of drugs can also be achieved resulting in negligible adverse effects and enhancing the effectiveness of medically administered drugs. Ceramics may also be potentially used for localized chemotherapy and bone-targeted drug delivery in the treatment of osteoporosis [18].

- **Diagnostic equipment**

Ceramics are significant in the development of diagnostic instruments in addition to being utilized for therapeutic purposes. For example, to produce and detect acoustic waves in medical ultrasonography, piezoelectric ceramics are used [4]. For rapid and precise detection of biomarkers, ceramic substrates are also utilized in biosensors. Ceramics are essential to medical imaging and diagnostics because of their enhanced electrical, thermal, and mechanical characteristics.

2 Different Ceramic Synthetic Techniques

With their wide range of characteristics, ceramics have long been considered a basic material for the fabrication of several advanced materials [17]. However, the synthesis of ceramics, which includes the production of solid, inorganic materials with precisely controlled characteristics, is a complex and flexible task at the same time. Several synthetic techniques have been explored over time, each with its advantages and disadvantages [59]. The variety of these methods allows us to engineer ceramics with accuracy and control, satisfying the demands of diverse sectors such as electronics, aerospace, healthcare, etc. Traditionally employed synthetic techniques like sintering and precipitation are combined with more advanced approaches like sol–gel, microwave-assisted synthesis, and hydrothermal routes [33, 54, 68]. The different synthetic routes used for ceramics are discussed below:

2.1 Sol–Gel Route

Synthesis of different nanostructures, particularly metal oxides, sol–gel route is widely employed chemical approach. In this route, the precursor (often a metal alkoxide) is dissolved in a solvent like water or alcohol, heated, and continuously stirred until it gels. The gel obtained during the hydrolysis is wet and must be properly dried to obtain the desired qualities. The obtained gels are pulverized and then heat treated using calcination after drying. The cost-effectiveness of the sol–gel process allows considerable control of materials compositions owing to low reaction temperature [6, 94]. This technique may be utilized to obtain transition between thin layers of metal oxides in a variety of applications. The synthesized materials via the sol–gel route are employed in a variation of optical, electrical, energy-related, surface engineering, biosensor, medicinal, and separation technologies (like chromatography) applications [78]. To obtain the homogeneous composition of ceramics from different oxide precursors, solution mixing followed by gelation was invented more than 30 years ago. The precursors used at first were organic, and afterward, mostly colloidal inorganic precursors [73, 75] were employed in the synthesis. Methods for producing abrasive grit, ceramic fibers, thin coatings, and nuclear fuel pellets have been developed using sol–gel. A typical flowchart for synthesis of ceramics via the sol–gel route is shown in Fig. 4.

The three main steps in the sol–gel route are: (i) dissolution of different oxides in solution using metal organic precursors; (ii) obtaining sol and allowing to gel to maintain chemical homogeneousness through dehydration; and (iii) formative after gelation into the desired shape. Some of the important applications of the sol–gel route are illustrated in Fig. 5.

During the mid-nineteenth century, there was a growing inclination towards employing sol–gel method to yield mineral ceramics and glass-like materials. This trend was initiated by *Ebelman* and *Graham*, who conducted extensive experiments on silica gels. During the 1950s, comprehensive research initiatives were carried out to investigate the ceramics synthesis and glass structures by solgel route [8]. The technique possessed the capability to synthesize a variety of mineral oxides, including TiO_2, SiO_2, and ZrO_2. Two distinct forms of dry gels, namely aerogels,

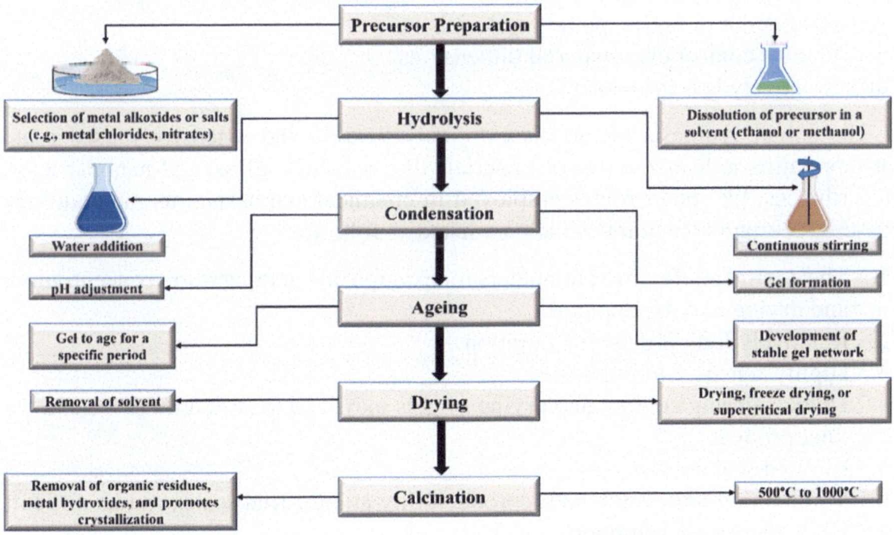

Fig. 4 Flowchart for synthesis of ceramics nanoparticles via sol–gel route

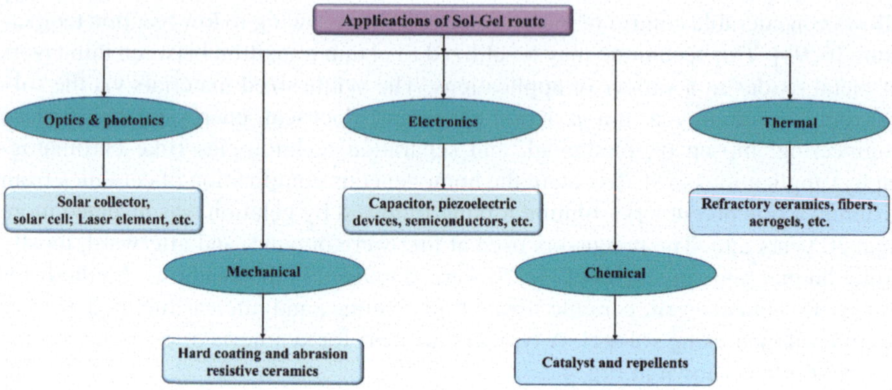

Fig. 5 Some possible applications of sol–gel route [77]

and xerogels, were also developed by the drying of wet gels. An aerogel is a type of gel that is classified as dry due to the removal of moisture. The preservation of the structure of the primary gel is mostly reliant on the method used for the removal of moisture from the gel. The main advantage of employing sol–gel techniques is listed below:

i. Simplicity of process
ii. High-purity end products
iii. High efficiency and production yields
iv. Complex and homogeneous morphologies can be obtained
v. Production of materials with modified physio-chemical properties
vi. Production of highly porous materials
vii. Precise control over material dimensions
viii. Relatively low investments

The solgel synthesis method is extremely flexible and extensively used route for fabricating a diverse range of materials like ceramics, glass, and nanoparticles. Nevertheless, like other routes employed in chemical synthesis, this methodology possesses certain constraints which are listed as follows:

1. The control of effective parameters for example pH, temperature, concentration, and drying may be required
2. Relatively long processing duration
3. Highly sensitive to impurities
4. Defects during cooling and drying of gels may lead to structural defects in the final product
5. Limited scalability
6. Maintaining uniformity and reproducibility at industrial scales can be difficult
7. High energy consumption
8. May involve the use of hazardous precursors, such as alkoxides or acids which require proper handling and safety measures

9. Sol–gel route may not be suitable for all types of materials or intricate structures.
10. The disposal of by-products may severely affect the environment effects if not managed properly

2.2 *Hydrothermal Route*

In recent years, a constantly expanding broad family of materials, like ceramics, that can be synthesized under moderate environments (T < 350 °C, P < 100 MPa), has triggered interest in hydrothermal synthesis. This has led to a nearly quadrupled number of research articles between 2000 and 2004 based on the hydrothermal synthesis of ceramics [87]. The hydrothermal route uses for single or heterogeneous phase processes in hydrated mediums at relatively high temperatures (T > 25 °C) and pressures (P > 100 kPa) [10]. The synthesis is often carried out at an autogenously maintained pressure, that leads to saturated vapour pressure of the solution at a particular composition and temperature. The conditions including 500 MPa and 1000 °C are the upper limits of hydrothermal synthesis [74]. A typical flowchart for synthesis of ceramics by hydrothermal route is shown in Fig. 6.

Hydrothermal route includes the controlled development of nanoscale materials using high-temperature and high-pressure aqueous solutions. Due to its capacity to synthesize nanoparticles with certain dimensions, morphology, and characteristics, the hydrothermal route has found multiple applications in numerous disciplines some of which are listed in Fig. 7.

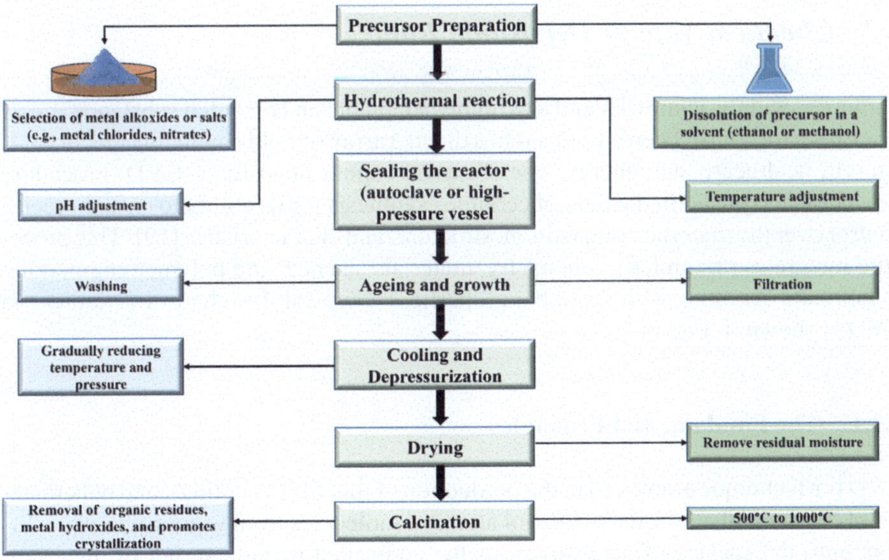

Fig. 6 Flow diagram for the synthesis of ceramics via hydrothermal route

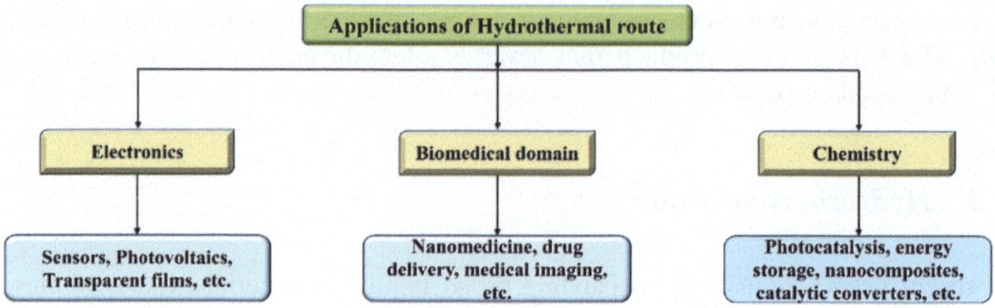

Fig. 7 Applications of hydrothermal route [32]

As compared to other synthetic techniques, hydrothermal route offers several advantages. The advantages of employing the hydrothermal route are listed below:

1. It is used for obtaining ceramics in all its forms, including powders, fibers, crystals, and coatings of polymers, etc.
2. Lower time and energy consumption
3. Post-treatments like calcination, mixing, and milling procedures are either not required or limited
4. Precipitation of preserved powders the from solution at a controlled rate with regularity of nucleation, development, and aged
5. Improved and controlled growth and morphology of crystals with low aggregation which is not possible with these processes

2.3 *Chemical Vapour Deposition Route*

Ceramics, widely acknowledged as a highly adaptable and essential substance across several sectors, have been employed in a diverse array of fields including electronics, aircraft, healthcare, and energy. The chemical vapour deposition (CVD) procedure is widely recognized in the field of ceramics synthesis for its ability to ensure precise control over the material composition, structure, and characteristics [19]. This procedure integrates principles of chemistry, materials science, and polymer engineering to fabricate ceramics with desirable properties. A typical flowchart of ceramics via CVD is shown in Fig. 8.

2.3.1 The Fundamental Principles

CVD is a technique employed in the production of thin films, coatings, and bulk materials [21]. It involves the deposition of atoms or molecules from vapour-phase precursors onto the surface of a substrate for the controlled manufacturing of materials. CVD enables the production of ceramic materials characterized by their remarkable

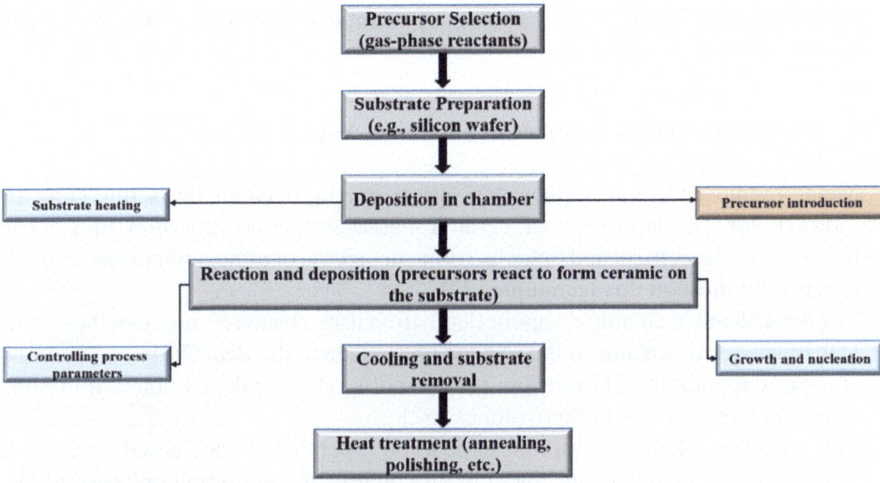

Fig. 8 Flowchart for the chemical vapour deposition-based synthesis of ceramics

purity, homogeneity, and accuracy. The procedure is dependent on a set of underlying principles:

1. The initial step involves the careful selection of precursors. These may exist as gaseous compounds, such as metalorganics or chlorides, vaporizable solid precursors, often in powder form, or liquid precursors, such as organometallic solutions [97].
2. Before deposition, it is the surface is thoroughly cleaned and the substrate is prepared. This is essential to ensure the appropriate adherence and consistency of the ceramic layer.
3. The control and optimization of temperature, pressure, and carrier gases (e.g., H_2, N_2, Ar) are crucial in obtaining the desired chemical reactions [44].
4. The process of precursor delivery involves the introduction of precursors into the reactor. In certain cases, it may be necessary to convert liquid precursors into vapours, whereas gaseous precursors and carrier gases are often combined and introduced into the reactor.
5. Chemical processes occur within the reactor, where the precursor molecules undergo transformations leading to the synthesis of ceramics. The process of deposition involves the formation of desired ceramics on the substrate [9].
6. To provide precise control over the thickness and quality of a film, it is necessary to monitor and adjust the parameters such as deposition time and growth rate.
7. Post-treatments may be employed to supplement the qualities of ceramic materials, depending on the application and desired properties [98].
8. The synthesized ceramic undergoes a range of characterization techniques, including X-ray diffraction, scanning electron microscopy, and different spectroscopic techniques to assess its physio-chemical, biological, mechanical, and

morphological characteristics. This is important to ensure that the material aligns with the necessary standards.

2.3.2 Variations and Techniques in Chemical Vapour Deposition

1. Metalorganic chemical vapour deposition is employed in the semiconductor industry for the deposition of ceramic-based semiconductor thin films. The production of electrical and optoelectronic devices is of utmost importance which is achieved through this technique [23].
2. Plasma-enhanced chemical vapour deposition uses plasma to augment the chemical processes occurring in the vapour phase, hence the deposition can be done at lower temperatures. This technique is employed in the deposition of thin films in microelectronics and photovoltaics [52].
3. Low-pressure chemical vapour deposition operates at decreased pressures, leading to significant enhancement in film quality and minimum impurities [35].
4. Atmospheric-pressure chemical vapour deposition is different from low-pressure chemical vapour deposition by its operation at atmospheric pressure, rendering it a comparatively basic and economically advantageous alternative. Its utilization is prevalent in the manufacturing a diverse array of ceramics, including oxide-based materials [50].
5. The process of hydriding and dehydriding-based CVD includes the use of hydrogen as a reactant, facilitating the production of carbides and nitrides.
6. Reactor designs can exhibit variations, with hot-wall reactors characterized by the presence of heated walls, whereas cold-wall reactors utilize external heating techniques.

2.3.3 Applications of Ceramics Synthesized via Chemical Vapour Deposition

1. Ceramics synthesized via CVD are widely utilized in the microelectronics sector [19]. These ceramics provide insulation, act as diffusion barriers, and even serve as piezoelectric substances in sensors and actuators.
2. CVD is employed in the production of optical coatings, waveguides, and semiconductor materials in the construction of lasers, light-emitting diodes, and photovoltaic devices [85].
3. Ceramics synthesized using CVD have found extensive applications in aircraft sectors owing to their capacity to withstand high temperatures [25]. It makes them well-suited to be used in engine components, thermal protection systems, and lightweight structural materials.
4. The chemical industry employs CVD in the manufacturing of catalysts, corrosion-resistant materials, and membranes in many chemical processes [3, 88]

5. The application of CVD-derived ceramics is prevalent in the biomedical domain, where they are employed as coatings on medical implants, bioactive ceramics, and drug delivery systems.
6. Ceramics fabricated using CVD find use in several energy-related fields such as solid oxide fuel cells, gas separation membranes, and thermal insulation for power production and energy storage systems.
7. Cutting tools and machining applications utilize hard and wear-resistant ceramics that are manufactured by CVD [26].
8. Ceramics produced via CVD have been used in photonic gadgets, fiber optic components, and optical coatings.

2.3.4 Challenges in chemical vapour deposition techniques

1. The availability of appropriate precursors is a crucial factor since specific ceramics may have restrained access to these precursors.
2. The design and operation of CVD reactors may be complex and need thorough regulation of several parameters.
3. Certain CVD methods employ precursors that are toxic or hazardous, hence giving rise to concern over their environmental impact and safety.
4. CVD can incur significant economic burdens as a result of the requirement for specialized apparatus and precursors.

2.4 Combustion-Based Synthesis of Ceramics

The synthesis of ceramics has been a significant area of academic examination and practical implementation for a prolonged period. Combustion-based synthesis, commonly referred to as self-propagating high-temperature synthesis, is a pioneering approach in the synthesis of ceramics [12]. This route has several benefits, including high efficiency, cost-effectiveness, etc. rendering it a convincing option for the fabrication of various ceramic morphologies. A typical flowchart for the combustion-based synthesis is shown in Fig. 9.

2.4.1 Principles of Combustion-Based Synthesis of Ceramics

Combustion-based synthesis is a relatively faster and highly exothermic route that relies upon the controlled burning of reactants [89]. The fundamental concepts underlying this route are as follows:

- **Exothermic reaction**

This depends on the chemically reactive processes which are exothermic. These reactions involve a fuel and an oxidizer, resulting in the release of a considerable

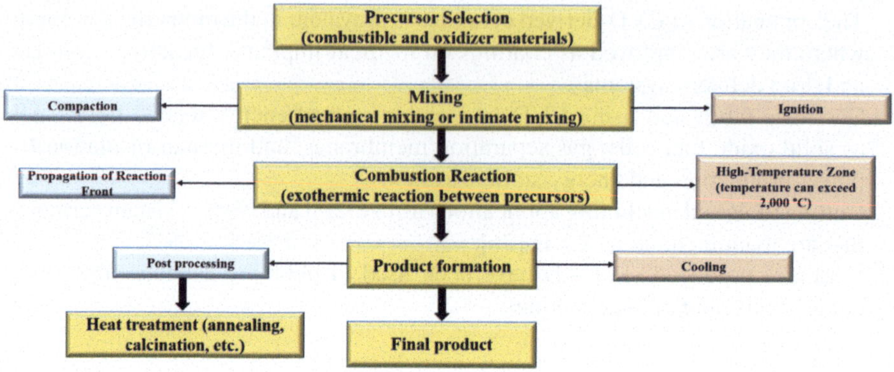

Fig. 9 Flowchart for the combustion-based synthesis of ceramics

quantity of thermal energy upon ignition [58]. The energy obtained upon combustion is utilized for aiding the synthesis of ceramics.

- **Elevated temperatures**

During the combustion process, temperatures can surpass the threshold of 1000 °C and in certain cases, even 2000 °C. The attainment of the appropriate phase transitions and crystal forms in the final ceramic product is dependent on these elevated temperatures [7].

- **Self-propagating reaction**

This phenomenon is observed when a chemical reaction can sustain and propagate itself without any external energy. Once started, the combustion is self-sustaining and can persist without external heat input.

2.4.2 Steps in Combustion-Based Synthesis

The process of combustion-based synthesis involves several sequential steps which are as follows:

- **Mixing of reactants**

The initial constituents, typically including a metallic powder (fuel) and an oxidizing agent, are combined to obtain a uniform powder mixture. The composition of finished ceramics is determined by the selection of precursors and their stoichiometry.

- **Ignition**

The composite powder is ignited by a heat source at an elevated temperature through an electrical discharge, or a laser. The ignition process serves as the first trigger to initiate the combustion reaction.

- **Propagation**

After ignition, the combustion reaction rapidly spreads throughout the mixture. The exothermic characteristic of the reaction maintains elevated temperatures as it propagates throughout the specimen.

- **Formation of products**

As the propagation of the combustion wave advances, it attains temperatures conducive to the formation of the desired ceramic phase(s). The ultimate ceramic product is obtained inside the reaction front and undergoes solidification as the reaction advances.

- **Cooling**

Once the combustion wave has traversed, the ceramic product undergoes cooling. The cooling process plays a crucial role in achieving stability and preventing the formation of undesired phases.

2.4.3 Advantages of Combustion-Based Synthesis

Combustion-based synthesis offers several benefits that render it a highly attractive option for ceramic production. Some of them are listed below:

- **Rapid reaction kinetics**

This makes the combustion-based synthesis be time-efficient approach as compared to the traditional heat treatments like sintering, roasting, calcination, etc.

- **Cost-effectiveness**

The rapid combustion-based synthesis facilitates the efficient fabrication of ceramics, rendering it a time-efficient approach in contrast to conventional sintering or sol–gel routes. The cost-effectiveness of the procedure is also enhanced by its simplicity and less complicated equipment required.

- **High yield and purity**

High temperatures generated during combustion promote the formation of pure ceramic phases, reducing the chances of impurities or secondary phases in the finally obtained product.

- **Stoichiometric control**

Controlled stoichiometry ensures precise management of reactant ratios in a chemical reaction to achieve desired results. The exact regulation of reactant stoichiometry enables the synthesis of ceramics with tailored compositions and characteristics.

- **Scalability**

Combustion-based synthesis has scalability, rendering it appropriate for both laboratory-scale investigation and industrial manufacturing as well.

2.5 *Microwave-Based Synthesis of Ceramics*

The utilization of microwave-based synthesis has been recognized as a groundbreaking method for the synthesis of ceramics, presenting a multitude of benefits including rapidity, energy conservation, consistency, specificity, and scalability [46]. This route has been implemented across several sectors, encompassing aircraft, electronics, healthcare, and environmental conservation. A typical flowchart for the synthesis of ceramics via a microwave-based route is shown in Fig. 10.

The microwave-based synthesis involves the interaction of microwaves with the precursors to enable the reaction between them [79]. The fundamental principles of this route are listed as follows:

- **Microwave energy**

Microwave energy refers to electromagnetic radiation with wavelengths ranging from 30 cm to 1 mm, falling within the microwave region of the electromagnetic spectrum. Microwaves encompass a spectrum of electromagnetic radiation, with frequencies from 300 MHz to 300 GHz.

- **Dielectric and uniform heating**

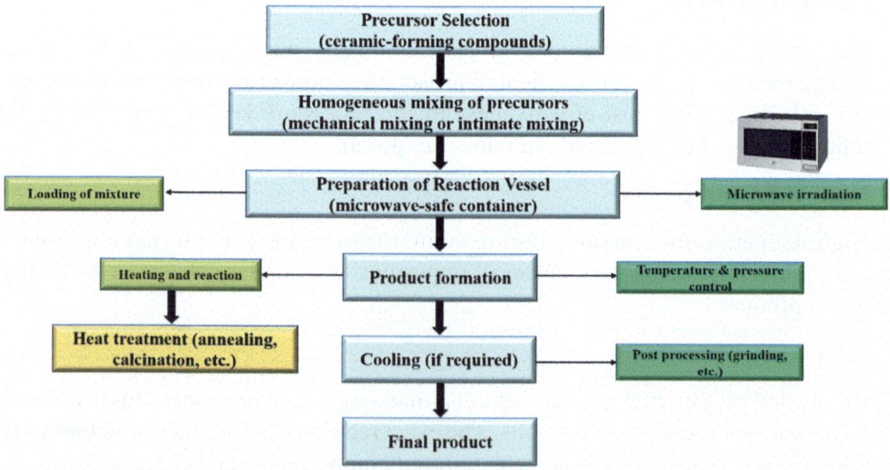

Fig. 10 Microwave-based synthesis of ceramics

Dielectric heating is a process in which an electric field is applied to a dielectric material, resulting in the generation of heat. In the field of ceramics, it is common for the precursor materials to consist of polar species. Consequently, the species are stimulated by the microwaves, resulting in efficient internal thermal energy generation.

- **Rapid kinetics**

Microwaves facilitate the homogeneous application of heat, hence resulting in the rapid reaction kinetics for ceramic synthesis. This is advantageous due to its ability to rapidly produce ceramic phases without the need for long exposure to high temperatures.

2.5.1 Steps Involved in Microwave-Based Synthesis of Ceramics

- **Precursor preparation**

The initial consisting of metal salts, metal oxides, or alternative ceramic precursors, are accurately blended to produce a uniform amalgamation. The blend is commonly denoted as the "precursor batch."

- **Loading of samples**

The initial precursor batch is placed into a microwave-safe container such as a ceramic crucible or a glass vial. To transmit the energy to the sample, the container must be transparent to microwaves. The container is positioned into the microwave cavity and then exposed to microwave radiation which triggers chemical reactions resulting in the synthesis of ceramic phases.

- **Monitoring of reactions**

The progression of the reaction may be calculated and analyzed by infrared spectroscopy, thermogravimetric analysis, X-ray diffraction, etc. These techniques aid in the guarantee of the formation of the desired ceramic phases.

- **Cooling**

The process of cooling is essential and involves the reduction of temperature and the collection of the desired product at the end. Cooling is employed to decrease the temperature of a substance or system. The container is removed from the microwave chamber and is allowed to be cooled. The ceramic product obtained is subsequently collected for further processing and characterization.

2.5.2 Applications of Microwave-Based Synthesis in Ceramics

The utilization of microwave-assisted synthesis in the field of ceramics includes a diverse array of industrial and research-based applications which are listed below:

- **Synthesis of advanced ceramics**

The utilization of microwave synthesis has been employed in the production of ceramics that exhibit enhanced characteristics like as high-temperature resistance, electrical conductivity, and mechanical strength. This enables the ceramics to be utilized in several industries such as aircraft, electronics, etc.

- **Nanomaterials**

Microwave-assisted synthesis has proven to be highly efficient in the production of nanomaterials, including ceramic nanoparticles and nanocomposites [14]. These materials can be widely employed in several applications including catalysis, drug delivery, and improved coatings.

- **Bio-ceramics**

Microwave-based routes can be employed for the synthesis of bio-ceramics which are extensively utilized in dental and orthopaedic implants. This route synthesis plays a significant role in facilitating the synthesis of biocompatible and architecturally stable ceramics due to its enhanced accuracy and control over the parameters [34].

- **Environmental catalysts**

The utilization of microwave-assisted synthesis in environmental aspects including catalytic converters for cars and pollution control systems has also gained a lot of importance in recent times. These catalysts play a crucial role in mitigating the pollutants and improving the air quality.

- **Energy storage materials**

Microwave-assisted techniques can be employed to manufacture ceramics to be utilized in energy storage applications including super capacitors and solid-state batteries. These materials play a significant role in enhancing energy efficiency and promoting environmental sustainability.

3 Mechanical and Biochemical Characterization of Ceramics

The use of ceramics in different domains like biomedical, mechanical engineering, etc. has experienced a notable rise due to its distinctive attributes such as exceptional mechanical strength and biocompatibility [49]. A comprehensive examination of the mechanical properties and biological interactions exhibited by ceramic-based biomaterials is essential for their effective utilization in medical devices and implants. A critical analysis of the mechanical characteristics with an emphasis on the *in-vitro* and *in-vivo* testing of ceramics along with their necessity for degradation and resorption investigations is of great interest.

3.1 Mechanical Properties of Ceramics

The mechanical properties of ceramics are a subject of significant interest and research in the field of materials science and engineering. These properties play a crucial role in determining the suitability and performance of ceramic biomaterials for various biomedical applications. The mechanical properties of ceramic biomaterials are essential qualities that govern their performance under different loading conditions and environmental scenarios. The applicability of ceramics for various biological applications can be strongly influenced by their mechanical characteristics.

- **Strength and toughness**

  One of the key aspects to consider in this context is the concept of strength. Ceramics possess appreciable compressive strength, rendering them well-suited for load-bearing applications, such as the use of bone implants [43]. Although ceramics possess high strength, they tend brittleness as well. Scientists are now engaged in efforts to augment the durability of ceramic biomaterials to mitigate the occurrence of early failures or fractures in ceramic-based medical implants.

- **Hardness**

  The concept of hardness is a fundamental property in materials science and engineering. It refers to the resistance of a material to deformation, indentation, etc. Ceramics possess intrinsic hardness, rendering them ideal in the context of medical equipment as they exhibit resistance to wear and abrasion as well [13].

- **Modulus of elasticity**

  The elastic modulus characterizes the stiffness or rigidity of a material. In the context of bone implants, the preference lies with ceramics that possess an elastic modulus comparable to that of natural bone [95].

- **Fracture behaviour**

  A careful examination of the fracture mechanisms shown by ceramics is crucial in the development of implants that are both biologically and mechanically safe. The primary objective is to mitigate the risk of sudden and unalarmed failure through crack initiation and propagation control.

- **Fatigue resistance**

  The ability to withstand fatigue is of utmost importance for materials subjected to repetitive loading, such as dental implants. Ceramic biomaterials are required to exhibit long-term durability by withstanding repetitive mechanical stresses without undergoing appreciable levels of plasticity.

- **Creep resistance**

It refers to the steady alteration of a material when subjected to a force over an extended period. Ceramic biomaterials in biomedical applications need a notable level of creep resistance to preserve their structural integrity within the human body.

- **In-vitro and in-vivo testing**

In-vitro testing may be done often in a controlled laboratory background. This type of testing involves the isolated cells, tissues, or organs to evaluate effects of various materials. *In-vivo* testing, on the other hand, involves experiments conducted within a living organism. The process of *in-vitro* testing involves the systematic assessment of the characteristics and performance of ceramic biomaterials inside the body of an organism or animal model. This test plays a vital role in evaluating the biocompatibility, mechanical properties, and degradation behaviour of ceramics. Several commonly used *in-vitro* assays include:

- Biocompatibility assay involves the evaluation of ceramics with biological systems, specifically cell cultures, through *in-vitro* investigations. These tests aid in assessing the potential of the substance to provoke adverse physiological reactions in the human body.
- Mechanical testing aims to evaluate the mechanical features of ceramics in a controlled environment. These include tensile strength, compression strength, and hardness tests, which are utilized to enumerate the response of ceramics to mechanical loads.
- Wear and friction testing is employed to replicate the wear and friction circumstances that ceramic implants can encounter within the human body. These tests assist in the development of implants to endure wearing and tearing over extended periods of usage.
- The chemical stability of ceramic biomaterials is assessed by *in-vitro* tests conducted in simulated body fluids, to confirm their resistance to degradation and prevention of the release of toxins.

The *in-vivo* testing requires the assessment of the functional characteristics and efficacy of ceramics when introduced into living organisms. These studies offer valuable insights into the interactions between materials and the human body, playing a critical role in evaluating their long-term safety and efficacy. Several commonly used *in-vivo* assays include:

- In animal studies, researchers employ animal models, including mice and bigger species, to evaluate the biocompatibility, mechanical stability, and tissue response associated with ceramic-based implants.
- The host-tissue reaction to the ceramic implants may include evaluation of several factors such as inflammation, tissue integration, toxin generation, and the processes involved in healing.
- The assessment of the long-term durability of ceramics about their capacity to sustain structural integrity and functionality over prolonged durations, is facilitated by *in-vivo* testing.

- Innovative medical imaging like MRI, and CT scan are utilized to monitor the behaviour and integration of ceramic implants within the human body.

3.1.1 Degradation and Resorption Studies

The knowledge of degradation and resorption mechanisms of ceramics within the human body is of ultimate importance in the advancement of biocompatible implants. The inclusion of degradation and resorption studies is integral to comprehensively assessing their efficacy for several biological applications.

- ***Degradation behaviour***

  In biomedical applications, ceramics must undergo regulated degradation [61]. The occurrence of excessive degradation may fail implants, whereas inadequate degradation may impede the process of tissue regeneration.

- ***Studies on resorption***

  Within the domain of biodegradable ceramics, resorption pertains to the progressive adaptation of the material through the inherent physiological mechanisms of the body. Its significance is particularly pronounced in various applications, such as bone transplants, where the ceramic scaffold is required to provisionally support and progressively substitute the tissues of the body. Resorption studies serve the purpose of determining the favourable resorption rate for therapeutics. Through the implementation of systematic degradation and resorption investigations, the characteristics of ceramics can be tuned to align with the intended results of biomedical treatments. These investigations aim to verify the intended function while mitigating the potential for unfavourable responses on the other hand.

4 Applications of Ceramics

4.1 Industry

The field of ceramics boasts an extended and recognized history that spans ages owing to its distinctive characteristics. In contrast to conventional ceramics, contemporary ones comprise a diverse array of refined materials that exhibit excellent properties and adaptability. These materials are widely utilized in many industries spanning numerous sectors. The applications of ceramics in different industries are listed below:

4.1.1 Electronics and Electrical Industry

The electronics and electrical industry agreements with the growth and distribution of electronics and electrical devices. The electronics and electrical sectors extensively use ceramics as a prominent material [62]. Ceramics possess exceptional electrical insulation, refractoriness, and low coefficient of thermal expansion. These attributes render ceramics highly essential elements in the production of electrical gadgets. The following are some of the significant applications in the electronics and electrical sector:

a. Ceramic insulators are employed in high-voltage transmission lines to mitigate the risk of electrical leakages, short circuits, and electrical failures [57]. These materials offer reliable thermal insulation even under severe conditions.
b. Ceramic capacitors serve as the necessary components for the storage and discharge of electrical energy when required [15]. These components provide stability, and high capacitance, and are frequently used in consumer electronics.
c. Several ceramics, including piezoelectrics, can produce an electrical voltage when exposed to mechanical stresses. Piezoelectric devices, such as sensors, actuators, and ultrasonic transducers are based on this characteristic [31].

4.1.2 Automotive Industry

a. Ceramics have been incorporated into the automobile sector, owing to their remarkable performance, efficiency, and safety [63].
b. Catalyst-coated ceramic substrates play a crucial role in mitigating detrimental emissions in internal combustion engines. These converters also play a crucial role in ensuring compliance with eco-friendly requirements [40].
c. Ceramic brake discs and pads offer several benefits including reduced mass, better thermal endurance, and higher braking efficacy, and are used in high-performance and luxury automobiles.
d. Spark plugs and bearings made from ceramics are used in automobiles to enhance the efficiency and durability of engines. These also exhibit resistance to elevated temperatures and mechanical abrasion.

4.1.3 Aerospace and Aviation

a. Ceramic matrix composites are used in gas turbine engines to withstand elevated temperatures and reduce the weight of the engine. This also improves the fuel economy and lifespan of critical automobile components [66].
b. Ceramic-based thermal barrier coatings protect the aircraft and spacecraft components against extreme levels of thermal agitation by the management of temperature differentials.

4.1.4 Medical and Healthcare

a. Dental ceramics, including zirconia-based materials, are used in dental prosthetics such as implants and crowns [24]. These provide patients with long-lasting benefits in terms of biocompatibility, aesthetics, and durability.
b. Orthopaedic implants for joint and bone repair commonly use ceramic biomaterials, including alumina and zirconia [38]. These offer mechanical robustness, resistance to wear, and compatibility with the biological tissues.
c. Ceramics find application in a diverse range of medical devices like X-ray tubes, ultrasound transducers, etc. owing to their enhanced electrical and thermal properties.

4.1.5 Chemical and Process Industries

a. Ceramics are utilized to line industrial furnaces and kilns to endure high temperatures and provide protection against both thermal and chemical corrosion [30]. They show a vital role in metal smelting, and glass making.
b. Several ceramics are utilized as catalysts to support various chemical processes. These catalysts offer a reliable foundation for facilitating chemical reactions in many industrial processes, such as petrochemical refining.

4.1.6 Energy Sector

a. Ceramics are widely utilized in the manufacturing of solar cells and photovoltaic panels [47].
b. Solid oxide fuel cells employ ceramic electrolytes to produce electrical energy through electrochemical reactions between hydrogen and oxygen [86]. The industrial applications of ceramics are shown in Fig. 11.

4.2 Biocompatibility and Biomedical Engineering

Ceramics have emerged as a prominent material in biomedical research owing to their distinctive characteristics making them suitable for diverse applications in medical devices, biomedical implants, and drug delivery [20]. The biocompatibility of ceramics determines their capacity to interact with biological systems without producing any harmful toxins [69]. By considering their biocompatibility, these materials show a vital role in the growth of implants, prosthetics, drug-delivery, and bone tissue engineering. To stick to severe demands of biomedical applications, biomaterials must exhibit the following distinct attributes:

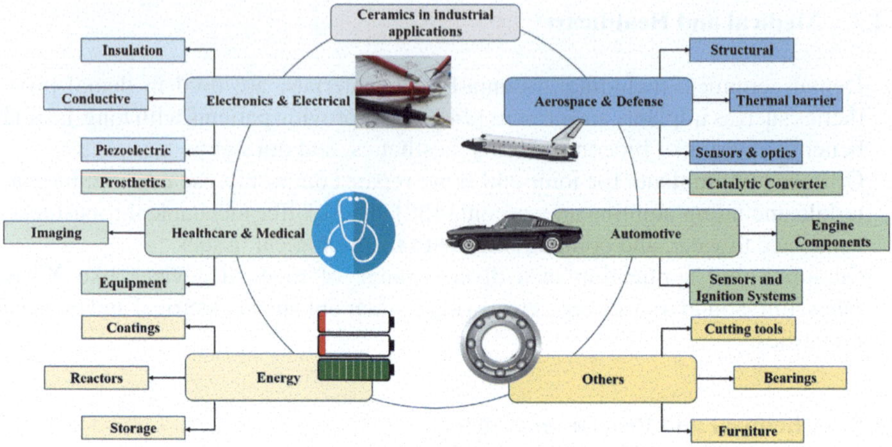

Fig. 11 Applications of ceramics in different industries

- **Biocompatibility**

Biocompatibility pertains to the capacity of a substance to engage with biological entities without inducing adverse effects. The avoidance of undesirable consequences, such as inflammation, toxicity, or immunological responses, is an important consideration in the development of ceramic-based biocompatible materials [51].

- **Mechanical properties**

Biomaterials must possess appreciable mechanical strength and stiffness. For instance, ceramic-based materials used as load-bearing implants must have sufficient mechanical strength and stiffness.

- **Surface properties**

The significance of surface characteristics in ceramic-based biomaterials governs their interactions with the biological molecules or cells, which ultimately determines the level of biocompatibility [16]. Surface modifications can be used to supplement cell adhesion, mitigate infections, and obtain the sustained release of drugs.

- **Degradation and bio-stability**

The degradation of biomaterials may be necessary such as in the case of temporary implants whereas long-term bio-stability may be required in the case of permanent implants. The concept of bioactivity refers to the ability of a substance or organism to have a biological effect or activity. It is important to note that degradation products must possess non-toxic properties and have ease of excretion from the human body. Bioactive materials possess the capacity to produce targeted biological reactions [96]. Certain ceramics can facilitate bone tissue regeneration by promoting the production of hydroxyapatite, a naturally occurring mineral in bone. The biomedical applications of ceramics are shown in Fig. 12.

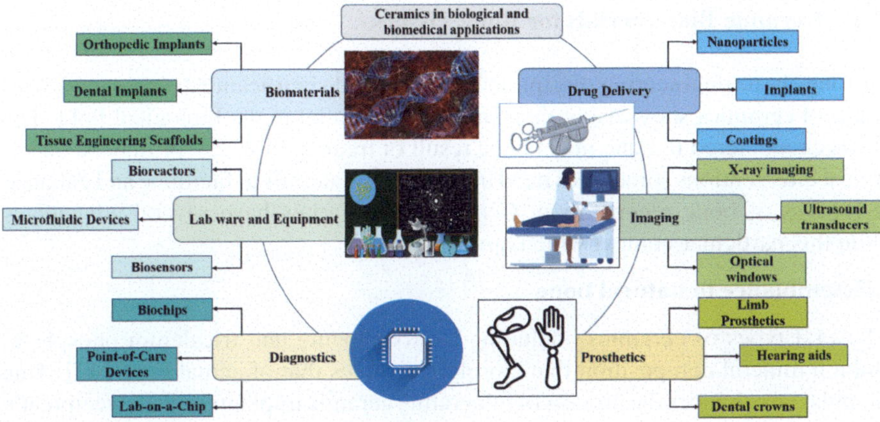

Fig. 12 Biological and biomedical applications of ceramics

4.2.1 Interaction of Ceramics with Biological Systems

Gaining a comprehensive understanding of relations among ceramics and biotic systems may be crucial to the efficiently utilize their full potential in medicine and biomedical applications.

- **Surface chemistry**

The correlation between the surface chemistry of ceramics and their interactions with biological systems is of significant importance. Protein adsorption, cell adhesion, tissue integration, etc. are all influenced by the composition and charge on their surface.

- **Osteoconductivity**

Numerous ceramic-based biomaterials demonstrate osteoconductivity, wherein they possess the proliferation and formation of the bone tissues [27]. Hydroxyapatite, a ceramic composed of calcium phosphate, is widely recognized for its osteoconductive properties, rendering it an appropriate option for bone formation.

- **Bioactivity**

It is a quality associated with particular ceramics, such as calcium phosphate. Through the development of a hydroxyapatite layer on their surface, the implants can link with the surrounding bone tissue promoting bone regeneration and integration.

- **Inflammatory response**

Ceramics are commonly linked to a reduced inflammatory response. Their biocompatibility may be prejudiced by their dimensions, morphologies, and its chemical compositions. The tendency of nanoscale ceramic particles to cause inflammation is lower as compared to the bigger particles.

4.2.2 Ceramic Biomaterials for Bone Tissues

The bone-tissue engineering discipline has witnessed significant advancements with the use of ceramics showcasing their immense potential in the biological field. The defects encountered in bone may be the result of trauma, diseases, or may be due to surgical interventions sometimes need implantable materials to facilitate and enhance the process of bone regeneration. Ceramic biomaterials have several advantages within this particular context which are as follows:

- **Resemblance to natural bone**

  Several types of ceramics, including hydroxyapatite and tricalcium phosphate, exhibit a mineral composition that closely resembles that of actual bone [41]. This resemblance enhances the process of integrating ceramic implants with the recipient's bone.

- **Porosity and scaffold design**

  Ceramic scaffolds can be designed with precise porosity and pore size distributions, hence aiding in the filtration of cells and blood vessels. These porous structures serve as a 3D scaffolding bone tissue formation [99].

- **Controlled drug delivery**

  This involves the specialized drug delivery that may release drugs in a sustained manner, ensuring optimal delivery of drugs. The different drug molecules can be incorporated into the ceramic-based scaffolds [56]. This mechanism facilitates the regulated discharge of growth factors or antimicrobial drugs, augmenting the regeneration potential of the implant.

- **Biodegradability**

  Certain ceramics possess biodegradability, wherein they undergo progressive resorption along with the development of bone tissues. Although in some cases, this particular characteristic removes the necessity for *in-vivo* surgical procedures to remove the implants. Some other potential applications of ceramics in the biomedical domain are as follows:

i. **Cancer therapy and bone regeneration**

Ceramic nanoparticles have been developed as nanocarriers for targeted drug delivery of chemotherapeutic drugs against cancer cell, resulting in a reduction in systemic toxicity and improvement in the therapeutic index [42]. The integration of nanoparticles into ceramic-based scaffolds offers the promising approach to process of bone regeneration.

ii. **Antimicrobial applications, medical imaging, and regenerative medicine**

Ceramic nanoparticles possessing antimicrobial characteristics can effectively address infections and hinder the formation of biofilms on medical implants [71, 91].

Ceramic nanoparticles can function as contrast agents in different imaging modalities, hence enabling the timely identification and precise diagnosis of diseases [36]. Regenerative medicine focuses on therapeutic strategies to repair, replace, regenerate damaged to the diseased tissues or organs. Ceramic nanoparticles have considerable potential within the realm of regenerative medicine, since they possess the ability to enhance tissue regeneration across diverse organs and tissues [29].

4.3 Mechanical Engineering, Material Science, and Nanotechnology

- **3D printing and additive manufacturing**

The emergence of 3D printing and additive manufacturing has initiated a novel phase in the field of ceramics. These have enabled the precise production of intricate ceramic structures with desired properties. Significant progress has already been made in biomedicine field with the expansion and utilization of 3-D printed scaffolds for the tissue engineering and medical implants tailored specifically to individual patients [28].

- **Multifunctional ceramic materials**

The incorporation of multifunctionality into ceramics represents a promising direction for future progress. For example, a ceramic-based implant has the potential to not only offer mechanical reinforcement but also administer drugs to promote tissue regeneration and mitigate the risk of infections. The utilization of multifunctional ceramics can provide novel opportunities for several biomedical applications, hence enhancing patient care and minimizing the necessity for additional treatments [100].

- **Nanotechnology and ceramic nanoparticles**

Ceramic nanoparticles have a wide range of potential applications, encompassing improved drug transport, advanced medical imaging, and scaffolds for tissue engineering. These nanoparticles can be accurately designed to be employed with biological systems hence resulting in treatments that are more focused and highly efficient [5]. Some widely used ceramic-based nanomaterials are shown in Fig. 13.

5 Challenges and Future Directions

Ceramics have demonstrated their immense value in several sectors owing to their extraordinary characteristics, such as tolerance towards elevated temperature, hardness, etc. In recent times, ceramics have achieved notable advancements in the biomedical field, particularly in the domains of medical implants and drug delivery for

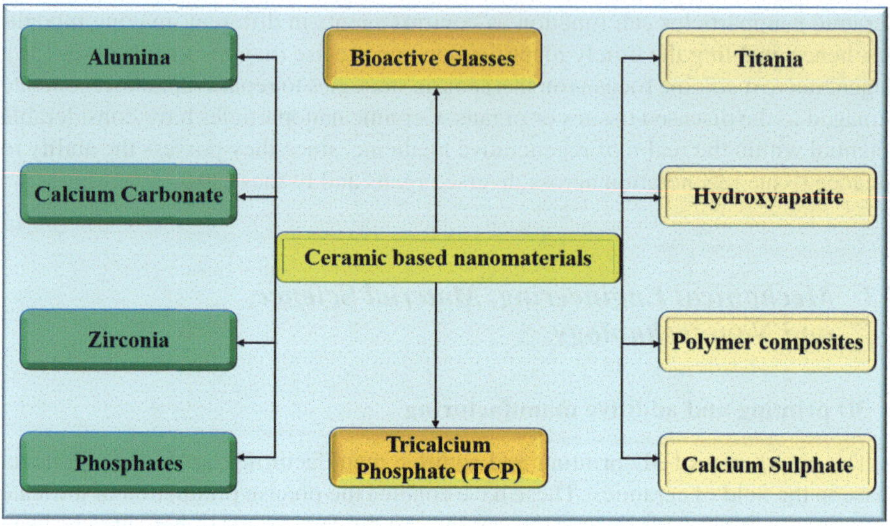

Fig. 13 Some ceramic-based nanomaterials [2]

biomedical applications. Nevertheless, similar to other disciplines, use of ceramics in of biomedicine field presents several difficulties and opportunities for further enhancement.

- **Addressing biocompatibility challenges**

The concept of biocompatibility holds significant significance in the context of biomedical applications involving ceramics, as it pertains to the capacity of material to interact with biological systems in a manner that avoids any detrimental effects. The consideration of biocompatibility problems is of utmost importance to guarantee the safe and efficient utilization of ceramics within healthcare environments.

i. **Inflammatory response**

One of the leading considerations in ceramics is biocompatibility which revolves around the possibility of provoking inflammatory reactions. Certain ceramics, when introduced into the human body, have the potential to cause an immunological response or provoke inflammation. To tackle this, investigations are carried out on the surface modifications and coatings that can effectively mitigate the inflammatory responses, hence enhancing their biocompatibility.

ii. **Toxicity**

Toxicity may arise due to the release of molecules, ions, or the breakdown of products from ceramics. Selecting ceramic compositions and production techniques that effectively reduce the release of harmful chemicals is of utmost importance. Further, biocompatibility testing is also vital to ascertain that ceramic materials must adhere to the established safety protocols.

iii. **Long-term bio-stability**

Long-term bio-stability mentions the capability of a biological material to maintain the structural and functional integrity over an extended period. Ceramic materials are required to sustain their structural integrity and biocompatibility for a prolonged duration when introduced into the human body. Current studies are dedicated to the advancement of ceramics that possess improved bio-stability, hence enabling their prolonged use.

iv. **Integration with tissues**

The attainment of a seamless integration between ceramics and host tissues is a substantial difficulty. Scientists are now investigating various surface modifications and scaffolding approaches to enhance tissue development and the overall biocompatibility of ceramic implants.

v. **Patient-specific considerations**

There can be substantial variations in the physiology and immunological response of different individuals. Customizing ceramic materials and designs to cater to the specific requirements of particular patients is a promising avenue for tackling the biocompatibility issues that arise in medical applications. The utilization of personalized medicine strategies is expected to witness an increase in its application within the field of ceramic implant technology shortly.

6 Conclusions

To summarize, the examination and analysis of ceramic synthesis in this discourse establishes a robust framework for the continuous progress in different fields of science, engineering, medicine, etc. The healthcare sector holds significant potential for the utilization of ceramics due to several advantages such as enhanced biocompatibility, tailored properties, enhanced mechanical strength, multi-functionality, etc. The landscape of ceramics is expected to undergo a significant transformation due to the emergence of technologies like as 3D printing, nanotechnology, and integration with digital health. Nevertheless, it is important to maintain a thorough awareness of the obstacles that may arise, including those related to biocompatibility and the sustainability of the environment. These problems highlight the significance of employing responsible and ethical procedures in the study and development of ceramics. The forthcoming trajectory presents promising prospects, and the convergence of ceramics and biology constitutes a domain conducive to ground-breaking advancements and outcomes. Through persistent devotion, effective teamwork, and a steadfast commitment to enhancing patient care, ceramics will undeniably play a pivotal role in crafting a healthier and more dynamic future for society as a whole.

References

1. Abusrewil, S.M., McLean, W., Scott, J.A.: The use of Bioceramics as root-end filling materials in periradicular surgery: a literature review. Saudi Dent. J. **30**(4), 273–282 (2018)
2. Balasubramanian, S., Gurumurthy, B., Balasubramanian, A.: Biomedical applications of ceramic nanomaterials: a review. Int. J. Pharm. Sci. Res. **8**(12), 4950–4959 (2017)
3. Behera, A., Mallick, P., Mohapatra, S.: Nanocoatings for anticorrosion: an introduction. In: Corrosion Protection at the Nanoscale, pp. 227–243. Elsevier (2020)
4. Berlincourt, D.: Piezoelectric ceramics: characteristics and applications. J. Acoust. Soc. Am. **70**(6), 1586–1595 (1981)
5. Boccaccini, A.R., Ma, P.X., Liverani, L.: Tissue Engineering using Ceramics and Polymers. Woodhead Publishing (2021)
6. Bokov, D., Turki Jalil, A., Chupradit, S., Suksatan, W., Javed Ansari, M., Shewael, I.H., Valiev, G.H., Kianfar, E.: Nanomaterial by sol-gel method: synthesis and application. Adv. Mater. Sci. Eng. **2021**, 1–21 (2021)
7. Bozo, C., Guilhaume, N., Garbowski, E., Primet, M.: Combustion of methane on CeO2–ZrO2 based catalysts. Catal. Today **59**(1–2), 33–45 (2000)
8. Brinker, C.J., Scherer, G.W.: Sol-Gel Science: The Physics and Chemistry of Sol-Gel Processing. Academic (2013)
9. Buzby, S., Franklin, S., Shah, S.I.: Synthesis of metal-oxide nanoparticles: gas-solid transformations. In: Synthesis, Properties, and Applications of Oxide Nanomaterials, vol. 119 (2007)
10. Byrappa, K., Yoshimura, M.: Handbook of Hydrothermal Technology. William Andrew Publishing. New York (2001)
11. Thomas, S.C., Kumar Mishra, P., Talegaonkar, S.: Ceramic nanoparticles: fabrication methods and applications in drug delivery. Curr. Pharm. Des. **21**(42), 6165–6188 (2015)
12. Cahill, J.T.: A study on the phase stability and diffusion behavior of alkaline-earth hexaborides. University of California, San Diego (2016)
13. Catledge, S.A., Fries, M.D., Vohra, Y.K., Lacefield, W.R., Lemons, J.E., Woodard, S., Venugopalanc, R.: Nanostructured ceramics for biomedical implants. J. Nanosci. Nanotechnol. **2**(3–4), 293–312 (2002)
14. Chang, Y.-C., Chiao, Y.-C., Hsu, P.-C.: Rapid microwave-assisted synthesis of ZnIn2S4 nanosheets for highly efficient photocatalytic hydrogen production. Nanomaterials **13**(13), 1957 (2023)
15. Chauhan, A., Patel, S., Vaish, R., Bowen, C.R.: Anti-ferroelectric ceramics for high energy density capacitors. Materials **8**(12), 8009–8031 (2015)
16. Chen, H., Yuan, L., Song, W., Wu, Z., Li, D.: Biocompatible polymer materials: role of protein–surface interactions. Prog. Polym. Sci. **33**(11), 1059–1087 (2008)
17. Chevalier, J., Gremillard, L.: Ceramics for medical applications: a picture for the next 20 years. J. Eur. Ceram. Soc. **29**(7), 1245–1255 (2009)
18. Chindamo, G., Sapino, S., Peira, E., Chirio, D., Gonzalez, M.C., Gallarate, M.: Bone diseases: current approach and future perspectives in drug delivery systems for bone targeted therapeutics. Nanomaterials **10**(5), 875 (2020)
19. Choy, K.: Chemical vapour deposition of coatings. Prog. Mater. Sci. **48**(2), 57–170 (2003)
20. Colilla, M., Manzano, M., Vallet-Regí, M.: Recent advances in ceramic implants as drug delivery systems for biomedical applications. Int. J. Nanomed. **3**(4), 403–414 (2008)
21. Creighton, J., Ho, P.: Introduction to chemical vapor deposition (CVD). Chem. Vap. Depos. **2**, 1–22 (2001)
22. D'Amato, R., Falconieri, M., Gagliardi, S., Popovici, E., Serra, E., Terranova, G., Borsella, E.: Synthesis of ceramic nanoparticles by laser pyrolysis: From research to applications. J. Anal. Appl. Pyrol. **104**, 461–469 (2013)
23. Dapkus, P.D.: Metalorganic chemical vapor deposition. Annu. Rev. Mater. Sci. **12**(1), 243–269 (1982)

24. Denry, I., Holloway, J.A.: Ceramics for dental applications: a review. Materials **3**(1), 351–368 (2010)
25. Dhanasekar, S., Ganesan, A.T., Rani, T.L., Vinjamuri, V.K., Rao, M.N., Shankar, E., Dharamvir, E., Kumar, P.S., Golie, W.M.: A comprehensive study of ceramic matrix composites for space applications. In: Advances in Materials Science and Engineering (2022)
26. Dobrzański, L.A., Pakuła, D.: Structure and properties of the wear resistant coatings obtained in the PVD and CVD processes on tool ceramics. Mater. Sci. Forum (2006)
27. Ducheyne, P., Qiu, Q.: Bioactive ceramics: the effect of surface reactivity on bone formation and bone cell function. Biomaterials **20**(23–24), 2287–2303 (1999)
28. Eckel, Z.C., Zhou, C., Martin, J.H., Jacobsen, A.J., Carter, W.B., Schaedler, T.A.: Additive manufacturing of polymer-derived ceramics. Science **351**(6268), 58–62 (2016)
29. Fathi-Achachelouei, M., Knopf-Marques, H., Ribeiro da Silva, C.E., Barthès, J., Bat, E., Tezcaner, A., Vrana, N.E.: Use of nanoparticles in tissue engineering and regenerative medicine [review]. Front. Bioeng. Biotechnol. **7**, 113 (2019). https://doi.org/10.3389/fbioe.2019.00113
30. Fauchais, P., Vardelle, A.: Thermal sprayed coatings used against corrosion and corrosive wear. Adv. Plasma Spray Appl. **10**, 34448 (2012)
31. Galassi, C.: Processing of porous ceramics: piezoelectric materials. J. Eur. Ceram. Soc. **26**(14), 2951–2958 (2006)
32. Gan, Y.X., Jayatissa, A.H., Yu, Z., Chen, X., Li, M.: Hydrothermal synthesis of nanomaterials. J. Nanomater. **2020**, 1–3 (2020)
33. Gavrilović, T.V., Jovanović, D.J., Dramićanin, M.D.: Synthesis of multifunctional inorganic materials: from micrometer to nanometer dimensions. In: Nanomaterials for green Energy, pp. 55–81. Elsevier (2018)
34. Gupta, S., Sharma, A.K., Agrawal, D.: Processing biomaterials using microwave energy and its futuristic scopes. In: Advances in Microwave Processing for Engineering Materials, pp. 191–211. CRC Press (2022)
35. Hersee, S., Duchemin, J.: Low-pressure chemical vapor deposition. Annu. Rev. Mater. Sci. **12**(1), 65–80 (1982)
36. Huang, H.-C., Barua, S., Sharma, G., Dey, S.K., Rege, K.: Inorganic nanoparticles for cancer imaging and therapy. J. Control. Release **155**(3), 344–357 (2011)
37. Jazayeri, H.E., Rodriguez-Romero, M., Razavi, M., Tahriri, M., Ganjawalla, K., Rasoulian-boroujeni, M., Malekoshoaraie, M.H., Khoshroo, K., Tayebi, L.: The cross-disciplinary emergence of 3D printed bioceramic scaffolds in orthopedic bioengineering. Ceram. Int. **44**(1), 1–9 (2018)
38. Jin, W., Chu, P.K.: Orthopedic implants. Encycl. Biomed. Eng. **1**(3), 425–439 (2019)
39. Kaur, M., Singh, K.: Review on titanium and titanium based alloys as biomaterials for orthopaedic applications. Mater. Sci. Eng. C **102**, 844–862 (2019)
40. Keane, M.: Ceramics for catalysis. J. Mater. Sci. **38**, 4661–4675 (2003)
41. Khan, A.F., Awais, M., Khan, A.S., Tabassum, S., Chaudhry, A.A., Rehman, I.U.: Raman spectroscopy of natural bone and synthetic apatites. Appl. Spectrosc. Rev. **48**(4), 329–355 (2013)
42. Khan, M.I., Hossain, M.I., Hossain, M.K., Rubel, M., Hossain, K., Mahfuz, A., Anik, M.I.: Recent progress in nanostructured smart drug delivery systems for cancer therapy: a review. ACS Appl. Bio Mater. **5**(3), 971–1012 (2022)
43. Khan, Y., Yaszemski, M.J., Mikos, A.G., Laurencin, C.T.: Tissue engineering of bone: material and matrix considerations. JBJS 90(Supplement_1), 36–42 (2008)
44. Kumar, S., Sahare, P., Kumar, S.: Optimization of the CVD parameters for ZnO nanorods growth: its photoluminescence and field emission properties. Mater. Res. Bull. **105**, 237–245 (2018)
45. Kurinjinathan, P., Arul, K.T., Ramya, J.R.: Cobalt ions doped bioactive ceramics for biosensor biomedical applications. Int. J. Curr. Res. Rev. **10**(21), 49–52 (2018)
46. Li, C., Zhang, Z., Sui, J., Jiang, X., Waterhouse, G.I., Zhang, Z., Yu, L.: Microwave-based synthesis of (NiCo) x/(MnO) y/C composites and their tunable wave absorption properties in the K band. Ceram. Int. **46**(7), 9353–9362 (2020)

47. Li, X., Li, P., Wu, Z., Luo, D., Yu, H.-Y., Lu, Z.-H.: Review and perspective of materials for flexible solar cells. Mater. Rep.: Energy **1**(1), 100001 (2021)
48. Luyten, J., Mullens, S., Cooymans, J., De Wilde, A., Thijs, I., Kemps, R.: Different methods to synthesize ceramic foams. J. Eur. Ceram. Soc. **29**(5), 829–832 (2009)
49. Mamidi, N., García, R.G., Martínez, J.D.H., Briones, C.M., Martinez Ramos, A.M., Tamez, M.F.L., Del Valle, B.G., Segura, F.J.M.: Recent advances in designing fibrous biomaterials for the domain of biomedical, clinical, and environmental applications. ACS Biomater. Sci. Eng. **8**(9), 3690–3716 (2022)
50. Manning, T.D., Parkin, I.P., Clark, R.J., Sheel, D., Pemble, M.E., Vernadou, D.: Intelligent window coatings: atmospheric pressure chemical vapour deposition of vanadium oxides. J. Mater. Chem. **12**(10), 2936–2939 (2002)
51. Markides, H., Rotherham, M., El Haj, A.: Biocompatibility and toxicity of magnetic nanoparticles in regenerative medicine. J. Nanomater. **2012**, 13–13 (2012)
52. Meyyappan, M.: A review of plasma enhanced chemical vapour deposition of carbon nanotubes. J. Phys. D Appl. Phys. **42**(21), 213001 (2009)
53. MH, S.: Bioactive Glasses in Dentistry: A Review (2015)
54. Mirzaei, A., Neri, G.: Microwave-assisted synthesis of metal oxide nanostructures for gas sensing application: a review. Sens. Actuators B Chem. **237**, 749–775 (2016)
55. Misra, K.P., Misra, R.: Advanced ceramics. In: Ceramic Science and Engineering, pp. 21–43. Elsevier (2022)
56. Mouriño, V., Boccaccini, A.R.: Bone tissue engineering therapeutics: controlled drug delivery in three-dimensional scaffolds. J. R. Soc. Interface. **7**(43), 209–227 (2010)
57. Moustafa, M.A., Mosalam, K.M.: Structural performance of porcelain and polymer post insulators in high voltage electrical switches. J. Perform. Constr. Facil. **30**(5), 04016002 (2016)
58. Mukasyan, A., Dinka, P.: Novel approaches to solution-combustion synthesis of nanomaterials. Int. J. Self Propag. High Temp. Synth. **16**, 23–35 (2007)
59. Ndolomingo, M.J., Bingwa, N., Meijboom, R.: Review of supported metal nanoparticles: synthesis methodologies, advantages and application as catalysts. J. Mater. Sci. **55**(15), 6195–6241 (2020)
60. Nigam, A., Pawar, S.: Structural, magnetic, and antimicrobial properties of zinc doped magnesium ferrite for drug delivery applications. Ceram. Int. **46**(4), 4058–4064 (2020)
61. O'brien, F.J.: Biomaterials & scaffolds for tissue engineering. Mater. Today **14**(3), 88–95 (2011)
62. Okada, A.: Automotive and industrial applications of structural ceramics in Japan. J. Eur. Ceram. Soc. **28**(5), 1097–1104 (2008)
63. Okada, A.: Ceramic technologies for automotive industry: Current status and perspectives. Mater. Sci. Eng. B **161**(1–3), 182–187 (2009)
64. Olajide, J.L., Desai, D., Ajibola, J., Adekoya, G.J., Daramola, O.O., Alaneme, K., Fasiku, V.O., Sadiku, E.R.: Nosocomial bacterial infection of orthopedic implants and antibiotic hydroxyapatite/silver-coated halloysite nanotube with improved structural integrity as potential prophylaxis. In: Antibiotic Materials in Healthcare, pp. 171–220. Elsevier (2020)
65. Patil, K.: Advanced ceramics: combustion synthesis and properties. Bull. Mater. Sci. **16**, 533–541 (1993)
66. Peter, P.I., Adekunle, A.A.: A review of ceramic/bio-based hybrid reinforced aluminium matrix composites. Cogent Eng. **7**(1), 1727167 (2020)
67. Pina, S., Rebelo, R., Correlo, V.M., Oliveira, J.M., Reis, R.L.: Bioceramics for osteochondral tissue engineering and regeneration. In: Osteochondral Tissue Engineering: Nanotechnology, Scaffolding-Related Developments and Translation, pp. 53–75 (2018)
68. Polychronopoulou, K., Zedan, A.F., Katsiotis, M., Baker, M., AlKhoori, A., AlQaradawi, S.Y., Hinder, S., AlHassan, S.: Rapid microwave assisted sol-gel synthesis of CeO2 and CexSm1-xO2 nanoparticle catalysts for CO oxidation. Mol. Catal. **428**, 41–55 (2017)
69. Raghavendra, G.M., Varaprasad, K., Jayaramudu, T.: Biomaterials: design, development and biomedical applications. In: Nanotechnology applications for tissue engineering, pp. 21–44. Elsevier (2015)

70. Rane, A. V., Kanny, K., Abitha, V., Thomas, S.: Methods for synthesis of nanoparticles and fabrication of nanocomposites. In: Synthesis of Inorganic Nanomaterials, pp. 121–139. Elsevier (2018)
71. Raphel, J., Holodniy, M., Goodman, S.B., Heilshorn, S.C.: Multifunctional coatings to simultaneously promote osseointegration and prevent infection of orthopaedic implants. Biomaterials **84**, 301–314 (2016)
72. Ravaglioli, A., Krajewski, A.: Bioceramics: Materials·Properties Applications. Springer Science & Business Media (1991)
73. Roy, R.: Ceramics by the solution-sol-gel route. Science **238**(4834), 1664–1669 (1987)
74. Roy, R.: Accelerating the kinetics of low-temperature inorganic syntheses. J. Solid State Chem. **111**(1), 11–17 (1994)
75. Roy, R., Agrawal, D., Alamo, J., Roy, R.: [CTP]: a new structural family of near-zero expansion ceramics. Mater. Res. Bull. **19**(4), 471–477 (1984)
76. Sáenz, A., Rivera, E., Brostow, W., Castaño, V.M.: Ceramic biomaterials: an introductory overview. J. Mater. Educ. **21**(5/6), 267–276 (1999)
77. Sakka, S.: Handbook of sol-gel science and technology. 1. In: Sol-Gel Processing, vol. 1. Springer Science & Business Media (2005)
78. Schmidt, H.: Chemistry of material preparation by the sol-gel process. J. Non-Cryst. Solids **100**(1–3), 51–64 (1988)
79. Schwenke, A.M., Hoeppener, S., Schubert, U.S.: Synthesis and modification of carbon nanomaterials utilizing microwave heating. Adv. Mater. **27**(28), 4113–4141 (2015)
80. Segal, D.: Chemical synthesis of ceramic materials. J. Mater. Chem. **7**(8), 1297–1305 (1997)
81. Shadjou, N., Hasanzadeh, M.: Application of graphene and mesoporous silica nanomaterials on the orthopaedic implants: recent advances. Mater. Technol. **31**(13), 806–811 (2016)
82. Shahidzadeh, M., Shabihi, P., Pourmortazavi, S.M.: Sonochemical preparation of copper (II) chromite nanocatalysts and particle size optimization via taguchi method. J. Inorg. Organomet. Polym. Mater. **25**, 986–994 (2015)
83. Sharmila, P., Tharayil, N.J.: Chemically synthesized nano composite (Zinc/Magnesium) Oxide for tunable band gap devices. IOP Conf. Ser.: Mater. Sci. Eng. (2015)
84. Song, X., Segura-Egea, J.J., Díaz-Cuenca, A.: Sol-gel technologies to obtain advanced bioceramics for dental therapeutics. Molecules **28**(19), 6967 (2023)
85. Soref, R.A.: Silicon-based optoelectronics. Proc. IEEE **81**(12), 1687–1706 (1993)
86. Stambouli, A.B., Traversa, E.: Solid oxide fuel cells (SOFCs): a review of an environmentally clean and efficient source of energy. Renew. Sustain. Energy Rev. **6**(5), 433–455 (2002)
87. Suchanek, W.L., Riman, R.E.: Hydrothermal synthesis of advanced ceramic powders. Adv. Sci. Technol. **45**, 184–193 (2006)
88. Vahlas, C., Caussat, B., Serp, P., Angelopoulos, G.N.: Principles and applications of CVD powder technology. Mater. Sci. Eng. R. Rep. **53**(1–2), 1–72 (2006)
89. Varma, A., Mukasyan, A.S., Rogachev, A.S., Manukyan, K.V.: Solution combustion synthesis of nanoscale materials. Chem. Rev. **116**(23), 14493–14586 (2016)
90. Varshney, S., Mishra, N., Gupta, M.: Progress in nanocellulose and its polymer based composites: a review on processing, characterization, and applications. Polym. Compos. **42**(8), 3660–3686 (2021)
91. Varshney, S., Nigam, A., Mishra, N., Pawar, S.: Microwave-assisted synthesis of magnesium oxide nanoflakes via green chemistry approach using Ficus Racemosa leaf extract: characterization and antibacterial activity. J. Korean Ceram. Soc. **60**(1), 62–74 (2023)
92. Varshney, S., Nigam, A., Pawar, S., Mishra, N.: Structural, optical, cytotoxic, and antimicrobial properties of amorphous silica nanoparticles synthesised via hybrid method for biomedical applications. Mater. Technol. **37**(10), 1504–1515 (2022)
93. Varshney, S., Nigam, A., Pawar, S.J., Mishra, N.: An overview on biomedical applications of versatile silica nanoparticles, synthesized via several chemical and biological routes: a review. Phosphorus Sulfur Silicon Relat. Elem. **197**(2), 72–88 (2022)

94. Varshney, S., Nigam, A., Singh, A., Samanta, S.K., Mishra, N., Tewari, R.: Antibacterial, structural, and mechanical properties of MgO/ZnO nanocomposites and its HA-based bioceramics; synthesized via physio-chemical route for biomedical applications. Mater. Technol. **37**(13), 2503–2516 (2022)
95. Venkatraman, S.K., Swamiappan, S.: Review on calcium-and magnesium-based silicates for bone tissue engineering applications. J. Biomed. Mater. Res. Part A **108**(7), 1546–1562 (2020)
96. Whitlow, J., Paul, A., Polini, A.: Bioactive materials: definitions and application in tissue engineering and regeneration therapy. In: Biocompatible Glasses: From Bone Regeneration to Cancer Treatment, pp. 1–17 (2016)
97. Yadav, A., Kumar, H., Sharma, R., Kumari, R.: Synthesis, processing, and applications of 2D (nano) materials: a sustainable approach. Surf. Interfaces **39**, 102925 (2023)
98. Yin, X., Guo, Y., Xie, H., Que, W., Kong, L.B.: Nickel oxide as efficient hole transport materials for perovskite solar cells. Solar Rrl **3**(5), 1900001 (2019)
99. Zhang, F., Li, Z., Xu, M., Wang, S., Li, N., Yang, J.: A review of 3D printed porous ceramics. J. Eur. Ceram. Soc. **42**(8), 3351–3373 (2022)
100. Zhang, L., Forgham, H., Shen, A., Wang, J., Zhu, J., Huang, X., Tang, S.-Y., Xu, C., Davis, T.P., Qiao, R.: Nanomaterial integrated 3D printing for biomedical applications. J. Mater. Chem. B **10**(37), 7473–7490 (2022)

Thin Film Preparation of Electro Ceramics

Suganeswaran Kandasamy, Gobinath Velu Kaliyannan, Raja Gunasekaran, Parameshwaran Rathinasamy, Nithyavathya Nagarajan, and Santhosh Sivaraj

Abstract Electro ceramics poses ability to conduct the electricity for various applications such as LEDs, microwave amplifiers, solar cells, high power switches, dielectric capacitors etc. In spite of mechanical, chemical and thermal stability, some ceramics hold excellent electrical, optical and mechanical properties. Thin film ceramic materials with negligible thickness were synthesized by increasing the temperature of non-metallic minerals especially for optical or electronic devices applications. Thin film electro ceramics can be deposited over substrates through various coating techniques such as spin coating, spray pyrolysis, pulsed laser deposition, metal organic chemical vapor deposition etc. Due to its high degree of adhesion, thin ceramic films found to remain intact with substrate surface. This chapter delivers the compilation of various techniques for the preparation of thin film electro ceramics.

Keywords Ceramics · Substrates · Chemical vapor deposition · Sputtering · Thermal spraying · Atomic layer deposition · Silicon carbide

S. Kandasamy (✉) · G. V. Kaliyannan · P. Rathinasamy · N. Nagarajan
Department of Mechatronics Engineering, Kongu Engineering College, Erode 638060, Tamil Nadu, India
e-mail: suganeswaran22@gmail.com

R. Gunasekaran
Department of Mechanical Engineering, Velalar College of Engineering and Technology, Erode 638012, Tamil Nadu, India

S. Sivaraj
Department of Robotics and Automation, Easwari Engineering College, Chennai 600089, India

U. Kumar (ed.), *Defects Engineering in Electroceramics for Energy Applications*, Engineering Materials, https://doi.org/10.1007/978-981-97-9018-0_6

1 Introduction

1.1 Ceramics

Ceramics are a highly advanced and diverse family of materials that offer a wide range of options for creating innovative outcomes. Ceramics have had a profound influence on our lives and interactions with the world, revolutionizing manufacturing processes and facilitating technical advancements in various domains. The emergence of ceramic thin films has led to the exploration of several new applications, expanding the potential for invention and exploration. For scientists, engineers, and enthusiasts of materials science, the study of ceramics offers a dynamic and constantly evolving landscape that holds great potential for numerous groundbreaking advancements in the future.

1.2 Ceramic Thin Films

The thickness of a thin film can range from a few micrometers to a few nanometers. Thin films find extensive utility across multiple uses, ranging from electronic gadgets to optical coatings. They possess such minimal thickness that they are frequently regarded as two-dimensional substances. The substance employed to construct a thin layer of ceramic determines its composition. The process of subjecting non-metallic minerals to extremely high temperatures result in the formation of ceramics. Subsequently, these minerals undergo a process of cooling and solidification, resulting in the formation of a material that possesses exceptional hardness, high resistance to heat, and resistance to corrosion. These characteristics are maintained even when ceramics are made into thin sheets with a thickness of less than a micrometer.

Ceramic thin coatings are mostly made using sputtering technique. The process includes ionizing a target substance with high-energy particles until its atoms evaporate and settle on the target substrate. This procedure allows designers to fabricate the exact layers required for producing coatings with optical properties and other parts that need to operate in adverse circumstances and endure significant pressures.

1.3 Thin Film Electro-Ceramics

The formation of oxide films is consistently achieved using several thin-film approaches, including evaporation, sputtering, laser ablation, chemical vapor deposition, and sol–gel methods. For these procedures to be used for deposition of multinary systems, the preparation parameters must be carefully controlled in order to create thin oxide layers that are stoichiometric [1]. For single oxides like SiO_2, TiO_2, and Ta_2O_5, a non-vacuum chemical deposition technique called the sol–gel thin film process has been under development since 1950. The thin-film technique is notable for its ability to deposit multicomponent oxide systems with different compositions and consistent dopant concentration [2–4]. The growing need for thin-film devices has also led to the advancement of sol–gel methods for dielectric thin films.

Hennings et al. [4] used a modified sol–gel technique to apply perovskite films with a regulated concentration of acceptor or donor dopants onto silicon substrates. This procedure relies on the breakdown of organometallic materials. Organometallic materials, such as metal alkoxides, metal carboxylates, or metal acetylacetonates, are dissolved in an organic solvent and applied onto different substrates, which includes Si-single-crystal substrates with a thin Pt layer on top, employing a spin coating or dip-coating technique.

At temperatures ranging from 500 to 700 °C, the organic molecules undergo decomposition, resulting in the formation of thin oxide layers. Perovskite films can be deposited by using many deposition and firing cycles, resulting in thicknesses ranging from a few micrometers to several micrometers. The thin layers produced with this sol–gel technique at low temperatures are characterized by a single phase and exhibit crystallization with the perovskite arrangement. The perovskite films were made and their grain size was assessed to be between 30 and 90 nm using X-ray diffraction analysis (XRD) and scanning electron microscopy (SEM).

The sol–gel thin film procedure is a chemical method that does not require a vacuum to create thin films. Thin film electro-ceramics are important components in various electrical and optical equipment that are now being produced or are under development. Ceramic films serve several functions in electronic components, such as providing mechanical support, acting as diffusion barriers, serving as intermediary layers in epitaxial development (buffer layers), and functioning as dielectrics [5]. Common dielectric materials used in integrated circuits include silica and silicon nitride. Ferroelectric ceramics are at present garnering significant attention due to their applications as non-volatile memory components, pyroelectric thermal imager detectors, actuators, and piezoelectric devices. The common materials used in this group consist of lead titanate doped with lanthanum and zirconium, also known as PLZT. Perovskite superconductors, such as $YBa_2Cu_3O_7$-x, are intricately connected. Thin film electro-ceramics are utilized in electrical gas sensor devices, such as tin oxide for the detection of carbon monoxide. Additionally, they are being combined with circuits that run on standard substrates. In optical uses, ceramic films can be classified as either active or passive, similar to their counterparts in electronic applications. Researchers are currently working on creating active ceramic films for non-linear optical equipment, including thin film lasers like Ti-doped sapphire and electro-optic switches like lithium niobate. Epitaxial or strongly orientated films are necessary for these applications. Indium-doped tin oxide is extensively utilized as the transparent electrode, whereas thin film electrochromic ceramics are now being researched for the purpose of creating “smart” windows. Commonly used materials include silica and titania arranged in multilayer stacks with precise thickness and homogeneity. Thin film electro-ceramics encompass a wide range of examples.

(i) **Ferroelectric lead titanate**

Ferroelectric materials have an inherent electric dipole moment without the need for an external electric field. In order to do that it is necessary for the crystal structure to be non-centrosymmetric and possess a polar axis. The stability of the ferroelectric phase is limited to temperatures below a critical point known as the Curie temperature (Tc). Above Tc, the crystal becomes centrosymmetric and exhibits electrical para magnetism. Ferroelectrics, similar to ferromagnetic materials, possess microstructural regions where the polar direction is consistent. When examining the relationship between polarization and applied field, hysteresis loops can be observed. An example of a ferroelectric oxide is the perovskite $BaTiO_3$, which exhibits a Curie temperature (TC) of 120 °C. At temperatures above the Curie temperature (TC), the crystal structure is cubic and exhibits paraelectric behavior. However, below TC, the Ti4+ ions undergo displacement from their center location within their coordination octahedron towards one of the six coordinating 0^2- ions. The outcome is the emergence of a dipole moment and a reduction in symmetry to tetragonal. The arrangement of perovskite is just one of several structures that can undergo this type of transformation. Additionally, the shift to the ferroelectric phase gives rise to several other beneficial characteristics, including piezoelectricity, pyroelectricity, electro-optical activity, and non-linear optical activity. Therefore, ferroelectric ceramics possess great appeal for their utilization in a diverse range of electrical, electro-mechanical, electro-optical, and non-linear optical systems. The diverse range of actions exhibited by these materials provides significant advantages, yet it can also lead to challenges as devices often aim to harness a certain property while minimizing undesired side effects. For instance, in a pyroelectric image sensor, a substantial pyroelectric coefficient is necessary. However, if the piezoelectric coefficient is likewise elevated, the device will exhibit sensitivity to stress and function as a microphone. Hence, while choosing the substance for a specific purpose, it is necessary to minimize undesired sensitivities while maximizing the desired ones. The competing criteria can be optimized by substituting suitable dopants into the fundamental crystal.

(ii) **Lead titanate thin films**

Various methods have been employed to produce thin ceramic films, and a majority of these approaches have been tested on ferroelectric ceramics. AEA Technology has conducted research on sol–gel process and reactive sputtering, that are currently the most commonly used methods for producing ferroelectric ceramics. Sol–gel processing involves applying a liquid colloidal dispersion (known as the sol) onto a substrate and allowing it to dry into a solid layer (known as the gel). This gel is then subjected to heat treatment to breakdown and densify it, resulting in the formation of the final ceramic material. The process of dehydrating lead acetate was initiated by subjecting it to reflux in methoxyethanol. Subsequently, the substance was treated with titanium isopropoxide, resulting in the formation of a presumed mixed alkoxide complex (compounds with the general formula R-0-M-0-M"-OR, where R represents an alkyl group) consisting of lead and titanium. The mixed alkoxide underwent partial hydrolysis upon the incorporation of dilute nitric acid, resulting in the formation of

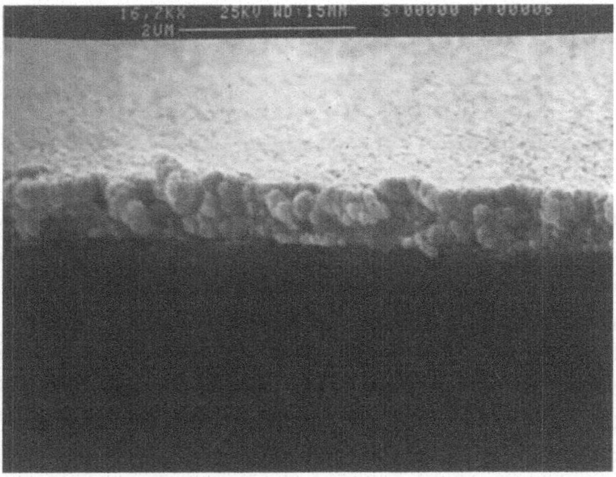

Fig. 1 SEM micrograph depicting the multicoating of a lead titanate film using the sol–gel process [5]

polymeric molecules dispersed in a solution. The dispersion was deposited onto platinum-coated silicon substrates using spin coating at a speed of 2000 revolutions per minute in a clean facility with a class 100 rating. The desiccated gel film was subsequently transformed into an oxide film using a heating process at either 600 or 700 °C for a duration of 15 min. Figure 1 displays the scanning electron micrograph of a film's cross section. The thickness of the ferroelectric film obtained from a single coating is around 200 nm. Therefore, films with a thickness of approximately 1 μm must be formed through the process of repeated coating.

Sputter deposition offers the benefit of being a dry method compared to solgel. However, it presents challenges in controlling the film's stoichiometry. To attain precise control over the Pb/Ti ratio in the film, we conducted a study on reactive sputter deposition. This involved employing two metal targets, one made of Pb and the other of Ti, which were independently regulated. The deposition process took place in an atmosphere including oxygen. The substrate holder may also be elevated to a temperature of 1000 °C, while concurrently applying a radio frequency biased signal to facilitate the bombardment of the developing film by low-energy ions originating from the cathode plasma. This bombardment facilitates the compaction of the film at temperatures that are lower compared to previous methods. The ambient gas pressure consisted of a mixture of 25% oxygen and argon, with a pressure of 5 millitorr. The ion source had the capability to either clean the substrate prior to deposition or serve as an alternative source of ions during the formation of the film. The ratio of Pb/Ti was closely regulated to be about equal to one by adjusting the power provided to each metal cathode.

2 Preparation of Thin Film Electro Ceramics

2.1 Preparation of Pb(Zr, Ti)O_3 Thin Films by Sol Gel Method

Figure 2 depicts the flow diagram outlining the process of preparing stock solutions and films employing a sol gel method. The selection of precursor chemicals and solvent for this procedure was determined by the following criteria: The compounds must possess a substantial amount of metal, exhibit a high level of solubility in the chosen solvent, undergo thermal decomposition without vaporizing, and demonstrate chemical compatibility with one another. The solvent must possess the appropriate boiling point, viscosity, and surface tension. The resulting solution should exhibit a significant amount of the required constituents and demonstrate stability under normal environmental conditions. The precursor materials employed were lead acetate ($Pb(CH_3COO)_{23}H_2O$), zirconium propoxide ($Zr(C_3H_7O)_4$), and titanium isopropoxide ($Ti[(CH_3)_2CHO)_4]$, taking into account these factors [6]. Solvents such as glacial acetic acid, distilled water, or propanol were utilized. Ethylene glycol was employed as an addition to mitigate cracking and enhance the surface uniformity of the films.

2.1.1 Preparation of Solution

The lead acetate was mixed in hot acetic acid at a ratio of 2 g of lead acetate per 1 ml of acetic acid. The solution was raised to a temperature of 10 degrees Celsius to eliminate water. The dehydrated solution was brought down to a temperature lower than 80 °C before to the sequential addition of the necessary amounts of zirconium propoxide and titanium isopropoxide. The addition of zirconium propoxide is necessary as it undergoes a reaction with acetic acid to produce a solution that is resistant to hydrolysis. Subsequently, this solution effectively inhibits the hydrolysis and condensation of titanium isopropoxide upon its addition. When titanium isopropoxide is introduced first, it undergoes a reaction with acetic acid resulting in the development of either mono or diacetylates. Additionally, condensation takes place leading to the production of polytitanyl acetylates. The solution was vigorously mixed with an ultrasonic cleaner till all the solid particles had completely dissolved. Subsequently, ethylene glycol was introduced at a ratio of 1 ml per 10 g of lead acetate in solution. Prior to the addition of glycol, it was necessary to ensure that the initial reaction had fully completed, as any remaining titanium isopropoxide and zirconium propoxide would combine with ethylene glycol to produce a solid substance through condensation. The mixture was diluted using either distilled water or propanol in order to modify the viscosity and surface tension of the mixture. Lower viscosity solutions led to the formation of thinner films. The stock solution given in Table 1a is a convenient method for creating PZT layers with a composition close to the morphotropic phase boundary.

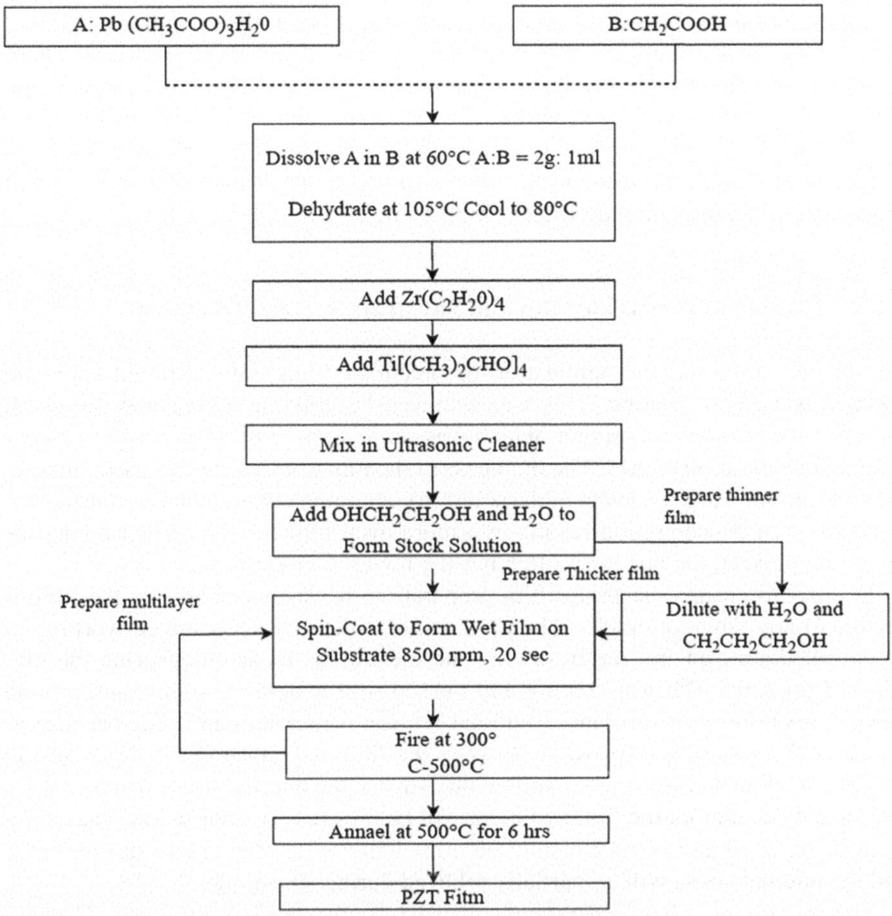

Fig. 2 Flow diagram for the process of preparing stock solutions and films by sol gel method

Table 1 Chemicals used for the stock solution for **a**) PZT film with Zr/Ti = 53.5/46.5 and **b** for alumina thin film

	Chemicals	Quantities
(a) PZT thin film		
Lead acetate	Pb $(CH_3COO)_2.3H_2O$	12.00 g
Acetic acid	CH_2COOH	6 ml
Zirconium propoxide	$Zr(C_3H_7O)_4$	5.54 g
Titanium isopropoxide	$Ti[(CH_3)_2CHO]_4$	2.18 g
Ethylene glycol	$OHCH_2CH_2OH$	1.2 ml
Distilled water	H_2O	6 ml
(b) Alumina thin film		
Aluminium isopropoxide	$Al[(CH_3)_2CHO]_3$	4.00 g
Nitric acid	HNO_3	5 ml
Ethylene glycol	$OHCH_2CH_2OH$	4 ml
Propanol	$CH_2CH_2CH_2OH$	4 ml

This solution can be used to produce a film that is precisely 1 μm thick. Thinner films were produced through incorporating a mixture of water and propanol, where the propanol helped to reduce the surface tension of the solution and enhance the capacity to evenly cover the substrate. The ultimate solution was purified through a membrane filter and preserved in a clean tight container. The substance exhibited stability and a moderate drying rate when exposed to air. When diluted, it formed films with a thickness ranging from 0.05 to 1 pm.

2.1.2 Coating of Precursor Film and Substrate Surface Treatment

The process involved the application of precursor films onto different substrate surfaces using spin coating. This was achieved by utilizing a Headway Research Incorporated photoresist spinner, which was set to a speed of 8500 revolutions per minute for a duration of 20 s. The thickness of the film was determined by regulating the concentration and viscosity of the solution containing organometallic substances. A decrease in concentration results in a more even film surface. When numerous layers are present, the last layer often has the lowest concentration.

In order to ensure successful film preparation, it was necessary for the film to be free of any faults or cracks. The production of the film necessitated working in a controlled atmosphere free from dust and meticulous focus on ensuring the film adhered properly. The wet coating and burned film undergo a significant volume change, resulting in a substantial internal tension. Cracking can occur because to a mismatch in thermal expansion between the film and substrate. If the adhesion between the film and substrate is sufficiently strong, the internal stress will be evenly distributed throughout the film and there will be no stress concentrations caused by areas of weak adhesion. As a result, the films will not develop cracks during firing and the internal stress will be partially relieved during annealing.

The substrates utilized encompassed fused quartz, borosilicate glass, Corning 7509 glass covered with a stannic oxide-based transparent conducting layer, mica, silicon wafers, and stainless-steel plates. Conductive glass facilitated the attainment of crack-free surface coatings, whereas the creation of even the thinnest film on fused quartz was challenging. Such substrates necessitated a specific surface treatment. The substrates underwent boiling in water and detergent, followed by a final rinsing in distilled water. Ultrasound and vapor degreasing in methanol effectively eliminated organic impurities. Next, the substrate was thoroughly washed in boiling filtered distilled water and dried using filtered nitrogen.

Before the PZT layer was placed onto the surface, a thin intermediate layer was used to improve the film's adherence to the substrate. The following layers were analyzed for this purpose: Al_2O_3, MgO, ZrO_2, and TiO_2. The compound Al_2O_3 exhibited superior characteristics, as it possessed chemical stability, transparency, and was readily manufacturable using sol gel processing. A solution of aluminum nitrate was diluted with a mixture of ethylene glycol and propanol to create the alumina precursor solution. A mixture of 4 g of aluminum isopropoxide and 5 ml of nitric acid was prepared. 4 ml of propanol and ethylene glycol were incorporated to the mixture after the process had finished. The final composition is outlined in Table 1b. The substrate was heated to a temperature ranging from 400 to 600 °C before the resulting mixture was spin coated onto it at 8500 rpm for 20 s. The Al2O3 film, as it was initially burned, did not exhibit any distinguishable x-ray diffraction peaks, suggesting that it had an amorphous structure. The alumina layers exhibited satisfactory clarity and transparency. The observation of the enhancement in film adhesion is noted.

2.1.3 Firing of Precursor Film

The fire process, which involves the conversion of an organic precursor film into an inorganic ceramic film through the pyrolysis of inorganic metallic compounds, is the crucial stage in the preparations. The firing sequence has an impact on various aspects of the film, including film breaking, structure of the crystal, size of grain, transparency, and roughness of the surface. The process is intricate, including a sequence of physicochemical processes. The procedure can be described by the following:

(i) The solvents undergo evaporation, resulting in a significant alteration in volume and the development of internal tension. This phenomenon takes place within the temperature range of ambient temperature to 250 °C.

(ii) Lead acetate undergoes dehydration, melting, and decomposition. During this phase, the dry film undergoes rehydration, causing the organic molecules to initiate decomposition. The internal stress is reduced, allowing for the ongoing volume change of the film. Upon reaching the last stage, the organic film undergoes a transformation into a fine blend of lead, titanium, zirconium oxides, and unbound carbon.

(iii) At elevated temperatures, the carbon that is not bound to other elements undergoes oxidation, resulting in the conversion of the mixture of oxides into a transparent and structureless PZT film. Knowing and managing these intersecting physicochemical interactions are crucial in the process of film preparation.

The elevated boiling point and substantial latent heat of vaporization of ethylene glycol increase the temperature at which the solution evaporates in stage (i) approaching the temperature at which it transforms into a molten state in stage (ii) This preserves the movement of atoms and drastically reduces the possibility of cracking. Increasing the firing temperature additionally decreases this latency. Films can be heated either by using an oven or by laying the material on a heated surface. In the latter instance, the fired film experiences fewer fractures due to the substrate temperature exceeding the film temperature and the subsequent rapid increase in temperature. As the wet film undergoes pyrolysis and solidifies on the heated expanded substrate, it tends to generate compressive or reduced tensile stress upon cooling. Nevertheless, there are physical constraints regarding the discharge rate and temperature. For thick films, the temperature range in which they can be heated without any negative effects is between 350 and 500 °C. As the firing temperature increases above this range, the films start to lose their transparency and become milky in appearance. The reason was the formation of porosity due to quick evaporation of solvents and the decomposition of ligands in the organometallic compounds.

2.1.4 Annealing

As-fired films are predominantly amorphous in nature. Annealing is performed with the aim of inducing a transformation in the amorphous structure, leading to the formation of the perovskite structure. The procedure involved utilizing a temperature–time profile where the annealing temperature was gradually increased, then held at a steady temperature for a maximum of 6 h, and thereafter decreased in 2-h increments until reaching room temperature. All films in the current investigation underwent annealing using this specific profile, which involved being held at a temperature of 600 °C for a duration of 6 h.

2.2 Preparation of Indium Tin Oxide (ITO) Thin Films Prepared by a Spin Coating Process

The primary materials utilized were tin (IV) isopropoxide, indium (III) acetylacetonate, and 2-methoxyethanol [7]. The weight ratio of Sn to In was set at 8 wt%. The weight loss and exothermic reaction characteristics of the precursor solution, which contains the sources of Indium (In) and Tin (Sn), were determined using thermogravimetric-differential thermal analysis (TG–DTA). For the film deposition, clean soda-lime silicate glass plates measuring 200 mm × 200 mm × 1 mm were utilized as substrates. The precursor solution, which contained sources of In and Sn, was applied to the substrate through a repeated procedure of spinning (500 rpm for 10 s and 2000 rpm for 20 s) and drying. This process was performed five times. The solvent in the combined solution was eliminated using the process of drying at a temperature of 150 °C for a duration of 10 min. The ITO thin films underwent a firing process in air at temperatures ranging from 450 to 600 °C for a duration of 30 min. Subsequently, they were cooled to room temperature and annealed at temperatures between 500 and 600 °C for another 30 min, this time under an atmosphere of Ar. Furthermore, the annealed ITO thin films underwent etching using 0.1 N HCl for a duration of 5–60 s.

2.3 *Preparation of Indium Tin Oxide Ultra-Thin Films Prepared by Jet Nebulizer Spray Pyrolysis Technique*

The jet nebulizer spray pyrolysis equipment as shown in (Fig. 3) comprises a jet nebulizer spraying unit, a substrate holder equipped with a heater, and an air compressor. In order to create thin films of tin doped indium oxide, a solution is prepared by dissolving indium (III) chloride (InC13) in 100 mL of double distilled water, resulting in a 0.4 M starting solution. Tin doping was carried out by introducing tin (II) chloride dihydrates ($SnCl_2{\cdot}2H_2O$) into the initial solution. To achieve a transparent and uniform solution, a small amount of acetic acid was added. The concentration of doping in the solution was systematically adjusted, ranging from 0 to 30 weight percent in intervals of 10 weight percent. The mixture was agitated at a uniform velocity for 1 h using a magnetic stirrer. Before the deposition, glass substrates of 1 square inch were sequentially cleaned with acetone, isopropyl alcohol, and distilled water for a duration of 15 min using an ultrasonicator.

The temperature of the substrate during each deposition was maintained at 500 °C under ambient air conditions. The solution that had been created was applied onto the heated substrate using a jet nebulizer. The solution was sprayed at a rate of 0.5 mL/min, with compressed air being used as the carrier gas. The nebulizer was positioned at a distance of 5 cm from the surface of the substrate. An X-ray diffractometer (XRD) equipped with the analytical system and Cu K1 radiation (k = 1.54056A) was used to investigate the structural properties of spray coated ITO films. The compounds Li_2CO_3, $BaCO_3$, La_2O_3 (previously dried at 900 °C for 12 h in the presence of air) and Ta_2O_5 were utilized in the production of the $Li_6BaLa_2Ta_2O_{12}$ material in large

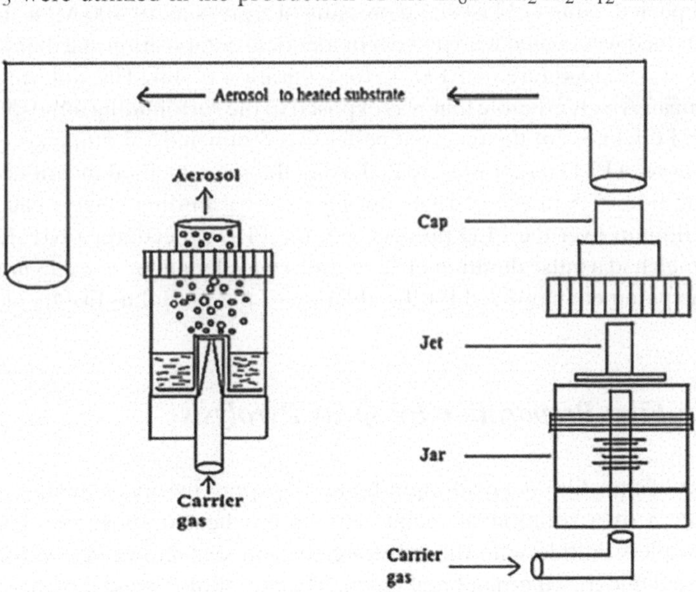

Fig. 3 Schematic diagram of jet nebulizer apparatus [8]

Table 2 PLD process parameters [8]

Gas atmosphere	Oxygen
Pressure (Pa)	5
Temperature (°C)	550
Laser fluence (j/cm^2)	2
Laser frequency (Hz)	10
Number of laser pulses	40,000
Target/substrate distance/mm	45

quantities. The initial components were combined in the correct proportion, along with an excess of 10 mol% of the lithium source, relative to the amount of lithium present (with an additional 5 mol% excess of Li_2CO_3), in order to account for any loss of lithium caused by volatile substances. Following the first manual blending process in an agate mortar, the raw materials underwent a 10-h ball milling process at 160 rpm in a zirconia grinding bowl containing zirconia spheres (diameter 3 mm) immersed in isopropanol. Subsequently, the powder underwent drying using a rotary evaporator and was then subjected to calcination in a Heraeus chamber furnace at a temperature of 750 °C for a duration of 6 h in air.

Following an additional ball milling process under identical conditions, the mixture was subjected to a temperature of 1000 °C for a duration of 12 h in an atmospheric air. The acquired substance was subjected to further milling, following the aforementioned procedure, and subsequently compressed into a pellet. First, the pellet was subjected to uniaxial pressing at a pressure of 12.5 MPa. Then, it was further compacted using cold isostatic pressing at a pressure of 400 MPa. The pellet that was obtained was coated with powder of identical composition and then subjected to sintering at a temperature of 1000 °C for a duration of 4 h. The sintering process took place in an Al_2O_3 crucible that was exposed to the surrounding atmosphere. The diameter and thickness of the received pellet was 8 mm and 3.2 mm respectively.

Furthermore, a PLD target was created using the same method mentioned earlier, with the addition of 5 mol% of Li_2O during the final milling stage to account for the loss of lithium over the PLD process. For the PLD procedure, a KrF-Laser was utilized, which had a pulse duration of 25 ns and a wavelength of $\lambda = 248$ nm. Table 2 provides the parameters utilized for the ablation of the $Li_6BaLa_2Ta_2O_{12}$ target.

2.4 *Thin Film Preparation by Spray Pyrolysis*

The process of thin-film deposition, utilizing the spray pyrolysis method, includes the application of a solution of metal salts onto a heated substrate as shown in (Fig. 4). Droplets collide with the substrate surface, expanding into a disc-shaped formation, and undergo thermal breakdown. The morphology and dimensions of the disk are contingent upon the droplet's momentum, volume, and the temperature of

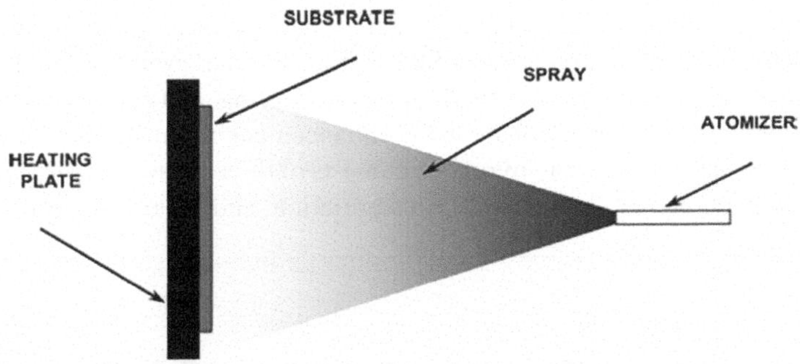

Fig. 4 Schematic representation of spray pyrolysis equipment [9]

the substrate. As a result, the film is often made up of metal salt disks that overlap and are transformed into oxides on the surface that is heated.

Spray pyrolysis encompasses a multitude of procedures that occur either concurrently or in a specific order. Key factors include aerosol formation and dispersion, solvent drying, droplet collision resulting in spreading, and precursor breakdown. All of the processes mentioned involve the deposition temperature, with the exception of aerosol generation. Hence, the substrate's surface temperature is the primary factor that governs the film's shape and characteristics. By elevating the temperature, the film's morphology can transition from a cracked to a porous microstructure. The deposition temperature has been identified as the most critical spray pyrolysis characteristic in numerous research. By altering the deposition temperature, the characteristics of deposited films can manipulate and regulated. In the case of zinc oxide films, this temperature adjustment directly impacts their optical and electrical properties [10]. The films with the lowest electrical resistance were created by depositing an aqueous solution of zinc acetate at a temperature of 490 °C. This process led to enhanced crystallinity. Additionally, films prepared at temperatures of 420 and 490 °C exhibited high transmission rates (90–95%) in the visible range. This phenomenon can be related to a reduction in the thickness of the film and an enhancement in the uniformity of its structure.

Indium oxide coatings doped with fluorine were analyzed in terms of their deposition temperature, dopant concentration, air flow rate, and film thickness in relation to their physical properties [11]. The deposition temperature has been determined to have a significant impact on the film structures. The degree of preferred (4 0 0) direction expands as the film thickness develops. Electrostatic spray deposition was utilized to deposit thin films of yttria-stabilized terbia doped with terbium [12]. The manipulation of the deposition variables and solution compositions allowed for precise control over the surface morphology. The film's morphology changed from an extremely dense to a significantly porous structure by elevating the deposition temperature. Thus far, only basic models regarding the mechanics of spray deposition and film development have been devised. The multitude of processes that take place during film production via spray pyrolysis, whether occurring sequentially or concurrently, are too complex to be accurately modelled in a simple manner. These include atomization of the precursor solution, transport and evaporation of droplets, distribution of the precursor salt onto the substrate, drying, and decomposition. With knowledge of these processes will aid in the enhancement of film quality. The process of thin-film deposition via spray pyrolysis consists of three primary stages: precursor solution atomization, resultant aerosol transportation, and precursor decomposition on the substrate.

2.5 Barium Titanate Thin Films by Chemical Solution Deposition

An equimolar mixture of titanium isopropoxide, mixed with glacial acetic acid, and acetylacetone, which was used as the chelating agent, was utilized to produce a precursor solution. The concentrated solution was reduced to a concentration of 0.3 M using methanol and diethanolamine (DEA). The use of diethanolamine enhances the process of strain relaxation during removal, hence promoting enhanced morphology [13]. After spin-coating 18 m thick bare copper foils with the precursor solution at 3000 rpm for 30 s, the foils were dried on heated plates at 250 °C for 7 min. The technique was iterated until the appropriate thickness was attained, with 3–4 layers being applied in this instance. The films were subjected to annealing at a temperature of 900 °C for a duration of 30 min. This process took place in an environment composed of a mixture of high-quality nitrogen and hydrogen. The forming gas used in this process was passed through water at a temperature of 25 °C. These circumstances are enough to create an atmosphere with a total pressure of one atmosphere and a partial pressure of oxygen (pO_2) of 10–9 atmospheres. The film thickness obtained after the firing process was 0.6 μm. A controlled oxidation anneal was conducted at a temperature of 600 °C in a vacuum environment, with a deliberate addition of oxygen at a total pressure of approximately 10–7 atm O_2, over a duration of 30 min. Platinum electrodes with a surface area of approximately 10–4 cm^2 and a thickness of 100 nm were created using DC magnetron sputtering, using a shadow mask for deposition.

Figure 5 displays an X-ray diffraction pattern of a standard BaTiO3 sample that was produced on copper foil. The pattern displays narrow peak widths, arbitrary alignment, and a minimal, even background noise level. The characteristics correlate more closely with meticulously produced ceramic material rather than conventionally produced thin films. Figure 5 demonstrates that the crystallization procedure does not yield any Cu_2O or CuO phases. The outcome is anticipated by thermodynamic computations of the $Cu–O_2Cu_2O$ system at a temperature of 900 °C and an oxygen partial pressure (pO_2) below 10–9 atm [15].

It is important to mention that the area detector utilized has a remarkably high sensitivity in identifying poor crystalline or low volume fraction second phases. Therefore, if there are no scattering events detected that cannot be attributed to Cu

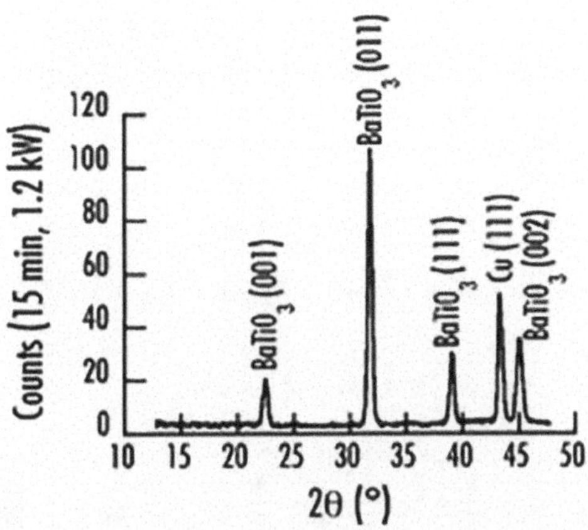

Fig. 5 XRD pattern of a 3 layer (~0.6 μm) $BaTiO_3$ film on 18 μm thick Cu-foil [14]

or $BaTiO_3$, it strongly suggests the presence of a sharp Cu/$BaTiO_3$ interface across a significant sampling area.

Figure 6a depicts a topographic image of the surface of a $BaTiO_3$ sheet, which was obtained by employing atomic force microscopy with a scan size of 5 μm by 5 μm. The observed surface microstructure comprises equiaxed granules, which vary in diameter from 0.1 to 0.3 μm, with a mean value of 0.2 μm. Figure 6b displays a transmission electron microscope (TEM) image taken under bright field conditions. This micrograph confirms the presence of equiaxed grain morphology over the whole thickness of the film. It is notable to observe that the $BaTiO_3$ layer appears to be nearly divided along a particle boundary [15, 16]. These findings indicate that crystallization takes place at two interfaces of the film. The growth continued until the separate fronts converge at the center of the film, and the influence of the number of organic layers applied is minimal. As a consequence, the film has an average thickness of two grains. It is important to mention that this observation contradicts previous analyses, which indicated that homogenous nucleation was the preferable process for the crystallization of $BaTiO_3$ thin films [5, 16]. The variation is believed to arise, at least in part, from the utilization of a strongly reducing environment for the extraction of organic matter and the formation of BT layers. The high-resolution transmission electron microscopy (TEM) analysis indicates the absence of any interfacial phases at the interface between $BaTiO_3$ and copper.

2.6 *Hydrothermal Synthesis of Thin Films of Barium Titanate Ceramic Nano-Tubes*

Commercially acquired polycrystalline Ti substrates made of metal with a purity of 99.99% and dimensions of $50 \times 50 \times 0.25\ mm^3$ were polished on one side to achieve a finish of 0.05 μm using standard metallography methods. The substrates were subsequently anodised using the conventional process. The anodization process

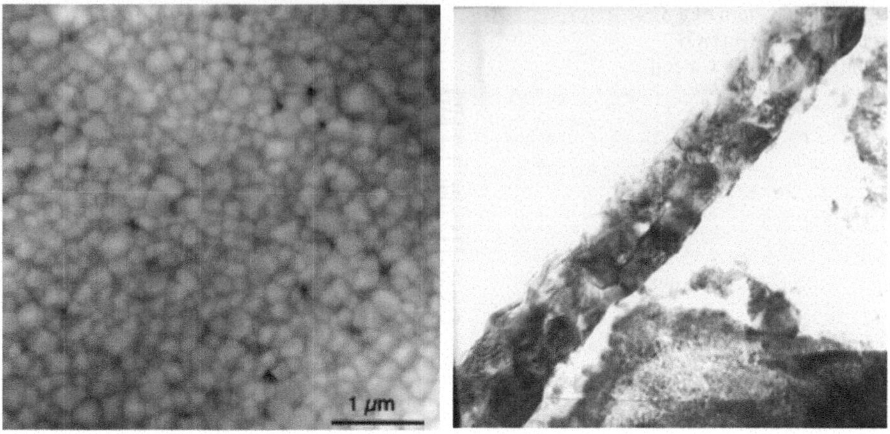

Fig. 6 **a** AFM surface image of a $BaTiO_3$ thin film deposited on Cu-foil and **b** TEM micrograph of a cross-sectioned $BaTiO_3$ thin film deposited on Cu-foil [14]

took place in a stirred aqueous bath containing a 1.5 wt% solution of hydrofluoric acid (HF) at a temperature of 16 °C. A direct-current power source was utilized to provide a voltage of 20 V over the anode, made of Ti-substrate, and the cathode, consisting of a Pt-sheet, with a separation distance of 10 mm, for a duration of 20–30 min. Distilled water was subsequently used to rinse the anodized substrates. The anodized substrates were promptly moved to a hydrothermal vessel, specifically a Teflon-lined stainless steel acid-digestion reactor or "bomb," which was filled with a 0.1 M $Ba(OH)_2$ (pH = 13.2) solution in water that was free of CO_2. The vessel was filled up to 70% of its 45-mL capacity. To prevent the introduction of settling powder or debris into the thin films, a holder was used to retain the substrates vertically in the center of the vessel. Subsequently, the vessel was subjected to a temperature of 200 °C in a forced-air convection oven for a duration of 80 min. The substrates were extracted from the vessel, washed, dried, and analyzed. Controlled hydrothermal studies were conducted on Ti metal substrates that were either as-polished or had a uniform, thin surface coating of titanium oxide (without any nano-tubes). The latter was accomplished by subjecting the as-polished Ti substrates to oxidation in a 1 M H_2SO_4 solution for a duration of 60 min. The aforementioned substrates underwent similar hydrothermal treatment. The X-ray diffraction (XRD) pattern of a substrate that underwent anodization and subsequent hydrothermal treatment reveals the existence of two different stages: alpha Ti metal substrate and crystalline $BaTiO_3$. Clamping the films onto the substrate alters their functional characteristics. This field has been the focus of numerous studies, encompassing both experimental and theoretical approaches. For instance, the response of piezoelectric materials is significantly affected by the altered boundary conditions. Specifically, the response perpendicular to the surface is considerably reduced compared to that of bulk ceramics, whilst the bending response of the sandwich arrangement (consisting of a substrate and a film) is amplified [17]. The pyroelectric response is altered due to the thermal expansion imbalance between the substrate and the film, leading to the addition of a piezoelectric component to the pyroelectric effect. Furthermore, the process of switching is affected by the substrate. Several investigations have demonstrated that the strength of non-180° switching is significantly reduced in thin films compared to the bulk material [18]. Stresses arise as the film is cooled down after sintering due to a difference in thermal expansion that exists between the substrate and film [19].

2.7 *ZrO_2 and SiC Ceramic Thin Films Prepared by Electrostatic Atomization*

The electrospray system as depicted in (Fig. 7) consists of a stainless-steel capillary with a precise inner diameter of 0.23 mm and an outside diameter of 0.51 mm. An elevated voltage of 4–6 kilovolts was given to the capillary. A concentric ring electrode, having an inner diameter of 15 mm, was positioned at a distance of 8 mm from the capillary tip. The electrode was linked to the reference voltage of zero volts. A metallic container, functioning as an electrode, was connected to the ground. The ceramic suspensions were introduced into the capillary at a rate of flow ranging from 0.5 to 1.5 ml/h, utilizing a Harvard 11 syringe pump. The silicon and Cr-Fe alloy substrate were electrically connected to the ground and placed 10 mm below the ring to capture droplets during the deposition process.

The initial and crucial step in electrospray deposition is the preparation of the ceramic suspension, as the characteristics of the suspension significantly impact electrostatic atomization. To achieve a stable cone-jet, it is necessary for the conductivity of the suspension to be within an appropriate range [10]. The study employed

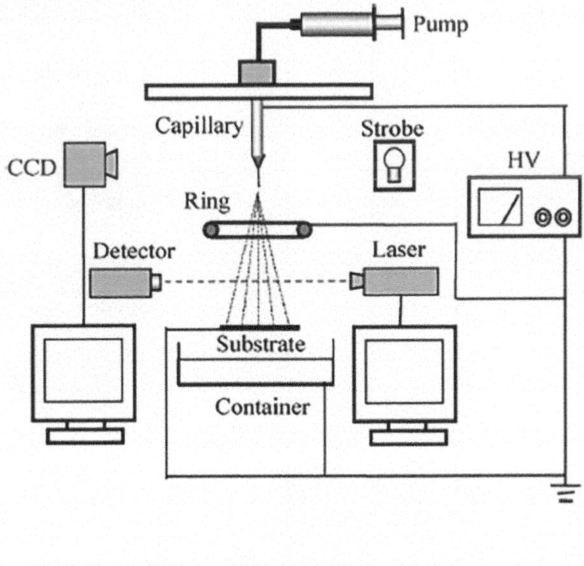

Fig. 7 Schematic representation of electrospray system [20]

HYY-3 zirconia powder, characterized by a typical particle size of 0.47 μm, and fine silicon carbide (SiC) powder with a mean particle size of 1.0 μm. The materials used in the research were fine silicon carbide (SiC) powder with an average particle size of 1.0 μm and HYY-3 zirconia powder, which typically has a particle size of 0.47 μm. A suspension of ZrO2 was produced by dispersing 5% by weight of ZrO2 powder in a solution of distilled water and ethanol (Aldrich) in a 1:1 volume ratio. The SiC suspensions were prepared by dispersing 1.3% by weight of SiC powder in pure ethanol. The suspensions underwent ultrasonic agitation and electrical stirring using a magnetic bar for a duration of 30 min each. The suspensions exhibited steady behavior without significant sedimentation for an extended period of time, providing sufficient duration for the completion of studies. To achieve the deposition of homogenous and compact thin films, it is necessary to first produce small droplets of ceramic suspensions. Aerosol can be formed in several modes, including dripping mode, fully-developed mode, cone-jet mode, and multi-jet mode, under the effect of varying flow rates and applied potentials. Out of these methods, the cone-jet mode of electrostatic atomization is the most appropriate technique to fulfill this criterion since it has the inherent ability to produce uniformly sized small droplets in a regulated manner. Figure 8 depicts a detailed analysis of a standard cone-jet formed under specific conditions: a flow rate of 0.8 ml/h, a potential of 4.5 kV, and the utilization of ZrO2 solution.

Based on the image, the diameter of the jet near the apex of the cone was approximately 4 μm. The jet's length was determined to be around 200 μm.

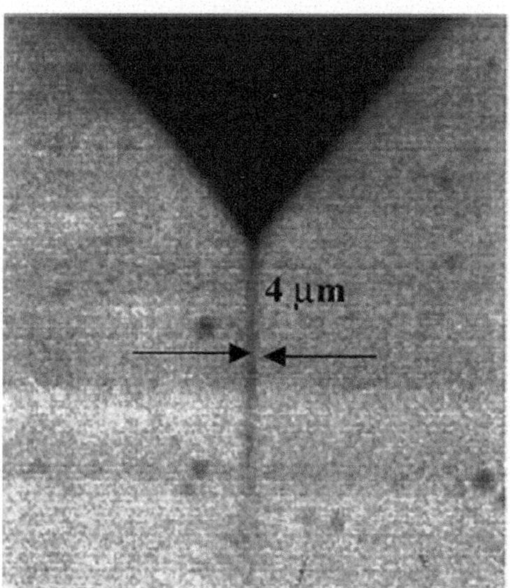

Fig. 8 CCD image shows a close examination of cone-jet [20]

3 Conclusion

This chapter discusses several procedures employed in the preparation of thin film electro-ceramics, including the sol–gel method, spin coating, spray pyrolysis, chemical solution deposition, hydrothermal synthesis, and Electrospray system. Different methods of preparing the substrate and applying a thin layer on its surface improve the compatibility, making it appropriate for a range of applications such as biomedical, electronics, and electrical sectors.

References

1. Saenger, K., Roy, R., Etzold, K., et al.: Lead zirconate titanate films produced by pulsed laser deposition. MRS Online Proc. Libr. (OPL). **200**, 115 (1990)
2. Chapin, L.N., Myers, S.A.: Microstructure characterization of ferroelectric thin films used in non-volatile memories-optical and scanning electron microscopy. MRS Online Proc. Libr. (OPL). **200**, 153 (1990)
3. Budd, K., Dey, S., Payne, D.: The effect of hydrolysis conditions on the characteristics of PbTiO3 gels and thin films. MRS Online Proc. Libr. (OPL) **72** (1986)
4. Hennings, D., Klee, M., Waser, R.: Advanced dielectrics: bulk ceramics and thin films. Adv. Mater. **3**(7–8), 334–340 (1991)
5. Atkinson, A., Moseley, P.: Thin film electroceramics. Appl. Surf. Sci. **65**, 212–219 (1993)
6. Yi, G., Wu, Z., Sayer, M.: Preparation of Pb (Zr, Ti) O3 thin films by sol gel processing: Electrical, optical, and electro-optic properties. J. Appl. Phys. **64**(5), 2717–2724 (1988)
7. Cho, H., Yun, Y.-H.: Characterization of indium tin oxide (ITO) thin films prepared by a sol–gel spin coating process. Ceram. Int. **37**(2), 615–619 (2011)
8. Reinacher, J., Berendts, S., Janek, J.: Preparation and electrical properties of garnet-type Li6BaLa2Ta2O12 lithium solid electrolyte thin films prepared by pulsed laser deposition. Solid State Ionics **258**, 1–7 (2014)
9. Perednis, D., Gauckler, L.J.: Thin film deposition using spray pyrolysis. J. Electroceram.Electroceram. **14**, 103–111 (2005)
10. Afify, H., Nasser, S., Demian, S.: Influence of substrate temperature on the structural, optical and electrical properties of ZnO thin films prepared by spray pyrolysis. J. Mater. Sci. Mater. Electron. **2**, 152–156 (1991)
11. Mirzapour, S., Rozati, S., Takwale, M., et al.: Influence of different process parameters on physical properties of fluorine dopes indium oxide thin films. J. Mater. Sci. **29**, 700–707 (1994)
12. Stelzer, N., Schoonman, J.: Synthesis of terbia-doped yttria-stabilized zirconia thin films by electrostatic spray deposition (ESD). J. Mater. Synth. Process. **4**, 429–438 (1996)
13. Dawley, J., Clem, P., Siegal, M., et al.: Improving sol-gel Yba2Cu3O7− δ film morphology using high-boiling-point solvents. J. Mater. Res. **17**(8), 1900–1903 (2002)
14. Ihlefeld, J., Laughlin, B., Hunt-Lowery, A., et al.: Copper compatible barium titanate thin films for embedded passives. J. Electroceram.Electroceram. **14**, 95–102 (2005)
15. Hoffmann, S., Waser, R.: Control of the morphology of CSD-prepared (Ba, Sr) TiO3 thin films. J. Eur. Ceram. Soc. **19**(6–7), 1339–1343 (1999)
16. Schwartz, R.W., Clem, P.G., Voigt, J.A., et al.: Control of microstructure and orientation in solution-deposited BaTiO3 and SrTiO3 thin films. J. Am. Ceram. Soc. **82**(9), 2359–2367 (1999)
17. Lefki, K., Dormans, G.: Measurement of piezoelectric coefficients of ferroelectric thin films. J. Appl. Phys. **76**(3), 1764–1767 (1994)
18. Trolier-McKinstry, S., Shepard, J., Jr., Lacey, J., et al.: Piezoelectricity in ferroelectric thin films: domain and stress issues. Ferroelectrics **206**(1), 381–392 (1998)

19. Setter, N.: Electroceramics: looking ahead. J. Eur. Ceram. Soc. **21**(10–11), 1279–1293 (2001)
20. Miao, P., Balachandran, W., Xiao, P.: Characterization of ZrO 2 and SiC ceramic thin films prepared by electrostatic atomization. J. Mater. Sci. **36**, 2925–2930 (2001)

Fundamental Characterization Techniques: Impedance and Modulus Spectroscopy

Uma Sharma, Vani Pawar, and Prabhakar Singh

Abstract Impedance spectroscopy (IS) is one of the effective experimental techniques employed for investigating the electrical characteristics of materials and systems over a broad range of frequencies. Its adaptability makes it invaluable in numerous fields, including electrochemistry, material science, electronics, and biophysics. This technique is used to study the dynamics of bound and mobile charges within the bulk or interfacial regions of diverse solid and liquid materials like ionic, mixed ionic-electronic, semiconducting, and dielectrics. By analyzing the electrical response over a range of frequencies, these techniques provide valuable insights into the materials' electrical properties, conductivity, dielectric behavior, and other relevant parameters. Modulus (M) is one of the important measured quantities related to impedance (Z) to study the small signal measurement with respect to frequency and known as *modulus spectroscopy*. These are interconnected like—

Complex impedance, $Z^* = Z' - iZ'' = \frac{1}{i\omega C_0 \varepsilon^*}$

Electric Modulus $M^* = M' + iM'' = 1/\varepsilon^*$, where, $\omega = 2\pi f$.

For the study of electro-ceramics complex Z^* and M^* are mostly used. In a polycrystalline ceramic, for the study of grains, grain boundaries, and electrodes, these can be represented by the equivalent circuit in which Resistors (R) and Capacitors (C) are connected in parallel. Consequently, in a polycrystalline electroceramic, both the properties, the dielectric and electrical, have contributions from bulk or grains, grain boundaries, and electrode polarization.

This chapter will commence by providing an introduction to the fundamentals of impedance and modulus spectroscopy, followed by an explanation of impedance in electrical circuits. Further, we'll discuss the Jonscher's Power law, Nyquist, and Bode plot representation of the impedance/modulus of an electrical circuit. In the concluding section of this chapter, we will delve into some examples that are closely related to both spectroscopy techniques.

U. Sharma · P. Singh (✉)
Department of Physics, Indian Institute of Technology (Banaras Hindu University), Varanasi, Varanasi 221005, India
e-mail: psingh.app@iitbhu.ac.in

V. Pawar
Center for Nano Science and Engineering, Indian Institute of Science, Bangalore, Karnataka 560012, India

U. Kumar (ed.), *Defects Engineering in Electroceramics for Energy Applications*, Engineering Materials, https://doi.org/10.1007/978-981-97-9018-0_7

Keywords Impedance spectroscopy · Modulus spectroscopy · Electroceramic

1 Introduction

Complex plane analysis is a widely recognized and powerful method that has been used extensively in studying dielectric and semiconducting properties of the materials. The impedance has frequently been designed as the AC impedance or complex impedance. It expresses the circuit's capability to resist the passage of electrical current. Impedance spectroscopy (IS) examines a specific material's electrical behavior and is subsequently analysed within the frequency domain [1].

A piece of ceramic consists of numerous tiny crystallites within grains that connect arbitrarily, forming grain boundaries. Ceramic assembly may be described as a grain-grain boundary-electrode system for the purpose of device application. These grain and grain boundary properties are distinctive and influenced by factors like heating–cooling rates, atmosphere, and initial ingredients during processing. Manipulating these variables allows for achieving the desired properties in the chosen material. In barrier layer capacitors (BLCs), the high permittivity could relate to the effects of the sample's area and the grain, grain boundaries related to the thin layer. The capacitance (C) of a single-layer device is represented as—

$$C = \varepsilon_o \varepsilon_r A \ell \tag{1.1}$$

where ε_r represents the dielectric material's relative permittivity, A is the area with thickness [2, 3].

The analysis of the electrical microstructure involves impedance spectroscopy (IS) to examine bulk properties, grain boundary, and sample electrodes. This method measures the material response to applied sinusoidal voltage at different frequencies. Various phenomena occur at distinct frequency ranges. Lattice vibrations occurred at the infrared frequency while long-range ionic conduction was observed at frequencies where ions hop due to variation in electric field, pausing at certain points before hopping again. Thin interfaces, such as those found at sample electrodes, grains, and grain boundaries have a capacitive effect that measures the polarization effectively. These interactions combine lattice polarization with ionic or electronic transport across the sample and are often described as the R–C elements that are connected in parallel or series to each other in an equivalent circuit. The frequency spectrum identifies these phenomena by examining the magnitude of RC products or relaxation time, τ, where $\omega\tau = 1$ serves as a key identifying factor [2, 4].

In IS, most instruments cover a wide frequency range (ω), typically ranging from 10^{-2} to 10^{8} Hz, enabling the study of relaxation times (τ), ranging between 10^{2} and 10^{-8} s. As many phenomena exhibit temperature dependency, accessing various conductivity spectrum regions necessitates temperature variation for comprehensive analysis. Specifically, effects at low frequencies with large τ can be observed at higher temperatures. At these higher temperatures, the mobility of charge carriers increases for their relaxation frequencies and falls within the measuring range. Similarly, easily occurring local hops can be investigated by lowering the temperature. So, characterizing materials through IS offers an investigation into their electrical microstructure. There are main four impedance functions that are used to study the electric/dielectric behavior of a system:

Impedance,

$$Z^* = Z' - iZ'', \tag{1.2}$$

Admittance,

$$Y^* = Y' + iY'' = 1/Z^*, \tag{1.3}$$

Modulus,

$$M^* = 1/\varepsilon^* = M' + iM'', \tag{1.4}$$

Complex dielectric permittivity,

$$\varepsilon^* = \varepsilon' - i\varepsilon'', \tag{1.5}$$

here C_o represents the capacitance of empty cell, $\omega = 2\pi f$ is angular frequency, Z', M', and Z'', M'', represents real and imaginary part of corresponding functions. The AC response of the material can be expressed by any of four formalisms [1, 5].

Impedance analysis can be performed via two different methods:

(i) Complex plane method
(ii) Spectroscopic method

They are valuable tools for investigating defects, microstructure, electrical conductivity, and surface chemistry in a wide range of materials. It covers diverse properties from dielectrics and ionic conductors to interfaces involving adsorbate and adsorbent interactions.

The relation among these parameters can be inscribed as follows-

$$Z' = \frac{D^2}{G}\left(1 + D^2\right) \tag{1.6}$$

$$Z'' = Z'/D \tag{1.7}$$

$$D = tan\delta \tag{1.8}$$

Z*, ***Y****, ***M**** are commonly used in the analysis of electronic ceramics. *G* and *D* are conductance and dielectric loss in relation to both frequency and temperature [5].

2 Theory

2.1 Impedance Theory

Unlike direct current (dc), alternating current (AC) can address all polarization processes by varying its frequency. Generally, impedance (Z) exhibits a linear response when measured across a wide frequency range, from a few millihertz to Megahertz and resistance under a small ∽1 mV fluctuation [4]. It is a complex quantity (Z^*) having a real part (Z') and imaginary part (Z'') of impedance (refer to Fig. 1b):

$$Z^* = Z' - iZ'' = 1/(i\omega C_o\varepsilon^*) \tag{2.1}$$

$$Z' = |Z|cos\theta \tag{2.2}$$

$$Z'' = |Z|sin\theta \tag{2.3}$$

with phase angle,

$$\theta = tan^{-1}\left(\frac{Z''}{Z'}\right) \tag{2.4}$$

$$|Z| = (Z'^2 + Z''^2)^{1/2} \tag{2.5}$$

here, $\omega = 2\pi f$ depicts angular frequency, $i = \surd - 1$, C_o stands for geometrical capacitance, and mathematically $C_o = \varepsilon_o a^2/d$, ε_o= 8.85 × 10^{-12}, a^2 represents the electrode surface area, and d is electrode spacing [6].

Figure 2 summarizes the basic examples of the complex impedance plane. Pure capacitance is represented as a point along the axis corresponding to the real part

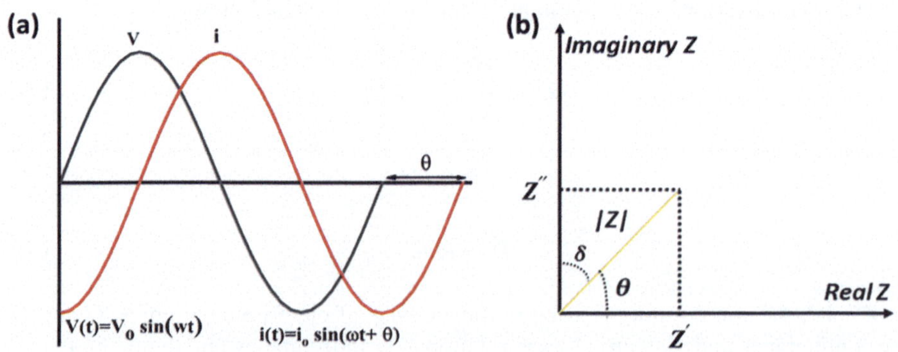

Fig. 1 **a** AC voltage and current **b** complex impedance plane

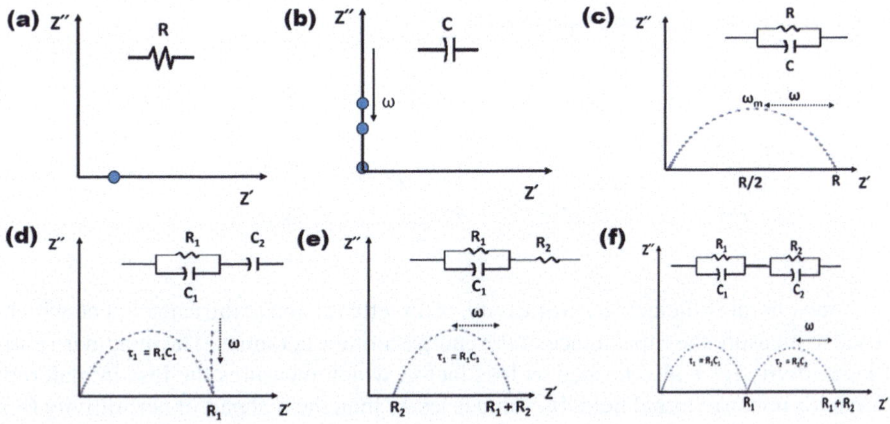

Fig. 2 Z' versus Z'' and their equivalent circuit for some simple examples

of impedance (Z'), with $Z' = R$. Conversely, pure capacitance observed as a linear response along the Z'' axis. When connecting resistance (R) and capacitance (C) in series, the impedance maintains a linear profile at $Z' = R$. However, in a parallel connection of R and C, known as the RC element, a semicircle appears positioned with its centre at *(R/2, 0)* along the Z' axis. The highest point on this semicircle is related with $\omega_m RC = 1$, where $RC = \tau$ represents the time constant or relaxation time, and ω_m denotes the angular frequency where the real part is equal to the imaginary part of impedance [3].

The relaxation frequency, denoted by $f_{max} = \omega_m/2\pi$.

Generally, in a polycrystalline sample, grain boundaries have more resistance and capacitance in comparison to the grain but the relaxation frequency of grain boundaries is of lesser value than grain. These properties provide the main difference between grains and grain boundaries. Thus, the RC element contributes as a fundamental unit in impedance fitting, as explained below [2].

To meet diverse analytical needs, the impedance complex quantity (Z^*) can be easily convert into other complex quantities like electric modulus (M^*), permittivity (ε^*), admittance (Y^*), complex conductivity (σ^*), and dielectric loss tangent ($tan\delta$). Like the modulus (M^*) can be derived from impedance (Z^*) as given below-

$$M^* = M' + iM'' = i\omega C_o Z^* = i\omega C_o(Z' - iZ'') = i\omega C_o Z' + \omega C_o Z'' \qquad (2.6)$$

$$tan\delta = M''/M' = Z'/Z'' = Y'/Y'' = \varepsilon''/\varepsilon' \qquad (2.7)$$

where, real part of modulus is $M' = \omega C_o Z''$, imaginary part of modulus is represented as $M'' = \omega C_o Z'$. All the comprehensive relations among complex quantities are summarized in Table 1.

Table 1 Relation among complex quantities [7]

	M^*	Z^*	Y^*	ε^*	σ^*
M^*	M^*	$i\omega C_o Z^*$	$i\omega C_o/Y^*$	$1/\varepsilon^*$	$i\omega C_o/\sigma^*$
Z^*	$M^*/i\omega C_o$	Z^*	$1/Y^*$	$1/i\omega C_o\varepsilon^*$	$\varepsilon_o/\sigma^* C_o$
Y^*	$i\omega C_o/M^*$	$1/Z^*$	Y^*	$i\omega C_o\varepsilon^*$	$\varepsilon_o\sigma^*/C_o$
ε^*	$1/M^*$	$1/i\omega C_o Z^*$	$Y^*/i\omega C_o$	ε^*	$\sigma^*/i\omega\varepsilon_o$
σ^*	$i\omega\varepsilon_o/M^*$	$\varepsilon_o/C_o Z^*$	$Y^*\varepsilon_o/C_o$	$i\omega\varepsilon_o\varepsilon^*$	σ^*

ε^*, the complex dielectric permittivity, is directly related to the capacitance which is used to measure the capacitance of the energy-storage ceramics. The imaginary part of permittivity (ε''), also termed as loss factor, which measures the loss in material after applying an external field. Its value is lesser than the real part of permittivity (ε') and always greater than zero. However, M^* and Z^* provide distinct perspectives on associated information. Hence, it becomes essential to weigh the strengths of each parameter when seeking comprehensive insights from impedance spectroscopy (IS).

Figure 3a depicts the instruments used in the laboratory experiments. Figure 3b illustrates a simplified impedance measurement setup comprising an impedance analyser, a computer, a sample fixture, and an external stimulus.

A small AC voltage is applied to the ceramic samples through the impedance analyser to observe their corresponding responses. Various unwanted losses are mitigated by the sample fixture, which also offers the required external inputs like temperature, pressure and atmosphere. Meanwhile, the computer controls these inputs to the fixture and collects data through an impedance analyser. An impedance spectrum is obtained for the ionic relaxation and conduction, ranging from low to high impedance (Z) and traverses a wider frequency range, ranging from low frequencies to approximately 10 MHz levels [2].

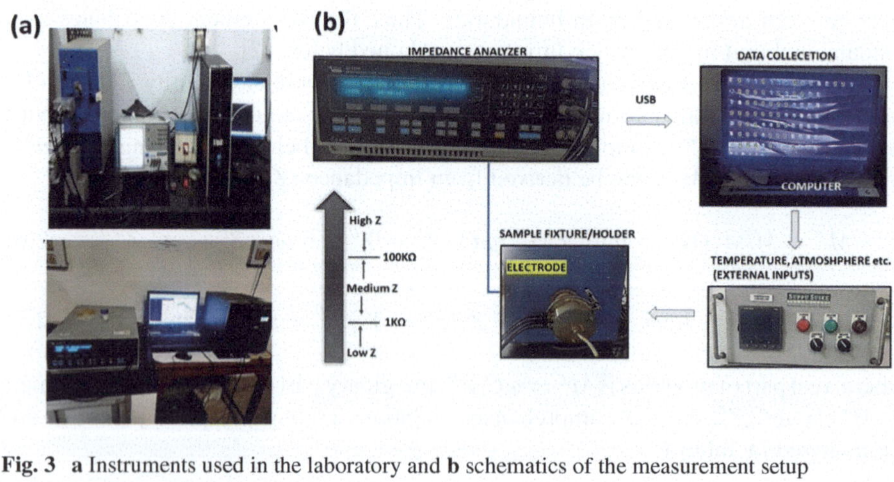

Fig. 3 **a** Instruments used in the laboratory and **b** schematics of the measurement setup

The frequency and time domains represent two important methods. Time domain technology gives the impedance values examines the step response within time parameters and translates it into impedance values, making it advantageous for acquiring ultralow-frequency responses. On the other hand, frequency domain technology captures the phase and amplitude of every frequency response using a sinusoidal signal, making it preferable for application in higher frequency ranges.

2.2 Impedance and Conductivity Types

Conduction is the primary mode of capacitive energy storage when charge carriers shift from short-range relaxation to long-range migration greater than 0.5 nm [2]. Electrons, holes, or charged ions contribute to both electronic and ionic conductivity. Utilizing temperature and frequency-dependent impedance for studying the conduction mechanism in energy-storage ceramics that considers both temperature and frequency variations in conductivity. Mathematically, the conductivity (σ^*) can be estimated from the impedance data as follows:

$$\sigma^*(\omega) = \frac{d}{[AZ^*(\omega)]} = \frac{Y^*(\omega)d}{A} \tag{2.8}$$

where A represents the electrode area and d is space between them.

The conduction mechanism can be understand with the help of both conduction and modulus spectra. The conductivity (σ) for polymers, glasses, nanocomposites, and solids can be described in terms of Jonscher's power law (JP law) which is also known as Almond-West equation [8]:

$$\sigma' = \sigma_{dc} + \sigma_{ac} = \sigma_{dc}[1 + (\frac{\nu}{\nu_h})^{\eta}] \tag{2.9}$$

where, σ' is the real part of conductivity, σ_{dc} is the frequency-independent conductivity, ν depicts the frequency, and ν_h is hopping frequency. Here, the power-law exponent is represented by η, and its value lies in the range of 0 to 1. Types of conductivity can be identified with the help of activation energy. The conductivity estimated through the impedance is combined form of ionic and electronic conductivities [9].

The samples' conduction activation energy can be determined by fitting the conductivity data using the Arrhenius equation:

$$\sigma_{dc} = \frac{\sigma_o}{T}\exp(-E_a/kT) \tag{2.10}$$

where σ_{dc} represents the frequency independent term, σ_o is the pre-exponent factor, and E_a is activation energy. This activation energy depicts the formation and migration of carriers [8]. This can be graphically obtained using above Eq. (2.10) refer to Fig. 4a and b. Here, Fig. 4a suggests that conductivity is frequency dependent

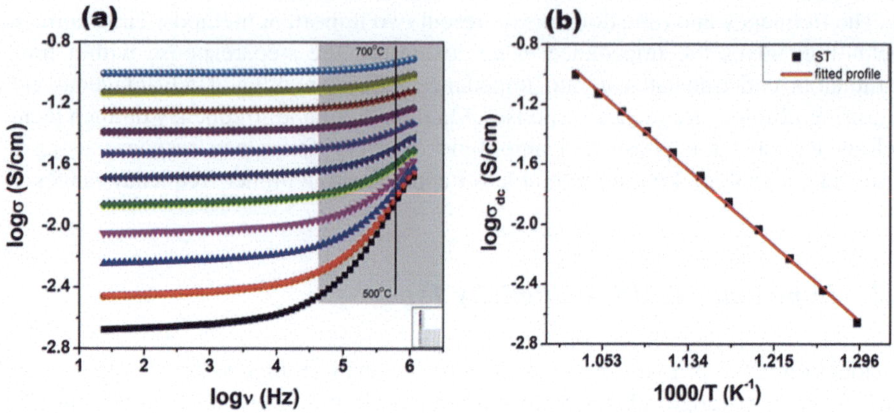

Fig. 4 **a** Log–log plot of conductivity vs frequency at distinct temperatures [10] **b** $log\sigma_{dc}$ versus 1000/T of $SrTiO_3$ [10]

and obeys the JP law as given in Eq. (2.9). From this, fitting parameters such as dc conductivity, hopping frequency, and the exponent can be obtained.

2.3 Impedance and Grain Boundary Relation

The ceramics used for the energy storage consists of a complex structure involving grains, grain boundaries, pores, and other elements. With the help of the equivalent circuit, any grain and grain boundary-electrode system can be explained, as depicted in Fig. 5a. The real and imaginary parts for the equivalent circuit can be estimated as:

$$Z' = \frac{R_b}{1 + (\omega R_b C_b)^2} + \frac{R_{gb}}{1 + \left(\omega R_{gb} C_{gb}\right)^2} + \frac{R_{el}}{1 + (\omega R_{el} C_{el})^2} \quad (2.11)$$

$$Z'' = \frac{\omega C_b R_b^2}{1 + (\omega R_b C_b)^2} + \frac{\omega C_{gb} R_{gb}^2}{1 + \left(\omega R_{gb} C_{gb}\right)^2} + \frac{\omega C_{el} R_{el}^2}{1 + (\omega R_{el} C_{el})^2} \quad (2.12)$$

Here, R_b, R_{gb}, R_{el} are resistance and C_b, C_{gb}, C_{el} are capacitance of grain, grain boundary, and the electrode, respectively [2]. But the main issue is that due to frequency dispersion, we cannot achieve a perfect fit of the impedance data. This has led to the introduction of impedance element called a constant phase element (CPE) which describes the fluctuation from the ideal impedance data. Mathematically, it can be expressed as—

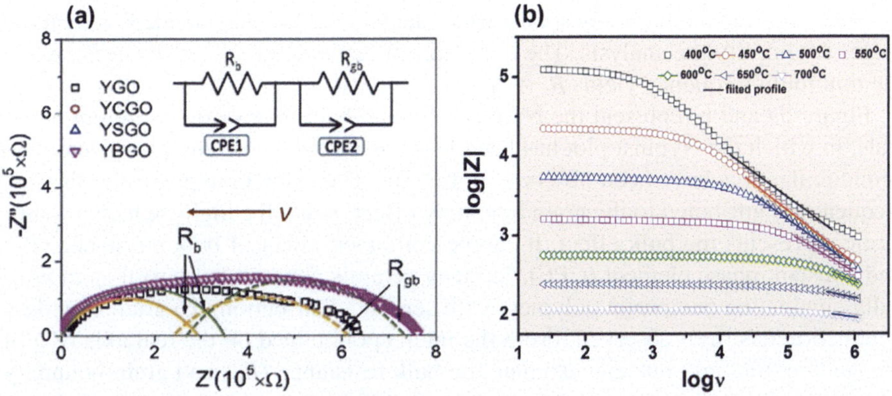

Fig. 5 **a** Nyquist plot and the equivalent circuit of YGO and doped YGO ceramic sample [13] **b** bode plot of Strontium titanate sample [10]

$$Q_{CPE} = Q_O(i\omega)^{1/n} \tag{2.13}$$

where n lies between 0 and 1 and Q_o and n are frequency-independent [11].

For $n = 0$, it shows pure resistive behavior, and for $n = 1$ shows pure capacitive nature.

Nyquist Plots—It is the most common way to represent impedance spectra, and to estimate the useful information from them. It is a complex-impedance plane representation, where frequency variation is studied with both the imaginary and real parts of impedance (refer to Fig. 5a). These types of plots are useful for identifying various characteristics features exhibited by a system like high and low real axis intercept with characteristic frequency of an arc, ω. The high-frequency intercept yields R_h, which are frequency independent contributions, called ohmic resistance. The low-frequency intercept results in $R_b + R_{gb}$, from which R_{gb} can be determined by subtracting R_b. The configuration of elements, like several semicircles or low-frequency tail at a 45°, gives possible indication about the governing kinetic or transport phenomena. These plots often contain several semicircles, where only a portion of a semicircle is seen. These semicircles are the characteristics of a single "time-constant" [12].

Bode Plot—It is another representation in which log frequency is plotted on the x-axis and modulus of impedance ($|Z|$) on the Y-axis (refer to Fig. 5b). Unlike the Nyquist plot, the bode plot does show frequency information.

The magnitude and phase angle are depicted as:

$$|Z| = \sqrt{Z'^2 + Z''^2} \tag{2.14}$$

$$\theta = tan^{-1}\left(\frac{Z''}{Z'}\right) \tag{2.15}$$

Since, $|Z|$ and ν commonly span a wide range of values, they are depicted on log scales to simplify the analysis. The maximum frequency for $|Z|$ results in R_b, while the minimum frequency yields $R_b + R_{gb}$.

Figure 5a and b represent the Nyquist plot and bode plot of two different materials, in which the Nyquist plot has been fitted using cole–cole equation. Here, two semicircular arcs have been observed in Fig. 5a. The semicircle present at the low frequency is attributed to the grain boundary effect, while the high-frequency semicircle represents the bulk effect. It can be combined effect of bulk resistance (R_b) and constant phase element (CPE). In many ceramic conductors as well as in ionically conducting composite polymers with ceramic distribution, the grain boundary conduction has been observed. From the intercept obtained on the real axis (Z') of the semicircular arc, one can estimate the bulk resistance (R_b) and grain boundary resistance (R_{gb}) values at various temperatures [13].

2.4 *Modulus Spectroscopy*

At the grain boundaries, the impedance modulus in the polycrystalline ceramics emphasizes the conductive process, while bulk effects in the frequency region dominate the electric modulus formalism. Valuable information can be derived with the help of modulus spectroscopy for distinguishing spectral elements of materials that share similar resistances but vary in capacitance. Another critical aspect of the electric modulus process is its ability to mitigate the influence of electrodes. Consequently, the complex electric modulus formalism has favored for these reasons. In the investigation of dielectric relaxation, studies were conducted using the complex modulus M^* formalism because it is an inverse form of the complex relative permittivity (ε^*). Figure 6 illustrates the variation in both the real (M') and imaginary (M'') part of modulus across distinct temperatures as a function of frequency [10].

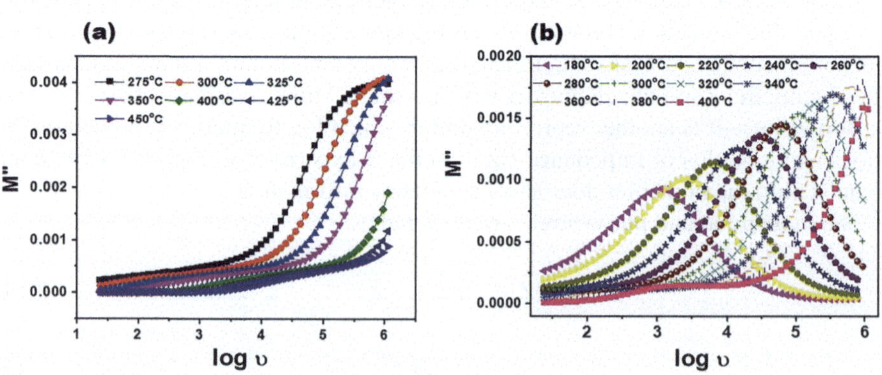

Fig. 6 **a** Real and **b** imaginary parts of the complex modulus (M' and M'') of $SrTiO_3$ on frequency [10]

Mathematically, M^* can be mathematically estimated from the ε^* or Z^*:

$$M^* = 1/\varepsilon^* = i\omega\varepsilon_0 Z^* A/d \tag{2.16}$$

Here, A is electrode area, d is sample thickness and ε^*,ε_0,Z^* represents the complex permittivity, permittivity in vacuum, and complex impedance, respectively. Here, Fig. 6a and b represent the change of frequency with the M' *(real part of modulus)* and M'' *(imaginary part of modulus)*, at various temperatures. At the lower frequencies the value of M' is nearly zero suggesting the effect of electrode polarization. Each peak corresponds to the relaxation with different time constants, and the irregular shape of peaks suggest the type of relaxation which is non-Debye type [10].

Comprehending both Debye-type and non-Debye type behavior is essential for analysing the dynamic properties of materials in polymer science, solid-state physics and materials engineering. Although the Debye type behavior refers to the response of material with a single relaxation time as predicted by Debye model [7, 14]—

$$\varepsilon(\omega) = \varepsilon_\infty + \frac{\varepsilon_s - \varepsilon_\infty}{1 + i\omega\tau} \tag{2.17}$$

here, ε_s is the static permittivity at lower frequencies ($\omega \to 0$) and ε_∞ is the permittivity at higher frequencies ($\omega \to \infty$). By selecting a range of relaxation times with appropriate strengths and frequencies, it becomes feasible to characterize the response of dielectric materials. Additionally, the range of relaxation times has been suggested as the source of the "constant phase elements" (CPEs) frequently observed in impedance studies of polymers, ceramic and solid state physics [6]. Information regarding modulus and dissipation factor can differentiate dielectric relaxation from long-range conductivity in frequency dependent plots [14].

By employing a combined study of impedance and modulus spectroscopy, it becomes possible to comprehend grain boundary and bulk phenomenon in polycrystalline ceramics. Here, semiconducting ceramics like $BaTiO_3$, which is electrically inhomogeneous material in which the grain-boundary resistance in many ways will dominate the impedance. Like in Fig. 7a there is no difference from ideal data at high frequencies [15]. For that case, Modulus spectroscopy or modulus formalism is an alternative for observing the response of grain boundary and bulk regions. It can be represented by equations discussed in previous sections.

In Fig. 7b, plots of imaginary part of M and Z with frequency, a single peak observed in Z'' in correspondence with the Fig. 7a but in the M'' two peaks identified. That one peak which is at the higher side of frequency depicts a bulk component of the sample. Thus, the data presented as Z'' and M' spectroscopic plots filtered the valuable and highlighted features of the sample. So, for the materials those are uniform and represented by more than one RC elements, M'' and Z'' spectroscopic plots shows a markable difference [15].

Complex modulus spectroscopy is used to understand the electrical properties of the material and studying other impacts present in the sample. Like Impedance spectroscopy, it is also an important and efficient technique to analyze and measure

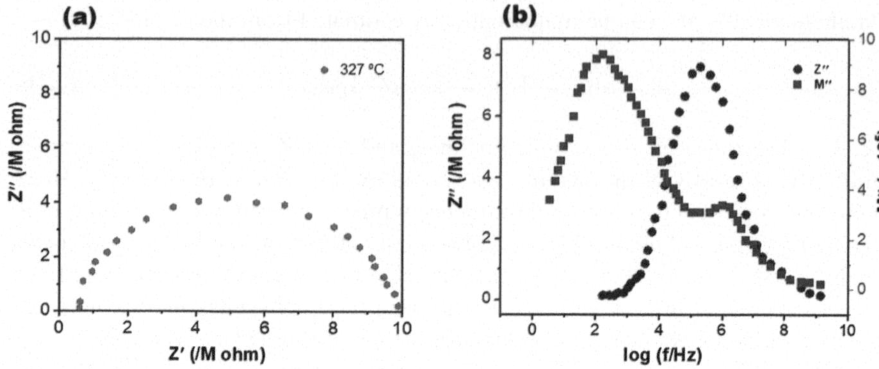

Fig. 7 **a** Nyquist plot, **b** M'' and Z'' plot of $BaTiO_3$ ceramics [15]

the various aspects of electrical transport phenomena such as conductivity relaxation time, ion hopping rate and many more [6].

3 Application

1. To identify whether the primary resistance in a material arises from bulk or grain boundary components.
2. To assess the quality and electrical uniformity of an electro-ceramic, often correlated with sintering/microstructure and its AC response.
3. To determine the individual resistances and capacitances of the component.

4 Conclusion

The fundamental principles, adaptability, and the practical applications of impedance and modulus spectroscopy in analysing electro-materials and solid-state devices have been examined. Within these materials and devices, impedance spectroscopy has provided a comprehensive electrical analysis of thin interfacial layers. These are present within the grain and grain boundaries and the layer in between the electrodes and electrolytes in solid state batteries. Generally, spectroscopic plots reveal the real or imaginary components of a complex formalism which is used within the complex plane plots. This overview of impedance and modulus spectroscopy illustrates the efficacy and versatility of the technique in characterizing an extensive range of materials and phenomena.

References

1. Macdonald, D.D.: Reflections on the history of electrochemical impedance spectroscopy **51**, 1376–1388 (2006)
2. Zhang, L., Pu, Y., Chen, M.: Complex impedance spectroscopy for capacitive energy-storage ceramics: a review and prospects **28** (2023)
3. Irvine, B.J.T.S., Sinclair, D.C., West, A.R.: Electroceramics : characterization by impedance spectroscopy **2**(3), 132–138 (1990)
4. Barsoukov, E., Macdonald, R., J.: Impedance spectroscopy theory, experiment, and applications, Second Edition. John Wiley & Sons (2005)
5. Pawar, V., Kumar, M., Dubey, P.K., Singh, M.K., Sinha, A.S.K., Singh, P.: Influence of synthesis route on structural, optical, and electrical properties of TiO2. Appl. Phys. A Mater. Sci. Process. **125**(9), 1–14 (2019)
6. Bredar, A.R.C., Chown, A.L., Burton, A.R., Farnum, B.H.: Electrochemical impedance spectroscopy of metal oxide electrodes for energy applications. ACS Appl. Energy Mater. **3**(1), 66–98 (2020)
7. Profile, S.E.E.: Impedance spectroscopy for emerging photovoltaics impedance spectroscopy for emerging photovoltaics, March 2019 (2019)
8. Singh, P., Raghvendra, Parkash, O., Kumar, D.: Scaling of low-temperature conductivity spectra of BaSn 1xNbx: Temperature and compositional-independent conductivity. Phys. Rev. B - Condens. Matter Mater. Phys. **84**(17), 1–6 (2011)
9. Zaidi, S.H., Jonscher, A.K.: Cite this article: Andrew K Jonscher. J. Phys. D Appl. Phys. To **32**, 57 (1999)
10. Sharma, U., Pawar, V., Singh, P.: Charge particle dynamics and electrochemical behaviour of SrTiO 3- δ as anode material for IT-SOFC applications. Int. J. Hydrog. Energy **52**, 1278–1289 (2024)
11. Morrison, F.D., Jung, D.J., Scott, J.F.: Constant-phase-element (CPE) modeling of ferroelectric random-access memory lead zirconate-titanate (PZT) capacitors. J. Appl. Phys. **101**(9) (2007)
12. Singh, P., Jha, P.K., Jha, P.A., Singh, P.: Science Direct Influence of sintering temperature on ion dynamics material for SOFC. Int. J. Hydrog. Energy **xxxx** (2019)
13. Singh, P., Pandey, R., Miruszewski, T., Dzierzgowski, K., Mielewczyk-Gryn, A., Singh, P.: Signature of oxide-ion conduction in alkaline-earth-metal-doped Y3GaO6. ACS Omega **5**(47), 30395–30404 (2020)
14. Gerhardt, R.: Impedance and dielectric revisited: distinguishing from long-range conductivity **55**(12), 1491–1506 (1995)
15. Sinclair, D.C.: Cerámica y Vidrio Characterization of electro-materials using ac impedance spectroscopy **65**, 55–66 (1995)

Defect Engineering for Tailoring Thermoelectric Properties of Electroceramics

Rajan Walia, Yogendra K. Gautam, Sagar Vikal, Ashwani Kumar, Paritosh Dubey, and Prashant Shahi

Abstract The field of thermoelectric materials has garnered significant attention due to its potential to revolutionize energy conversion and harvesting technologies. Delving into the fascinating realm of utilizing defect engineering strategies to enhance and tailor the thermoelectric properties of Electroceramic materials, this book explores the intricate relationship between crystal defects and thermoelectric performance. This chapter defines various thermometric, electroceramic, and 2D materials. It provides a comprehensive overview of defect types, their origins, and their influence on material behavior. With the aid of computational simulations and state-of-the-art experimental methods, defect engineering strategies are unveiled, enabling precise control over material properties. This chapter highlights the harnessing of defect engineering principles to unlock the full potential of electroceramics thermoelectric materials. The chapter provides a detailed study of the impact of defect Engineering on tailoring the thermometric properties of electroceramics, thermoelectric devices, and energy conversion technologies.

Keywords Thermoelectric materials · Electroceramics · Thermometric properties · Crystal defects · Defect engineering · Energy conversion

---

R. Walia
Department of Physics, University of Allahabad, Prayagraj 211002, Uttar Pradesh, India

Y. K. Gautam (✉) · S. Vikal
Smart Materials and Sensor Laboratory, Department of Physics, Ch. Charan Singh University, Meerut 250004, Uttar Pradesh, India
e-mail: ykg.iitr@gmail.com

A. Kumar (✉)
Departemnt of Physics, Regional Institute of Education (NCERT), Bhubaneswar 751022, Odisha, India
e-mail: ashraj@ncert.nic.in

P. Dubey
CSIR-National Metallurgical Laboratory, Jamshedpur 831007, India

P. Shahi
Department of Physics, DDU Gorakhpur University, Gorakhpur 273009, Uttar Pradesh, India

U. Kumar (ed.), *Defects Engineering in Electroceramics for Energy Applications*, Engineering Materials, https://doi.org/10.1007/978-981-97-9018-0_8

1 Introduction

1.1 Thermoelectric Materials

One of the most important concerns at the recent times is to reduce the production of electricity through fossil fuels and generate it through the green technology like thermoelectric materials, solar power plants, Hydro power plants etc. Thermoelectric technology is a promising solution in which waste heat energy can be converted into electricity in a fully environment-friendly manner. "Thermoelectric materials are a class of materials that can convert heat energy into electrical energy and vice versa through a phenomenon called the Seebeck effect." This means that when there is a temperature gradient across these materials, a voltage difference is generated, which can be used to produce electrical power or, conversely, applying an electrical current to these materials can create a temperature difference (Peltier effect).The performance of a thermoelectric material is decided by the Seebeck coefficient (S), electrical conductivity (σ), thermal conductivity (κ_t), and their variations with temperature (T) and is quantified by the figure of merit (ZT), which is given as [51]

$$ZT = \left(\frac{S^2\sigma}{\kappa_t}\right)T = \left(\frac{PF}{\kappa_t}\right)T$$

The value of this dimensionless figure of merit parameter determines the efficiency of a TEG. The signature of an efficient thermoelectric (TE) material is the high Seebeck coefficient, electrical conductivity, and low thermal conductivity. Nevertheless, controlling these three parameters independently is the most challenging part of designing an efficient TE material.

1.2 Electroceramics

Ceramic materials are a broad class of inorganic, non-metallic compounds that are known for their exceptional thermal, electrical, and mechanical properties. They are typically made from a combination of metal and non-metal elements and are processed at high temperatures to achieve desired shape and hardness [4]. Ceramics have a wide range of applications due to their unique characteristics. Ceramic materials are generally brittle, hard, insulating and corrosion resistant.

Electroceramics are a subcategory of advanced ceramics that are specifically engineered to exhibit unique electrical properties. They are designed to have controlled electrical behaviour, making them valuable for a variety of electrical and electronic applications. Some of the key electrical properties of electroceramics include ferroelectricity, piezoelectricity, dielectric properties, and ionic conductivity. Electroceramics are crucial in the development of modern electronics and electrical devices. Their unique electrical properties make them well-suited for a wide range of applications, from sensors and actuators to energy conversion and storage.

Ceramic and non-ceramic thermoelectric materials can be classified into several categories based on their chemical composition and properties. These classifications help researchers and engineers choose the most suitable materials for specific thermoelectric applications. Some common classes of thermoelectric materials include:

1.2.1 Inorganic Thermoelectric Materials

(a) **Bismuth Telluride (Bi_2Te_3):**

Bismuth telluride is one of the most widely used thermoelectric materials. It is a compound composed of bismuth (Bi) and tellurium (Te) elements, forming a layered crystal structure. The material's unique hexagonal lattice arrangement allows for efficient charge carrier transport, contributing to its outstanding thermoelectric performance. The layered structure is a crucial factor in promoting the material's electronic and thermal properties, making it an ideal candidate for thermoelectric applications. For Bi_2Te_3, the Seebeck coefficient is relatively high, especially in the temperature range around room temperature [8], making it suitable for practical thermoelectric applications. Bi_2Te_3 and its alloys with Sb_2Te_3 and Bi_2Se_3 etc. have the highest known efficiency [15] near room temperature which makes them one of the most important materials for thermoelectric applications.

The high ZT value of Bismuth Telluride arises from its unique electronic band structure and the ability to optimize charge carrier mobility while minimizing thermal conductivity. While Bismuth Telluride has shown immense promise, there are still challenges to address. One of the primary challenges is improving ZT values at higher temperatures. Researchers are actively exploring methods such as nanostructuring and alloying to enhance the material's performance and widen its applicability [41].

(b) **Lead Telluride (PbTe):**

Lead telluride is another efficient thermoelectric material, particularly in the mid-temperature range. It is commonly used for power generation in applications like radioisotope thermoelectric generators (RTGs). This compound belongs to the family of transition metal tellurides. Palladium telluride, represented by the chemical formula PdTe, is a binary compound composed of the transition metal palladium (Pd) and the chalcogen tellurium (Te). This compound exists in various phases, with the most studied being PdTe and Pd_3Te_4. The crystal structure of PdTe depends on the specific phase and conditions of synthesis [9]. In its most stable form, PdTe adopts a layered structure, similar to other transition metal tellurides. The layered structure of PdTe is crucial for its electroceramic applications, as it influences the material's electronic, thermal, and mechanical properties. This layered arrangement provides a foundation for controlling charge transport and electrochemical processes within the material [43].

PdTe demonstrates a high degree of ionic and electronic conductivity, a key attribute for electroceramic materials. The movement of ions and electrons within the crystal lattice allows for efficient electrochemical processes, making PdTe suitable for various applications. The electrochemical stability of PdTe is vital for its use in electronic devices and components. The material's ability to withstand repeated cycles of oxidation and reduction is crucial for longevity and reliability in electroceramic applications

(c) **Skutterudites**:

Skutterudite materials have a complex crystal structure and can exhibit high thermoelectric performance, especially at high temperatures. They are being explored for high-temperature thermoelectric applications. They typically belong to the general formula $A_xB_yC_z$ with A, B and C being metallic elements, often cobalt, iron or nickel and x, y, z representing the stoichiometry [34]. A defining feature of skutterudites is their crystal structure, characterized by a three-dimensional framework of corner-sharing polyhedra, forming cage-like structures that encapsulate guest atoms or molecules. This unique crystal structure is central to the thermoelectric properties of skutterudites, as it facilitates the manipulation of charge carriers and phonons, contributing to their efficiency in thermoelectric conversion. Skutterudites offer the advantage of a tunable band gap, allowing for optimization of thermoelectric performance over a range of temperatures [47].

Skutterudites play a pivotal role in the development of efficient thermoelectric generators. These generators convert waste heat from industrial processes or other sources into electrical power, showcasing the potential for skutterudites in enhancing energy efficiency. With the automotive industry seeking sustainable solutions, skutterudites are being explored for waste heat recovery in vehicles. Their ability to efficiently convert exhaust heat into electricity holds promise for improving fuel efficiency. While skutterudites exhibit outstanding thermoelectric properties, challenges such as cost, scalability, and the need for stable and environmentally friendly materials persist. Researchers are actively working to address these challenges through innovative synthesis methods, including nanostructuring and alloying, to enhance thermoelectric performance and reduce costs. The future prospects of skutterudites in the realm of electroceramics are promising. Ongoing research efforts aim to further optimize their properties, expand their applications, and contribute to the development of sustainable energy solutions [47].

(d) **Half-Heusler Compounds**:

Half-Heusler compounds, a fascinating class of intermetallic compounds, have garnered significant attention in recent years due to their exceptional thermoelectric properties and potential applications in energy conversion. As electroceramics, Half-Heusler compounds have emerged as key contenders in the quest for more efficient and sustainable energy technologies. Half-Heusler compounds belong to a family of intermetallic alloys with the chemical formula XYZ where X, Y and Z represent metallic elements. Common combinations include NiTiSn, TiCoSb, and

ZrNiSn [32]. The term "Half-Heusler" arises from the fact that the crystal structure of these compounds is related to the full Heusler structure but lacks one of the three interpenetrating fcc lattices. The crystal structure of Half-Heusler compounds enables high electrical conductivity and low thermal conductivity, resulting in an elevated thermoelectric efficiency [32]. Half-Heusler compounds are integral to the development of advanced thermoelectric generators. The band gap of Half-Heusler compounds can be tuned by adjusting the composition, allowing for optimization across different temperature ranges for enhanced thermoelectric performance. The robust mechanical stability of Half-Heusler compounds is advantageous for their application in practical devices, ensuring durability and longevity. The robust nature of Half-Heusler compounds makes them suitable for power generation in harsh environments, such as in space exploration missions or remote locations where traditional power sources may be impractical.

1.2.2 Organic Thermoelectric Materials

Organic thermoelectric materials encompass a diverse range of compounds, including polymers and small organic molecules. These materials exhibit unique properties that make them attractive for thermoelectric applications. Here are some examples of organic thermoelectric materials:

Poly (3,4-Ethylenedioxythiophene): Poly (Styrene Sulfonate) (PEDOT: PSS)

It is a conductive polymer blend commonly used in organic electronics. It has been explored for thermoelectric applications due to its high electrical conductivity and solution processability. PEDOT: PSS is a composite material composed of two main components: poly(3,4-ethylenedioxythiophene) (PEDOT) and poly (styrene sulfonate) (PSS). PEDOT, a conjugated polymer, provides electrical conductivity, while PSS serves as a stabilizing and dispersing agent. The combination of these components results in a transparent, conductive, and processable material. The polymer backbone of PEDOT: PSS is formed by the repeating units of ethylene dioxy thiophene, which contributes to its conductivity and electronic properties. The poly (styrene sulfonate) component surrounds and stabilizes the PEDOT chains, preventing them from agglomerating and improving the material's processability. PEDOT: PSS exhibits flexibility, which is particularly important for applications in flexible and stretchable electronics. Its ability to conform to various shapes and substrates makes it suitable for emerging technologies like wearable devices. Applications of PEDOT: PSS as an Electro ceramic include transparent conductive film in applications like touch screens, displays, and solar cells, as a hole transport layer in organic photovoltaic devices, as an electrode in organic light-emitting diodes (OLEDs) and other organic electronic devices and sensors [14].

Polyaniline (PANI)

PANI stands as a captivating example of an organic electroceramic, demonstrating a unique combination of electrical conductivity, processability, and versatility. This conductive polymer has garnered significant attention in the field of materials science and electronics due to its tunable properties and diverse applications. Polyaniline is a conducting polymer derived from the chemical polymerization of aniline monomers. Its structure is composed of alternating single and double-bonded nitrogen atoms, giving rise to a conjugated polymer backbone. The polymerization process results in various oxidation states of PANI, leading to different colors, including green, blue, and black, corresponding to the emeraldine, polaronic, and per nigraniline forms, respectively. The versatility of PANI arises from its ability to exist in different oxidation states, allowing for easy control of its electrical and optical properties through doping and de-doping processes.

Polyaniline exhibits electrical conductivity, a property that places it among the family of conductive polymers [14]. The degree of conductivity depends on the oxidation state, making it a versatile material for a range of electronic applications. PANI is soluble in various solvents, allowing for solution processing techniques such as spin-coating and inkjet printing. This processability facilitates the fabrication of thin films, coatings, and intricate structures on different substrates. The conjugated structure of PANI contributes to its unique electronic properties. This structure allows for the delocalization of π electrons along the polymer chain, contributing to its electrical conductivity and semiconducting behavior. PANI's properties can be tuned by controlling its oxidation state through chemical doping or electrochemical methods. This tunability is advantageous for tailoring the material's characteristics for specific applications [45].

Applications of Polyaniline as an Organic Electroceramic include sensors, energy storage devices, anticorrosive coatings, conducting inks and coatings.

Polythiophenes

Polythiophenes are a class of conjugated polymers derived from the polymerization of thiophene monomers. Thiophene is a heterocyclic compound containing a sulfur atom, and when polymerized, it forms a conjugated backbone with alternating single and double bonds. The resulting polythiophene structure allows for delocalization of π electrons, imparting unique electronic and optoelectronic properties to the material. Polythiophenes can be modified and functionalized, providing a versatile platform for tailoring their properties for specific applications. Polythiophenes are intrinsically conductive due to the delocalization of π electrons along the conjugated backbone. This conductivity can be enhanced through various methods, including doping or chemical modifications. One of the advantageous features of polythiophenes is their solution processability. They can be easily dissolved in common solvents, enabling the deposition of thin films and coatings using techniques like spin-coating and inkjet printing. The polymer backbone of polythiophenes provides flexibility, making them suitable for applications in flexible and stretchable electronics. This property is crucial for the development of wearable devices and flexible displays. Applications of Polythiophenes as Electroceramics include Organic Photovoltaics (OPVs), Light-Emitting Diodes (LEDs), and Field - Effect Transistors (FETs), Sensors etc.

Poly(3-hexylthiophene) (P3HT)

P3HT is a conjugated polymer derived from the polymerization of 3-hexylthiophene monomers. The polymer chain consists of repeating units of hexyl side chains and thiophene rings. The conjugated backbone, formed by the alternating single and double bonds in the thiophene rings, facilitates π-electron delocalization, imparting semiconducting properties to the polymer. The hexyl side chains contribute to the solubility and processability of P3HT, allowing for the formation of thin films and coatings through solution-based techniques. P3HT is a semiconducting polymer, making it suitable for use in various electronic devices, including organic field-effect transistors (OFETs) and solar cells. While not as highly conductive as metals, P3HT exhibits electrical conductivity within the semiconducting range. Its conductivity can be modified through doping or by blending with other materials [13]. P3HT demonstrates strong optical absorption in the visible range of the electromagnetic spectrum. This property is crucial for its applications in organic photovoltaics, where the polymer absorbs sunlight to generate electrical energy. The hexyl side chains in P3HT contribute to its solubility in common organic solvents, allowing for solution-based processing methods like spin-coating and inkjet printing. This processability is advantageous for the fabrication of thin films and electronic devices [12].

Applications of P3HT as an Electroceramic include Organic Photovoltaics (OPVs), Organic Field-Effect Transistors (OFETs), Thin-Film Transistors (TFTs) and Conductive Coatings etc.

Fullerene Derivatives

Fullerene derivatives are molecular compounds derived from fullerene structures, with C60 as the most well-known representative. These derivatives often involve modifications of the fullerene cage, such as adding functional groups, substituting atoms, or encapsulating other molecules within the cage. Common fullerene derivatives include C70, C84, and others with appended functional groups like -OH, -NH_2, or -COOH. The unique structure of fullerene derivatives, characterized by a hollow sphere of carbon atoms arranged in a hexagonal pattern, imparts exceptional electronic and optical properties to these molecules.

Fullerene derivatives possess a conjugated π-electron system, allowing for efficient electron delocalization. This property is crucial for their application in various electronic devices. Fullerene derivatives exhibit versatility in their chemical and physical properties [26], making them amenable to functionalization and modification. This versatility enables the tailoring of their properties for specific applications. Many fullerene derivatives are soluble in common organic solvents, facilitating solution processing methods in thin-film fabrication and other applications. Depending on the specific fullerene derivative and its structure, some compounds may exhibit electrical conductivity. This property is essential for their use in electronic and optoelectronic devices.

Applications of Fullerene Derivatives as Electroceramics include Organic Photovoltaics (OPVs), Organic Light-Emitting Diodes (OLEDs), Thin-Film Transistors (TFTs) and sensors etc. Researchers aim to enhance the electrical conductivity of fullerene derivatives by exploring novel synthesis methods, doping strategies, and structural modifications to improve their performance in electronic devices [22].

1.3 Electroceramics-Based Nanocomposites

Nanocomposite materials encompass a diverse range of compositions and applications in the field of electroceramics. Here are some examples of nanocomposite materials used in various electroceramic applications:

1.3.1 Carbon Nanotube (CNT) Reinforced Ceramics

Carbon Nanotube (CNT) reinforced ceramics represent a ground-breaking class of materials that seamlessly blend the exceptional properties of ceramics with the unique attributes of carbon nanotubes. This synergistic combination has led to the development of electroceramics with enhanced mechanical strength, thermal conductivity, and electrical properties. In this article, we will explore the composition, properties, fabrication methods, applications, and future prospects of CNT-reinforced ceramics, shedding light on their transformative role in the realm of advanced electronic materials. CNT-reinforced ceramics are composite materials composed of a ceramic matrix, typically oxides such as alumina or zirconia, intertwined with carbon nanotubes. Carbon nanotubes, allotropes of carbon with cylindrical nanostructures, serve as reinforcing elements within the ceramic matrix [21]. The unique one-dimensional structure of CNTs, combined with their exceptional mechanical and electrical properties, imparts remarkable enhancements to the overall performance of the composite material.

The incorporation of CNTs into the ceramic matrix significantly enhances the mechanical strength and toughness of the composite material. This is attributed to the reinforcing nature of CNTs, which act as structural elements, preventing crack propagation and improving fracture resistance. CNTs, known for their high thermal conductivity, contribute to improved heat dissipation within the ceramic matrix. This property is crucial for applications in electronic devices and components operating under high-temperature conditions. The addition of CNTs introduces electrical conductivity to the traditionally insulating ceramics. This property is advantageous for applications in electronic devices, sensors, and conductive coatings. The dielectric properties of CNT-reinforced ceramics can be tailored by adjusting the CNT concentration and orientation within the matrix [39]. This flexibility makes them suitable for applications in capacitors and electronic components.

CNT-reinforced ceramics find applications in the fabrication of structural components for electronic devices, where enhanced mechanical strength and thermal conductivity are crucial for reliability and performance. Due to their improved thermal conductivity, CNT-reinforced ceramics are employed in thermal management systems for electronic devices, ensuring efficient heat dissipation and preventing overheating. The electrical conductivity of CNT-reinforced ceramics makes them suitable for applications in sensors and actuators, where their piezoresistive properties can be exploited for sensing and actuation purposes. CNT-reinforced ceramics are used to develop conductive coatings for electromagnetic interference (EMI) shielding, antistatic applications, and protective coatings for electronic devices. Research is focused on developing advanced techniques for functionalizing CNTs to enhance their compatibility with ceramic matrices, improving dispersion and reinforcing effects.

1.3.2 Graphene Oxide (GO)-Based Ceramics

GO-based ceramics are composite materials formed by embedding graphene oxide sheets within a ceramic matrix, typically composed of oxides such as silicon carbide (SiC), alumina (Al_2O_3), or zirconia (ZrO_2). Graphene oxide, a derivative of graphene, is characterized by oxygen-containing functional groups on its surface, imparting hydrophilic properties and facilitating its dispersion within polar ceramic matrices. The unique two-dimensional structure of graphene oxide contributes to the exceptional properties of the resulting composite material. The incorporation of graphene oxide into ceramics significantly enhances their mechanical strength and fracture toughness. The two-dimensional structure of graphene oxide sheets acts as a reinforcing agent, preventing crack propagation and improving the overall structural integrity of the material. Graphene oxide, known for its high thermal conductivity, enhances the thermal properties of ceramics. This is crucial for applications in electronic devices and components where efficient heat dissipation is essential. While ceramics are traditionally insulators, the addition of graphene oxide imparts electrical conductivity to the composite. This property is advantageous for applications in sensors, conductive coatings, and electronic components. GO-based ceramics exhibit tailored dielectric properties, making them suitable for use in capacitors, sensors, and electronic components where specific electrical characteristics are required.

GO-based ceramics find applications in the fabrication of structural components for electronic devices. The enhanced mechanical strength and thermal conductivity make them valuable for ensuring reliability and performance. Due to their improved thermal conductivity, GO-based ceramics are employed in thermal management systems for electronic devices, preventing overheating and ensuring efficient heat dissipation [17]. The electrical conductivity of GO-based ceramics makes them suitable for applications in sensors and actuators. The piezoresistive properties of the material are exploited for sensing and actuation purposes. GO-based ceramics are used to develop conductive coatings for electromagnetic interference (EMI) shielding, antistatic applications, and protective coatings for electronic devices. Research is focused on developing advanced reduction techniques to achieve a well-controlled reduction of graphene oxide during the fabrication process, ensuring optimal electrical conductivity.

1.3.3 Metal Oxide Nanoparticle-Embedded Polymer Composites

Metal oxide nanoparticle-embedded polymer composites consist of a polymeric matrix, such as polyvinylidene fluoride (PVDF) or polyaniline, interspersed with metal oxide nanoparticles. The choice of metal oxide and polymer depends on the desired properties and applications of the composite. Common metal oxides include titanium dioxide (TiO_2), zinc oxide (ZnO), and iron oxide (Fe_2O_3). The resulting composite structure is characterized by a uniform dispersion of metal oxide nanoparticles within the polymer matrix. The addition of metal oxide nanoparticles imparts electrical conductivity to the traditionally insulating polymer matrix. This property is crucial for applications in flexible electronics, sensors, and conductive coatings. Metal oxide nanoparticles serve as reinforcing agents, enhancing the mechanical strength and toughness of the polymer composite. This is beneficial for structural applications and devices requiring mechanical robustness. Metal oxide nanoparticles contribute to tailored dielectric properties, making these composites suitable for applications in capacitors, sensors, and other electronic components. Certain metal oxides, such as ZnO, exhibit piezoelectric properties. When embedded in polymers, these composites can be utilized in sensors and actuators, converting mechanical strain into electrical signals. Metal oxide nanoparticles can improve the thermal stability of polymer composites, expanding their applicability in high-temperature environments and electronic devices [27].

Applications of Metal Oxide Nanoparticle-Embedded Polymer Composites as Electroceramics include Flexible Electronics, Sensors and Actuators, Energy Storage Devices and Conductive Coatings etc.

1.3.4 Barium Titanate Nanoparticle-Polymer Composites

Barium titanate nanoparticle-polymer composites are composed of a polymeric matrix, often polyvinylidene fluoride (PVDF) or other suitable polymers, incorporating barium titanate nanoparticles. Barium titanate ($BaTiO_3$) is a ferroelectric ceramic with remarkable dielectric properties. The incorporation of $BaTiO_3$ nanoparticles into the polymer matrix creates a composite material with enhanced dielectric constant, piezoelectricity, and tailored electrical conductivity. The composite structure is characterized by a uniform dispersion of barium titanate nanoparticles within the polymer matrix. Barium titanate nanoparticles contribute to a significant increase in the dielectric constant of the composite material. This property is crucial for applications in capacitors, sensors, and other electronic components.

The inherent piezoelectric properties of barium titanate are transferred to the composite, allowing it to convert mechanical stress into electrical signals. This feature is exploited in sensors, actuators, and energy harvesting devices. The addition of barium titanate nanoparticles imparts tailored electrical conductivity to the polymer matrix. While polymers are typically insulating, the composite material exhibits improved conductivity suitable for certain electronic applications. Barium titanate nanoparticles act as reinforcing agents, enhancing the mechanical strength and toughness of the polymer matrix [1]. This is beneficial for applications in structural components and devices requiring mechanical robustness.

Applications of Barium Titanate Nanoparticle-Polymer Composites as Electroceramics include Capacitors, Sensors and Actuators, Flexible Electronics and Energy Harvesting.

1.3.5 Zinc Oxide Nanowire-Embedded Ceramic Composites:

Zinc oxide nanowire-embedded ceramic composites are composed of a ceramic matrix, often alumina (Al_2O_3) or another suitable ceramic material, in which zinc oxide nanowires are uniformly dispersed. Zinc oxide (ZnO) nanowires, with their unique one-dimensional structure, act as reinforcing agents within the ceramic matrix [29]. This incorporation enhances the overall properties of the composite material, particularly in terms of electrical conductivity, mechanical strength, and piezoelectricity. The resulting composite structure exhibits a well-distributed network of zinc oxide nanowires within the ceramic matrix. The addition of zinc oxide nanowires imparts electrical conductivity to the traditionally insulating ceramic matrix. This property is pivotal for applications in electronic devices, sensors, and conductive coatings. Zinc oxide nanowires act as reinforcing agents, significantly enhancing the mechanical strength and toughness of the ceramic composite. This improved strength is crucial for structural applications and devices requiring resistance to mechanical stress. Zinc oxide is inherently piezoelectric, and the nanowires embedded in the ceramic matrix retain this property. The composite material can convert mechanical energy into electrical signals, making it suitable for sensors, actuators, and energy harvesting devices. The incorporation of zinc oxide nanowires can enhance the thermal stability of the ceramic composite, making it suitable for applications in high-temperature environments.

Applications of Zinc Oxide Nanowire-Embedded Ceramic Composites as Electroceramics include Sensors and Actuators, Energy Harvesting Devices, Flexible Electronics and Conductive Coatings etc.

1.3.6 Lead Zirconate Titanate (PZT) Nanoparticle-Polymer Composites

PZT nanoparticle-polymer composites consist of a polymeric matrix, often polyvinylidene fluoride (PVDF) or other suitable polymers, incorporating PZT nanoparticles. PZT, a ferroelectric ceramic, is composed of lead, zirconium, and titanium oxides. The inclusion of PZT nanoparticles within the polymer matrix enhances the overall properties of the composite material. The resulting structure showcases a uniform dispersion of PZT nanoparticles within the polymer matrix. The incorporation of PZT nanoparticles imparts piezoelectric properties to the composite material [11]. This enables the conversion of mechanical energy into electrical signals, making these composites highly suitable for sensors, actuators, and energy harvesting applications. PZT nanoparticles contribute to an increase in the dielectric constant of the composite material. This property is pivotal for applications in capacitors, sensors, and electronic components. While polymers are typically insulating, the addition of PZT nanoparticles enhances the electrical conductivity of the composite. This feature is advantageous for applications in electronic devices, conductive coatings, and flexible electronics. PZT nanoparticles act as reinforcing agents within the polymer matrix, enhancing the mechanical strength and toughness of the composite. This improved mechanical stability is crucial for structural components and devices subjected to mechanical stress. Applications of PZT Nanoparticle-Polymer Composites as Electroceramics include Sensors and Actuators, Energy Harvesting Devices, Flexible Electronics and Medical Imaging etc.

1.3.7 Ferroelectric Polymer Nanocomposites

Ferroelectric polymer nanocomposites are crafted from a polymer matrix, often centered around ferroelectric polymers like poly (vinylidene fluoride) (PVDF) or its copolymers. This matrix is interwoven with nanoscale ferroelectric additives, such as lead zirconate titanate (PZT), barium titanate ($BaTiO_3$), or other ferroelectric ceramics. The marriage of ferroelectric polymers with nanoparticles results in a composite material with synergistic properties, laying the foundation for versatile applications in electronic devices. The defining characteristic of these nanocomposites is their ferroelectric behavior. The incorporation of ferroelectric polymers and nanoparticles allows these materials to exhibit spontaneous electric polarization, making them suitable for non-volatile memory devices and ferroelectric capacitors. Ferroelectric polymer nanocomposites often display piezoelectric properties. This enables them to convert mechanical energy into electrical signals and vice versa, rendering them ideal for sensors, actuators, and energy harvesting applications. The addition of ferroelectric nanoparticles significantly enhances the dielectric constant of the composite material. This property is vital for applications in capacitors, sensors, and other electronic components [48]. Ferroelectric polymer nanocomposites can exhibit tailored electrical conductivity, depending on the choice of ferroelectric polymers and nanoparticles. This feature is advantageous for applications in flexible electronics, conductive coatings, and electromagnetic interference (EMI) shielding.

Applications of Ferroelectric Polymer Nanocomposites in Electroceramics include Non-volatile Memory Devices, Sensors and Actuators, Energy Harvesting Devices: Capacitors and Flexible Electronics etc.

1.3.8 Silver Nanoparticle-Embedded Ceramic Composites

Silver nanoparticle-embedded ceramic composites consist of a ceramic matrix, often alumina (Al_2O_3) or another suitable ceramic material, with dispersed silver nanoparticles. The integration of silver nanoparticles into the ceramic matrix is carefully engineered to achieve a uniform distribution. The resulting composite structure exhibits a well-balanced synergy between the insulating ceramic matrix and the conductive silver nanoparticles. The incorporation of silver nanoparticles imparts electrical conductivity to the traditionally insulating ceramic matrix. This property is pivotal for applications in electronic devices, conductive coatings, and other technologies requiring efficient charge transport. Silver is an excellent conductor of heat. The addition of silver nanoparticles enhances the thermal conductivity of the ceramic matrix, making the composite material suitable for applications in thermal management [19], heat sinks, and electronic packaging. Ceramic matrices are known for their high mechanical strength. The inclusion of silver nanoparticles reinforces the material, contributing to improved mechanical properties, such as toughness and fracture resistance. The electrical conductivity of silver nanoparticles enables the composite to effectively shield against electromagnetic interference. This property is valuable in electronic devices to prevent unwanted electromagnetic signals.

Applications of Silver Nanoparticle-Embedded Ceramic Composites in Electroceramics include Electronic Devices, Conductive Coatings, Thermal Management, Biomedical Devices and EMI shielding etc.

1.3.9 Perovskite Nanoparticle-Embedded Thin Films

Perovskite nanoparticle-embedded thin films are constructed from a thin film matrix, often composed of a semiconducting material or polymer, embedded with perovskite nanoparticles. Perovskite materials typically follow the ABX3 crystal structure, where A, B, and X represent cations, anions, and halide ions, respectively. In the context of thin films, the perovskite nanoparticles are dispersed or integrated within the thin film matrix to create a composite structure. The resulting material showcases a synergistic combination of the desirable properties of perovskite materials and the thin film's inherent advantages, such as flexibility and scalability. Perovskite materials are renowned for their excellent optoelectronic properties, including high light absorption coefficients and charge carrier mobility. The incorporation of perovskite nanoparticles into thin films enhances their ability to absorb light, making them suitable for applications in solar cells and photodetectors. The bandgap of perovskite materials can be easily tuned by altering the composition, allowing for the customization of the thin film's optical and electrical properties. This tunability is crucial for optimizing the performance of devices like solar cells. Perovskite materials exhibit high charge carrier mobility, facilitating efficient charge transport within the thin film. This property is essential for achieving high-performance electronic devices and sensors. Thin films, by nature, are flexible and conformable to various surfaces. The integration of perovskite nanoparticles further enhances the flexibility of the composite material [16], enabling its application in flexible electronics and wearable devices. Perovskite materials often exhibit semiconductor behavior, allowing for the creation of p-n junctions within the thin film. This property is advantageous for constructing electronic devices such as transistors and diodes.

Applications of Perovskite Nanoparticle-Embedded Thin Films in Electroceramics include Solar Cells, Photodetectors, Flexible Electronics, Sensors and Light-Emitting Diodes (LEDs) etc.

1.3.10 Polymer-Ceramic Nanocomposites for Flexible Electronics

Polymer-ceramic nanocomposites are composed of a polymer matrix, often based on flexible and stretchable polymers like polydimethylsiloxane (PDMS) or polyurethane, embedded with ceramic nanoparticles. The ceramic phase may include materials such as alumina (Al_2O_3), zirconia (ZrO_2), or other ceramic oxides. The integration of ceramics into the polymer matrix is engineered to achieve a uniform dispersion of nanoparticles, creating a composite material with enhanced mechanical strength and electrical properties. The polymer matrix imparts flexibility and stretchability to the nanocomposite, enabling its use in flexible electronic devices that can conform to irregular shapes and withstand deformation. Ceramic nanoparticles act as reinforcing agents within the polymer matrix, enhancing the overall mechanical strength, toughness, and durability of the composite material. The inclusion of ceramic nanoparticles contributes to improved thermal conductivity, making polymer-ceramic nanocomposites suitable for applications in thermal management and heat dissipation in electronic devices [31]. While polymers are typically insulating, the addition of ceramic nanoparticles enhances the electrical conductivity of the composite. This feature is crucial for the development of conductive paths in flexible electronic circuits. The dielectric properties of the nanocomposite are influenced by the ceramic phase, making it suitable for use in capacitors, sensors, and other electronic components.

Applications of Polymer-Ceramic Nanocomposites in electroceramics include Flexible Electronic Circuits, Wearable Devices, Flexible Displays, Stretchable Electronics and Energy Harvesting Devices etc.

1.4 2D Materials

Two-dimensional (2D) materials, such as graphene and transition metal dichalcogenides, are being studied for their thermoelectric properties. Their ultrathin nature can lead to unique electronic and thermal properties that are advantageous for thermoelectric applications.

1.4.1 Graphene

Graphene, a single layer of carbon atoms arranged in a hexagonal lattice, has emerged as a revolutionary material with extraordinary electronic, mechanical, and thermal properties. While traditionally not classified as an electroceramic, graphene's versatile characteristics and applications in electronics make it a noteworthy contender in the realm of electroceramics. Graphene is renowned for its unparalleled electrical conductivity (10^6S cm^{-1}) at room temperature due to its high electron mobility [5] surpassing traditional semiconductors and conductors. This property makes it a crucial component in electronic applications. Its maximum Seebeck coefficient value has been reported to be about 80 mV K^{-1}. The thermal conductivity of graphene is in the 4840–5300 W mK^{-1} range at room temperature [41]. Extensive theoretical and experimental endeavors have been undertaken to investigate the thermoelectric characteristics of graphene [6]. With exceptional mechanical strength and flexibility, graphene is one of the strongest materials known. It can withstand deformation without losing its structural integrity, making it ideal for flexible electronic devices. Despite its exceptional electrical conductivity, graphene remains transparent. This unique combination of conductivity and transparency is pivotal for applications in optoelectronics. Graphene exhibits outstanding thermal conductivity, allowing it to efficiently dissipate heat. This property is advantageous in applications where heat management is crucial.

While graphene is not a traditional electroceramic, its integration with ceramics has shown promising results in enhancing various properties:

1.4.2 Ceramic Composites

Graphene can be incorporated into ceramic matrices to form composites with improved mechanical strength, electrical conductivity, and thermal properties. These composites find applications in electronic packaging and structural materials.

1.4.3 Dielectric Applications

Functionalized graphene can be used as additives in dielectric materials to enhance their electrical properties. The resulting composites exhibit improved dielectric constants and reduced dielectric loss.

1.4.4 Thermal Management

Graphene-based ceramics are explored for thermal management applications in electronics. These ceramics with embedded graphene provide efficient heat dissipation, crucial for maintaining the stability and performance of electronic devices.

1.4.5 Sensors and Actuators

Graphene-ceramic composites are investigated for use in sensors and actuators due to their synergistic combination of electrical, mechanical, and thermal properties.

Transition Metal Dichalcogenides (TMDs)

Transition Metal Dichalcogenides (TMDs), a class of layered materials with a distinctive two-dimensional structure, have gained prominence in recent years due to their fascinating electronic, optical, and mechanical properties.

Molybdenum Disulfide (MoS_2)

It is a member of the transition metal dichalcogenide (TMD) family and has emerged as a promising material in the field of thermoelectricity. While traditionally not categorized as an electroceramic, MoS_2's unique electronic and thermal properties make it a captivating subject in the exploration of advanced thermoelectric materials. MoS_2 exhibits a layered structure, consisting of molybdenum atoms sandwiched between two layers of sulfur atoms. This arrangement facilitates the exfoliation of MoS_2 into thin, two-dimensional sheets. MoS_2 is a semiconductor with a tunable bandgap [42], allowing for precise control of its electrical conductivity by adjusting the number of layers or through doping. MoS_2 displays anisotropic thermal conductivity, meaning it conducts heat differently along different crystallographic directions. This property is advantageous for thermoelectric applications. MoS_2 exhibits a high thermoelectric power factor, a crucial parameter for efficient thermoelectric energy conversion. The Seebeck coefficient of MoS_2 is favorable for thermoelectric applications, signifying its ability to generate an electric voltage in response to a temperature gradient. MoS_2's anisotropic thermal conductivity enables efficient heat transport along specific directions, contributing to improved thermoelectric performance. The electrical conductivity of MoS_2 can be tailored through doping or layer manipulation, allowing for optimization of its thermoelectric properties. Ongoing research aims to optimize the thermoelectric performance of MoS_2 by enhancing its electrical conductivity and reducing thermal conductivity through strategies like nanostructuring. Developing scalable and cost-effective manufacturing methods for MoS_2-based thermoelectric devices is crucial for their practical implementation on a larger scale. Ensuring the stability and reliability of MoS_2-based thermoelectric devices over extended periods is essential for their real-world applications, especially in harsh environments.

Tungsten Diselenide (WSe_2)

WSe_2 possesses a layered atomic structure, consisting of tungsten atoms sandwiched between two layers of selenium atoms. This layered arrangement allows for the exfoliation of WSe_2 into thin, two-dimensional sheets. WSe_2 exhibits semiconducting properties with a tunable bandgap, enabling precise control over its electrical conductivity through adjustments in the number of layers or via doping. Similar to other TMDs, WSe_2 displays anisotropic thermal conductivity, allowing for efficient heat transport along specific crystallographic directions. WSe_2 demonstrates a high thermoelectric power factor, a critical parameter for effective thermoelectric energy conversion. WSe_2 exhibits a favorable Seebeck coefficient, indicating its ability to generate an electric voltage in response to a temperature gradient. The anisotropic nature of WSe_2's thermal conductivity contributes to enhanced thermoelectric performance, as it allows efficient control over heat transport. WSe_2's electrical conductivity can be tailored through doping or layer manipulation, providing flexibility in optimizing its thermoelectric properties [38].

It's important to note that the choice of thermoelectric material depends on the specific requirements of the application, such as the desired operating temperature range and efficiency. Researchers are continually working to discover and engineer new materials and improve existing ones to enhance the performance of thermoelectric devices for various practical uses.

2 Defects Engineering and Thermoelectric Materials

Defect engineering plays a crucial role in improving the thermoelectric performance of materials. Thermoelectric materials are designed to efficiently convert heat into electricity or generate a temperature difference when an electric current is applied. The efficiency of this energy conversion is determined by a property called the thermoelectric figure of merit (ZT), which is directly affected by various types of defects in the material's crystal structure. Here's how defects engineering is related to thermoelectric materials:

2.1 Point Defects

Introduction of point defects, such as vacancies, interstitials, or substitutional impurities, can affect the electrical and thermal conductivity of thermoelectric materials. Careful engineering of these defects can lead to improved thermoelectric performance. For example, introducing impurities can enhance the electrical conductivity while scattering phonons (vibrational lattice defects), leading to lower thermal conductivity [35].

2.2 Nano Structuring

Creating nanoscale defects and structures within thermoelectric materials, like nano inclusions or grain boundaries, can enhance phonon scattering, reducing thermal conductivity [18]. This results in better ZT values as it minimizes heat loss.

2.3 Charge Carriers

Defect engineering can be used to control the concentration and mobility of charge carriers (either electrons or holes) in the material [25]. This influences electrical conductivity and Seebeck coefficient, key factors in ZT optimization.

2.4 Doping

Controlled doping with certain elements can help optimize the electronic band structure, improve carrier concentration, and reduce thermal conductivity, all of which can enhance the thermoelectric properties [7].

2.5 Texture and Grain Boundaries

The manipulation of the material's texture and grain boundaries can also lead to improved thermoelectric properties. These features can enhance phonon scattering and control the mobility of charge carriers [37].

2.6 *Complex Structures*

Engineering complex or hierarchical structures can create multiple scattering centres for phonons, further reducing thermal conductivity while preserving electrical conductivity. Defect engineering is a critical aspect of developing high-performance thermoelectric materials. Researchers work to optimize materials by carefully introducing and controlling defects to enhance their thermoelectric properties, ultimately increasing the efficiency of converting waste heat into usable electrical power. The goal is to improve the thermoelectric figure of merit (ZT), making these materials more practical and efficient for various applications, including waste heat recovery, power generation, and solid-state cooling and heating systems. Further details of defect engineering in electroceramic materials in view of their thermoelectric applications will be discussed in subsequent sections [18].

3 Types of Defects

Ideal crystals represent the epitome of perfection, achieving a state of flawlessness through the precise repetition of a periodic unit cell. These crystalline structures, often referred to as perfect or ideal, serve as a theoretical benchmark for understanding the fundamental principles of crystallography. However, the reality of crystalline materials introduces a captivating complexity, as even the most meticulously arranged crystals deviate from perfection at a local level, giving rise to what we term as defects. These defects, classified into various dimensions, add a fascinating layer to the study of crystallography. Dimensionless point defects, for instance, represent minute imperfections in the lattice structure where an atom is missing or occupies an irregular position. One-dimensional defects, known as dislocations, are disruptions along a line in the crystal lattice. Moving up the scale, we encounter two-dimensional defects, such as grain boundaries, stacking faults, and twin boundaries, disrupting the regularity of crystal planes. Finally, three-dimensional defects manifest as precipitates or inclusions, adding a further layer of intricacy to the crystal structure. The influence of these defects on material properties is profound and diverse. Certain properties, like Young's modulus and the dilation coefficient, find their roots in the underlying crystal structure, making them relatively impervious to the presence of defects [35]. On the other hand, properties such as stress behaviour under external forces, the diffusion phenomenon during heat treatment, and the electric conductivity of materials are significantly influenced by the nature and concentration of defects within the crystal lattice [30].

3.1 Point Defects (Vacancies, Interstitials, Substitutions)

Dimensionless defects, signifying their lack of dimensionality, exist on a scale comparable to the interatomic distance within a crystal lattice [33]. This category encompasses three distinct types of point defects, each contributing to the nuanced intricacies of crystal imperfections. These defects-vacancies, interstitials, and substitutions-play a pivotal role in shaping the dynamic nature of crystalline structures and are elucidated in the accompanying diagram for clarity. Vacancies denote the absence of an atom at a lattice site, creating a void within the otherwise ordered structure. Interstitials involve atoms occupying positions between regular lattice sites, disrupting the seamless arrangement. Substitutions, on the other hand, entail the replacement of existing lattice atoms with foreign elements, introducing an additional layer of complexity to the crystalline matrix. The introduction of point defects induces local distortions within the crystalline lattice, setting off a cascade of effects. This perturbation generates elastic stress and strain fields in the volume of the crystal surrounding the defect. The intricate interplay of these stress and strain fields becomes a crucial determinant in understanding the complex behaviours and properties exhibited by crystalline materials, as they contribute to the localized environment that differs from the idealized, defect-free state.

3.2 Line and Planar Defects (Dislocations, Grain Boundaries)

When a crystal undergoes deformation beyond its elastic domain, distinct lines emerge on the surface, known as slip traces.

These traces correspond to small ledges formed by slip along crystalline planes, as illustrated below. They result from the movement of linear defects within the crystal, specifically dislocations, under the influence of mechanical load [28]. Two primary types of dislocations exist: edge dislocations and screw dislocations. In practical scenarios, dislocations often exhibit both edge and screw characteristics simultaneously, termed as mixed dislocations. These dislocations, forming curves or tortuous curves, can be decomposed into edge and screw components at any point along their line.

3.3 Edge Dislocations

Geometrically, an edge dislocation arises from introducing an atomic half-plane into a perfect crystal. The dislocation's location is defined as the limit of the additional half-plane within the otherwise pristine crystal. This creates a deformation similar to that induced by adding an extra atomic plane to the upper part of the crystal, leading to compression in the upper half-crystal and expansion in the lower half-crystal [36].

3.4 Screw Dislocations

A screw dislocation can be visualized as the result of making a notch in the crystal and sliding one edge of this notch away from the other by one interatomic distance. This motion transforms successive atomic planes into a helical surface [36].

3.5 Dislocation Movement

Plastic deformation arises from the propagation of dislocations. To conceptualize their movement, consider envisioning a heavy carpet that needs to be moved along the ground. This movement can occur by either pulling the carpet to make it slide or creating a wave motion at one edge and propagating it across the carpet. The former corresponds to slip along a compact plane, while the latter mirrors the propagation of dislocations in crystals.

3.6 Elastic Properties of Dislocations

In the elastic realm, dislocations contribute to stored energy in the crystal, and they are the focal points of specific stress fields. Elastic interactions between dislocations occur, influencing their behaviour and creating a complex network within the crystal lattice.

3.7 Stress Fields Associated with Dislocations

Stress fields adjacent to dislocations result in specific configurations, as illustrated for edge and screw dislocations. These fields are characterized by elastic strains and energy associated with the dislocation core.

3.8 Dislocation Energy

Creating a dislocation demands energy, encompassing both the elastic energy from stress and strain fields and the core energy. This energy input is crucial for understanding dislocation behaviour, and it plays a role in the deformation and mechanical properties of the crystal.

3.9 Force Acting on a Dislocation

External stress fields can induce dislocation movement within their slip system. The force acting on a dislocation, defined by the Peach and Koehler equation, depends on the applied stress field.

3.10 Elastic Interaction Between Dislocations

Every dislocation within a crystal exerts a force on other dislocations, leading to elastic interactions. Understanding these interactions is vital for predicting material behaviour and responses to external stresses.

3.11 Multiplication of Dislocations

Dislocations can multiply through mechanisms like the Frank-Read mechanism, where a dislocation segment forms a loop through iterative processes. The multiplication of dislocations contributes to material hardening and can lead to localized stresses, potentially resulting in micro-cracks and crystal fracture.

3.12 Dissociation of Dislocations

Perfect dislocations can undergo dissociation into partial dislocations, creating bidimensional defects or stacking faults. Differences in mechanical behaviour among materials can be attributed to the ease or difficulty of dislocation dissociation, with stacking fault energy being a crucial parameter.

3.13 Observation of Dislocations

Transmission Electron Microscopy (TEM) allows for the observation of dislocations in thin, electron-transparent specimens, revealing contrast due to the strain field's effect on electron diffraction.

Understanding the behaviour and characteristics of dislocations is fundamental in unravelling the mechanical properties and deformation mechanisms of crystalline materials.

3.14 Extended Defects (Stacking Faults, Twin Boundaries)

Extended defects in crystalline materials, such as stacking faults and twin boundaries, add a layer of complexity to the perfect lattice structure. These defects involve deviations from the regular arrangement of atoms in more than one dimension, introducing unique characteristics and influencing material properties [52].

3.15 Stacking Faults

A stacking fault occurs when there is a disruption in the regular stacking sequence of crystal planes. Instead of a continuous, ordered stacking of planes, there is a deviation, leading to a mismatch in the arrangement. Stacking faults can create regions with different stacking sequences within the crystal lattice. These defects play a crucial role in influencing material properties, including mechanical strength, electrical conductivity, and even magnetic behavior [52].

3.16 Twin Boundaries

Twin boundaries are planes within a crystal where the atomic arrangement mirrors that of another portion of the crystal. This mirror-image relationship results in twin domains, with each domain exhibiting a distinct crystal orientation. Twinning can occur during crystal growth or in response to external factors. Twin boundaries have significant implications for material behaviour, affecting mechanical properties, optical characteristics, and even influencing the response to external stresses.

Understanding extended defects like stacking faults and twin boundaries is essential in comprehending the diverse range of behaviours exhibited by crystalline materials. These defects not only contribute to the richness of crystallography but also play a pivotal role in tailoring material properties for specific applications.

4 Influence of Defects on Electrical Properties

The influence of defects on the electrical properties of materials is a multifaceted and pivotal aspect within the realm of materials science [10]. Defects, spanning from minute point defects to more extensive disruptions like dislocations and grain boundaries, wield a significant impact on how materials conduct, transport charges, and respond to electrical stimuli. At the foundational level, point defects such as vacancies and substitutions introduce localized energy levels within the crystal lattice, altering the conductivity and charge transport characteristics. As we delve into more complex defects like dislocations and grain boundaries, the story becomes even richer. Dislocations, for instance, can act as both charge carriers and trapping sites, influencing overall conductivity. Intentional introduction of impurities, known as doping, is a common strategy in semiconductor technology, where defects are manipulated to create charge carriers or modify the conductivity type. This interplay of defects is particularly critical in semiconductor devices, where the distribution and density of defects directly impact the performance of electronic components. In emerging materials and nanoscale structures, defect engineering becomes a deliberate strategy for tailoring electronic properties, offering both challenges and opportunities for innovation. Understanding the intricate relationship between defects and electrical properties is not only a scientific pursuit but a practical necessity for advancing electronic technologies and designing materials with tailored functionalities. This chapter explores the diverse manifestations of defects in the electrical landscape of materials, shedding light on their role in shaping the future of materials science and technology.

5 Relationship Between Carrier Concentration, Carrier Mobility, and Electrical Conductivity

The relationship between carrier concentration, carrier mobility, and electrical conductivity forms a fundamental framework for understanding the electrical behaviour of materials. This intricate interplay is pivotal in determining how efficiently a material can conduct electric current, making it a cornerstone concept in the realm of semiconductor physics and electronic device engineering.

5.1 Carrier Concentration

Carrier concentration refers to the number of charge carriers, either electrons or holes, within a given volume of a material. In semiconductors, the carrier concentration is a key parameter influenced by factors such as doping, temperature, and external stimuli. Doping introduces impurities to intentionally modify the carrier concentration, thereby altering the material's electrical properties.

5.2 Carrier Mobility

Carrier mobility is a measure of how easily charge carriers can move through a material in response to an applied electric field. It is influenced by factors such as crystal structure, temperature, and the presence of defects. High carrier mobility is desirable for efficient charge transport in electronic devices. Mobility is often expressed in units of cm^2/Vs and is a crucial parameter in semiconductor device design.

5.3 Electrical Conductivity

Electrical conductivity, represented by the symbol σ, quantifies a material's ability to conduct electric current. It is directly influenced by both carrier concentration and mobility. The relationship is expressed by the formula:

$$\sigma = \mathrm{q} \cdot \mathrm{n} \cdot \mu$$

where: σ is the electrical conductivity, q is the charge of the carrier (elementary charge), n is the carrier concentration, μ is the carrier mobility.

This formula underscores the importance of both carrier concentration and mobility in determining the overall electrical conductivity of a material. A higher carrier concentration increases the number of charge carriers available for conduction, while higher carrier mobility enhances their ability to move through the material.

5.4 Temperature Dependence

The relationship between carrier concentration, mobility, and conductivity is temperature-dependent. Elevated temperatures can influence the concentration of thermally generated carriers and impact the mobility of charge carriers through lattice scattering mechanisms. Understanding these temperature dependencies is crucial for predicting and optimizing the electrical performance of materials over a range of operating conditions.

In essence, the intricate relationship between carrier concentration, mobility, and electrical conductivity provides a foundational understanding of how materials conduct electricity. This understanding is not only essential for designing and optimizing semiconductor devices but also plays a key role in advancing technologies ranging from integrated circuits to solar cells and beyond.

5.5 Impact of Point Defects on Charge Carrier Scattering and Mobility

The influence of point defects on the scattering of charge carriers and their mobility is a critical aspect in the study of semiconductor materials. Understanding how point defects, such as vacancies, interstitial, and substitutions, affect the movement of charge carriers is essential for optimizing the performance of electronic devices. These defects act as scattering centers, disrupting the otherwise smooth trajectory of charge carriers and influencing their overall mobility within the material. Investigating the impact of point defects on charge carrier scattering provides valuable insights into the material's electronic behaviour, offering a foundation for defect engineering strategies aimed at enhancing carrier mobility and, consequently, the efficiency of semiconductor devices [23].

5.6 The Role of Grain Boundaries in Affecting Carrier Transport

The section delves into the intricate role of grain boundaries in shaping carrier transport within semiconductor materials. Grain boundaries, interfaces between crystalline grains, present both challenges and opportunities in the realm of charge carrier movement. These boundaries disrupt the crystal lattice, introducing scattering mechanisms that hinder the smooth flow of charge carriers and consequently reduce their mobility [37]. The discussion encompasses the impact of grain boundaries on carrier scattering, emphasizing the need to comprehend their nature and effects on material conductivity. Despite these challenges, the chapter explores the potential for engineering carrier transport by strategically modifying grain boundary properties. Techniques such as grain boundary engineering and controlled material processing conditions are scrutinized for their efficacy in optimizing carrier mobility. Practical implications for semiconductor devices are underscored, emphasizing the importance of minimizing the adverse effects of grain boundaries for enhanced device performance. Furthermore, the chapter addresses recent advancements in material characterization techniques, providing a detailed exploration of grain boundaries and their intricate interplay with carrier transport. This comprehensive examination contributes valuable insights to the broader understanding of semiconductor materials, laying the foundation for informed strategies in electronic technology applications.

6 Engineering Defects to Enhance the Electrical Performance

The process of deliberately engineering defects to enhance electrical performance stands as a transformative concept within materials science and engineering. This approach involves the strategic introduction and manipulation of defects in a material to achieve specific improvements in its electrical properties. Defects, such as vacancies, interstitials, or dopants, play a crucial role in influencing charge transport and conductivity within semiconductors.

In this paradigm, the chapter explores how controlled defect engineering can be employed to optimize electrical performance. By tailoring the type, concentration, and distribution of defects, researchers and engineers aim to fine-tune the material's electronic characteristics. This includes modifying carrier concentration, enhancing carrier mobility, and influencing other key parameters that directly impact electrical conductivity.

The discussion encompasses various strategies for defect engineering, including precision doping, lattice modifications, and the creation of defect-rich structures. Each strategy is dissected for its efficacy in achieving specific electrical enhancements. The chapter also delves into the intricate interplay between defects and their influence on band structure, energy levels, and carrier dynamics, offering a comprehensive understanding of the underlying mechanisms.

Practical applications of defect engineering in semiconductor devices, sensors, and electronic components are highlighted, showcasing how this approach translates into tangible improvements in device performance. Moreover, the chapter explores the evolving landscape of emerging materials, such as two-dimensional structures and nanomaterials, where defect engineering plays a pivotal role in tailoring electrical properties.

The synthesis of theoretical frameworks and experimental methodologies for defect engineering is addressed, providing a holistic view of the subject. Advanced characterization techniques, computational simulations, and innovative fabrication processes are discussed as integral components of a successful defect engineering strategy.

Ultimately, the chapter underscores the transformative potential of defect engineering in advancing materials for electronic and optoelectronic applications. By harnessing the controllable introduction of defects, researchers can unlock new frontiers in materials design, pushing the boundaries of electrical performance and contributing to the evolution of cutting-edge technologies [7].

7 Influence of Defects on Electrical Properties of Electroceramics

7.1 Strategies to Improve the Electrical Conductivity

Electrical conductivity refers to a material's ability to conduct electric current. It is directly related to both carrier mobility (μ) and carrier concentration (n), which can be expressed through the formula: $\sigma = ne\mu$. A high-quality thermoelectric (TE) material typically possesses an electrical conductivity of approximately 10^4 S/m. There are several factors that can contribute to the improvement of electrical conductivity in a material:

7.2 Carrier Concentration

In order to create a good TE (thermoelectric) material, it is crucial that the carrier concentration falls within a specific range, which is typically between 10^{19} and 10^{21} cm^{-1}. There are two methods to optimize carrier concentration: tuning intrinsic defects and extrinsic doping.

Extrinsic doping is the process of adding suitable elements from the periodic table to the material to adjust its carrier concentration. It is considered to be one of the main methods to achieve this, as evidenced by the data shown in Fig. 1a–c. However, determining the appropriate dopant can be difficult due to several factors, such as the solubility limit and efficiency of the dopant. This can result in lower TE performance than what is theoretically predicted for certain materials. Therefore, it is important to carefully choose the dopant to optimize the TE performance of the material.

Intrinsic defects such as antisite, vacancies, and interstitials can also be used to optimize carrier concentration, as shown in Fig. 1d–f. Compared to extrinsic doping, intrinsic defects are more difficult to control and optimize. However, they are still considered to be a viable method for fine-tuning the TE performance of a material.

In addition to extrinsic doping and intrinsic defects, carrier concentration can also be manipulated by magnetic transitions and temperature-dependent solubility, as demonstrated in Fig. 1g–h. By taking advantage of these methods, researchers can create TE materials with improved performance, which can be useful in a variety of applications.

7.3 Carrier Mobility

In the realm of electrical conductivity, both carrier mobility and carrier concentration play equally important roles. The carrier mobility is determined by a number of

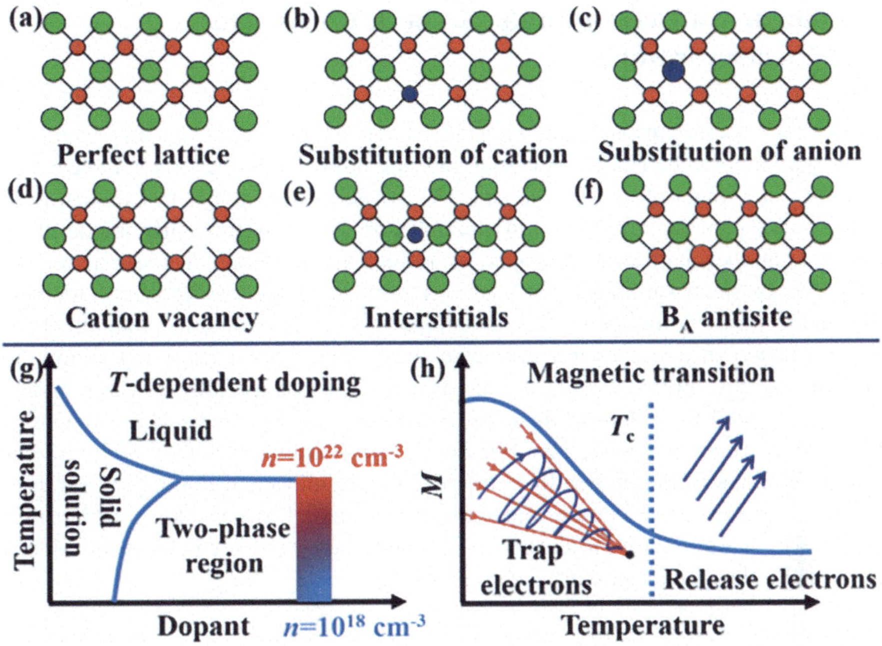

Fig. 1 **(a)** perfect lattice **(b)** substitution of cation **(c)** substitution of anion **(d)** cation vacancy **(e)** interstitials **(f)** antistites **(g)** temperature dependent doping **(h)** magnetic transition [49]

factors, including the material's electronic structure and various scattering mechanisms such as alloying scattering, acoustic phonon scattering, polar optical phonon scattering, and ionized impurity scattering. Additionally, elastic scattering carrier mobility can be expressed through a simple parabolic band assumption. Understanding these factors is crucial in optimizing the performance of thermoelectric for various technological applications. The carrier mobility is explained using the following expression:

$$\mu = \frac{e\tau}{m_I^*}$$

where m_I^* and τ represent charge carriers' effective mass and scattering time. The scattering time is related to the effective mass of density-of-states (DOS) (m_d^*), carrier energy (E), and temperature (T), which is related to the scattering mechanisms, as shown in the equation below

$$\tau \propto E^r T^S m_d^{*t}$$

Specifically, t = 1/ 2, r = 3/2, and s = 0 for ionized impurity scattering, while for acoustic phonon scattering, s = -1, r = -1/2, and t = − 3/2, respectively [20].

8 Influence of Defects on the Seebeck Coefficient of Electroceramics

The Seebeck coefficient, also known as thermoelectric power or thermopower, is a crucial parameter in thermoelectric materials, including electroceramics. It measures the ability of a material to generate an electric voltage in response to a temperature gradient. The presence of defects in electroceramics can significantly impact their electrical and thermal transport properties, influencing the Seebeck coefficient [34].

Defects in electroceramics can take various forms, such as vacancies, interstitials, dislocations, grain boundaries, and dopants. Each type of defect can affect the electronic band structure, charge carrier mobility, and phonon scattering within the material, thereby influencing its thermoelectric performance [50]. Several studies have investigated the relationship between defects and the Seebeck coefficient in different electroceramic materials. For example, research may focus on the role of grain boundaries in polycrystalline ceramics or the impact of doping on the thermoelectric properties of certain materials. It is essential to consider both intrinsic defects, which are inherent to the crystal structure, and extrinsic defects introduced intentionally through processing techniques [46].

9 Role of Point Defects in Seebeck Coefficient Modification

The role of point defects in modifying the Seebeck coefficient represents a critical and fascinating aspect in the realm of electroceramics for thermoelectric applications. Point defects, which include vacancies, interstitials, and dopants, play a pivotal role in shaping the electronic structure of materials, ultimately influencing their thermoelectric properties. These defects act as active participants in the charge transport mechanisms, impacting the movement of charge carriers and, consequently, the Seebeck coefficient.

In the intricate interplay between charge carriers and point defects, the Seebeck coefficient emerges as a key parameter that is particularly sensitive to the presence and type of defects. The introduction of point defects into the crystal lattice can lead to adjustments in the band structure, affecting the energy landscape through which charge carriers navigate. This, in turn, influences the mobility of charge carriers and modulates their scattering behaviour, thereby altering the material's thermoelectric response.

Understanding the mechanisms through which point defects impact the Seebeck coefficient involves delving into intricate concepts [47]. Defect-induced energy levels, for example, serve as stepping stones for carriers within the material, influencing their movement and contributing to changes in the Seebeck coefficient. Moreover, the interactions between charge carriers and defects can lead to complex phenomena such as energy filtering and selective scattering, further adding layers of intricacy to the modification process.

Noteworthy studies in this field explore the diverse avenues through which point defect engineering can be harnessed to enhance the thermoelectric performance of electroceramics. These studies often leverage both experimental and computational approaches to tailor specific point defects for optimal Seebeck coefficient outcomes. By elucidating the underlying mechanisms and providing insightful case studies, researchers aim to contribute to the broader understanding of how point defects can be strategically employed to tailor the Seebeck coefficient in electroceramics [24]. These endeavours not only deepen our theoretical understanding but also open up practical avenues for advancements in thermoelectric technology, with the potential for more efficient energy conversion and utilization.

10 Impact of Microstructural Defects on Seebeck Coefficient

The impact of microstructural defects on the Seebeck coefficient is a crucial consideration in the study and engineering of electroceramics for thermoelectric applications. Microstructural defects, encompassing grain boundaries, dislocations, and other structural irregularities, significantly influence the transport of charge carriers and, consequently, the thermoelectric performance of materials. Grain boundaries, acting as barriers for charge carriers, can lead to increased scattering events, affecting their mobility and contributing to changes in the Seebeck coefficient. Dislocations, on the other hand, introduce strain fields that can alter the electronic structure of the material, further influencing the thermoelectric properties. Understanding the intricate relationship between microstructural defects and the Seebeck coefficient requires a comprehensive exploration of how these defects modify the energy landscape within the material and interact with charge carriers [24].

Researchers delve into experimental investigations and computational simulations to unravel the complex interplay between microstructural defects and the thermoelectric response. Strategies for tailoring microstructural defects emerge as essential focal points, aiming to optimize the Seebeck coefficient for enhanced thermoelectric efficiency. As we navigate the complexities of microstructural engineering, a deeper understanding of how these defects impact the Seebeck coefficient not only advances fundamental knowledge but also guides the development of materials with tailored microstructures for improved thermoelectric performance. The pursuit of optimized microstructural design holds promising implications for the broader field of energy conversion, offering pathways for the development of more efficient and sustainable thermoelectric devices.

11 Defect-Induced Phonon Scattering and Seebeck Coefficient Enhancement

The intricate interplay between defect-induced phonon scattering and the enhancement of the Seebeck coefficient stands as a pivotal aspect in the optimization of thermoelectric materials for electroceramic applications. Phonons, as carriers of heat, contribute significantly to the thermoelectric performance of materials. The deliberate introduction of defects, such as point defects, dislocations, or grain boundaries, acts as a strategic approach to manipulate phonon transport. In this context, defect-induced phonon scattering becomes a crucial mechanism for modifying the thermal conductivity of the material. By impeding the smooth propagation of phonons, defects effectively reduce the material's thermal conductivity, enhancing its thermoelectric efficiency. Simultaneously, the impact of defect-induced phonon scattering on the Seebeck coefficient is substantial. The modification of phonon behavior influences the interaction between charge carriers and lattice vibrations, leading to alterations in the electronic transport properties. Researchers investigate various theoretical models to understand the nuanced effects of defect-induced phonon scattering on the Seebeck coefficient. The exploration often involves computational simulations and experimental studies to quantify the influence of different defects on the thermoelectric response. Strategies focusing on defect-induced phonon scattering aim to strike a delicate balance, as excessive scattering may adversely affect electrical conductivity. Hence, researchers seek optimal defect configurations and densities to achieve a harmonious enhancement of the Seebeck coefficient without compromising electrical transport. This multifaceted approach to defect engineering underscores the importance of considering both electronic and thermal aspects in the pursuit of highly efficient thermoelectric materials. The ongoing research in defect-induced phonon scattering not only contributes to the fundamental understanding of thermoelectric phenomena but also holds promise for the development of materials with tailored thermal and electrical properties, paving the way for more effective energy conversion technologies.

12 Defect Density and Distribution: Optimizing Seebeck Coefficient Performance

The optimization of the Seebeck coefficient performance in electroceramics critically hinges on the strategic control of defect density and distribution within the material. Defects, including vacancies, interstitials, and dopants, play a pivotal role in influencing the electronic structure and transport properties of materials. Understanding how defect density and distribution impact the Seebeck coefficient involves a delicate balance, as these factors affect charge carrier mobility, scattering mechanisms, and the overall thermoelectric response. Researchers are engaged in comprehensive studies to explore the nuanced effects of varying defect densities on the Seebeck coefficient. Higher defect densities can lead to increased scattering of charge carriers, thereby modifying the electronic transport properties and influencing the thermoelectric performance. However, finding the optimal defect density is crucial, as excessively high densities may introduce detrimental effects, such as reduced carrier mobility or increased phonon scattering. Equally important is the consideration of defect distribution throughout the material. Uniform distribution versus clustering of defects can have distinct consequences for the Seebeck coefficient. Researchers investigate experimental and computational approaches to tailor defect distribution, seeking to optimize charge carrier pathways and minimize the impact of localized charge carrier scattering. Strategies for optimizing defect density and distribution involve a multidisciplinary approach, combining theoretical models, advanced characterization techniques, and materials synthesis methods. By gaining a deeper understanding of the interplay between defects and the Seebeck coefficient, researchers aim to design materials with precisely engineered defect landscapes for enhanced thermoelectric performance. This avenue of research not only contributes to the fundamental understanding of defect engineering but also holds promise for the development of advanced materials with improved thermoelectric properties, thereby advancing the field of energy conversion and utilization [24, 47].

13 Defects and Thermal Conductivity

13.1 Strategies to Reduce the Thermal Conductivity

The cumulative thermal conductivity comprises both lattice thermal conductivity and electronic thermal conductivity. Electronic thermal conductivity hinges on electrical conductivity and can be evaluated using the Wiedemann–Franz law, denoted as $k_{ele} = L\sigma T$, where L represents the Lorentz number, σ electrical conductivity, and T temperature. The value of L lies in the range of (1.6–2.5) × 10^{-8} $W\Omega K^{-2}$. So, for a good thermoelectric material, optimum carrier concentration is required. Furthermore, the lattice thermal conductivity is established upon classical kinetic theory and is outlined by the subsequent equation

$$k_{lat} = \frac{1}{3} C_v l v_s$$

Here l represents mean free path of phonons, C_v stands for specific heat at constant volume, and v_s signifies the average velocity of sound. At temperatures below 40 K, the number of excited phonons is minimal and their wavelengths are extended, resulting in the prevalence of the Debye T^3 law for C_v in lieu of phonon scattering as the dominant contributor to lattice thermal conductivity. As the temperature surpasses the Debye temperature, C_V approaches the classical value of 3R, influenced by the phonon–phonon scattering-induced mean free path of the phonons.

Phonons are basically the vibrational waves having various wavelengths that arises due to the displacement of atoms from their equilibrium position. Lattice thermal conductivity hinges on these phonons, as they bear the primary heat-carrying role and are connected to a range of wavelengths and mean free paths. Diverse crystal defects, including nanoprecipitates, point defects, dislocations, and grain boundaries, can effectively impede the propagation of phonons, as depicted in Fig. 2.

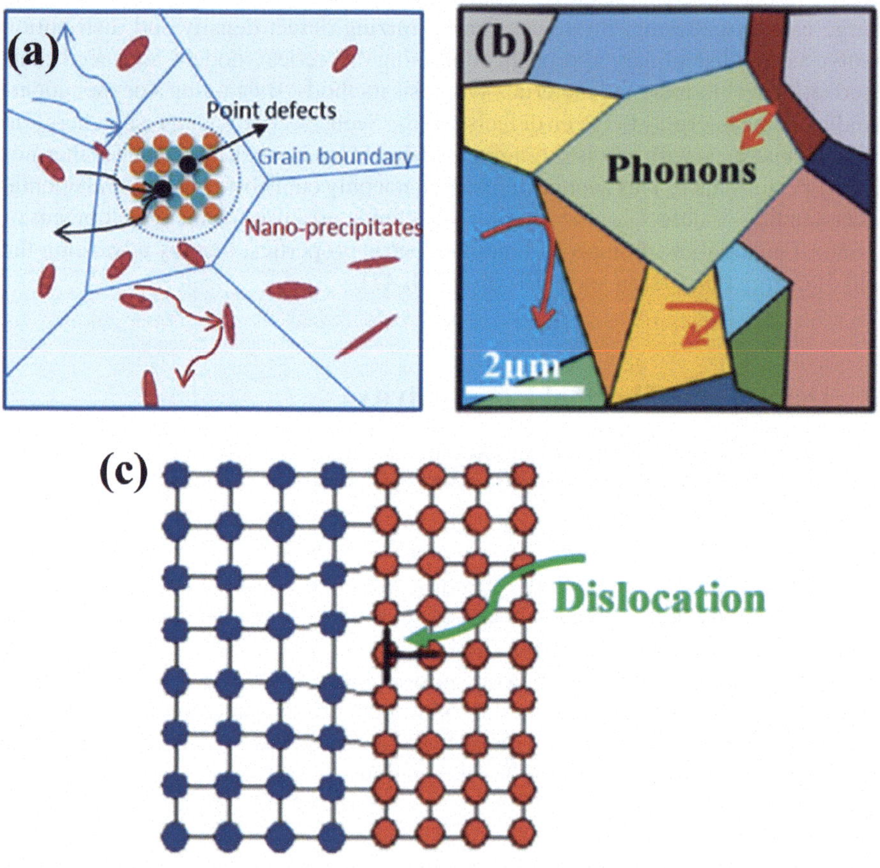

Fig. 2 Phonon scattering by **a** point defects, nano-precipitates and grain boundaries, **b** mesoscale structuring, **c** dislocations [2, 3, 34, 40, 44]

13.2 Point Defects

Point defects are intentionally introduced to induce imperfections within the lattice, typically occurring over a length scale of one interatomic distance. This is achieved by incorporating vacancies, interstitials, and substitutions, consequently leading to a reduction in lattice thermal conductivity.

13.3 Nano Structuring

It is a method to form single phase nanoparticles or a solid solution having nanoscale precipitate embedded in the matrix. These nanoscale precipitate and the grain boundaries can act as a barrier to scatter the phonons thereby reducing the lattice thermal conductivity.

13.4 Dislocations

Dislocations which can be introduced by plastic deformation during the process of liquid phase compaction are utilized to diminish lattice thermal conductivity by causing scattering of phonons with medium frequencies.

13.5 Mesoscale Structuring

Mesoscale structuring is used to scatter long wavelength phonons. Mesoscale structuring is a method of creating mesoparticles ranging from 100 nm to 5 μm. These fine-grained powders are then sintered which enables boundary scattering within the intergrain region. The long mean free path phonons are restricted to propagate due to finite size of the grains. Hence, the long wavelength phonons are scattered by Mesoscale grain boundaries.

14 Defect Chemistry and Thermoelectric Efficiency

Defect chemistry plays a pivotal role in shaping the thermoelectric efficiency of materials, offering a promising avenue for enhancing energy conversion technologies. In the realm of thermoelectric materials, where the conversion of waste heat into electricity is paramount, defects introduce opportunities for tuning electronic and thermal transport properties. The deliberate introduction of point defects, such as vacancies or dopants, can significantly influence the carrier concentration and mobility, impacting the Seebeck coefficient and electrical conductivity. Simultaneously, defects act as scattering centres for phonons, reducing thermal conductivity and thereby enhancing the thermoelectric figure of merit (ZT) [2]. Understanding the intricate interplay between defect chemistry and thermoelectric properties enables researchers to tailor materials for optimal performance. Advances in computational methods, such as ab initio calculations and modeling techniques, facilitate the prediction and design of materials with specific defect configurations to achieve superior thermoelectric efficiency. Harnessing the potential of defect engineering opens new vistas for the development of sustainable and efficient energy conversion devices, contributing to the ongoing quest for cleaner and more effective power generation technologies.

14.1 Theoretical Models for Studying the Effects of Defects

Theoretical models elucidating the influence of defect chemistry on thermoelectric properties are essential tools in the quest to optimize materials for efficient energy conversion. One such model is the point defect scattering model, which focuses on the interaction between charge carriers and defects, such as vacancies or dopants. This model mathematically describes the impact of defects on carrier mobility and the Seebeck coefficient, offering insights into how specific defect types and concentrations affect electrical conductivity and thermopower. Additionally, the Callaway model extends the understanding to thermal transport by considering phonon scattering mechanisms induced by defects, providing a framework to analyse how defects influence thermal conductivity. The incorporation of these theoretical models into the broader framework of transport equations allows researchers to predict and optimize thermoelectric properties by strategically manipulating defect concentrations. Computational methods, such as density functional theory, contribute to these models by providing a microscopic understanding of the electronic and vibrational interactions involved in defect-mediated changes. As researchers refine and expand these theoretical models, they pave the way for more precise tailoring of materials with enhanced thermoelectric efficiency through controlled defect engineering.

14.2 Defects as a Means to Optimize Carrier Concentration and Electronic Band Structure

Defects serve as a compelling avenue for the deliberate optimization of carrier concentration and the electronic band structure in materials, offering a nuanced approach to enhancing their thermoelectric performance. Point defects, such as vacancies or dopants, play a pivotal role in influencing the density of charge carriers within a material. Through the intentional introduction of these defects, researchers can precisely control the concentration of electrons or holes, thereby modulating the electrical conductivity. This manipulation of carrier concentration is pivotal for tailoring the thermoelectric properties of a material. Moreover, defects contribute to the modification of the electronic band structure, influencing the energy levels available to charge carriers. Dopants, for instance, introduce additional electronic states within the bandgap, affecting the Seebeck coefficient and electrical conductivity. By strategically engineering defects, researchers can tailor the electronic band structure to enhance the thermopower, or the ability of a material to convert a temperature gradient into an electric voltage. The optimization of carrier concentration and electronic band structure through defect engineering is a powerful strategy to improve the overall thermoelectric efficiency of a material. This approach allows for a fine-tuned adjustment of electrical transport properties, striking a delicate balance between electrical conductivity and thermopower to maximize the thermoelectric figure of merit (ZT). As researchers delve deeper into understanding and manipulating these defect-mediated effects, they pave the way for the development of advanced materials with superior thermoelectric capabilities, holding great promise for sustainable energy harvesting and conversion technologies.

14.3 Trade-Offs Between Enhanced Electrical Properties and Decreased Thermal Conductivity

The optimization of thermoelectric materials often involves navigating intricate trade-offs between enhancing electrical properties and suppressing thermal conductivity. This delicate balance is a central consideration in the quest for materials with superior thermoelectric efficiency. On one hand, the intentional introduction of defects, such as vacancies, dopants, or grain boundaries, can substantially enhance electrical conductivity by influencing the charge carrier concentration and mobility. This is a crucial aspect as higher electrical conductivity contributes to improved thermoelectric performance. However, the same defects that boost electrical properties often act as scattering centres for phonons, impeding the smooth transmission of heat. This phenomenon leads to a decrease in thermal conductivity. While this reduction is desirable for enhancing the thermoelectric figure of merit (ZT), which is a key indicator of overall efficiency, it creates a trade-off scenario [2]. Enhancing electrical conductivity might inadvertently compromise thermal conductivity, and vice versa. Striking an optimal balance between these opposing trends is essential. Researchers must carefully engineer defects to maximize electrical properties without excessively sacrificing thermal conductivity. This delicate compromise is particularly crucial in thermoelectric materials, where the overall efficiency is directly tied to both the electrical and thermal transport properties. The pursuit of materials with minimal trade-offs, or even synergistic improvements in both electrical and thermal properties, remains a critical challenge and an area of active research in the field of thermoelectric materials science. Achieving this delicate balance holds the key to unlocking the full potential of thermoelectric materials for sustainable energy harvesting and conversion applications.

15 Strategies for Defect Control and Material Design

Efficient strategies for defect control and material design play a pivotal role in tailoring the properties of materials for specific applications, ranging from electronics to energy conversion. One approach involves precise manipulation of the synthesis conditions to control defect types and concentrations. By carefully adjusting parameters such as temperature, pressure, and reaction time, researchers can influence defect formation and distribution, allowing for a targeted design of material structures. Additionally, advanced doping techniques offer a powerful tool for introducing controlled defects into materials, enabling the tuning of electronic properties to meet desired specifications. In the realm of computational materials science, modelling and simulation techniques provide valuable insights into defect behaviour and their impact on material performance. Density functional theory and other simulation methods allow researchers to predict the influence of specific defects on electronic, optical, and thermal properties, aiding in the rational design of materials with tailored functionalities. Combining experimental and computational approaches facilitates a comprehensive understanding of defect engineering, empowering researchers to design materials with enhanced properties and address challenges in various technological domains. These strategies not only pave the way for the development of advanced materials with improved performance but also contribute to the broader field of materials science and engineering.

15.1 Defect Engineering Techniques: Doping, Alloying, Controlled Sintering

Defect engineering techniques encompass a variety of sophisticated methods aimed at intentionally introducing and manipulating defects in materials to tailor their properties. Doping, a widely employed technique, involves the deliberate introduction of specific atoms or ions into a material's crystal lattice to modify its electronic and structural characteristics. By carefully selecting dopants and controlling their concentration, researchers can influence charge carrier density, band structure, and conductivity. Alloying is another powerful defect engineering strategy that involves incorporating foreign elements into a material to form a solid solution. This alters the material's properties by introducing lattice distortions, changing the electronic structure, or creating new phases. Alloying provides a versatile means to tailor mechanical, thermal, and electrical properties, enabling the design of materials with enhanced performance for specific applications.

Controlled sintering, or thermal processing, is a crucial technique for defect engineering, particularly in ceramics and other materials with high-temperature processing requirements. During sintering, the material undergoes controlled heating and cooling, influencing defect formation, grain growth, and overall microstructure. By adjusting sintering parameters such as temperature, pressure, and dwell time, researchers can precisely control defect populations, grain sizes, and distribution, thus influencing the material's mechanical and thermal properties.

Combining these defect engineering techniques allows for a synergistic approach to material design. For instance, controlled doping can be complemented by alloying to fine-tune electronic properties, while controlled sintering aids in optimizing the material's structural integrity. The judicious integration of these techniques provides a comprehensive toolbox for tailoring materials across a broad spectrum of applications, from electronics to energy conversion and beyond. This multidimensional approach to defect engineering empowers researchers to achieve materials with unprecedented and customizable properties, meeting the demands of diverse technological challenges.

15.2 Computational Methods for Predicting and Optimizing Defect Configurations

Computational methods have emerged as indispensable tools for predicting and optimizing defect configurations in materials, offering an in-depth understanding of the intricate interplay between defects and material properties. One prominent approach is density functional theory (DFT), a quantum mechanical method that calculates electronic structure and properties with high accuracy. DFT enables researchers to explore the energetics and electronic effects of various defect configurations, shedding light on how defects influence the material's overall behaviour. Molecular dynamics (MD) simulations extend the computational arsenal by providing insights into the dynamic aspects of defect interactions. MD simulations allow researchers to study the temporal evolution of defects, examining their diffusion, aggregation, and reaction kinetics. This temporal perspective is crucial for understanding how defects evolve over time and influence material stability. Machine learning techniques have also found application in defect prediction and optimization. By training models on large datasets of known defect structures and their corresponding properties, machine learning algorithms can predict the impact of specific defects on material performance. This approach accelerates defect screening processes, helping researchers identify promising defect configurations for further experimental investigation. Furthermore, ab initio thermodynamics is employed to assess the stability and prevalence of different defect configurations under varying conditions. This method considers the temperature, pressure, and chemical potential to predict the thermodynamically favourable defect states, aiding in the identification of stable configurations that could be experimentally achievable. The integration of these computational methods allows researchers to explore vast configurational spaces, predict the most favourable defect structures, and optimize material performance. This computational synergy not only expedites the materials discovery process but also guides experimental efforts by providing valuable insights into the underlying mechanisms governing defect-induced changes in materials. As computational techniques continue to advance, their role in defect engineering will likely expand, facilitating the design of materials with tailored properties for specific applications.

15.3 Case Studies Highlighting Successful Defect-Controlled Thermoelectric Material.

Several studies have successfully demonstrated the impact of defect engineering on the thermoelectric properties of materials, showcasing the potential for controlled defects to enhance energy conversion efficiency. One notable example involves the use of point defects in silicon-based materials. Researchers have employed intentional doping with specific elements to introduce defects, effectively tuning the charge carrier concentration and enhancing electrical conductivity. This controlled defect strategy has shown promise in improving the thermoelectric efficiency of silicon, traditionally not considered a strong thermoelectric material. In another study, lead telluride (PbTe), a renowned thermoelectric material, has been subject to defect engineering to further enhance its performance. By introducing specific dopants, such as sodium or silver, researchers manipulated the material's electronic band structure, optimizing the balance between electrical conductivity and Seebeck coefficient. This deliberate defect control led to a substantial increase in the thermoelectric figure of merit (ZT), reflecting a notable improvement in the material's ability to convert waste heat into electricity. Additionally, research has focused on complex oxides, such as strontium titanate ($SrTiO_3$), where controlled oxygen vacancies serve as effective point defects. These vacancies influence both electronic and thermal transport properties, offering a unique platform for tailoring thermoelectric performance. Studies involving $SrTiO_3$ have demonstrated the feasibility of defect-controlled strategies to enhance ZT, emphasizing the importance of precise defect engineering in oxide-based thermoelectric materials [58]. Furthermore, advancements in nanostructuring techniques have allowed researchers to create defect-rich nanomaterials. For instance, nanostructured silicon germanium (SiGe) alloys with grain boundaries and interfaces have exhibited enhanced thermoelectric properties owing to the increased scattering of phonons at defect sites. These studies highlight the significance of defect engineering in nanomaterials to achieve superior thermoelectric efficiency. Collectively, these studies underscore the success of defect-controlled strategies in improving the thermoelectric properties of diverse materials. By manipulating defects, researchers have demonstrated the ability to fine-tune electronic and thermal transport properties, paving the way for the development of more efficient thermoelectric materials for sustainable energy applications.

16 Integration of Defect-Controlled Electroceramics into Practical Thermoelectric Devices

The ultimate goal of defect engineering in electroceramics is to translate fundamental insights into practical applications. Integrating defect-controlled electroceramics into thermoelectric devices poses a unique set of challenges, including scalability, stability, and compatibility with existing technologies. Overcoming these challenges requires a collaborative effort between materials scientists, engineers, and device designers. Successful integration will not only contribute to the realization of efficient thermoelectric devices for waste heat recovery but also foster advancements in sustainable energy solutions. As we navigate these challenges and explore new research frontiers, the field of defect-controlled electroceramics holds great potential for revolutionizing the landscape of thermoelectric technology.

17 Potential Impact of Defect-Controlled Electroceramics on Energy Conversion Technologies

Looking forward, the potential impact of defect-controlled electroceramics on energy conversion technologies is profound. The exploration of emerging research areas, such as the use of two-dimensional materials and nanostructuring techniques, presents exciting possibilities for manipulating defects at the nanoscale. This not only deepens our fundamental understanding of defect engineering but also opens new vistas for practical applications.

As we envision the integration of defect-controlled electroceramics into thermoelectric devices, the anticipated outcome is a transformative shift in energy conversion technologies. The marriage of fundamental scientific insights with technological innovation holds the promise of highly efficient and sustainable thermoelectric devices, contributing significantly to the global pursuit of clean and renewable energy solutions. The study of defects in electroceramics stands not only as a scientific endeavour but as a driving force toward a more energy-efficient and environmentally conscious future. In this synthesis of knowledge and innovation, the strategic engineering of defects emerges as a powerful tool for crafting a new era in energy conversion and utilization.

18 Challenges and Future Directions

The study of defects in electroceramics has significantly advanced our understanding of thermoelectric phenomena, yet several challenges persist in comprehending and controlling defects in these materials. Achieving a comprehensive understanding of the intricate interplay between different types of defects, such as point defects, grain boundaries, and dislocations, remains a formidable task. The complex nature of defect interactions with charge carriers and lattice vibrations demands further exploration to unveil the underlying mechanisms governing thermoelectric behaviour.

One of the persisting challenges lies in unravelling the intricate relationships between defect types and their influence on various thermoelectric parameters. Additionally, the development of precise and scalable techniques for controlling defect density and distribution is essential for achieving reproducibility and consistency in material performance. Addressing these challenges will not only deepen our fundamental understanding of defect engineering but also pave the way for the intentional design of electroceramics with tailored thermoelectric properties.

19 Conclusion

In summary, our journey through the exploration of defects in electroceramics has illuminated their profound significance in modulating thermoelectric properties, laying the foundation for advancements in energy conversion technologies. The deliberate engineering of defects, encompassing a variety of forms from point defects to microstructural irregularities, emerges as a transformative approach to fine-tune the electronic and thermal transport properties of materials.

As evidenced by the chapters on the influence of point defects, microstructural defects, and defect-induced phonon scattering on the Seebeck coefficient, it is evident that defects are not mere imperfections but active agents shaping the thermoelectric landscape. Point defects alter the electronic structure, microstructural defects affect charge carrier mobility, and defect-induced phonon scattering impacts thermal conductivity. The intricate dance between these defects becomes the key to achieving optimal thermoelectric performance.

Acknowledgements Author Yogendra K. Gautam, thank to Government of Uttar Pradesh for Research grant under Research and Development Scheme [No. 23/2023/680/Sattar-4-2023-001-4(59)/2022] and CCS University, Meerut [DEV/URGS/2022-23/39].

References

1. Anno, H., Yamaguchi, K., Nakabayashi, T., Kurokawa, H., Akagi, F., Hojo, M., Toshima, N.: Thermoelectric properties of conducting polyaniline/BaTiO3 nanoparticle composite films. IOP Conf. Ser.: Mater. Sci. Eng. **18**(SYMPOSIUM 9C) (2011)
2. Aswal, D.K., Basu, R., Singh, A.: Key issues in development of thermoelectric power generators: High figure-of-merit materials and their highly conducting interfaces with metallic interconnects. In: Energy Conversion and Management, vol. 114, pp. 50–67. Elsevier (2016)
3. Baláž, P., Baláž, M., Achimovičová, M., Bujňáková, Z., Dutková, E.: Chalcogenide mechanochemistry in materials science: insight into synthesis and applications (a review). J. Mater. Sci. **52**(20), 11851–11890 (2017)
4. Carter, C.B., Norton, M.G.: Ceramic Materials Science and Engineering (2007)
5. Dey, A., Bajpai, O.P., Sikder, A.K., Chattopadhyay, S., Shafeeuulla Khan, M.A.: Recent advances in CNT/graphene based thermoelectric polymer nanocomposite: A proficient move towards waste energy harvesting. Renew. Sustain. Energy Rev. **53**, 653–671) (2016)
6. Dragoman, D., Dragoman, M.: Giant thermoelectric effect in graphene. Appl. Phys. Lett. **91**(20) (2007)
7. Feng, Y., Wu, J., Chi, Q., Li, W., Yu, Y., Fei, W.: Defects and aliovalent doping engineering in electroceramics. Chem. Rev. **120**(3), 1710–1787. American Chemical Society (2020)
8. Goldsmid, H.J.: Bismuth telluride and its alloys as materials for thermoelectric generation. Materials 7(4), 2577–2592. MDPI AG (2014)
9. Hao, X., Chen, X., Zhou, X., Zhang, L., Tao, J., Wang, C., Wu, T., Dai, W.: Performance optimization for PbTe-based thermoelectric materials: Front. Energy Res. **9**. Frontiers Media S.A. (2021)
10. Hassen, A.B., Rhouma, F.I.H., Daoudi, M., Dhahri, J., Zaidi, M., Abdelmoula, N.: Influence of defect on the electrical and optical properties of A-site non-stoichiometry $Ca_{0.67}La_{0.22\text{-}0.11}Ti_{(1-x)}Cr_xO_{3-\delta}$ perovskite. RSC Adv. **9**(34), 19285–19296 (2019)
11. Huang, H.L., Cao, G.Z., & Shen, I.Y.: Hydrothermal synthesis of lead zirconate titanate (PZT or $Pb(Zr_{0.52}Ti_{0.48})O_3$) nano-particles using controlled ramping and cooling rates. Sens. Actuators A: Phys. **214**, 111–119 (2014)
12. Imae, I., Koumoto, T., Harima, Y.: Thermoelectric properties of polythiophenes partially substituted by ethylenedioxy groups. Polymer **144**, 43–50 (2018)
13. Jiang, X.Y., Zhang, Q.K., Deng, S.P., Zhou, B., Wang, B., Chen, Z.Q., Qi, N., Tang, X.F.: Enhanced thermoelectric performance of polythiophene/carbon nanotube-based composites. J. Electron. Mater. **49**(4), 2371–2380 (2018)
14. Jin Bae, E., Hun Kang, Y., Jang, K.S., Yun Cho, S.: Enhancement of thermoelectric properties of PEDOT:PSS and tellurium-PEDOT:PSS hybrid composites by simple chemical treatment. Sci. Reports **6** (2016)
15. Kadel, K., Kumari, L., Li, W.Z., Huang, J.Y., Provencio, P.P.: Synthesis and thermoelectric properties of Bi_2Se_3 nanostructures. Nanoscale Res. Lett. **6**(1), 1–7 (2011)
16. Karim, A.M.M.T., Khan, M.K.R., Hossain, M.S.: Evaluation of the thermoelectric properties and thermal conductivity of $CH_3NH_3PbI_{3-x}Cl_x$ thin films prepared by chemical routes. ACS Omega **6**(26), 16775–16782 (2021)
17. Khoso, N.A., Jiao, X., Guangyu, X., Tian, S., Wang, J.: Enhanced thermoelectric performance of graphene based nanocomposite coated self-powered wearable e-textiles for energy harvesting from human body heat. RSC Adv. **11**(27), 16675–16687 (2021)
18. Li, H., Su, X., Tang, X., Zhang, Q., Uher, C., Snyder, G.J., Aydemir, U.: Grain boundary engineering with nano-scale InSb producing high performance $In_xCe_yCo_4Sb_{12+z}$ skutterudite thermoelectrics. J. Mater. **3**(4), 273–279 (2017)
19. Li, J., Cheng, R., Cheng, Z., Duan, C., Wang, B., Zeng, J., Xu, J., Tian, X., Chen, H., Gao, W., Chen, K.: Silver-nanoparticle-embedded hybrid nanopaper with significant thermal conductivity enhancement. ACS Appl. Mater. Interfaces **13**(30), 36171–36181 (2021)

20. Li, J., Sui, J., Pei, Y., Barreteau, C., Berardan, D., Dragoe, N., Cai, W., He, J., Zhao, L.D.: A high thermoelectric figure of merit ZT > 1 in Ba heavily doped BiCuSeO oxyselenides. Energy Environ. Sci. **5**(9), 8543–8547 (2012)
21. Li, Y.N., Wu, P., Zhang, S.P., Pei, Y.L., Yang, J.G., Chen, S., Wang, L.: Effect of carbon nanotubes addition on thermoelectric properties of $Ca_3Co_4O_9$ ceramics. Chin. Phys. B **31**(4) (2022)
22. Liu, J., Qiu, L., Shao, S.: Emerging electronic applications of fullerene derivatives: An era beyond OPV. J. Mater. Chem.C **9**(45), 16143–16163. Royal Society of Chemistry (2021)
23. Lordi, V., Erhart, P., Åberg, D.: Charge carrier scattering by defects in semiconductors. Phys. Rev. B – Condens. Matter Mater. Phys. **81**(23) (2010)
24. Massetti, M., Jiao, F., Ferguson, A.J., Zhao, D., Wijeratne, K., Würger, A., Blackburn, J. L., Crispin, X., Fabiano, S.: Unconventional thermoelectric materials for energy harvesting and sensing applications. Chem. Rev. **121**(20), 12465–12547. American Chemical Society (2021)
25. Musiienko, A., Yang, F., Gries, T.W., Frasca, C., Friedrich, D., Al-Ashouri, A., Sağlamkaya, E., Lang, F., Kojda, D., Huang, Y.T., Stacchini, V., Hoye, R.L.Z., Ahmadi, M., Kanak, A., Abate, A.: Resolving electron and hole transport properties in semiconductor materials by constant light-induced magneto transport. Nat. Commun. **15**(1) (2024)
26. Nadtochiy, A., Kozachenko, V., Korotchenkov, O., Schlosser, V.: Nickel-fullerene nanocomposites as thermoelectric materials. Nanomaterials **12**(7) (2022)
27. Nandihalli, N., Liu, C.J., Mori, T.: Polymer based thermoelectric nanocomposite materials and devices: fabrication and characteristics. Nano Energy **78** (2020)
28. Ok, K.M., Ohishi, Y., Muta, H., Kurosaki, K., Yamanaka, S.: Effect of point and planar defects on thermal conductivity of TiO_{2-x}. J. Am. Ceram. Soc. **101**(1), 334–346 (2018)
29. Pashkevich, A.V., Fedotov, A.K., Poddenezhny, E.N., Bliznyuk, L.A., Fedotova, J.A., Basov, N.A., Kharchanka, A.A., Zukowski, P., Koltunowicz, T.N., Korolik, O.V., Fedotova, V.V.: Structure, electric and thermoelectric properties of binary ZnO-based ceramics doped with Fe and Co. J. Alloys Compd. **895** (2022)
30. Pham, V.T., Fang, T.H.: Effects of temperature and intrinsic structural defects on mechanical properties and thermal conductivities of InSe monolayers. Sci. Reports **10**(1) (2020)
31. Rodrigues-Marinho, T., Correia, V., Tubio, C.R., Ares-Pernas, A., Abad, M.J., Lanceros-Méndez, S., Costa, P.: Flexible thermoelectric energy harvesting system based on polymer composites. Chem. Eng. J. **473** (2023)
32. Rogl, G., Rogl, P.F.: Development of thermoelectric half-heusler alloys over the past 25 years. Crystals **13**(7) (2023). Multidisciplinary Digital Publishing Institute (MDPI)
33. Rounds, R., Sarkar, B., Alden, D., Guo, Q., Klump, A., Hartmann, C., Nagashima, T., Kirste, R., Franke, A., Bickermann, M., Kumagai, Y., Sitar, Z., Collazo, R.: The influence of point defects on the thermal conductivity of AlN crystals. J. Appl. Phys. **123**(18) (2018)
34. Rowe, D.M.: Thermoelectrics Handbook: Macro to Nano. CRC/Taylor & Francis (2006)
35. Scott, E.A., Hattar, K., Rost, C.M., Gaskins, J.T., Fazli, M., Ganski, C., Li, C., Bai, T., Wang, Y., Esfarjani, K., Goorsky, M., & Hopkins, P.E.: Phonon scattering effects from point and extended defects on thermal conductivity studied via ion irradiation of crystals with self-impurities, Phys. Rev. Mater. **2**(9) (2018)
36. Termentzidis, K., Isaiev, M., Salnikova, A., Belabbas, I., Lacroix, D., Kioseoglou, J.: Impact of screw and edge dislocations on the thermal conductivity of individual nanowires and bulk GaN: a molecular dynamics study. Phys. Chem. Chem. Phys. **20**(7), 5159–5172 (2018)
37. Vladimirov, I., Kühn, M., Geßner, T., May, F., Weitz, R.T.: Energy barriers at grain boundaries dominate charge carrier transport in an electron-conductive organic semiconductor. Sci. Reports **8**(1) (2018)
38. Wang, J., Xie, F., Cao, X. H., An, S. C., Zhou, W. X., Tang, L. M., & Chen, K. Q.: Excellent Thermoelectric Properties in monolayer WSe_2 Nanoribbons due to Ultralow Phonon Thermal Conductivity, Scientific Reports, 7(2017)
39. Wei, J., Fan, Y., Zhao, L., Xue, F., Hao, L., Zhang, Q.: Thermoelectric properties of carbon nanotube reinforced cement-based composites fabricated by compression shear. Ceram. Int. **44**(6), 5829–5833 (2018). Elsevier Ltd.

40. Wei, J., Yang, L., Ma, Z., Song, P., Zhang, M., Ma, J., Yang, F., Wang, X.: Review of current high-ZT thermoelectric materials. J. Mater. Sci. **55**(27), 12642–12704 (2020). Springer
41. Witting, I.T., Chasapis, T.C., Ricci, F., Peters, M., Heinz, N.A., Hautier, G., Snyder, G.J.: The thermoelectric properties of bismuth telluride. Adv. Electron. Mater. **5**(6) (2019). Blackwell Publishing Ltd.
42. Wu, J., Liu, Y., Liu, Y., Cai, Y., Zhao, Y., Ng, H.K., Watanabe, K., Taniguchi, T., Zhang, G., Qiu, C.-W., Chi, D., Castro Neto, A.H., Thong, J.T.L., Loh, K.P., Hippalgaonkar, K.: Large enhancement of thermoelectric performance in MoS_2/h-BN heterostructure due to vacancy-induced band hybridization. PNAS **117**, 13929–13936 (2020)
43. Xiao, Y., Zhao, L.D.: Charge and phonon transport in PbTe-based thermoelectric materials. NPJ Quantum Mater. **3**(1) (2018). Nature Publishing Group
44. Yan, X., Liu, W., Wang, H., Chen, S., Shiomi, J., Esfarjani, K., Wang, H., Wang, D., Chen, G., Ren, Z.: Stronger phonon scattering by larger differences in atomic mass and size in p-type half-heuslers $Hf_{1-x}Ti_xCoSb_{0.8}Sn_{0.2}$. Energy Environ.Sci. **5**(6), 7543–7548 (2012)
45. Yin, S., Lu, W., Wu, X., Luo, Q., Wang, E., Guo, C.Y.: Enhancing thermoelectric performance of polyaniline/single-walled carbon nanotube composites via dimethyl sulfoxide-mediated electropolymerization. ACS Appl. Mater. Interfaces **13**(3), 3930–3936 (2021)
46. Zebarjadi, M., Esfarjani, K., Dresselhaus, M.S., Ren, Z.F., Chen, G.: Perspectives on thermoelectrics: from fundamentals to device applications. Energy Environ. Sci. **5**(1), 5147–5162) (2012)
47. Zevalkink, A., Smiadak, D.M., Blackburn, J.L., Ferguson, A.J., Chabinyc, M.L., Delaire, O., Wang, J., Kovnir, K., Martin, J., Schelhas, L.T., Sparks, T. D., Kang, S.D., Dylla, M.T., Snyder, G.J., Ortiz, B.R., Toberer, E.S.: A practical field guide to thermoelectrics: Fundamentals, synthesis, and characterization. Appl. Phys. Rev. 5(2) (2018). American Institute of Physics Inc.
48. Zhang, G., Li, Q., Gu, H., Jiang, S., Han, K., Gadinski, M.R., Haque, M.A., Zhang, Q., Wang, Q.: Ferroelectric polymer nanocomposites for room-temperature electrocaloric refrigeration. Adv. Mater. **27**(8), 1450–1454 (2015)
49. Zhang, L.: Investigation of tin chalcogenides thermoelectric materials for energy conversion (2018)
50. Zhao, C., Li, Z., Fan, T., Xiao, C., Xie, Y.: Defects engineering with multiple dimensions in thermoelectric materials. Research (2020)
51. Zheng, X.F., Liu, C.X., Yan, Y.Y., Wang, Q.: A review of thermoelectrics research - recent developments and potentials for sustainable and renewable energy applications. Renew. Sustain. Energy Rev. **32**, 486–503 (2014)
52. Zimbone, M., Sarikov, A., Bongiorno, C., Marzegalli, A., Scuderi, V., Calabretta, C., Miglio, L., La Via, F.: Extended defects in 3C-SiC: stacking faults, threading partial dislocations, and inverted domain boundaries. Acta Mater. **213** (2021)

Role of Electroceramics in Renewable Energy Technologies

Pinki Singh, Nisha Shankhwar, and Satyendra Singh

Abstract The global warming has become one of the major concerns of today's era, carbon dioxide emissions being the largest contributors amongst all. The technologies employing use of renewable resources viz. sun, water, wind etc. can play a vital role in the reduction of greenhouse gas emissions. In order to circumvent the usage of fossil fuels, which is still the major source for power generation and is amongst the largest CO_2-emitting sectors, it is essential to integrate the technologies associated with energy-harvesting and energy storage devices. These devices with high energy/power density play a significant role in transport electrification and portable electronics. The conventional dielectric capacitors have garnered larger interest due to higher power density and their faster charge–discharge speed making them suitable for pulsed power applications. The smaller remanent and larger maximum polarization with high electric breakdown strength of dielectric ceramics are the keys to its high energy density. The higher values of dielectric breakdown temperature of electroceramics than its counterparts make them feasible alternative for high-temperature applications. In this chapter, an overview of the role of electroceramics, the historical evolutions and future projections in the renewable energy technology sector have been discussed. A brief description on different types of ceramics, the principles of energy storage in electroceramics and the strategies for enhancement of energy storage properties is presented for more understanding.

Keywords Electroceramics · Energy-storage · Energy–density · Power-density · Dielectric breakdown

1 Introduction to Electroceramics in Renewable Energy

To combat with the climate crisis and achieve Paris agreement goals of net-zero carbon emissions along with reduction of the global warming to 1.50 °C, it is essential that anthropogenic carbon dioxide emissions must decline ~45% by 2030 [1]. This

P. Singh (✉) · N. Shankhwar · S. Singh
Special Centre for Nanoscience, Jawaharlal Nehru University, New Delhi, India
e-mail: pinki9386@gmail.com

U. Kumar (ed.), *Defects Engineering in Electroceramics for Energy Applications*,
Engineering Materials, https://doi.org/10.1007/978-981-97-9018-0_9

strongly highlights the urge for development of technologies related to renewable resources such as sun, biomass, hydrothermal, wind, etc. Despite of so many attempts being taken to increase the subsequent deployment of renewable energies, a lot of challenges are being faced to replace the fossil fuels-based power generation system which are a major source of CO_2 emissions, due to the epochal properties of renewables. To overcome the limitation caused due to this intermittent nature of renewables, and to shift our reliability on fossil fuels simultaneously, it is essential to integrate the technologies related to energy-harvesting with devices for energy storage.

The increased demand for the development of sustainable renewable technologies has led the researchers to focus on energy harvesting from various sources to limit the usage of batteries or to extend their lifespan. Energy harvesting is defined as the collection of various forms of energy from ambient sources (solar, thermal, mechanical, wind, etc.) and converting them to electrical energy using a specific material or transduction mechanism.

Energy storage devices play a pivotal role in sustainable renewable technologies which includes applications in particular transport electrification, pulsed power applications and more specialized applications viz. active armors and heart defibrillators. The technologies to store energies such as batteries, fuel cells, electrochemical and electrostatic capacitors are already available [2]. The drawbacks such as limited energy storage capacities, power efficiency and limited lifespan of conventional energy sources such as batteries has led to the recent developments of portable, remote and low-power devices introduced as non-conventional sources from past few decades. These energy storage devices eliminate the need of frequent recharging with extended lifespan.

The electrostatic or dielectric capacitors are passive components that store electrical energy with faster charge/discharge performances (~μs) leading to high power density (~10^4–10^5 W/kg) as compared to batteries and fuel cells [3, 4]. These capacitors also provide several other advantages such as higher temperature/frequency stability, increased lifespan, fatigue resistance and better manufacturing cost making them useful for practical applications.

The newly developed electroceramics are used as energy harvesting materials which facilitate unique conversion mechanism for energies which plays an essential role in renewable energy technology.

1.1 Defining Electroceramics

Traditionally, ceramic materials have been focused due to their incomparable mechanical properties and thermal stability. In addition to these properties, nowadays, their unique optical, electrical and magnetic properties have garnered significant attention of scientific community due to their applications in several technologies that include energy storage and conversion, electronics, communications and automation. The materials with these specific properties are termed as electroceramics.

With the growth of modern-day technologies, electroceramics have been developed in various subclasses i.e., ferroelectrics, ferrites, solid electrolytes, piezoelectrics and semiconducting oxides. The properties of devices based on electroceramics are heavily dependent on their processing, structure, chemistry and device physics.

1.2 Importance of Renewable Energy

The renewable energy is generally extracted from natural energy sources which do not deplete the earth's natural resources and replenish in time less than human lifespan. These energy sources are available everywhere in one form or the another and are inexhaustible. The most common examples of renewable energy resources are sun, tidal waves, biomass, rain, wind, green hydrogen and geothermal energy. These renewable energy resources cause negligible or minimal damage to the environment.

The fossil fuels or non-renewable energy sources such as coal, oil and natural gases are available in limited quantities and can run out in few centuries as they are formed via natural processes which replenish in millions of years. The extraction and production of these fossil fuels involve processes which release severe green house gases and many health affecting particles.

1.3 Scope of the Chapter

The electroceramics have been of utmost importance in modern day technologies of energy storage and conversion, sensors, automation, electronics, information storage etc. due to their unique optical, electrical and magnetic properties, extraordinary thermal, mechanical and chemical stability. The processing of electroceramics can be done to obtain them in various forms such as powders, composites, porous or dense materials, solid state devices and thin films.

They exhibit versatile properties that include tunable dielectric behavior, varying conductivities from insulating, metallic to fast ionic, ferromagnetic, pyro-, ferro- and piezoelectric properties, and variable electro- to nonlinear optical properties. As discussed in the earlier section, the deployment of electroceramics in energy conversion and storage technology offers a wide spectrum for renewable energy technology. The present chapter is dedicated to discussing the role of electroceramics in various technologies concerning energy storage and conversion, the principles behind the energy storage and efficient ways to enhance these properties that makes them a great contributor in renewable energy technology.

2 Electroceramics and Energy Storage

The electrostatic ceramic capacitors provide major technological characteristics in energy storage properties such as high energy density, high power density, better conversion efficiencies, wider temperature stability, etc. The principals involved in energy storage and key parameters affecting the properties are discussed in the next section.

2.1 Fundamentals of Energy Storage in Capacitors

A simple electrostatic capacitor is a set of two parallel conducting plates with insulator separation which gets polarized when subjected to an electric field (Fig. 1). This behavior characterizes a dielectric material. The ability to store the charge, termed as the capacitance, C, for an ideal capacitor is defined as the ratio of stored charges on each conducting plate, Q, to the voltage applied, V as given by

$$C = \frac{Q}{V}$$

From the viewpoint of Gauss's law, the value of V can be written as

$$V = \frac{Qd}{\varepsilon A}$$

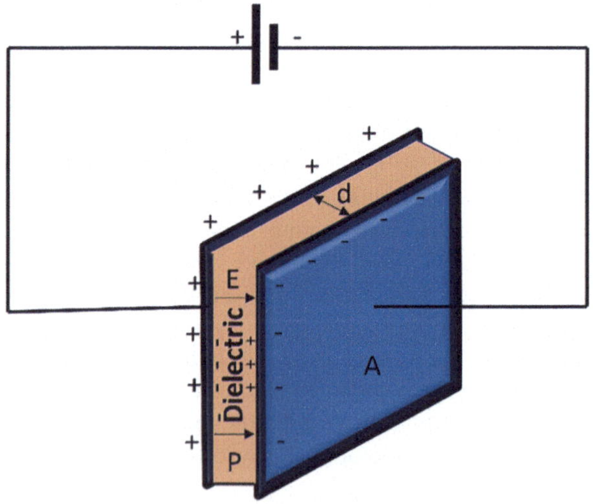

Fig. 1 Schematic diagram of a simple electrostatic capacitor

where d, ε and A are the separation distance between two conducting plates, the dielectric permittivity and are of two parallel conducting plates, respectively. Using this value of V in equation for C, we obtain

$$C = \frac{A}{\varepsilon d}(F)$$

From the above equation, it can be inferred that the capacity to charge a dielectric capacitor, and thus to store energy is directly associated with dielectric value of the insulating material between the metallic plates.

The electrostatic energy storage capacity under applied electric field, E, for dielectric capacitors are generally expressed as

$$W_{total} = \int_{0}^{P_{max}} EdP$$

$$W_{rec} = \int_{P_r}^{P_{max}} EdP$$

where W_{total} and W_{rec} represent the total energy density and recoverable energy densities, respectively. The term P denotes the electric displacement, P_{max} shows maximum electric displacement and P_r indicates the remnant electric displacement. Further, the efficiency (η) for the electrostatic capacitors can be calculated as

$$\eta = \frac{W_{rec}}{W_{total}} \times 100\%$$

The key parameters used to optimize energy density of electroceramics include microstructural features like grain size, porosity, density, core–shell structure and the secondary phases. The intrinsic electrical response, such as band gap, distribution of different resistive and conductive elements also plays decisive roles in optimizing the energy density.

2.2 *Role of Electroceramics in Energy Storage*

The potential of ceramic capacitors to be used in the energy storage applications is evident back from Jaffe's work, Clevite Corp., U.S.A. in 1960 [5]. Later, nearly after a decade, two researchers from Sprague Electric Company named Burn and Smyth studied the performance of $SrTiO_3$ and $BaTiO_3$ for energy storage for larger electric fields till 400 kV/cm [6]. Before that time, the quantitative analysis of the energy storage performance for these materials were bounded to comparatively smaller electric fields (<150 kV/cm). This limitation was due to the lower dielectric breakdown strength (BDS) of the available ceramics. It was then realized that the thinner dielectric ceramic layers can provide maximum energy density due to reduced probability of defects occurrence (pores, microcracks or voids are regarded as defects) which are leading causes of DBS. In 1990s, Love, again from Sprague Electric Company, took up the studies of ceramic capacitors for energy storage enhancement and developed empirical design principles to obtain it [7].

The commercially available C0G-type capacitors are developed from linear dielectrics with smaller relative permittivity (ε_r) which can obtain energy storage up to 1 J/cm^3 due to its intrinsic high BDS. The importance of BDS becomes much prominent for the capacitors fabricated using high ε_r, viz X7R-type capacitors. The thickness of dielectric layer is strongly correlated with the BDS, which can be understood by the fact that just reducing the thickness to half, there is a three-fold increase in the energy storage performance of the material. It was then proposed by Love that the materials with intermediate ε_r are much suitable for capacitors in energy storage applications rather than high ε_r materials due to their greater BDS values. To study the trend, Yang et al. compiled the available BDS data of several dielectric materials with different thickness from the literatures and concluded that decay is inversely proportional to (t)*0.5 [8]. In 1996, Fletcher et al. discovered that ferroelectric materials with their Curie temperature adjusted to operate the material in paraelectric regime to show small zero-field εr, exhibit larger values of energy storage densities [9]. This approach to enhance the energy storage density has been explored by Jaffe in 1961. Ogihara, in 2009, reported the fabrication of energy storage devices by making the use of weakly coupled relaxors due to extraordinary stability of their ε_r over wider range of temperatures [10]. One of the recent approaches to enhance energy storage performance is to fabricate chemically heterogeneous ceramics with electrically homogeneous properties.

2.3 Types of Electroceramic Materials

The electroceramics can be classified into different types based on their characteristics. The most used types of electroceramics are dielectric ceramics, ferroelectric ceramics, piezoelectric ceramics and ionic conductors. Dielectric ceramics show properties of higher electrical insulation whereas properties of switchable spontaneous polarization are shown by ferroelectric ceramics. The piezoelectric ceramics can convert mechanical stress to electrical signals. The ionic conductors are useful in applications like batteries and fuel cells by enabling the movement of ions across the specimen.

3 Electroceramics in Solar Energy Applications

The solar energy systems are a major contributor in the renewable energy technology. Due to their distinctive structural, thermal, and electrical properties, the contribution of electroceramics in solar energy applications is very extensive. They can be incorporated in various components of solar devices which leads to enhanced efficiency and improved performances of the solar energy systems as illustrated in Fig. 2. The key areas for implementation of electroceramics in solar energy applications are listed below.

Solar Batteries and Solar Energy Storage Systems

The various components associated with storage of solar power in solar energy storage systems are built up of electroceramics. The solid-state Li-ion batteries are one of the best examples of use of electroceramics in solar energy storage systems [11]. The use of electroceramics can effectively improve the solar batteries performance, their safety and increase their lifespan which is essential to store the excess solar power to be used in periods with low or no sunlight.

Solar Thermal Collectors

The thermal collectors in solar devices are used to collect and capture the heat produced by sun and transform them into thermal energy useful for water heating

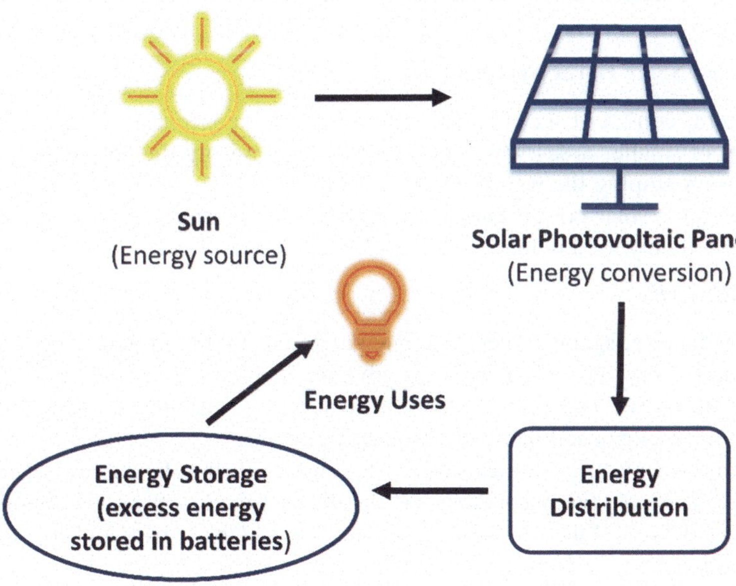

Fig. 2 Pictorial representation of different solar energy applications of electroceramics

and space heating. The components of thermal collectors are made up of electroceramic materials that possess higher thermal conductivity to improve the heat transfer efficiency.

Perovskite Solar Cells

They are the most promising and advancing technology in photovoltaics in recent days. The perovskite-based electroceramics are used in the cells as an active layer leading to conversion of sunlight in electricity. The perovskite solar cells are highly efficient and very cost-effective which makes them a major contributor in solar energy research and development [12].

Piezoelectric Ceramics for Solar Tracking Systems

The devices used to track the maximum sunlight exposure direction of the solar panels and orient them in the preferable direction are known as solar tracking systems. These systems involve the implementation of piezoelectric ceramics leading to generation of mechanical motion from the electrical signals. These devices are helpful in improving the efficiency of solar panels by enabling them to always face the sun [13].

Transparent Conductive Oxides

The transparent conductive materials which are generally used to fabricate the transparent coatings on the solar panels. This allows the light to pass through it while allowing the PV cells to generate an efficient amount of electricity. A well-known transparent conductive oxide material used in these applications is Indium Tin oxide (ITO).

Concentrated Solar Power Systems

These systems consist of set of lenses and mirrors that concentrate larger amounts of sunlight in smaller areas which generate heat by raising the temperature. The solar receivers employ the use of electroceramic materials which traps and stores this concentrated solar power. Due to their ability to withstand high temperatures, they can be suitably used in thermal energy storage applications.

Photovoltaic Cells

The photovoltaic cells require electroceramic materials for their fabrication, that can convert solar energy into electricity. They are implemented to encapsulate and passivate solar cell layers. This protects them from several environmental degradations, enhances their lifespan and increases their efficiency.

To summarize, electroceramics play a very crucial role in solar energy generation and storage. This effectively enhances the durability, functionality and efficiency of solar energy devices. Further research advances are being carried out to improve solar technology.

4 Electroceramics in Fuel Cells

The device used to generate electricity, water or heat by combination of hydrogen and oxygen is termed as fuel cell. There are various types of fuel cells which are categorized depending on the electrolyte or the fuel they employ for their operation. The fuel cells that use ceramics or solid oxides to generate electricity directly using fuel are classified as solid oxide fuel cells (SOFCs). The ceramic fuel cell stack consists of an anode, a cathode, the electrolyte, and the interconnect as their main components [14]. The planar and cross-sectional design of solid oxide fuel cells are represented in Fig. 3. Each of these parts have significant contribution in the working of fuel cell which require certain specifications such as relevant electrical conductivity, identical thermal expansion co-efficient, chemically compatible with different parts and working conditions stability. The anode and cathode should allow transport of gases at reaction sites; thus, they must be porous in nature. They can be composed of electronic and mixed ionic conductors whereas the electrolyte and the interconnect strictly must be ionic conductor and electronic conductor, respectively.

The ceramic fuel cells such as solid-oxide fuel cells use electrolytes composed of oxide ion conductors which needs to be operated at usually higher temperatures of 700- 1000 °C [15]. The latest solid fuel cell technology (SOFCs) includes functional materials (such as strontium substituted lanthanum manganite) as choice of materials for cathode and Ni-yttria stabilized zirconia (YSZ) cermet as anode. YSZ is material choice for electrolyte which behaves as a good oxide ionic conductor beyond 800 °C. This property of YSZ raises the operation temperature of fuel cells based on YSZ up to 1000 °C which causes the cell degradation [16]. This limitation can be overcome by reducing the temperature of operation to 600–800 °C as in the case of intermediate temperature solid oxide fuel cells (ITSOFCs) and become cost-efficient since cheaper materials for construction can be used [17]. The doped ceria are most promising electrolyte material for ITSOFCs where Gadolinium is the most studied doping element in Ceria. The proton conducting ceramics for ITSOFCs applications, due to smaller size of protons as compared to oxide ion, are also being investigated

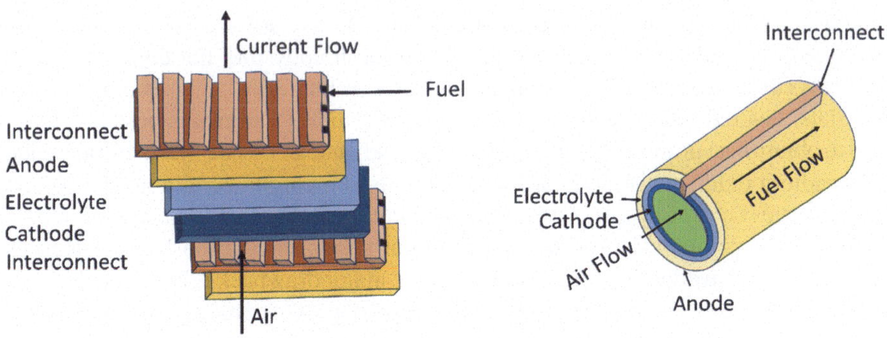

Fig. 3 Planar and cross-sectional design of solid-oxide fuel cells (SOFCs)

where higher conductivity for low temperatures is anticipated. Perovskite structures based on barium show protonic conductivity that approaches the better oxide ion conductivity at the low operation temperature of ITSOFCs. Due to higher ionic conductivity and catalytic activity of rare earth based nicolates, ferrites and cobaltite, they are considered a superior material choice as cathode for ITSOFCs. The structural support in anode is made up of cermets as it matches the thermal expansion co-efficient of anode and the electrolyte. In case of high temperature solid oxide fuel cells (HTSOFCs), a ceramic or a specific nickel or chromium alloys are used as interconnect which makes them highly expensive while ITSOFCs use cost-effective ferritic stainless steel alloys.

From the above discussion, we can conclude that electroceramics play a very important role in power generation, making them remarkable contributor in renewable energy technology.

5 Batteries and Electroceramics

Like fuel cells, battery is also an electrochemical cell that generates electricity from chemical reactions. The fuel cells need a constant supply of fuel for their operation, whereas the batteries have stored chemical energy which can be extracted in form of electrical energy on demand.

An extensive literature on batteries are available for different types which include lead acid [18], Na/$NiCl_2$ [19], Ni/metal hydride [20] and Lithium-ion based batteries [21], each having several advantages and disadvantages. Several advances are being done on Li-ion batteries to make them much safer, less expensive, enhanced durability and improved performances. The Li-ion batteries typically have graphite and $LiCoO_2$ as electrode materials, due to their layered structures. The use of $LiCoO_2$ is not seemingly much effective as its layered structure is not-so stable at the time of delithiation where it can decompose to release oxygen. The process of 'thermal runaway' can occur due to escalated exothermic reactions at higher temperatures. Due to these drawbacks, it is necessary to replace the anode, cathode and electrolyte materials for Li-ion batteries where carbon, $LiCoO_2$ and organic Li-ion conducting materials are used for respective roles. The morphology and the composition of materials can cause a significant impact on the performance of batteries. For cathode materials, layered oxides with mixtures of cobalt, nickel and manganese are being used to obtain best properties with minimized disadvantages. The other materials such as Vanadium oxide are olivines are also being used owing to certain limitations on the properties such as cycling behavior, conductivity and capacity. In case of anodic materials, the commonly used ones are lithium-alloys and carbon-based materials due to their easy availability and low cost but the major drawback is low capacity. The carbon nanotubes some novel graphite varieties are used to enhance the capacity but are expensive due to high processing charges. The capacity, mechanical stability and performance of the anodic material can be enhanced using alloys or intermetallic compounds. The recent approaches include alloying the nanocrystalline materials (containing Zn, Mg, Bi, Al, etc.) with non-alloying stabilization matrix (that contain Cu, Fe and Co or Ni), eutectic mixture composed of Al and Al_3Ni, Si_5Li_{22}, amorphous carbon coated quantum dots. The materials to be used for electrolytes in the lithium-ion battery applications should be able to withstand the high temperature and existing voltage, and longer shelf life. The different types of electrolytes used are polymers, solvent-based liquids and solid-state electrolytes. In case of liquid electrolytes, there is always a competition between electrochemical performance and flammability. The liquid electrolytes based on complexes of 2-methyl tetrahydrofuran and boron trifluoride (BF_3) are reported to show higher ionic conductivity and oxidatively stable. The electrolytes made up of ceramic nanoparticles based composite polymers are observed to show higher conductivity and resistance to oxidation at higher voltages. The solid electrolytes comprise of ceramic glasses and Li-ion conducting crystals which reduce the chances of thermal runaway. The lesser mobility of lithium ions at lower temperature leads to their poor performances. In addition, they require high temperature treatments and special deposition techniques to get desired behavior which increases their cost. The focus is to obtain suitable material which can manage unavoidable changes in volume of active electrode materials to prevent cracks or fractures. The processing of desired materials is focused on packing density that can expose maximum active material, access the electrolyte through open pores and to ensure electronic continuity for charge exchange into the current collector.

6 Ceramic Capacitors and Supercapacitors

The modern-day electronics mostly uses multilayered ceramic capacitors (MLCCs) as passive component due to their compactness and high capacitance. The ceramic capacitors are also used in distinctive forms other than MLCCs in technological applications [1]. They are extensively used in devices which perform at high voltages and high temperatures. Due to their abundant compositions and diverse dielectric properties, ceramic capacitors are ubiquitous in various extreme conditions. On the basis of their withstanding temperature and dielectric behavior they can be categorized in three different classes based on their temperature characteristics over ranges of specified temperature. These factors decide the type of dielectric material that can be used for a particular application. A typical dielectric ceramic capacitor is shown in Fig. 4. The dielectric ceramics that show no significant effect on dissipation factor and capacitance with respect to voltage, frequency and time along with highly stable dielectric characteristics are termed as class I dielectrics. Class II dielectrics show higher volumetric efficiency than class I dielectrics while being stable in the defined limits. The dissipation factors and capacitance are dependent on voltage, frequency and time. The dielectrics classified as class 3 are the ones having less accuracy, stability, non-linear temperature characteristics and show high losses. They have been considered outdated and are of no use in the current electronic market.

The class I dielectrics are used in applications involving high temperature such as power electronics or deep well drilling where low conductivity and low loss are

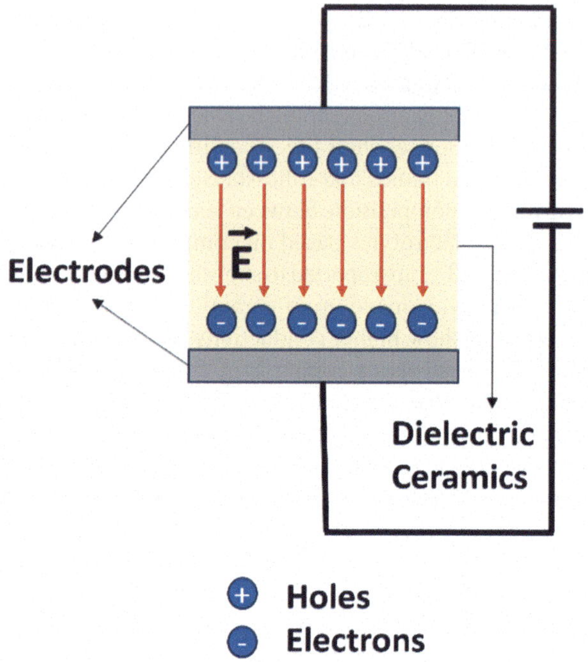

Fig. 4 A dielectric ceramic capacitor

the key components for their role. Amongst class II dielectrics, $BaTiO_3$ was the first ferroelectric material to be discovered with the highest permittivity at the time of discovery. Even after more than seven decades of its findings, it is still the most important base material for the electronics industries in MLCCs.

The double-layered capacitors, termed as supercapacitors, possess higher capacitance as their power and energy densities are higher than the normal capacitors. They are used in energy storage devices and are helpful in achieving voltage stability. They are environment friendly as they can release energy in shorter duration without causing any harm to the environment while scrapping and production.

7 Piezoelectric Materials for Energy Harvesting

The term 'energy harvesting' refers to conversion of different energies from various ambient sources (solar, wind, mechanical, fluid flow, thermal, etc.) in form of electrical energies with the help of certain material or via transduction mechanism [22]. The piezoelectric materials are being explored as non-conventional sources of power for low-power and small-sized electronic devices that typically contributes to piezoelectric energy harvesting. To eliminate the issue of limited lifetime, periodic recharging and batteries replacement in remote, portable and implantable devices, the ambient energies in form of thermal energy, solar energy and vibrational energy can be found. Among all the available energy sources, the vibrational energy is persistent in natural as well as manmade structures due to the surroundings or the operational conditions. As compared to the solar and thermal energy harvesters, the piezoelectric harvesters generally operate in much low-level energies, thus, they are useful when solar and thermal energies are not present. The piezoelectric generators are advantageous over different methods due to their efficiency preservation in reduced scale, functionality at higher frequencies, inherent transduction capacity and higher power density. A piezoelectric harvester is attached to host structures for effective transfer of vibratory energy to the harvester. The operating scale for piezoelectric harvester is at microwatts to milliwatts, that can supply energy to the low-power requirement electronics such as portable electronics, biomedical devices implanted inside the body and wireless sensor nodes. Although many materials with piezoelectricity have been discovered, the perovskite material lead zirconium titanate (PZT) often doped with lanthanum or niobium. The piezoceramics are widely used in sensors and actuators due to their easy operation to be able to output large voltages (~50–100 V) in absence of bias voltages via direct coupling. The piezoceramics utilizes direct and inverse piezoelectric effect in application as energy harvester. The direct piezoelectric effect refers to generation of surface charge on application of pressure whereas the converse piezoelectric effect refers to variation of length due to application of electrical field (shown in Fig. 5). As an application in energy harvesting, the effect of direct polarization is utilized where the electrical charge is induced due to the applied force. In the absence of an applied field, the electric dipoles inside the piezoelectric materials are randomly aligned in between the two electrodes. When an external electrical field is applied on the energy harvester, the electric dipoles align themselves in the direction of field. When the external force is absent, the device remains in an equilibrium state, thus providing zero signal. The compressive force applied in vertical direction leads to polarization of materials due to compressive strain that induces piezoelectric potential in between electrodes. This leads to generation of an output signal. When the applied field is released, reverse piezoelectric potential is induced due to development of a slight tensile force.

The use of piezoelectric ceramics can efficiently minimize the costs related to battery replacement where they act as autonomous and permanent power source for electronic devices placed in remote areas or embedded in structures. The recent trends of low-power electronic devices utilize piezoelectric energy harvesting making a remarkable contribution in renewable energy technologies.

8 Electroceramics in Hydrogen Production

This section provides an overview of the role of electroceramics in hydrogen production via water electrolysis. It discusses the key electroceramic materials used as proton-conducting membranes in electrolyzers and their impact on the efficiency, durability, and scalability of hydrogen generation systems.

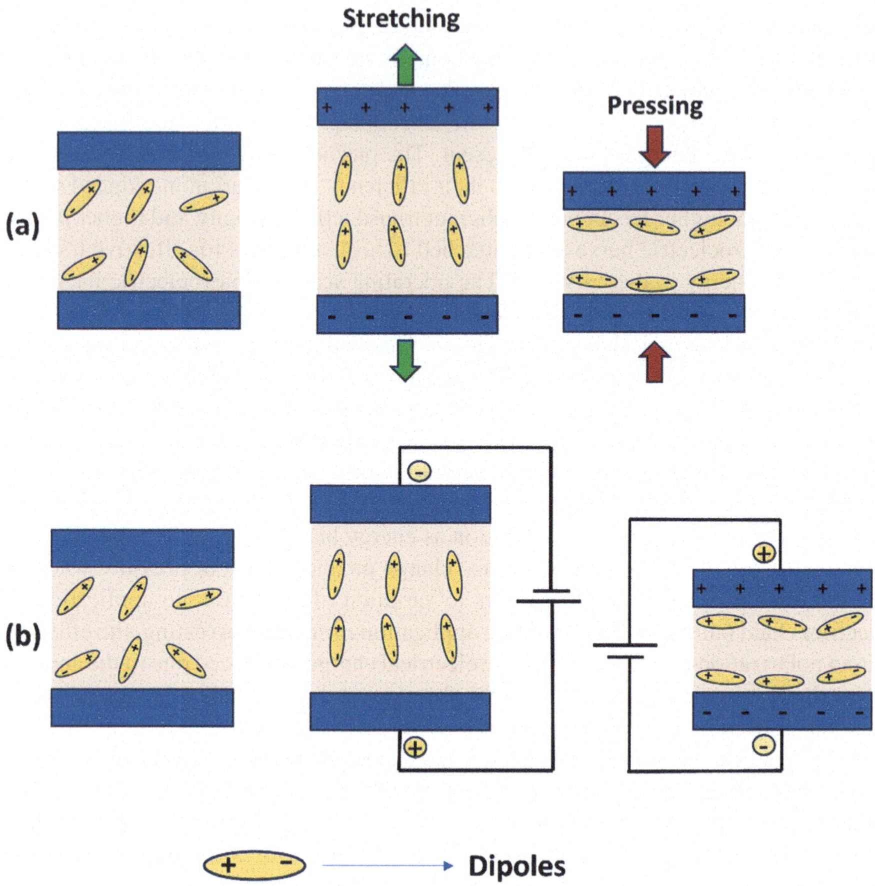

Fig. 5 Schematic illustration of **a** direct piezoelectric effect, and **b** converse piezoelectric effect

Hydrogen emerges as a pivotal player in the quest for sustainable energy solutions, heralded as a clean and versatile energy carrier with the potential to revolutionize various sectors. Its significance lies in its ability to serve as a zero-emission fuel for transportation, a storage medium for renewable energy, and a feedstock for industrial processes, promising a pathway towards decarbonization and energy transition.

One of the most promising methods for large-scale hydrogen production is water electrolysis, a process wherein water molecules are split into hydrogen and oxygen gases using an electric current [23]. Electrolysis offers a direct and efficient means of harnessing renewable energy sources, such as solar and wind power, to generate hydrogen without greenhouse gas emissions or other pollutants.

In the realm of electrolysis, electroceramics emerge as key enablers of efficiency and performance. These specialized ceramic materials possess unique properties that make them indispensable for electrolysis systems. Electroceramics are primarily employed as proton-conducting membranes, responsible for facilitating the transport of protons while preventing the mixing of hydrogen and oxygen gases.

The role of electroceramics in electrolysis extends beyond mere proton conduction; they also contribute to improving the overall efficiency, durability, and scalability of electrolysis technologies. By leveraging the high proton conductivity and chemical stability of electroceramic membranes, electrolyzers can operate at lower temperatures and pressures, reducing energy consumption and extending equipment lifespan.

In essence, electroceramics act as the backbone of electrolysis systems, offering a robust foundation upon which sustainable hydrogen production can thrive. Their integration into electrolysis technologies underscores a commitment to innovation and progress in the pursuit of clean energy solutions.

Solid Oxide Electrolyzers

Solid oxide electrolyzers, that makes use a solid ceramic material as the electrolyte to selectively conduct negatively charged oxygen ions (O_2-) at elevated temperatures, generate hydrogen in a slightly different mechanism [24].

The steam generated at the cathode combines with electrons from external circuit to form hydrogen gas and negatively charged oxygen ions. The oxygen ions permeate through the solid ceramic membrane and react to form oxygen gas at the anode and generate electrons for the external circuit.

Solid oxide electrolyzers have higher operation temperature that is high enough for the solid oxide membranes to function properly (about 700°–800 °C, compared to PEM electrolyzers, which operate at 70°–90 °C, and commercial alkaline electrolyzers, typically operates at less than 100 °C). Advanced lab-scale solid oxide electrolyzers based on proton-conducting ceramic electrolytes are showing promise for lowering the operating temperature to 500°–600 °C. The solid oxide electrolyzers can effectively use the available heat at these elevated temperatures (from various sources, including nuclear energy) to reduce the amount of electrical energy needed to produce hydrogen from water [24].

8.1 Electroceramic Materials for Proton-Conducting Membranes

Electroceramic materials used in hydrogen production are selected based on their ability to conduct protons effectively while exhibiting suitable mechanical, chemical, and thermal stability under electrolysis conditions. These materials play a critical role in separating hydrogen and oxygen gases, thereby enabling efficient and selective hydrogen generation.

Types of Electroceramic Materials:

a. Perovskite Oxides: Perovskite-based electroceramics, such as $BaZrO_3$, $BaCeO_3$, and $SrCeO_3$, are widely studied for their high proton conductivity and chemical stability. These materials exhibit desirable properties for proton transport, including high oxygen ion vacancy concentrations and suitable crystal structures conducive to fast diffusion pathways.
b. Doped Zirconia: Zirconia-based electroceramics, such as yttria-stabilized zirconia (YSZ) and gadolinia-doped ceria (GDC), are commonly used as electrolyte materials in solid oxide fuel cells (SOFCs) and electrolysis cells due to their high oxygen ion conductivity [23]. These materials can also conduct protons under certain conditions, making them promising candidates for proton-conducting membranes in electrolysis applications.

Properties Required for Effective Proton Conduction:

a. High Ionic Conductivity: Electroceramic membranes must exhibit high proton conductivity to facilitate rapid proton transport across the electrolyte interface, enabling efficient hydrogen production.
b. Chemical Stability: Electroceramic materials should be chemically stable in both reducing and oxidizing environments encountered during electrolysis, ensuring long-term durability and performance.
c. Mechanical Strength: Proton-conducting membranes must possess sufficient mechanical strength to withstand mechanical stresses and thermal cycling associated with electrolysis operations.

Examples of Electroceramics and Their Suitability for Proton Transport:

a. Perovskite Oxides: Perovskite-based electroceramics demonstrate excellent proton conductivity due to their intrinsic crystal structure and high oxygen ion vacancy concentrations. These materials offer promising prospects for proton-conducting membranes in water electrolysis systems.
b. Doped Zirconia: Yttria-stabilized zirconia (YSZ) and gadolinia-doped ceria (GDC) are extensively researched for their proton-conducting properties in addition to their well-established oxygen ion conductivity. Their chemical stability and compatibility with various electrode materials make them attractive candidates for electrolysis applications.

In summary, electroceramic materials such as perovskite oxides and doped zirconia exhibit promising characteristics for proton conduction, making them essential components of proton-conducting membranes in hydrogen production via water electrolysis. Continued research and development in electroceramics are crucial for advancing the efficiency and scalability of electrolysis technologies for sustainable hydrogen generation.

8.2 *Electroceramics in Electrolysis Systems*

Integration of Electroceramic Membranes into Electrolyzers

Electroceramic membranes are integrated into electrolyzers as the key component responsible for separating hydrogen and oxygen gases during water electrolysis. These membranes are typically sandwiched between electrode layers, forming the electrolyte compartment where proton transport occurs while preventing gas crossover. Integration methods involve careful sealing to ensure gas-tight operation and optimization of electrode–electrolyte interfaces for efficient electrochemical reactions.

Design Considerations and Optimization Strategies

Electrolysis cell design must consider factors such as membrane thickness, porosity, and surface area to maximize proton conductivity and minimize ohmic losses. Optimization strategies include tailoring the composition and microstructure of electroceramic membranes to enhance proton transport properties. Electrode design optimization is crucial for achieving high catalytic activity and selectivity, ensuring efficient hydrogen and oxygen evolution reactions at lower overpotentials.

Performance Metrics and Challenges

Performance metrics for electroceramic-based electro lysis technologies include hydrogen production rate, energy efficiency, and durability. Challenges in commercializing electroceramic-based electrolysis technologies include high material and manufacturing costs, limited scalability of production processes, and reliability issues. Proton conductivity degradation over time, electrode poisoning, and membrane degradation due to harsh operating conditions are additional challenges that must be addressed to ensure long term performance and viability.

In summary, the successful integration of electroceramic membranes into electrolysis systems requires careful design optimization and consideration of performance metrics and challenges. Addressing these challenges is essential for the commercialization and widespread adoption of electroceramic-based electrolysis technologies for sustainable hydrogen production.

The overview of contribution of electroceramics in various renewable energy technologies is summarized in form of table for comparison of their role that highlights the efficiency, sustainability and reliability in different applications (Table 1).

9 Environmental Benefits of Electroceramics

The primary concern of mankind, nowadays, is environmental sustainability. The main focus of human beings has shifted to contribute in environmental protection by any means ranging from home-gardening to proper disposal of wastes from high-end chemical factories and industries. The chemical industries produce chemicals

Table 1 Compares the role of electroceramics in various renewable energy technologies

Renewable energy technology	Role of electroceramics	Description
Solar photovoltaics	Electroceramics are used in photovoltaic cells to enhance light absorption, improve charge separation, and increase energy conversion efficiency. Materials like perovskite-based ceramics are explored for their potential in next-generation solar cells	Electroceramic materials are crucial for improving the efficiency and performance of solar panels, thus making solar energy more competitive and widely adopted
Fuel cells	Electroceramic materials serve as electrolytes and electrodes in solid oxide fuel cells (SOFCs), proton exchange membrane fuel cells (PEMFCs), and other types of fuel cells. Ceramics like yttria-stabilized zirconia (YSZ) and perovskite oxides are used for their high ionic conductivity and chemical stability at elevated temperatures	Electroceramics are fundamental components in fuel cell systems, facilitating the electrochemical conversion of chemical energy into electrical energy with high efficiency, low emissions, and versatility for various applications, including stationary power generation and transportation
Energy storage (batteries)	Electroceramics are employed in the development of solid-state batteries, where ceramic electrolytes replace liquid electrolytes, offering higher energy density, improved safety, and longer lifespan. Ceramic separators are also used to enhance the performance and safety of lithium-ion batteries	Electroceramics facilitate advancements in battery technology, enabling higher energy storage capacities, faster charging rates, and safer operation, thus supporting the integration of renewable energy sources into the grid and promoting energy storage solutions for intermittent renewables

(continued)

Table 1 (continued)

Renewable energy technology	Role of electroceramics	Description
Capacitors and supercapacitors	Electroceramics serve as essential dielectric materials in capacitors, enabling high energy storage densities and low losses. In supercapacitors, electroceramic electrodes enhance charge storage capacity through high surface area and efficient charge transfer mechanisms, facilitating rapid energy storage and release for diverse applications, from electronics to renewable energy systems	Electroceramics play a vital role in capacitors and supercapacitors by providing high-performance dielectric materials and electrodes. In capacitors, they enable efficient energy storage with low losses, while in supercapacitors, they enhance charge storage capacity and facilitate rapid energy storage and release for various applications
Hydrogen production	Electroceramic membranes are employed in water electrolysis processes for hydrogen production, where ceramic proton-conducting membranes enable efficient proton transport while separating hydrogen and oxygen gases	Electroceramics play a crucial role in advancing hydrogen production technologies, offering high selectivity, durability, and scalability for electrolysis systems powered by renewable energy sources, thereby supporting the transition to a hydrogen economy and enabling sustainable energy storage and distribution

which are used directly or indirectly in our life that affect environmental protection. The sustainability of electroceramics is vastly used for environment protection and has direct applications in conventional products such as culinary designs, ornaments or buildings. The advanced ceramics, developed in recent years, are optimized for creation of new components that can significantly contribute to enhanced energy storage, environmental engineering, reduce air pollution, create green buildings and manage energy production.

The characteristics of electroceramics that make them suitable for environment-friendly technologies are:

1. The electroceramics are stable at higher temperatures even above 1000 °C leading to their applications in solar energy conversion, photovoltaics and various power plant engine systems.
2. They are highly resistant to thermal shock, making them suitable for thermal plants.
3. Electroceramics can resist the oxidizing conditions and withstand the corrosion causing solutions as well as gases which makes them ideal in industries related to water processing and treatment.
4. The ceramics have longer life due to their high wear resistance.
5. The excellent electrical and thermal insulating capacities

The ceramic materials like glass or metals are helpful in protection of the environment and helpful in regeneration of the destroyed ecosystems. The technological uses of these elements for this specified purpose are categorized as environmental engineering. One of the most popular building materials is ceramic tiles. The recycled old ceramics are being used to obtain ceramic tiles leading to safe environment with less dumping.

10 Ceramics in Clean Energy Production

The ceramics play an important role in green energy production technologies that involve nuclear energy, chemical energy and energy harvesting. The nuclear plants make use of ceramic for role of electrolytes and insulators. The thermal coatings of ceramics in gas turbines and the diesel engines enhances their operational efficiency, thus increasing the lifetime of operation and improved fuel efficiency. Moreover, the ceramics are also implemented via different kinds of sensors like gas sensors, water analyzing sensors, H_2 and CO sensors.

11 Conclusions

In the realm of renewable energy technology, electroceramics emerge as pivotal components, offering a diverse array of properties and applications that drive innovation and efficiency. Throughout this chapter, we have explored the multifaceted role of electroceramics in various renewable energy systems, from solar power to fuel cells, batteries, and capacitors.

One of the most significant contributions of electroceramics lies in their ability to enhance the performance and efficiency of renewable energy devices. For instance, in solar panels, electroceramic materials such as perovskites have shown remarkable potential in improving light absorption and charge transport, leading to higher conversion efficiencies. Similarly, in fuel cells, electroceramic electrolytes enable the efficient conversion of chemical energy into electrical energy, powering a range of applications from vehicles to stationary power generation. Moreover, electroceramics play a crucial role in energy storage solutions, addressing the intermittent nature of renewable energy sources. Advanced ceramic-based batteries and capacitors offer high energy densities, fast charging capabilities, and long cycle lifetimes, making them ideal candidates for storing renewable energy and balancing the grid.

As we look to the future, the importance of electroceramics in renewable energy technology is only expected to grow. Continued research and development efforts hold the promise of further optimizing electroceramic materials for enhanced performance and sustainability. Additionally, innovations in manufacturing techniques and scalability will drive down costs, making renewable energy solutions more accessible and widespread. In essence, the integration of electroceramics into renewable energy technologies represents a crucial step towards achieving a cleaner, more sustainable energy future. By leveraging the unique properties of electroceramic materials, we can overcome challenges, increase efficiency, and reduce our reliance on fossil fuels, ultimately mitigating climate change and preserving the planet for future generations.

To summarize, in this chapter, the importance and role of electroceramics in sustainable energy has been discussed. The electroceramics, in modern day technologies, are a crucial part in renewable energy by its implementation in different fields as various roles due to their ability to scale the properties.

References

1. Wang, G., Lu, Z., Li, Y., Li, L., Ji, H., Feteira, A., Zhou, D., Wang, D., Zhang, S., Reaney, I.M.: Electroceramics for high-energy density capacitors: current status and future perspectives. Chem. Rev. Am. Chem. Soc. 6124–6172 (2021). https://doi.org/10.1021/acs.chemrev.0c01264
2. Yao, Z., Song, Z., Hao, H., Yu, Z., Cao, M., Zhang, S., Lanagan, M.T., Liu, H.: Homogeneous/Inhomogeneous-structured dielectrics and their energy-storage performances. Adv. Mater. (2017). Wiley-VCH Verlag. https://doi.org/10.1002/adma.201601727
3. Sherrill, S.A., Banerjee, P., Rubloff, G.W., Lee, S.B.: High to ultra-high power electrical energy storage. Phys. Chem. Chem. Phys. 20714–20723 (2011). https://doi.org/10.1039/c1cp22659b
4. Li, Q., Liu, F., Yang, T., Gadinski, M.R., Zhang, G., Chen, L.Q., Wang, Q.: Sandwich-structured polymer nanocomposites with high energy density and great charge-discharge efficiency at elevated temperatures. Proc. Natl. Acad. Sci. U S A **113**(36), 9995–10000 (2016). https://doi.org/10.1073/pnas.1603792113
5. Jaffe: Antiferroelectric Ceramics with Field-Enforced Tran-Sitions: A New Nonlinear Circuit Element* (1961)
6. Burn, I., Smyth, D.M.: Energy Storage in Ceramic Dielectrics, vol. 7 (1972)
7. Love, G.R.: Energy Storage in Ceramic Dielectrics (1990)
8. Yang, L., Kong, X., Li, F., Hao, H., Cheng, Z., Liu, H., Li, J.F., Zhang, S.: Perovskite lead-free dielectrics for energy storage applications. Prog. Mater. Sci. 72–108 (2019). Elsevier Ltd. https://doi.org/10.1016/j.pmatsci.2018.12.005
9. Fletcher, N.H., Hilton, A.D., Ricketts, B.W.: Optimization of Energy Storage Density in Ceramic Capacitors, vol. 29 (1996). http://iopscience.iop.org/0022-3727/29/1/037
10. Ogihara, H., Randall, C.A., Trolier-Mckinstry, S.: High-energy density capacitors utilizing 0.7 BaTiO3-0.3 BiScO3 ceramics. J. Am. Ceram. Soc. **92**(8), 1719–1724 (2009). https://doi.org/10.1111/j.1551-2916.2009.03104.x
11. Guo, Y., Wu, S., He, Y.B., Kang, F., Chen, L., Li, H., Yang, Q.H.: Solid-state lithium batteries: safety and prospects. eScience 138–163 (2022) . KeAi Communications Co. https://doi.org/10.1016/j.esci.2022.02.008
12. Correa-Baena, J.-P., Saliba, M., Buonassisi, T., Grätzel, M., Abate, A., Tress, W., Hagfeldt, A.: Promises and challenges of perovskite solar cells. https://www.science.org
13. Sathiyamoorthy, S., Bharathi, N.: Hybrid energy harvesting using piezoelectric materials, automatic rotational solar panel, vertical axis wind turbine. Procedia Eng. **38**, 843–852 (2012). Elsevier Ltd. https://doi.org/10.1016/j.proeng.2012.06.106
14. Bharadwaj, S.R., Varma, S., Wani, B.N.: Electroceramics for fuel cells, batteries and sensors. In: Functional Materials: Preparation, Processing and Applications, pp. 639–674. Elsevier (2011). https://doi.org/10.1016/B978-0-12-385142-0.00016-7
15. Jiang, S.P.: Development of lanthanum strontium manganite perovskite cathode materials of solid oxide fuel cells: a review. J. Mater. Sci. **43**(21), 6799–6833 (2008). https://doi.org/10.1007/s10853-008-2966-6

16. Jacobson, A.J., Yu, C., Gong, W.: Thin-film solid oxide fuel cell (SOFC). Materials (2012). https://doi.org/10.1016/9780857096371.3.478
17. Brett, D.J.L., Atkinson, A., Brandon, N.P., Skinner, S.J.: Intermediate temperature solid oxide fuel cells. Chem. Soc. Rev. **37**(8), 1568–1578 (2008). https://doi.org/10.1039/b612060c
18. May, G. J., Davidson, A., Monahov, B.: Lead batteries for utility energy storage: a review. J. Energy Stor. 145–157 (2018). Elsevier Ltd. https://doi.org/10.1016/j.est.2017.11.008
19. Delmas, C.: Sodium and sodium-ion batteries: 50 years of research. Adv. Energy Mater. (2018). Wiley-VCH Verlag. https://doi.org/10.1002/aenm.201703137
20. Dhar, S.K., Ovshinsky, S.R., Gifford, P.R., Corrigan, D.A., Fetcenko, M.A., Venkatesan, S.: Nickel/Metal Hydride Technology for Consumer and Electric Vehicle Batteries-a Review and up-Date, vol. 65 (1997)
21. Bashir, T., Ismail, S.A., Song, Y., Irfan, R.M., Yang, S., Zhou, S., Zhao, J., Gao, L.: A review of the energy storage aspects of chemical elements for lithium-ion based batteries. Energy Mater. **1**(2), 100019 (2022). https://doi.org/10.20517/energymater.2021.20
22. Mahapatra, S. Das; Mohapatra, P.C., Aria, A.I., Christie, G., Mishra, Y.K., Hofmann, S., Thakur, V.K.: Piezoelectric materials for energy harvesting and sensing applications: roadmap for future smart materials. Adv. Sci. (2021). John Wiley and Sons Inc. https://doi.org/10.1002/advs.202100864
23. Alamiery, A.: Advancements in materials for hydrogen production: a review of cutting-edge technologies. ChemPhysMater 64–73 (2024). Elsevier B.V. https://doi.org/10.1016/j.chphma.2023.09.002
24. Robles, J.O., Almaraz, S.D.L., Azzaro-Pantel, C.: Hydrogen supply chain design: key technological components and sustainable assessment. In: Hydrogen Supply Chain: Design, Deployment and Operation, pp. 37–79. Elsevier (2018). https://doi.org/10.1016/B978-0-12-811197-0.00002-6

Structural Perspective on Multifunctional Oxide Materials

Harshit Agarwal

Abstract The global demand for technological advancement frequently involves discovering novel, exotic, multifunctional materials. Due to their broad spectrum of properties, including ferroelectricity, magnetism, magnetoresistance, and superconductivity, complex oxides are one of the most alluring research topics. These functional electrical and magnetic materials are crucial to modern science and technology. For example, ferroelectrics, materials with spontaneous electric polarization that can be changed by an applied electric field, are frequently utilized as tunable capacitors and are used as the basis for ferroelectric random-access memory in computers. On the other hand, ferromagnetic materials are the materials that are typically used for encoding and storing data, such as in hard drives. Multiferroic materials may allow the fabrication of multiple control devices, such as electric fields, electric field-controlled magnetic data storage, and energy storage devices. Rare earth manganites are a perfect example of multiferroic materials with unique structural properties that change their physical properties. Therefore, studies on functional oxide materials and compounds with multiple degrees of complexity are essential. The present article is focused on the structural aspects of these multifunctional oxide materials, which are vital to understanding the correlation to the physical properties of such exotic materials.

Keywords Multifunctional oxide materials · Structural properties · X-ray diffraction · Multiferroics

1 Introduction

Technological advancements in materials science have unanimously dealt with novel multifunctional compounds [1–3]. Complex oxides are quite possibly one of the most appealing research areas because of their amusing range of properties like ferroelectricity [4], magnetism [5], magnetoresistance [6], superconductivity [7], etc. These

H. Agarwal (✉)
Laboratory of Crystallography, University of Bayreuth, 95444 Bayreuth, Germany
e-mail: harshit.physics@gmail.com

U. Kumar (ed.), *Defects Engineering in Electroceramics for Energy Applications*, Engineering Materials, https://doi.org/10.1007/978-981-97-9018-0_10

functional electronic and magnetic materials are vital to cutting-edge technology [8]. In particular, ferroelectric materials are fundamental to ferroelectric random-access memory in computers, exhibit spontaneous electric polarization that may be switched by an applied electric field, and are commonly employed as tunable capacitors [9]. On the other hand, ferromagnetic materials are most commonly employed for data recording and storage, such as in hard drives [10]. The study of complex oxide materials aims at the compounds that display more than one type of ordering functionality, like ferromagnetism, ferroelectricity, ferrotoroidicity, and ferroelasticity [11, 12]. These multifunctional complex oxides cover a wide range of properties from insulating to semiconducting or metallic, exhibiting phenomena such as high-temperature superconductivity, multiferroics, and colossal magnetoresistance due to solid correlation in charge, spin, lattice, and orbitals [13, 14]. Materials in which more than one ferroic order coexists simultaneously are termed multiferroics [15, 16]. Multiferroic materials have the potential to facilitate the creation of various control devices, including energy storage and magnetic data storage devices that are controlled by an electric field. The present chapter follows this path, introducing the structural aspects of new promising oxide materials. These multifunctional oxide materials have become the focus of recent developments in various research fields, such as Magnetism, Electronics, Energy storage applications, etc. Quasicrystalline materials [17, 18], Spinel materials [19], Perovskites [20], Super-alloys, Shape memory Alloys, and High Entropy Alloys are some types of examples of multifunctional materials.

Perovskites are a unique class of materials that exhibit a number of fascinating characteristics, including superconductivity, charge ordering, colossal magnetoresistance, ferroelectricity, and an efficiency in absorption of light that is highly promising for the conversion of solar energy and other technologies [21]. Perovskite originally refers to calcium titanium oxide $CaTiO_3$, discovered by Gustav Rose (1839) who named it for the significant work by Lev Perosvki. Perovskites are compounds with the generic chemical formula ABX_3 that initially were identified by Victor Goldschmidt. Magnetic atoms can be took place in the perovskite structure at both positions A and B. A distinct method for the matching spins to interact allows for the creation of alternative long-range magnetic ordering, including non-collinear structures [22].

Perovskites are typically ceramic materials with a generic formula that combine a metal with a non-metal, such as oxygen. ABO_3, in which transition metal ions occupy the B sites and trivalent or divalent alkali, alkaline, or rare earth metal ions occupy the A sites. Six oxygen atoms in an octahedral coordination are centered around the B cation. The perfect perovskite has a simple cubic symmetric ion configuration. For example—$SrTiO_3$ is an ideal cubic perovskite with a lattice parameter, a = 3.905 Å, and space group *Pm-3m*. The Sr atoms are at the middle of the cube, and the Ti atoms are at the corners. The oxygen is positioned at the centers of the twelve cube edges, resulting in three-dimensional, infinitely-extending corner-shared TiO_6 octahedra strings. The six equal Ti–O bonds at 1.952 Å and 90° angles make up the ideal TiO_6 octahedra. At 2.761 Å, twelve equally spaced oxygen atoms encircle each Sr atom. The explicit formulation of the structural formula includes the coordination number of every ion as well as local symmetry. The ideal perovskite is cubic but can be distorted to rhombohedral, orthorhombic, hexagonal, or monoclinic structures, depending on the tolerance factor. Almost all the perovskite materials have a tolerance factor between 0.75 and 1 [22, 23]. Figure 1 shows the multifunctional properties of perovskite materials.

With weak-ferromagnetic (wFM) components, the majority of perovskites frequently generate antiferromagnetic ordering. Numerous studies on perovskites are motivated by the potential for wFM in insulators in order to find room-temperature multiferroic materials. The ability of perovskites to exhibit spiral magnetic ordering that disrupt inversion symmetry and produce ferroelectricity is another fascinating property of the material [24]. Since this kind of multiferroicity by construction implies strong magnetoelectric linkages, this second process is also gaining a lot of interest. For applications in spintronics and related fields, these properties are particularly attractive. Thus, perovskite oxide materials have a strong correlation between their existing crystal structure and their physical and magnetic properties. Therefore, these materials exhibit the proper knowledge of atomic arrangements to understand the microscopic origin of the physical phenomenon. The present chapter emphasizes the structural aspects of multifunctional oxide materials and their correlation to physical and magnetic properties.

Fig. 1 Perovskite materials as multifunctional materials

2 Multiferroics as Multifunctional Materials

Modern technology explores novel techniques to achieve high downsizing of integrated devices, combining characteristics into multifunctional materials. Multiferroic materials, which combine two or more ferroic orders, provide prospective applications in the developing disciplines of spintronics, optoelectronics, and data storage in this context, in addition to their fascinating physics. Multiferroics express materials with different ferroic ordering or various properties simultaneously in the same phase. From the perspective of multifunctionality, multiferroic materials are the best option. Figure 2 shows the effect of coupling in multiple properties.

The two types of magneto-electric coupling that are discussed in this chapter are direct and indirect coupling, and they both relate to the structural aspect of multifunctional materials. The class of multiferroics known as direct magneto-electric coupling

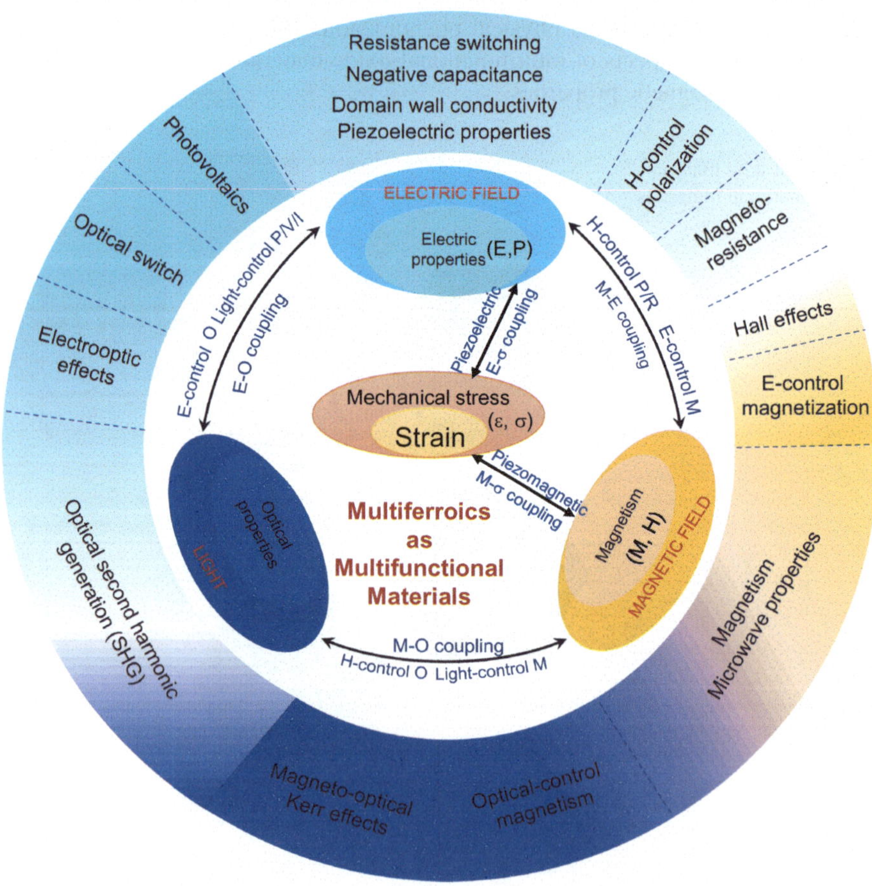

Fig. 2 Coupling in various ferroic ordering that originates multifunctional properties in multiferroic materials

is characterized by the direct application of an electric or magnetic field, which alters the system's electric polarization or magnetization, respectively [25]. This phenomenon is specific to a family of multiferroic materials known as "single-phase" multiferroics. Within elasto-magnetoelectric multiferroics, or composite multiferroics, there is a second class of the magneto-electric coupling phenomenon that is referred to as the indirect magneto-electric effect. When the indirect magneto-electric effect occurs, the electric polarization or magnetization are not directly affected by the application of a magnetic or electric field, respectively. Rather, they undergo indirect modification through a magneto-electric interaction mediated by strain [26].

The term "single-phase multiferroic materials" refers to chemically isotropic homogeneous compounds in which the electric and magnetic order states coexist within the material at a specific temperature or pressure. However, strong symmetry criteria are needed for the occurrence of magnetic and electric order states in the same phase. Naturally, dielectric oxides are solely single-phase multiferroic materials since one of the ferroic components is the electrically ordered phase, requiring for dielectric materials. Single-phase multiferroics are classified into two categories: Type-I and Type-II. Their different physical mechanisms of electric polarization constitute the primary connection between these two types [27].

2.1 *Type-I Single-Phase Multiferroics*

A material in which the electric and magnetic phases coexist simultaneously within the same compound is the first class of single-phasic multiferroics. Nevertheless, they come from distinct microscopic sources and are essentially independent to each other. The spontaneous polarization P in these multiferroic materials is frequently rather large, and ferroelectricity tends to appear at higher temperatures than magnetism (of order 10–100 $\mu C/cm^2$). Examples: $BiFeO_3$ (T_{FE} ~ 1100 K, $T_N = 643$ K, P ~ 90$\mu C/cm^2$) and $YMnO_3$ (T_{FE} ~ 914 K, $T_N = 76$ K, P ~ 6 $\mu C/cm^2$). The type I multiferroics are generally good ferroelectrics. These multiferroics' magnetic and ferroelectric transition temperatures usually happen above room temperature. The ferroelectric and magnetic transition temperatures of these multiferroics typically occur above room temperature. The weak interaction between ferroelectricity and magnetism is a disadvantage of these multiferroics. Depending on the mechanism of ferroelectricity, the following are the different subclasses of type-I multiferroics:

(i) **Multiferroic Perovskites**:

In this class of multiferroic materials, magnetic and ferroelectric perovskites exist, but the coexistence of Magnetism and ferroelectricity seems to disappear. All ferroelectric perovskites have transition metal ions with an empty d-shell, although partially filled d shells of a transition metal are a necessary condition for magnetism. Ferroelectricity in magnetic perovskites is inhibited by the presence of d-electrons on the d^n configuration of magnetic transition metal. The condition is referred as the "d^0 vs d^n problem", which may be overcome by mixing d^0 and d^n ions. The best example of this class type is $BaTiO_3$, where the polarization occurs due to the non-magnetic Ti ion's empty d-orbitals. This causes the Ti to shift in the crystal structure, leading to polarization. This mechanism does not allow for any magnetism in the system, as it only occurs because the d-orbitals are empty. Such types of ferroelectrics are Proper ferroelectrics, where the primary order parameter is the ferroelectric distortion.

(ii) **Lone Pair Multiferroics**:

The formation of ferroelectricity in certain multiferroics, such as $BiFeO_3$, $BiMnO_3$, and $PbVO_3$, is mostly dependent on Bi^{3+} and Pb^{2+}. These ions' two outer 6s2 electrons are not involved in any chemical bonding. They are called lone pairs and lead to high polarizability, which is mandatory for ferroelectricity in the classical analog. One can more microscopically describe the origin of ferroelectricity in these multiferroics by ordering lone pairs (with a certain admixture of p orbitals) in one direction. A single material can have many ferroic orders coexisting in it because of the magnetism caused by B-site transition metals like Fe^{3+} and Mn^{3+}. In the category of proper ferroelectrics, which are also comprised of these materials, lone pair electrons are the source of ferroelectricity.

(iii) **Charge Ordered Multiferroics**:

Charge ordering, which is frequently seen in transition metal compounds, particularly those that officially incorporate transition metal ions with differing valence, is the critical process that might result in ferroelectricity and type-I multiferroicity in the material. These ions form a polar arrangement, which causes improper ferroelectricity (i.e., no ionic displacement). The main order parameter in the case of improper ferroelectricity is not the ferroelectric distortion but rather a different kind of phase change, such as a structural change or magnetic ordering. In a charge-ordered system, the coexistence of inequivalent bonds and multiple valances leads to ferroelectricity. A coexisting magnetic order can be established if the magnetic ions are present, coupled to ferroelectricity. The charges in $LuFe_2O_4$ and $Pr_{0.5}Ca_{0.5}MnO_3$ arrange nonsymmetrically, causing the material to become electrically polarized.

(iv) **Geometric Ferroelectricity**:

In such types of multiferroics, the ferroelectricity originates due to geometrical tilting in the crystal structure. The ferroelectricity in $YMnO_3$ has nothing to do with the magnetic ordering of Mn^{3+} but happens due to the tilting of the practically rigid MnO_5 block. The geometric polarization generation mechanism in $YMnO_3$ describes tilting a severe MnO_5 block with a magnetic Mn remaining at the center. The system becomes ferroelectric as a result of the oxygen ions moving closer to the small Y ions due to this tilting, which also forms Y–O bonds into dipoles.

2.2 *Type-II Single-Phase Multiferroics*

A solid magneto-electric coupling is proposed in Type-II single-phase multiferroic materials, where magnetism induces the ferroelectric order. The polarization of these materials is, however, often significantly less ($10^{-2}\mu C/cm^2$) compared to type I multiferroics. Type-II single-phase multiferroics exhibit ferroelectricity in a magnetically ordered state; in $TbMnO_3$, for instance, magnetic order is present at $T_{N1} = 41$ K and changes to $T_{N2} = 28$ K at a lower temperature.

(i) **Spiral Type II Multiferroics**

This subgroup includes the majority of type II multiferroics that are currently known to exist. Such multiferroics belong to improper ferroelectric multiferroic materials where the magnetic ordering induces electrical polarization. The discovery of this multiferroic class has encouraged the recent revival of research in this field. Such structures are often incommensurate with the underlying crystal lattice, and they present a subclass of spin-density wave structures having different types [28]. They can be sinusoidal, with spins perpendicular or parallel to the wave vector of the spin density wave. Or they can be helicoidal (spiral) of two types: proper screw, with spins rotating in the plane perpendicular to the wave vector, or cycloidal, with the spin rotation plain containing the wave vector. Different types of conical structures may also exist. A magnetic spiral or cycloidal state is one type of magnetic ordering that can lead to an electrical polarization. In such states, the inverse Dzyaloshinskii-Moriya interaction between two neighboring spins can cause a slight displacement to a neighboring O ion, inducing polarization. Ferroelectricity is combined with a spiraling magnetic phase, and most of the cycloid type is observed in $MnWO_{4,}$ $Ni_3V_2O_{6,}$ and $TbMnO_3$ [29]. In the prominent example of spiral type multiferroic of $TbMnO_3$, where all spin points in one direction below $T_{N1} = 41$ K, the magnetic structure is a sinusoidal spin density wave while the size of the local moment fluctuates periodically in space. The Mn spins sweep out a cycloid magnetic ordering $T_{N2} = 28$ K [30]. Magnetic frustration is the origin of spiral magnetic ordering in such insulators. As a result, frustrated systems typically exhibit type-II multiferroicity.

(ii) **Type II Multiferroics with Collinear Magnetic Structure**

In the second class of magnetically driven ferroelectric materials, the ferroelectricity is observed in collinear magnetic structures. All of the magnetic moments in these materials aligned along a specific axis without the spin–orbit interaction being necessary. Due to exchange striction, these materials may exhibit polarization because the magnetic coupling fluctuates with the locations of the atoms. In complicated magnetic structures, the exchange interaction can cause a displacement of the magnetic ions if it reduces the system's energy. This displacement of ions in a non-symmetric way leads to electrical polarization in the materials [27].

3 Rare Earth Manganites

Multiferroic materials are the critical candidate of material science for developing novel devices and technologies using their coupled Magnetism and ferroelectric properties. Rare-earth manganites ($RMnO_3$) are a promising example of multiferroic oxide materials where the manganese ions are linked to oxygen ions with rare-earth (R = La − Lu)/alkali earth ions (AE = Sr, Ca, Ba) sitting in the gap of Mn and O ions having the $RMnO_3$ configuration. Several emergent phenomena in these manganites, including the multiferroicity and colossal magnetoresistance, arise due to a strong correlation between spin, charge, and orbital degree of freedom. The enormous magnetoresistance (CMR) behavior was first observed in doped $RMnO_3$ [La–Ca–Mn–O films at $T_N = 77$ in an applied magnetic field of 5 Tesla]. Many systems, including ferromagnetic metals and superlattices of magnetic and non-magnetic layers, have been shown to exhibit magnetoresistance (MR), which is the relative change in electrical resistivity with an externally applied magnetic field. However, the most considerable MR effect is observed in doped rare-earth manganites $R_{1-x}A_xMnO_3$ ($A = Ca^{2+}$, Sr^{2+}). A thousand-fold change in resistivity refers to 'colossal magnetoresistance.' The initial CMR was reported in a single-crystalline thin film of $La_{0.67}Ca_{0.33}MnO_3$ near the ferromagnetic transition temperature. Moreover, a sharp increase in MR at low applied fields was observed in polycrystalline $La_{0.67}Sr_{0.33}MnO_3$ [13].

Manganites represent several interesting phenomena like Jahn–teller distortion, super-exchange, double-exchange mechanism, half metallicity, etc. Technologically important discoveries like giant magnetoresistance, giant magneto-capacitance, and geometric ferroelectricity were observed in manganites. Generally, $RMnO_3$ possesses orthorhombic perovskite structures for larger rare-earth ions (R = La-Dy) with Pnma symmetry and a non-centrosymmetric hexagonal structure for the smaller rare-earth ions (R = Ho-Lu, Y) having the space group $P6_3cm$. The complex crystal structure and its unrevealed features make these classes of materials essential to explore. Studying the structural behavior in Rare earth substituted manganites at low temperatures with a magnetic field is still an exciting field of research.

4 Properties of Rare Earth Manganites

4.1 Crystal Field Theory

When an atom or ion is immersed in a solid, the electric field originating from the nearby atoms is referred to as the crystal field, and the Coulomb interaction of the atom's electronic charge distribution with the surrounding charges is known as the crystal field interaction. The degeneracies of electron orbital states, typically d and f-orbitals, break due to the static electric field created by a neighboring charge

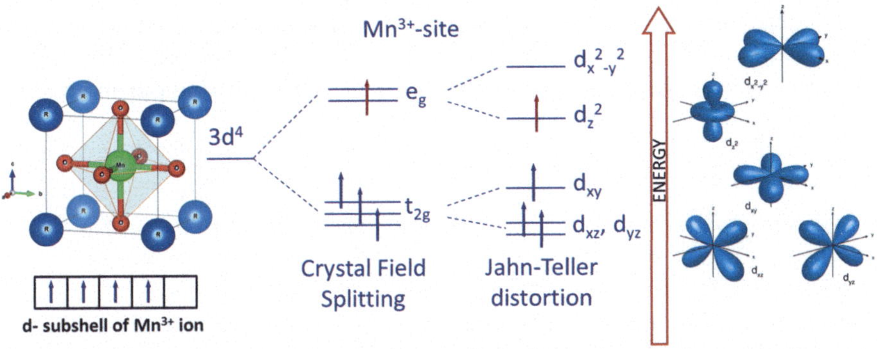

Fig. 3 Schematic of crystal field splitting and Jahn Teller distortion

distribution. This phenomenon is explained by the Crystal field theory, which was discovered in 1930 by Hans Bethe and J. H. Van Vleck [31] (Fig. 3).

An interaction between a transition metal and surrounding ions, also known as ligands, arises from the attraction between the positively charged metal cation and the negatively charged electrons of the ligand. Metals are considered electron acceptors, and these ligands are considered electron donors. The theory is established by considering energy changes of five degenerate d-orbitals. The electrons in d-orbitals and those in the ligand repel each other due to repulsion between like charges. As a result, the d-orbitals will split in energy because the d-electron locations closer to the ligands will have more energy than those farther away. The following factors exaggerate the crystal field splitting:

- The nature of the metal ion.
- The oxidation state of metal. A higher oxidation state leads to a more extensive splitting.
- The arrangement of the ligands around the metal ion.
- The nature of the ligands surrounding the metal ion.

To explain bonding, orbital organization, and other properties of coordination complexes, the "Ligand Field Theory" (LFT) is followed by the "Crystal Field Theory" (CFT). It is an example of how transition metal complexes can be studied using molecular orbital theory. Though most explanations start with describing octahedral complexes, where six ligands coordinate to metal, the LFT analysis is extremely dependent on the complex's geometry. This theory explains the loss of degeneracy of metal d-orbitals in transition metal complexes. Let us examine the scenario where six ligands form an octahedral complex. In this example, the metal ion is at the origin and the ligands are symmetrically positioned along the axes of a Cartesian coordinate system. The d-orbital breaks into sets of 3 low-energy orbitals in the octahedral environment, where six ligands create an octahedron surrounding the metal ion: d_{xy}, d_{xz}, d_{yz} and 2 orbitals of high energy: d_{z^2}, $d_{x^2-y^2}$. The orbitals lying along the axes d_{z^2} and $d_{x^2-y^2}$ will be more strongly repelled than the orbitals with lobes directed between the axes (d_{xy}, d_{xz}, and d_{yz}). The d-orbitals are thus split into two sets with the d_{z^2} and $d_{x^2-y^2}$ at higher energy than the other three. The energy difference between these 2 sets of d-orbitals is called **the splitting parameter** Δ_o. T field strength of the ligand determines the magnitude of Δ_o. We refer to the higher energy orbitals as e_g and the three lower energy orbitals as t_{2g} together. The $d_z{}^2$ is the combination of $d_z{}^2{}_{-y}{}^2$ and $d_z{}^2{}_{-x}{}^2$. It can also be written as $d_{2z^2-x^2-y^2} = d_{2z^2-(x^2+y^2)}$

$$\text{As } r^2 = x^2 + y^2 + z^2, d_{2z^2-x^2-y^2} = d_{2z^2-(r^2+z^2)} = d_{3z^2-r^2}$$

In dz^2, there is no independent existence, it has the average properties of the two $d_{z^2-y^2}$ and $d_{z^2-x^2}$. Both have high electron density along the z-axis; the subscript orbital has a significant fraction of the electron density concentrated along the same axis. Since one of the component wave function $d_{z^2-x^2}$ has lobes along the x-axis and the other $d_{z^2-y^2}$ along the y axis, the resultant d_{z^2} orbital has a torus of electron density in the x–y plane [32].

Crystal Field Stabilization Energy:

The stability that arises from adding a transition metal ion to the crystal field created by a group of ligands is known as the crystal field stabilization energy. It is caused by the fact that some of the split d-orbitals in a ligand field have lower energies than before with respect to the barycenter, a spherical field where all five d-orbitals are degenerate. If the splitting energy of the d-orbitals in an octahedral environment is Δ_o, the three t_{2g} orbitals are stabilized relative to the barycenter by 2/5 Δ_o, and the e_g orbitals are destabilized by 3/5 Δ_o.

4.2 Jahn–Teller Distortion

The Jahn Teller effect occurs in materials with a partial electronic occupation in t_{2g} or e_g symmetric classes. In order to explain the instability of a nonlinear system in a degenerate energy state, Herman Jahn and Edward Teller first postulated this phenomenon in 1937. They suggested that in an electronically degenerate energy state, a nonlinear system undergoes distortion that eliminates the degeneracy by reducing symmetry and, thus, energy. It is claimed that in the electrically degenerate state, the orbitals are asymmetrically occupied and acquire additional energy. Therefore, by lowering the molecule's overall symmetry—a process known as **Jahn–Teller distortion**—the system attempts to eliminate this extra energy. The existence of many degenerate orbitals for an electron is represented by an electrically degenerate state. The degenerate orbitals in this scenario are occupied asymmetrically. For example, in an octahedral symmetry, when three t_{2g} orbitals of the same energy are available to occupy the e^-, the d^1 configuration is considered to be an electronically degenerate state. In contrast, the d^3 configuration in octahedral symmetry is non-degenerate and symmetric. Putting two e^- in one orbital is impossible, which is against Hund's rule of maximum multiplicity [33].

The majority of cases of the Jahn–Teller distortion are found in octahedral settings. Theoretically, all the configurations except d^3, d^8, d^{10}, and low spin d^6 configurations are possible for exhibiting electronic degeneracy in octahedral symmetry. However, considerable distortion is usually observed in high spin d^4, low spin d^7, and d^9 configurations in the octahedral environment. The reason for this is that the energy gain is significantly higher and the Jahn Teller distortion is usually large for asymmetrically occupied e_g orbitals, which are directed towards the ligands.

Consider an Mn^{3+} ion with a d^4 configuration. After the field splitting, 3 electrons of Mn^{3+} occupy the t_{2g} set and one electron in the e_g set, which makes the Mn^{3+} ion Jahn Teller active. Two e_g orbitals split, raising the energy of one at the expense of the other termed as Jahn Teller distortion, followed by the octahedron distortion along the z-axis. The octahedron elongates along the z-axis and compresses along the x and y directions. The elongation and compression profoundly affect the energy states of the d orbitals—$d_z{}^2$ will be lowered in energy than $d_x{}^2{}_{-y}{}^2$ orbital; similarly, the d_{xz}, d_{yz} orbitals will have lower energy than d_{xy}. At this point, two types of energy will be taken into consideration: the electrical energy and the lattice strain energy caused by distortion. The single electron of the e_g set occupies the $d_z{}^2$ level, which lowers the electronic energy, outweighing the lattice strain energy thereby reducing the energy of the material structure. Due to the direct impact of lattice distortion on the electronic configuration and atomic arrangement, this distortion can have an impact on the material's properties, including electrical conductivity and magnetism.

The degeneracy of orbitals can be eliminated by lowering the symmetry of the molecule. This can be achieved by either elongation of bonds along the z-axis (z-out distortion) or by compression (shortening) of the bonds along the z-axis (z-in distortion). In the case of Z-out distortion, the energies of d-orbitals with z-factor (i.e., $d_z{}^2$, d_{xz}, d_{yz}) are lowered since the bonds along the z-axis are elongated. This is the most preferred distortion and occurs in most cases, mainly when the degeneracy occurs at the e_g level. Usually, the octahedral d^2, d^4 high spin, d^7 low spin, d^8 low spin, and d^9 configuration show the z-out distortion. In the case of Z—in distortion, the energies of orbitals with z-factor are increased since the bonds along the z-axis are shortened. This type of distortion is observed in the case of the octahedral d^1 configuration. The only electron will now occupy the d_{xy} orbital with lower energy.

4.3 Spin Interaction and Exchange Mechanism

The interactions between magnetic moments produce magnetic order in a material. The interactions between neighboring spins S_i and S_j are described by the **Heisenberg model**, with the exchange Hamiltonian given by [34]

$$\widehat{H}_J = -2\sum_{i<j} J_{ij}\widehat{S}_i.\widehat{S}_j$$

The value J_{ij} is called the exchange constant and describes the nature of the interaction between S_i and S_j. If J_{ij} is positive, neighboring spins favor a parallel alignment, and the system is ferromagnetic. If J_{ij} is negative, neighboring spins prefer an anti-parallel alignment, and the system is antiferromagnetic. Let the spin interact with external field B. If the field is inhomogeneous, the force will act on the spin, and the effect of the field is to align the spin. The energy of these spins inside the field is—μ.B. So, the interaction between the magnetic field and spin is generally called **Zeeman Interaction**. Spins also interact among themselves, and the interaction between these spins can be dipolar or exchange **interaction**. Exchange interaction could be of different types– Direct exchange interaction, indirect exchange interaction, and hyperfine interaction. This exchange interaction is essential as it gives rise to ferromagnetism, antiferromagnetism, and ferrimagnetism. When the spins interact with the lattice, the interaction is called Spin–orbit interaction. Now, all three interactions (Zeeman, Exchange, Spin–Orbit) can interact with each other. There is also an interaction between spin and phonon, which gives rise to **Magneto-optic effects**. As lattices can oscillate in a quantized fashion, these quantized oscillations are called phonons. Spin can also interact with phonons, giving rise to **magneto-acoustic impacts**. Exchange interaction results from electrostatic Coulombic attraction/repulsion and symmetrization postulates related to long-range magnetic ordering. There could be different types of flavors of the exchange interaction, which are as follows [35].

4.3.1 Direct Exchange Interaction

If the electron on neighboring magnetic atoms interacts directly with an exchange interaction without an intermediary electron, the interaction is known as a direct exchange. Direct exchange cannot be an essential mechanism in controlling magnetic properties because there is insufficient natural overlap between neighboring magnetic orbitals. This is a very short-range effect.

4.3.2 Indirect Exchange Interaction

Double Exchange Mechanism

The double exchange mechanism, which promotes ferromagnetic ordering, typically occurs between ions in various oxidation states. In 1951, C. Zener explained why the best ferromagnetic material displays the lowest resistivity in his theory of indirect magnetic exchange between 3d atoms by introducing the concept of a double exchange mechanism (DE) [36]. This theory was applied to the perovskite manganites in order to explain the value of the zero-temperature saturation magnetization, which is equivalent to the sum of all the unpaired electron spins, and the significant correlation between conductivity and ferromagnetism. Double exchange theory was utilized to account for ferromagnetic ordering when doped manganites were produced.

Two Mn ions with different valence states are coupled by oxygen, resulting in this magnetic ordering. Consider Mn^{3+} and Mn^{4+} ions. An electron transfer from Mn^{3+} to Mn^{4+} occurs due to a simultaneous double transfer of an electron from Mn^{3+} to O^{2-} and then from O^{2-} to Mn^{4+}. If the magnetic moments of the two Mn ion t_{2g} states are parallel, then such an electron transfer can occur; if not, Hund's rule would be broken. Hund's rule demands the occupancy of electrons with the same spins in the d orbitals of Mn^{3+} and Mn^{4+} ions.

The process known as the double exchange produces a ferromagnetic ordering between Mn ions by lowering the energy state as a result of the double transfer of electrons. Both ferromagnetism and metallicity are explained by spin-dependent electron transfer between Mn ions with parallel spin alignment in a double-exchange mechanism [35].

According to the double exchange theory, electrons will move between species more quickly when they are not required to change their spin direction in order to comply with Hund's rule when they adopt a new species. The kinetic energy is decreased by the capacity to localize or hop. Therefore, the ferromagnetic alignment of the nearby transition metal ions may result from the overall energy savings.

Super-Exchange Interaction

If the spins or magnetic moment are not directly overlapping but in between, they have some non-magnetic ion (as O^{2-}), and exchange interaction is mediated through these non-magnetic ions (which do not have magnetic dipole moment). Such a type of interaction is called **Super exchange interaction**. When two next-to-nearest neighbor cations coupled strongly through a non-magnetic anion, it is known as super-exchange and represents the exchange mechanism of the insulator. An anti-ferromagnetic ordering between the identical valence states of Mn ions is usually triggered by the super-exchange mechanism. For two Mn^{2+} ions, it is energetically favorable for the valence electrons on the Mn^{2+} and O^{2-} ions to undergo covalent bonding. The p orbital of the O^{2-} ion has a filled shell of electrons, and for the Mn–O–Mn hybridization to occur, a transfer of electron with a particular spin from the O^{2-} ion onto the Mn^{2+} ion is required, followed by the transferred electron of opposite spin to other Mn^{2+} ions. Such a process results in an antiferromagnetic alignment between the Mn^{2+} ions. The sign of the exchange integral guides a superexchange mechanism, J_{ij} in the Heisenberg Hamiltonian, which depends on the oxygen–metal bond angle and the d electron configuration on the transition metal. Goodenough-Kanamori-Anderson (GKA) rules and presented in Fig. 4 which state that:

1. If both neighbouring Mn ions point their empty orbitals (hybridized lattice orbital) towards O^{2-} then the Mn–Mn separation is small, and coupling between Mn ions is antiferromagnetically coupled.

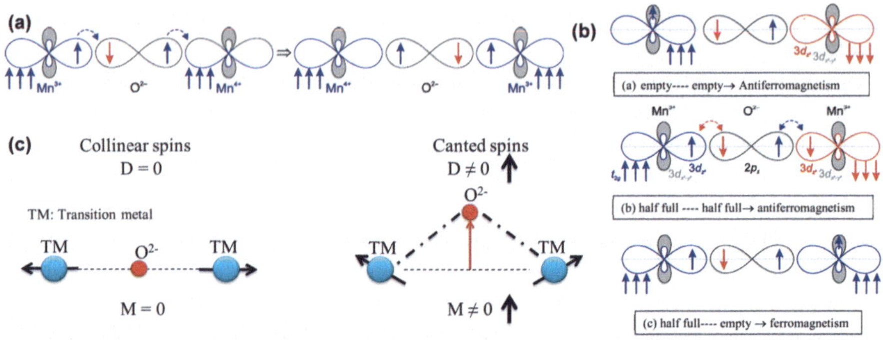

Fig. 4 Interaction in manganites **a** double exchange, **b** superexchange, **c** D-M interaction

2. If both neighbouring Mn^{3+} ions point their filled atomic orbitals towards O^{2-} then the Mn–O–Mn separation is largest, and the exchange interaction is antiferromagnetic.
3. If one of the two neighbouring Mn^{3+} ions point its occupied orbital (atomic orbit) towards O^{2-} and the other points its unoccupied orbital, then the Mn–O–Mn bond length will be large, and the two manganese ions will be Ferromagnetically coupled.

According to Goodenough-Kanamori's (GK) rules, it is generally concluded that 180° metal–oxygen–metal angles between identical metal ions with both orbitals, either filled or empty, lead to antiferromagnetic interactions. In contrast, 90° angles lead to ferromagnetism. Rare-earth manganites having the $R^{3+}Mn^{3+}O^{2-}{}_3$ configuration are insulators in nature where the exchange interaction in Mn–O–Mn occurs due to super-exchange interaction [37, 38].

Anisotropic Exchange Interaction (Dzyaloshinsky-Moria Interaction)

Dzyaloshinsky-Moria (D-M) Interaction is a process similar to superexchange where the intermediate process occurs via spin–orbit interaction rather than an Oxygen ion. Anisotropic exchange interactions, sometimes referred to as D-M interactions, are exchange interactions that occur between the excited state of a magnetic ion and the ground state of the next ion. In 1960, the concept was developed by Dzyaloshinskii to explain weak ferromagnetism. He added an asymmetrical term based on the symmetries, and Moria discovered that the spin–orbit coupling plays a role in the process underlying this interaction. DM interaction is induced by a lack of inversion symmetry of the compound and a robust spin–orbit coupling. When acting between two spins S_1 and S_2, it leads to a term in the Hamiltonian H_{DM} equal to,

$$\widehat{H}_{DM} = \vec{D}.\left(\vec{S}_1 \times \vec{S}_2\right), \text{ where } \vec{D} \sim \lambda(\vec{a} \times \vec{r}_{12})$$

The D-M vector $\overrightarrow{D}$ is proportional to the spin–orbit coupling constant and depends upon the position of the oxygen ion 'a' between two magnetic transition metal ions [34].

Two atomic spins interacting with a nearby atom that has a high spin–orbit coupling give rise to D-M interaction. This interaction tends to align S_1 and S_2 perpendicular against $\overrightarrow{D}$ and against each other which results in the small canting of magnetic moments. It causes an antiferromagnet's net moment to become weak. Therefore, the spin–orbit interaction in one of the spins S_1 and S_2 may result in weak anisotropy in the exchange in the structures where the line connecting their magnetic ions lacks inversion symmetry. This interaction is known as the D-M contribution. The direction of canting of spins is determined by the sign of $\overrightarrow{D}$ [39].

RKKY Interaction (Indirect Exchange in Metals)

In metals, conduction electrons, also known as itinerant electrons, can mediate the exchange interaction between magnetic ions. The spin of a localized magnetic moment polarizes the conduction electrons, and this polarization relates to a nearby localized magnetic moment that is *r* distant. Since there is no direct coupling between magnetic moments, the exchange contact is therefore indirect. It's referred to as **RKKY interaction** (or itinerant exchange interaction). The initial letters of the surnames of the people who discovered the effect—Roderman, Kittel, Kasuya, and Yosida—led to the creation of the given name RKKY. The coupling takes the form of an r-dependent exchange interaction J_{RKKY} (r) given by $J_{RKKY}(r) \propto \{\cos(2k_F r)\}/r^3$ at large r (assuming a spherical Fermi surface at radius k_F). There is no consistent dependence of the long-range interaction on the separation of the magnetic moments. Thus, it can be either ferromagnetic or antiferromagnetic depending on the separation.

4.4 Charge Ordering

Transition metal oxides and organic solids are strongly correlated materials where charge ordering (CO) transitions are primarily observed. When electron-to-electron interaction precedes the hopping term, the CO phase transition occurs. This results in creating an ordered superlattice and charge imbalance at various places. Symmetry breaking occurs together with the CO transition, and in certain situations, it stabilizes the ferroelectricity in the systems. The CO was usually driven by electron–electron and electron-lattice correlation effects. In the case of the CO manganites, due to the JT-active nature of the Mn^{3+} ions, the CO is accompanied by the distortion of the oxygen octahedron surrounding the Mn^{3+}. The tendency of this structural distortion leads to the so-called cooperative JT distortions. This will reduce the crystal symmetry. Thus, the orbital order gives rise to the anisotropy of the electron-transfer interaction. The phenomenon of CO is extended to explain the colossal magnetoresistance behavior in various manganites. The close interplay between manganite structural, magnetic, and electronic states makes them suitable for studying multiple physical phenomena, including colossal magnetoresistance, formation of CO patterns, and magneto-structural transitions. The magnetic field that induces melting of the CO state leads to a considerable resistance drop with a magnetic field.

5 Crystallographic Aspects of Rare Earth Manganites

The structure of $RMnO_3$ is almost a cubic perovskite consists of R ion at (0,0,0), Mn ion at (1/2,1/2,1/2), and three O ions at (0,1/2,1/2), (1/2,0,1/2), and (1/2,1/2,0). The structure is stabilized by the 12-fold and 6-fold oxygen coordination of the A and Mn ions, respectively [40]. R is of fixed valence +3, and O is of valence 2^-. So, the charge neutrality condition forces Mn to be in a +3-valence state. When an alkaline earth metal ion (AE = Ca, Sr, Ba) in a +2-valence state is doped in A site of $RMnO_3$, the charge neutrality condition leads to a mixed valent state of Mn, i.e. an ionic configuration is of the form $R^{3+}{}_{1-x}AE^{2+}{}_{x}Mn^{3+}{}_{1-x}Mn^{4+}{}_{x}\,O^{2-}{}_{3}$.

Another way to see the structure is as a stacking of two planes, the R-O and Mn-O_2, on top of each other. The manganite structure is determined by the tolerance factor "t," which is influenced by the ionic radii of R and doped AE. When the tolerance factor is near to unity, the cubic system is stable; any significant deviation from unity produces completely distinct crystal structures, such as hexagonal, orthorhombic, etc. Variation of the ionic radii at A sites of $RMnO_3$ can result in buckling or distortion of MnO_6 octahedra from cubic phase to several lower symmetry structures having reduced coordination numbers for R and Mn ions. Magnetic ordering in these materials arises from the spin ordering of adjacent Mn ions mediated by the hopping of electrons from Mn^{+3} to Mn^{+4} ions. Doped manganites can be further categorized as having a large bandwidth, intermediate bandwidth, and lower bandwidth manganites based on the electron hopping amplitude, also known as bandwidth.

In most perovskites, ferroelectricity reflects the property of spontaneous electric polarization which can be reversed by applying an external electric field [16]. Ferroelectric materials have a wide range of technical uses due to their ferroelectric phenomena as well as piezoelectric, pyroelectric, and electro-optic properties. Examine the traditional perovskite structure identified by ABO_3, which consists of an octahedron of negatively charged oxygen ions surrounding a central positive B-ion, a transition metal element. Ferroelectric order would arise from a shift in the B-site ion location, which would break the inversion symmetry and induce a dipole moment. When a system undergoes a structural phase transition—a transition from a high to a low symmetry state, for example, from cubic to tetragonal—such modifications may take place. $BaTiO_3$ is among the most widely recognized examples of a proper ferroelectric. The majority of ferroelectric perovskites feature an empty d electron shell on the B-site atom, which allows covalent bonding with the *p* orbitals of the oxygen atoms. Ferroelectricity can also occur due to lone pairs of electrons on the outer shell of the A-site atom, which is highly susceptible to polarization. Thus, the crystal must not possess the center of symmetry for being a ferroelectric material. Their lattice structure governs the symmetry of the crystals. Despite the fact that nature contains thousands of crystals, the symmetry elements allow for the grouping of all crystals into 230 space groups. The 32 crystallographic point groups with 14 Bravais lattices that correspond to one of the seven crystal systems combine to form the crystallographic space groups.

Based on the existence of an inversion center, the 32-point groups are divided into two categories: (a) crystals having a center of symmetry and (b) crystals that do not possess a center of symmetry. Crystals with the center of symmetry (labeled centrosymmetric) include 11-point groups and do not show polarity. The remaining 21-point groups do not have a center of symmetry (i.e., non-centrosymmetric). All non-centrosymmetric point groups, except point group 432, show the piezoelectric effect along the unique directional axes. A crystal having no center of symmetry possesses one or more crystallographically individual directional axes.

The non-centrosymmetric space groups are further classified into polar and chiral types due to the lack an inversion center. Without any roto-inversion symmetry components, a space group is said to be chiral. Roto-inversion is a rotation followed by inversion; for example, a mirror reflection corresponds to a two-fold roto-inversion. Chiral space groups must, therefore, only contain (purely) rotational and translational symmetry. These arise from the crystal point groups 1, 2, 3, 4, 6, 222, 422, 622, 32, 23, and 432. Chiral molecules, such as proteins, crystallize in chiral space groups. The term ‘polar’ is often used for space groups that are neither centrosymmetric nor chiral. However, the time ‘polar’ is more correctly used for any space group containing a unique anisotropic axis. These occur in point groups 1, 2, 3, 4, 6, m, mm2, 3 m, 4 mm, and 6 mm. Thus, some chiral space groups are also polar. Out of the twenty-point groups that show the piezoelectric effect, ten polar point groups have only one unique direction axis, and such crystals show spontaneous polarization. Thus, considering symmetry restrictions, all ferroelectric materials are pyroelectric. However, not all pyroelectric materials are ferroelectric. Since all pyroelectric materials are piezoelectric, ferroelectric materials are inherently piezoelectric [41].

The primary interaction in orthorhombic perovskites ABO3 compounds with *Pbmn* symmetry is the isotropic exchange interaction between nearest-neighboring spins. At this level of interaction, four types of collinear ordering can arise depending on the relative sign of these interactions because the crystal unit cell contains four magnetic B atoms. Following are the magnetic structures having collinear magnetic ordering (also shown in Fig. 5a–d).

1. **F-type**, with all the spins pointing in the same direction (FM ordering), $F = S_1 + S_2 + S_3 + S_4$,
2. **A-type**, with spins pointing in opposite directions in consecutive planes (AFM order of FM planes), $A = S_1 - S_2 - S_3 + S_4$,
3. **C-type**, with spins pointing in opposite directions in consecutive lines (AFM order of FM chains), $C = S_1 + S_2 - S_3 - S_4$,

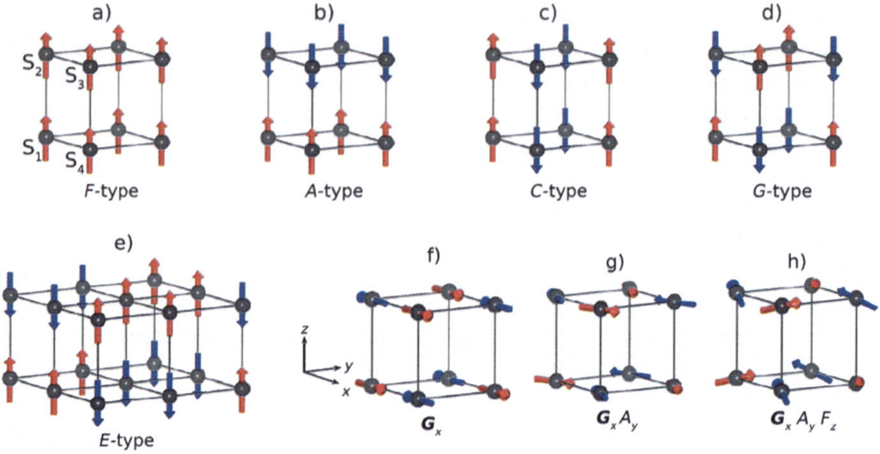

Fig. 5 Schematic view of **a** *F*-type, **b**, *A*-type, **c** *C*-type, **d** *G*-type, and **e** *E*-type collinear magnetic orders that can be constructed in cubic perovskites ABX_3 (for the sake of illustration, spins are directed along the *z*-axis). **f**, **g** and **h** show the G_x, G_xA_y, and $G_xA_yF_z$ non-collinear orders [42]

4. **G-type**, with nearest–neighboring spins pointing in opposite directions ('full' AFM ordering). $G = S_1 - S_2 + S_3 - S_4$.

In terms of the cubic lattice with one spin per unit cell, these orders are associated with the propagation vectors q = (0, 0, 0), (0, 0, 1/2), (0, 1/2, 1/2), and (1/2, 1/2, 1/2) respectively. The notation proposed by "Bertaut" is widely used to characterize the magnetic structure of perovskites. [43]. The magnetic structure is designated in Bertaut's notation by indicating the kind of order seen along each principal direction as $M_xM'_yM''_z$. As an example, the representation of a basic G-type structure with spins parallel to the x-direction is G_x (see Fig. 5f). If there is an additional A-type component along the y-direction in this arrangement, the structure is G_xA_y (see Fig. 5g). Moreover, if there is an additional F-type component along z, the entire structure is indicated as $G_xA_yF_z$ (see Fig. 5h). E-type AFM ordering is another kind of collinear ordering that is expected to cause an electric polarization that breaks inversion symmetry [42].

Further to these orders, the perovskite structure can hold spiral magnetism as well. In particular, this is observed in the case of orthorhombic manganites ($RMnO_3$), where isotropic exchange interactions compete with one another to produce this result. Rare-earth manganites have non-collinear magnetic ordering that disrupt the inversion symmetry and cause an electric polarization, both of which are necessary for the material's multiferroicity.

Rietveld Refinement Analysis for Magnetic Structure:

Neutron diffraction is the determination of the magnetic structure that determines the directions and absolute magnitude of atomic moments in a magnetically ordered material. Neutron diffraction patterns consist of both nuclear and magnetic reflection.

To describe the magnetic structures, it is required to relate them to the underlying crystal structures with a formalism of magnetic ordering propagation vectors. In the present thesis, the magnetic structure refinement of rare earth manganites has been done using the Fullprof software, for which the propagation vector has been observed using the k-search program. The irreducible representation for a particular space group can be calculated using the BASIREPS [45] program. The magnetic structure is plotted in FP studio and VESTA [46].

It is always important to observe the magnetic phase transition temperature of the material using magnetization studies before collecting neutron powder diffraction data (NPD). The first experimental study of neutron powder diffraction was done by Shull, who observed the NPD pattern at room temperature (293 K) and at 80 K, which is well below the magnetic ordering temperature of MnO (Tc = 120 K) [44]. At lower temperatures, extra antiferromagnetic reflections were observed, which showed that the ordered magnetic cell was more significant than the chemical cell shown in Fig. 6.

Suppose new magnetic reflections are generated in the neutron diffraction pattern below the magnetic ordering temperature. In that case, it gives the signature of antiferromagnetic ordering in the material. If there is an enhancement in the intensity of reflections, it provides the signature of the material's ferromagnetic or ferrimagnetic type of order. Different magnetic structures for different propagation vector can be seen in Fig. 7.

Neutron powder diffraction patterns and the observation of magnetic structure can be done using JANA2020 [47] and FullProf software. The neutron diffraction pattern in Fullprof is based on group theory, including the irreducible representation for a particular space group. However, in JANA2020, the magnetic structure is refined using the magnetic superspace group approach. The irreps approach in Fullprof lacks the assignment of a space (or superspace) group symmetry and the

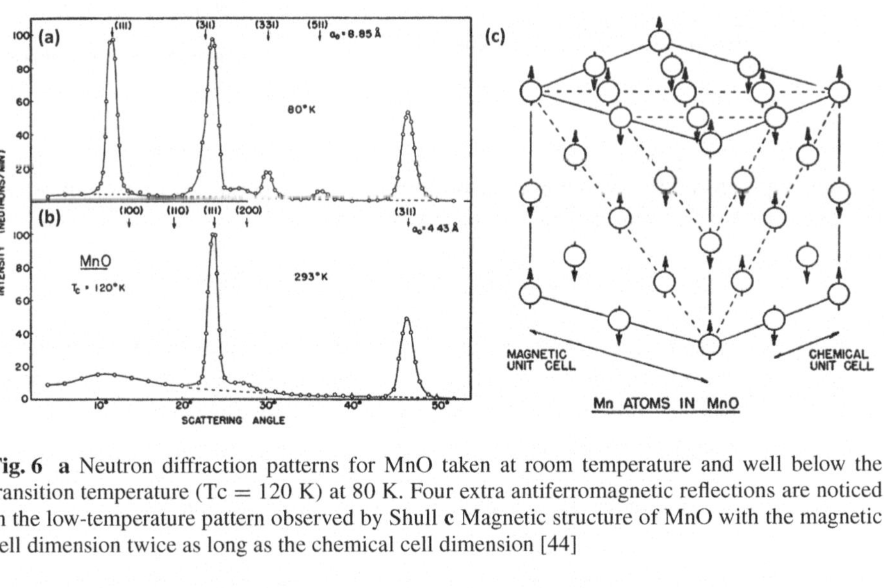

Fig. 6 **a** Neutron diffraction patterns for MnO taken at room temperature and well below the transition temperature (Tc = 120 K) at 80 K. Four extra antiferromagnetic reflections are noticed in the low-temperature pattern observed by Shull **c** Magnetic structure of MnO with the magnetic cell dimension twice as long as the chemical cell dimension [44]

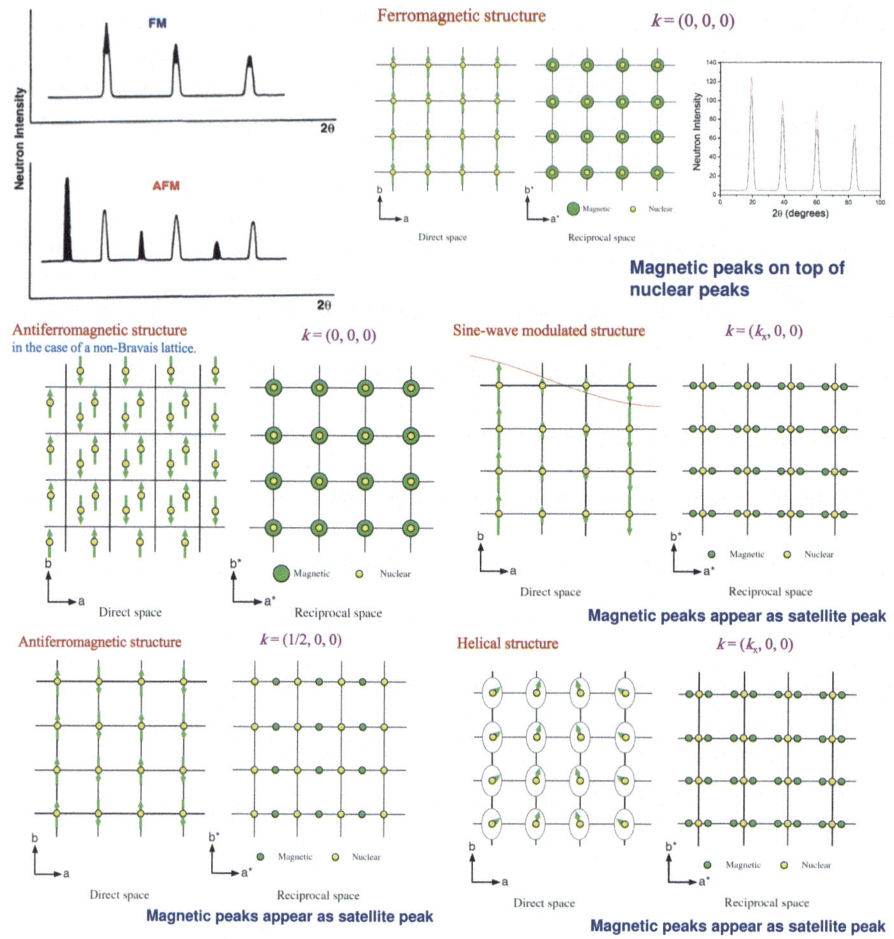

Fig. 7 Different magnetic structures for different propagation vector

point group symmetry for incommensurate magnetic structures. For example, the low-temperature phase of $TbMnO_3$ can be written as,

$$Pbnm1' \rightarrow^{(T_N)} Pbnm1'(0\beta 0)s00s \rightarrow^{(T_C)} Pbn2_1 1'(0\beta 0)s00s$$

The super-space concept is just a mathematical construct. It allows us to interpret the equations characterizing incommensurate structures and their symmetry as analogous to ordinary crystallography for a structure with lattice periodicity but in a (3 + 1)-dimension mathematical space.

Table 1 shows the magnetic super-space group of different magnetic materials with incommensurate magnetic k-vectors [48]. Representation analysis using group

Table 1 Magnetic superspace group of different multiferroic materials with incommensurate magnetic k-vectors

Material	Parent S.G.	k-vector	Magnetic superspace group	Magnetic point group
$TbMnO_3$	*Pbnm*	(0, 0.27, 0)	*Pbn2₁1' (0β0)s00s*	*mm21'*
$MnWO_4$	*P2/c*	(−0.214, 0, 0.457)	*X21' (α0γ)0 s*	*21'*
$SrFeO_3$	*Pm-3m*	(0.129, 0.129, 0.129)	*R321' (00γ)t0s*	*321'*
$CaCr_2O_4$	*Pbnm*	(0, 0, 0.477)	*Pbnm1' (00γ)s00s*	*mmm1'*
$DyMn_6Ge_6$	*P6/mmm*	(0, 0, 0.1651)	*P62'2' (00γ)h00*	*62'2'*

theory of magnetic structures is not generally equivalent to using magnetic superspace group symmetry (i.e., giving an irrep is not equal to providing the system's magnetic space (super-space) group). Both methods are reliable for observing the magnetic structure. Bilbao Crystallographic Server, ISOTROPY Suite, JANA2020 and Fullprof are the source for understanding the magnetic structure of the materials.

6 Conclusion

Multifunctional oxides are one of the most appealing research areas due to their rich range of properties. A recent trend of these multiferroic oxide materials is approaching to topological properties and opto-electronic properties for which a deep understanding of structural aspects is essential. This chapter briefly explains rare earth manganite materials and how the structure is necessary for describing the magnetic and physical properties of the materials. The super-space group approach for understanding the magnetic system is needed to explore new materials.

References

1. Uberuaga, B.P.: Complex oxides: intricate disorder. Nat. Mater. **15**(5), 496–497 (2016)
2. Zubko, P., Gariglio, S., Gabay, M., Ghosez, P., Triscone, J.M.: Interface physics in complex oxide heterostructures. Annu. Rev. Condens. Matter Phys. **2**(1), 141–165 (2011)
3. Schouwink, P., et al.,: Structure and properties of complex hydride perovskite materials. Nat. Commun. **5** (2014)
4. Bousquet, E., et al.: Improper ferroelectricity in perovskite oxide artificial superlattices. Nature **452**(7188), 732–736 (2008)
5. Emori, S., Bauer, U., Ahn, S.M., Martinez, E., Beach, G.S.D.: Current-driven dynamics of chiral ferromagnetic domain walls. Nat. Mater. **12**(7), 611–616 (2013)
6. Chen, Y.T., et al.: Theory of spin Hall magnetoresistance. Phys. Rev. B - Condens. Matter Mater. Phys. **87**(14) (2013)
7. Sato, M., Ando, Y.: Topological superconductors: a review. Reports Prog. Phys. **80**(7) (2017)

8. Rondinelli, J.M., May, S.J., Freeland, J.W.: Control of octahedral connectivity in perovskite oxide heterostructures: an emerging route to multifunctional materials discovery. MRS Bull. **37**(3), 261–270 (2012)
9. Garcia, V., Bibes, M.: Ferroelectric tunnel junctions for information storage and processing. Nat. Commun. **5** (2014)
10. Scheunert, G., Heinonen, O., Hardeman, R., Lapicki, A., Gubbins, M., Bowman, R.M.: A review of high magnetic moment thin films for microscale and nanotechnology applications. Appl. Phys. Rev. **3**(1) (2016)
11. Lorenz, M., et al.: The 2016 oxide electronic materials and oxide interfaces roadmap. J. Phys. D: Appl. Phys. **49**(43) (2016)
12. Saparov, B., Mitzi, D.B.: Organic-inorganic perovskites: structural versatility for functional materials design. Chem. Rev. **116**(7), 4558–4596 (2016)
13. Morosan, E., Natelson, D., Nevidomskyy, A.H., Si, Q.: Strongly correlated materials. Adv. Mater. **24**(36), 4896–4923 (2012)
14. Zhou, Y., et al.: Strongly correlated perovskite fuel cells. Nature **534**(7606), 231–234 (2016)
15. Schmid, H.: Some symmetry aspects of ferroics and single phase multiferroics. J. Phys. Condens. Matter **20**(43) (2008)
16. Schmid, H.: Multi-ferroic magnetoelectrics. Ferroelectrics **162**(1), 317–338 (1994)
17. Levine, D., Steinhardt, P.J.: Quasicrystals: a new class of ordered structures. Phys. Rev. Lett. (1984)
18. Keys, A.S., Glotzer, S.C.: How do quasicrystals grow? Phys. Rev. Lett. **99**(23) (2007)
19. Kim, J.H., Pieczonka, N.P.W., Yang, L.: Challenges and approaches for high-voltage spinel lithium-ion batteries. ChemPhysChem. (2014)
20. Xu, J., Cui, Y., Xu, H.: The effect of Fe content on dielectric and relaxation characteristics in $TbFe_xMn_{1-x}O_3$. Phys. B Condens. Matter **545**, 402–407 (2018)
21. Zhu, J., et al.: Perovskite oxides: Preparation, characterizations, and applications in heterogeneous catalysis. ACS Catal. **4**(9), 2917–2940 (2014)
22. Peña, M.A., Fierro, J.L.G.: Chemical structures and performance of perovskite oxides. Chem. Rev. **101**(7), 1981–2017 (2001)
23. Young, J., Lalkiya, P., Rondinelli, J.M.: Design of noncentrosymmetric perovskites from centric and acentric basic building units. J. Mater. Chem. C **4**(18), 4016–4027 (2016)
24. Katsura, H., Nagaosa, N., Balatsky, A.V.: Spin current and magnetoelectric effect in noncollinear magnets. Phys. Rev. Lett. **95**(5) (2005)
25. Fiebig, M.: Revival of the magnetoelectric effect. J. Phys. D: Appl. Phys. **38**(8) (2005)
26. Vopson, M.M.: Fundamentals of multiferroic materials and their possible applications. Crit. Rev. Solid State Mater. Sci. **40**(4), 223–250 (2015)
27. Khomskii, D.: Classifying multiferroics: mechanisms and effects. Physics (College. Park. Md) **2** (2009)
28. Tokura, Y., Seki, S.: Multiferroics with spiral spin orders. Adv. Mater. **22**(14), 1554–1565 (2010)
29. Agarwal, H., Yadav, P., Lalla, N.P., Alonso, J.A., Srivastava, O.N., Shaz, M.A.: Structural correlation of magneto-electric coupling in polycrystalline $TbMnO_3$ at low temperature. J. Alloys Compd. **806**, 510–519 (2019)
30. Kimura, T., Goto, T., Shintani, H., Ishizaka, K., Arima, T., Tokura, Y.: Magnetic control of ferroelectric polarization. Nature **426**(6962), 55–58 (2003)
31. Johnson, R.C.: A simple approach to crystal field theory. J. Chem. Educ. **42**(3), 147–148 (1965)
32. Deskin, W.A.: Inorganic chemistry: Principles of structure and reactivity (Huheey, James E.). J. Chem. Educ. **50**(7), A379 (1973)
33. Sawada, H., Morikawa, Y., Terakura, K.: Jahn-Teller distortion and magnetic structures. Phys. Rev. B - Condens. Matter Mater. Phys. **56**(19), 12154–12160 (1997)
34. Blundell, S., Thouless, D.: Magnetism in condensed matter. Am. J. Phys. **71**(1), 94–95 (2003)
35. Coey, J.M.D.: Magnetism and Magnetic Materials, vol. 9780521816 (2010)
36. Zener, C.: Interaction between the d shells in the transition metals. Phys. Rev. **81**(3), 440–444 (1951)

37. Anderson, P.W.: Antiferromagnetism. Theory of superexchange interaction. Phys. Rev. **79**(2), 350–356 (1950)
38. Goodenough, J.: Goodenough-Kanamori rule. Scholarpedia **3**(10), 7382 (2008)
39. Hu, C.D.: The Dzyaloshinskii-Moriya interaction in metals. J. Phys. Condens. Matter **24**(8) (2012)
40. Coey, J.M.D., Viret, M., Von Molnár, S.: Mixed-valence manganites. Adv. Phys. **58**(6), 571–697 (2009)
41. Mostovoy, M.: Ferroelectricity in spiral magnets. Phys. Rev. Lett. **96**(6) (2006)
42. Bousquet, E., Cano, A., Non-collinear magnetism in multiferroic perovskites. J. Phys. Condens. Matter **28**(12) (2016)
43. Bertaut, E.F.: Representation analysis of magnetic structures. Acta Crystallogr. Sect. A **24**(1), 217–231 (1968)
44. Rodríguez-Carvajal, J.: BasIreps: a program for calculating irreducible representation of little groups and basis functions of polar and axial vector properties. Solid State Phenom. (2004)
45. Momma, K., Izumi, F.: VESTA 3 for three-dimensional visualization of crystal, volumetric and morphology data. J. Appl. Crystallogr. **44**(6), 1272–1276 (2011)
46. Shull, C.G., Strauser, W.A., Wollan, E.O.: Neutron diffraction by paramagnetic and antiferromagnetic substances. Phys. Rev. **83**(2), 333–345 (1951)
47. Petříček, V., Palatinus, L., Plášil, J., Dušek, M.: Jana 2020 - a new version of the crystallographic computing system Jana. Zeitschrift fur Krist. - Cryst. Mater. **238**(7–8), 271–282 (2023)
48. Perez-Mato, J.M., Ribeiro, J.L., Petricek, V., Aroyo, M.I.: Magnetic superspace groups and symmetry constraints in incommensurate magnetic phases. J. Phys. Condens. Matter **24**, 163201 (2012)

Langmuir–Blodgett Films: A Unique Approach Towards Monomolecular Assemblies

Subrata Deb and Gautam Chel

Abstract Thin films of few nanometers are the sources of high expectations as being useful components in many practical and commercial applications such as sensors, detectors, and electronic circuit components. The possibility to synthesize organic molecules almost without limitations with desired structures and functionality in conjunction with a sophisticated thin film deposition technology enables the production of electrically, optically, and biologically active components on a nanometer scale. An organic thin film can be deposited on a solid substrate by various techniques such as *thermal evaporation, solvent casting, spin coating, dip coating, plasma polymerization, sputtering, electro-deposition, molecular beam epitaxy, adsorption from solution, solvent casting, lithography, Langmuir–Blodgett (LB) technique, Layer by Layer self-assembly,* etc. The scientific interest in the fabrication and characterization of ultrathin films prepared by the Langmuir–Blodgett (LB) technique has earned considerable attention in recent years as it offers an elegant method of organizing molecules at the microscopic level. It is one of the most promising and reliable methods for the preparation of films of nanometer order thickness as it offers precise control over monolayer and multilayer growth, homogeneous deposition over a large area, and the possibility to make multi-structure LB films with varying layer composition. This technique also allows the forming of thin films on almost any kind of substrate. Moreover, these films' optical, electrical, and other aggregating behaviors can be changed markedly by varying LB parameters, which may be extremely useful for a specific technological application. Hence, a detailed investigation of these films is extremely necessary, and a comparative analysis with other techniques of thin film depositions allows one to optimize different parameters in such films. These studies already opened promising possibilities for the fabrication of various devices in the field of molecular electronics. It is also important to study the spectroscopic characteristics of these films to understand the basic physics involved in such monomolecular assemblies of molecules. In this chapter, we highlighted the basic

S. Deb (✉)
Department of Physics, Women's College, Agartala, Tripura 799001, India
e-mail: subrata.tu@gmail.com

G. Chel
Department of Chemistry, Women's College, Agartala, Tripura 799001, India

U. Kumar (ed.), *Defects Engineering in Electroceramics for Energy Applications*,
Engineering Materials, https://doi.org/10.1007/978-981-97-9018-0_11

details in a very simple manner regarding the formation of thin films by following the Langmuir–Blodgett (LB) technique along with various aspects of linked mechanisms in order to study such a system of great importance.

1 Introduction

The interpretation and measurement of electromagnetic radiation absorption and emission during the transition of ions, molecules, or atoms between energy levels is known as spectroscopy. When atoms, ions, or molecules transition from one energy level to another, electromagnetic radiation is absorbed and emitted. This process is measured and interpreted in spectroscopy. In summary, because modern instruments with data processing capabilities are readily available, spectroscopy is a vital tool for physicists, chemists, and biologists. Using these techniques, one can look at the fundamental photochemistry and photophysics of both excited and ground-state molecules. The presence of highly sophisticated tunable and high-intensity laser beam as a monochromatic and coherent light source gives a unique opportunity to study the various happenings related to radiative and non-radiative processes. Various aspects of these radiative or non-radiative transitions are found to depend on the rigidity, nature, the shape of molecules, and also on the environment. Once more, only quantum mechanics could account for the mechanisms holding molecules of atoms together, the valence bond and molecular orbital ways of chemical bonding, and the limitations of spectroscopic transitions between specific pairings of levels.

High expectations are placed on thin films of a few nanometers in thickness because they are expected to be valuable parts of numerous real-world and industrial applications, including sensors, detectors, and electronic circuit components. When combined with advanced thin film deposition techniques, the nearly limitless ability to synthesize organic molecules with desired architectures and functions allows for the creation of physiologically, optically, and electrically active nanoscale components.

Many methods, including solvent casting, thermal evaporation, dip coating, spin coating, plasma polymerization, electro-deposition, sputtering, adsorption from solution, molecular beam epitaxy, lithography, solvent casting, LB (Langmuir–Blodgett) approach, layer-by-layer self-assembly, etc., can be utilized by depositing an organic thin film on a solid substrate. The scientific interest in the fabrications and characterizations of ultrathin films prepared by LB the technique has earned considerable attention in recent years [1–5] as it offers an elegant method of organizing molecules at the microscopic level. The LB approach is the most a reliable and promising method for the preparation of films of nanometer order thickness as it offers precise control over thickness, film homogeneous deposition over a large area and option to make multi-structure LB films with a composition of varying layers. In addition, this technique allows the thin film deposition on almost any kind of substrates. Furthermore, electrical, optical and other physicochemical properties of these films can be changed markedly by varying different LB parameters, which might be extremely suitable for specific technological applications. Therefore, from the perspective of their numerous practical uses, in-depth research of these films is essential. A comparative study of these films with those fabricated by some other techniques is also very essential for the optimization of different parameters in thin films. These studies have already opened promising possibilities for the fabrication of various devices in field of molecular electronics. The preparation of ultrathin films by the LB approach and their photophysical, morphological, electrical, and other properties are fundamental importance as they have broad applications in various molecular electronic devices [6–8] optoelectronic devices [9–12] etc. It is also important to study the spectroscopic characteristics of these films to understand the basic physics involved and to check whether they have any interesting photophysical and electrical properties based on which they may be utilized properly in the many fields of application. In the current chapter, considerable attention was made to the preparation of ultrathin films of some non-amphiphilic organic molecules having interesting photophysical and electrical properties. Monolayers of these molecules at gas–liquid (air–water) interface has been researched extensively as a 2-dimensional model system to get an idea about the miscibility or immiscibility of the binary components. The thermodynamic the behaviour of the floating monolayer and also the interactions nature involved among the constituent molecules have been highlighted in these studies. The most celebrated feature of Langmuir–Blodgett (LB) technique, which enables the transfer of spread monolayer from the air–water interface with precise control was also extensively investigated. The detailed spectroscopic characterizations of these films have been also studied to highlight the aggregation nature of molecules in the LB films. In addition, the spectroscopic along aggregating properties of these films were also compared with their respective solution and microcrystal form (bulk). The dynamics of the excited state of the molecules in LB films constrained geometry were also examined. The conventional and traditional UV–Vis Absorption and steady-state Fluorescence Spectroscopic methods may be employed for the photophysical characterizations of these films. In some cases, the traditional structural studies method which is, SEM (Scanning Electron Microscopy) was utilized to get an idea about the nature of aggregating species along with also to imagine the domain structure in the LB films.

This chapter aimed to deliver a brief overview regarding the basic things needed to understand the various experimental techniques and prospects of this pioneering method of thin film deposition.

2 Langmuir and Langmuir Blodgett Films

2.1 Historical Development

The development of LB films was the result of creativity, intuition, hard effort, and discoveries. The progress has persisted for a century, and the global scientific community continues to be deeply interested in monolayer research today. Let's start with a quick definition of "monolayer" before delving into the history itself. Particles stacked side by side, without any on top of one another, form a monolayer. The monolayer characteristics are often governed by the orientation of the molecules. These particles may be in the form of chemicals, proteins, viruses, nano- or micro-colloids, live cells, etc. Monolayers are sometimes referred to as thin films, ultra-thin films, or films.

Going back to historical observations, it was once noted that waves would become smoother when barrels of oil were dropped into rough seawater. This spreading of oil across the surface of the water was the first observed example of a molecular film of organic material [1].

2.2 Scientific Interest in Langmuir–Blodgett Films

The first serious observation of this was carried out by American politician Benjamin Franklin in 1774, when he deposited a spoon of oil on an agitated pond, and noticed that the rough waves were calmed on a determined surface [13].

> At length at Clapman where there is, on the common, a large pond, which I observed to be one day very rough with the wind, I fetched out a cruet of oil, and dropped a little of it on the water. I saw it spread itself with surprising swiftness upon the surface, the oil, though not more than a teaspoonful, produced an instant calm over a space several yards square, which spread amazingly and extended itself gradually until it reached the leeside, making all that quarter of the pond, perhaps half an acre, as smooth as a looking glass

The robust scientific findings of Rayleigh [14], who computed the thickness of the oil films to be 10–20 angstroms—which is near to more precise measurements of the thickness of monomolecular films of such substances today—followed his perceptive observations and pointed questions. A few years later, in 1882, Miss Agnes Pockels began methodical studies to determine the maximum area that one oil drop could cover. She determined the thickness of the oil film and, consequently, every oil molecule's surface area by knowing the volume of the oil drop and surface occupied. Her experimental results were published in the reputed journal Nature [15–18]. Scientists began looking into ways to compress the oil monolayer using a movable barrier—basically, a trough made of hydrophobic material—after Miss Pockets discovered the monolayers. The barrier was moved forward, gathering the molecules first and then compressing them, once they had been dispersed across the water's surface. Scientists noticed an intriguing occurrence during this time that varied according to the type of oil utilized. At a specific pressure, a few of them displayed a sort of transition condition. The proper explanation of this transition was first given by Devaux [19].

Irving Langmuir investigated how molecules behaved with respect to pressure and area on an aqueous surface [20–24]. The hydrocarbon chain length was found to have no bearing on the regions occupied by molecules like alcohols, esters, and acids, indicating that only the hydrophilic head groups have submerged in the subphase. The transferring of the film method onto the solid substrates and subsequently constructing multi-layer films was discovered by Katherine Blodgett [25–27], who collaborated with Langmuir on researching the characteristics of floating monolayers. Her work culminated in a 1937 [27] paper with Langmuir and a 1939 [24] paper on anti-reflection coatings. Her discoveries were published in 1934 and 1935 [25–27].

In several useful devices, LB films have been employed. The first were step thickness gauges and the coatings of anti-reflection already mentioned, both created by Blodgett. Following the start of World War II, interest in Langmuir–Blodgett films declined. This trend persisted until the 1960s when Möbius and Kuhn [28–31] demonstrated how accurate supermolecular structures could be created using monolayers. They demonstrated the fluorescence and quenching of dye molecules bonded to completely saturated fatty acids under UV light using the Langmuir–Blodgett technique. This work and publication of Gaines's "*Insoluble Monolayers at Liquid Gas Interfaces*" in 1966 [32] introduced a revival of interest in this field. Numerous potential uses of Langmuir–Blodgett films are presently being explored by research groups worldwide in the domains of electronics [33], optics [34, 35], molecular electronics [6–8], and biotechnology [36].

The next section gives a brief idea regarding the details of t various important terms related to the Langmuir and LB films.

2.3 Air–Water Interface

The boundary surface among 2 phases (such as the air–water interface) or 2 dissimilar immiscible layers is referred to as an interface (, oil–water interface). Any interface can be approximately assumed as two-dimensional as the thickness is on the order of a few molecular diameters. The abrupt change in composition and density at the interface, which raises the amount of excess free energy, is its most intriguing feature. The air–water (gas–liquid) interface is taken into consideration in current research because of its dipole moment, high dielectric constant, and increased propensity to create a hydrogen bond, all of which have a significant impact on the interface's characteristics. In addition, the water has a remarkably high value of surface tension of about 70×10^{-3} N/m at room temperature. Thus, the prominent ordering in the molecules of water is caused by the existence of strong intermolecular forces of attraction [37–39]. The electrostatic characteristics of the interface can also be significantly impacted by the higher ordering of water molecules with the atom of oxygen pointed toward the air.

Surface tension is the unit of measurement for the interfacial free energy and is defined as

$$\gamma = \left(\frac{\partial G}{\partial A}\right)_{T,P,n}$$

where G represents Gibb's free energy, A indicates the surface area at constant temperature T, P and n are the pressure and composition respectively.

Although, in recent investigations, the air–water interface has been given preference there are also some other interfaces such as a liquid–liquid interface.

2.4 Surfactants

Surfactants, which belong to a broader family of chemicals, are highly relevant both in terms of technology and biology. These compounds typically comprise a hydrophilic component (soluble in water) along with a hydrophobic component (insoluble in water). The association behavior of surfactants in solution (vesicles, bilayers, micelles, and so on) and their buildup at interfaces (water/oil or oil/water) are caused by their amphiphilic character. Hydrocarbon or fluorocarbon chains often make up the hydrophobic portion, whereas polar groups make up the hydrophilic portion $\left(-OH, -COOH, -NH_3^+, -PO_4^{3-} \text{ etc.}\right)$.

Ionic and non-ionic surfactants are categorized based on the type of hydrophilic head group present in surfactants. Once more, the ionic surfactants are further divided into cationic and anionic surfactants based on the type of charge (anion or cation) present in this portion. Semipolar and single-bond surfactants are two categories of non-ionic surfactants.

The affinity of surfactants for surfaces and their association behavior in solution is found by the chemical and physical characteristics of hydrophilic and hydrophobic groups, correspondingly. In this regard, the hydrocarbon moiety's shape and size as well as the hydrophilic head group's size, charge, and hydration are crucial. Numerous self-assembled structures were seen at the interfaces as well as in bulk, depending on how these attributes are balanced. The association's primary motivator is the system's decreased free energy. As a result, a surfactant's surface tension decreases when it accumulates at the air–water interface when it comes into contact with water.

It's critical to define hydrophobic and hydrophilic interactions precisely in this situation. When a small quantity of such a molecule is spread on the surface of water then as per entropy consideration, some molecules are dispersed on the surface of water. This increases the interactions of the polar molecules with the polar head groups and with each other, hence reducing the internal energy of the system for any configuration. Thus, the complete system behaves in such a way that it shows the 'like-dissolves-like' principle. These forces can result in a variety of intriguing structures, which are covered in more detail in the section that follows, depending

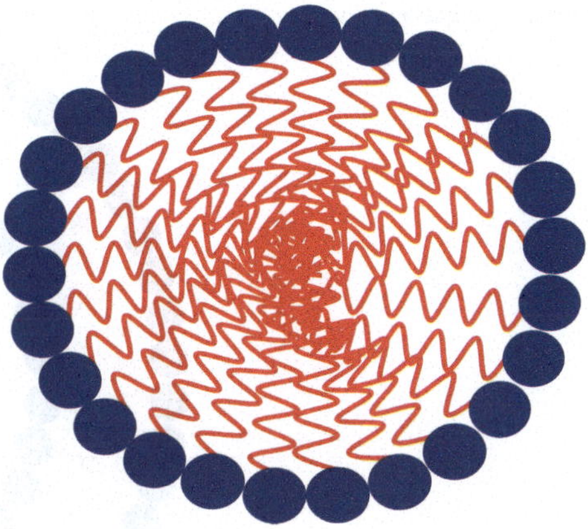

Fig. 1 Formation of micelle

on the characteristics of the particular parameters controlling the behavior of the system.

2.4.1 Micelle

A micelle is an aggregate of amphiphilic molecules organized in a polar solvent, such as water, with the non-polar groups gathered at the center of the combined and the polar groups facing outward. A micelle formed by the amphiphilic molecules is shown in Fig. 1.

The term *critical micelle concentration* (*c. m. c.*) refers to the specific concentration for many systems at which these micellar structures begin to develop. With polar groups facing inward and non-polar groups facing outward, these structures can also form in non-polar solvents.

2.4.2 Vesicle

It is composed of a spherically shaped, closed structure with walls composed of amphiphilic molecular bilayers. They are both found in and contain solvents, as Fig. 2 illustrates. In this case, the hydrophobic parts of all the molecules interact with the similar parts of the other molecules. That is the development of vesicles at the air–water interface is due to hydrophobic-hydrophobic interactions.

2.4.3 Monolayer

As illustrated in Fig. 3, a monolayer of these molecules may form at the liquid–gas interface such that their polar groups come into contact with the polar liquid.

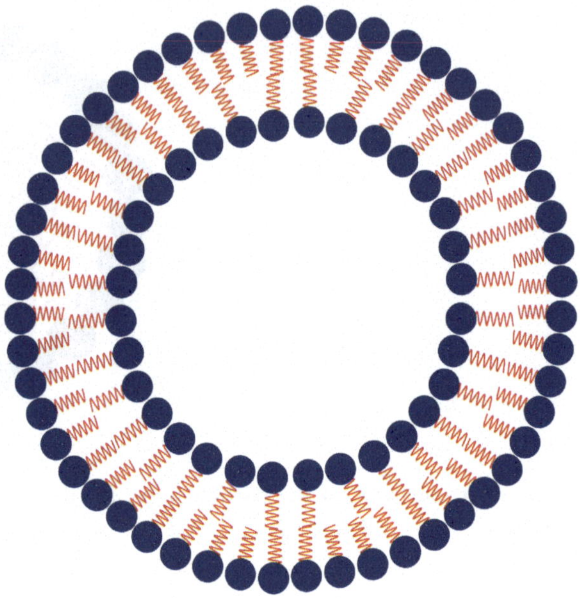

Fig. 2 Formation of vesicle

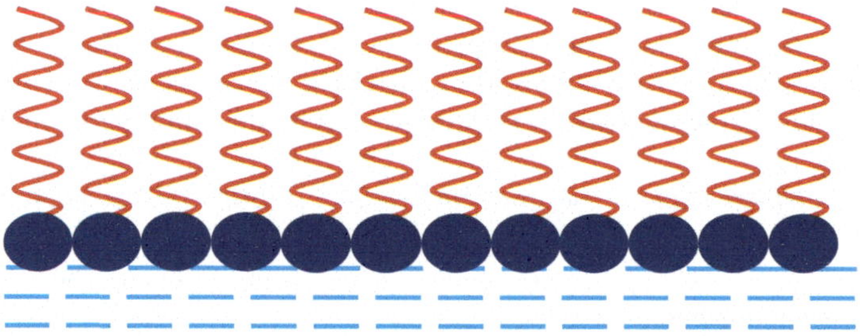

Fig. 3 Formation of Langmuir monolayer

2.5 Materials Suitable for LB Film Deposition

By definition, the production of a floating monolayer needs amphiphilic molecules to be trapped at the interface of water in the process of LB. An amphiphilic molecule typically consists of a water-soluble hydrophilic head group, such –COOH, and a water-insoluble hydrophobic portion, which is typically a long alkyl chain.

Generally speaking, substances like alcohol and long-chain fatty acids are utilized to prepare the films. The material is repelled from water by the hydrocarbon moiety of the molecule, which contains $-CH_2$. On the other hand, the polar head groups (–COOH, –OH,) have enough affinity for the water molecule to anchor the molecule in the aqueous sub-phase. For example, in SA, the long hydrocarbon tail ($-C_{17}H_{35}$) is hydrophobic and preferentially resides in the air and another part of the molecule i.e., the head-group –COOH is hydrophilic and preferentially immersed in water. Some of the typical LB-compatible materials are arachidic acid (AA), stearic acid (SA), zinc (arachidate)$_2$, etc. Figure 4 shows the structural formula of some amphiphilic molecules.

In this connection, it should be noted that the equilibrium between the hydrophilic head strength and the alkyl chain length determines an amphiphilic molecule's solubility in water. Figure 5 depicts the direction of an amphiphilic fatty acid molecule dispersed at the interface of water.

To become LB compatible, the hydrophilic portion with intriguing physical features (such as semi-conducting, conducting, gas sensing, lasing action, pyroelectric, etc.) is typically joined to an extended hydrophobic chain (alkyl chain). A polymeric backbone can be added to a molecule to further increase its LB compatibility. Nevertheless, both of these procedures need synthesis expertise and are quite expensive and time-consuming. Furthermore, the purification of these synthetic materials is a very skilled and time-consuming process.

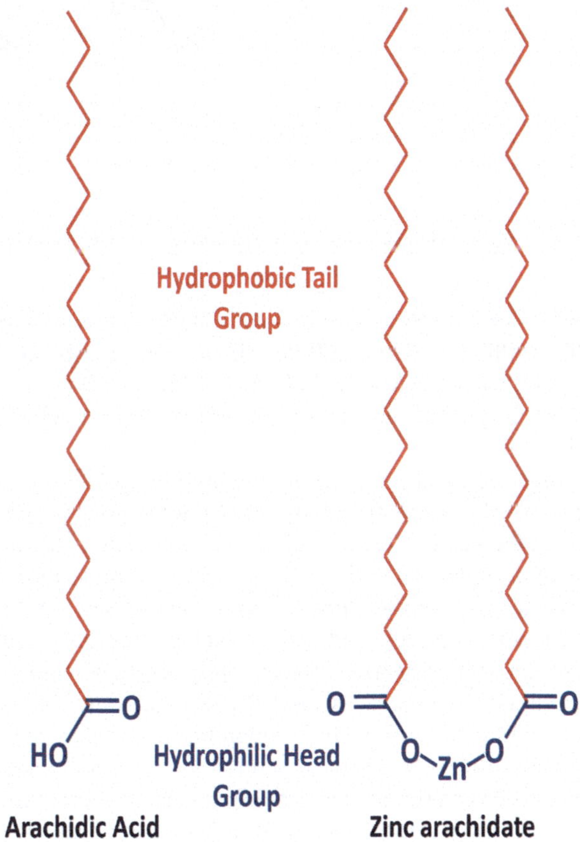

Fig. 4 Structural formula of some amphiphilic molecules

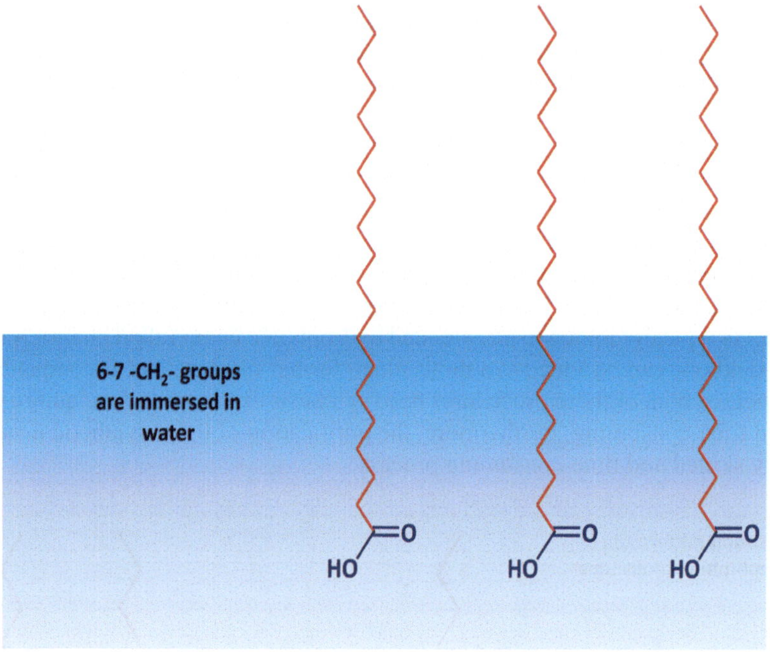

Fig. 5 Orientation of amphiphilic fatty acid molecule at the air–water interface

2.6 Recent Trends in Selection of Materials

According to recent studies in this area, some non-amphiphilic molecules—that is, ones without any hydrophobic chain, when combined with either "an inert polymer matrix, such as PMMA [Polymethyl Methacrylate] or PSt [Polystyrene], or a long-chain fatty acid, such as SA (Stearic Acid) or arachidic acid, result in excellent LB films [40–45].

In these situations, the non-amphiphilic molecule is supported in its organization at the interface of water by long-chain fatty acids or polymers acting as building blocks. From the perspective of various technological applications, studies on the development of organized molecular assemblies of these molecules in films of LB and their subsequent classifications are important because these non-amphiphilic molecules are readily available and with broad varieties with various interesting photophysical and electrical properties. Additionally, it was noted that the spectroscopic and aggregating characteristics of the LB films containing these non-amphiphilic molecules are similar to those of their amphiphilic counterparts [40–45]. Another significant feature of these kinds of mixed LB films is that" materials' optical along electrical properties can occasionally be significantly altered upon incorporation into specific inert matrices with constrained geometries. From the perspective of their technical uses, thorough analyses of these films are crucial. Additionally, it is required for the optimization of certain characteristics in certain materials' thin films, which could result in a few particular technological applications.

3 Surface Pressure

The discrepancies in the environments of the molecules on the surface and those in the bulk give rise to an excess free energy at the interface of water. Surface tension is a useful unit of measurement for this interfacial free energy.

The pure water surface tension is typically 70×10^{-3} N/m, which is remarkably higher when compared to other liquids. As a result, water is an excellent subphase for monolayer investigations. Surface tension decreases when an impurity is added to the water's surface because the force of cohesion decreases. The drop in surface tension could be evaluated in relation to this level of reference and yield a negative number if we use the pure water's surface tension as the zero-level reference. This negative value, or drop in surface tension, is commonly denoted as a rise in positive surface pressure in the LB approach.

Surface pressure and tension have similar units and are equivalent numerically. In contrast to surface tension, which is displayed along the negative Y-axis, surface pressure is represented along the positive Y-axis. The next section provides a discussion of the specific method used to detect surface pressure.

3.1 Quantitative Measure of Surface Pressure

The $ST_{inst.}$ ("Instantaneous Surface Tension") of a water surface decreases while a monolayer is compressed on it. This decrease in surface tension is typically interpreted in the LB approach as a rise in surface pressure. In addition, surface tension and surface pressure both are numerically equal.

$$\text{Hence, } ST_{inst.} + SP = \text{constant}$$

This constant being the absolute ST of the liquid—72.8×10^{-3} N/m for water.

By submerging an object, typically a plate, ring, or cylinder known as a Wilhelmy plate (as shown in Fig. 6) across the interface of air–water and assessing the force this object experiences, one can determine the surface pressure, also known as surface tension, of the water surface. Think about a plate that is suspended over the water–air interface.

The forces operating on the plate include buoyancy from displaced water acting upwards, and gravity and surface tension acting downward into the water. If the plate is filled with water to a depth of d and has dimensions t, w, and l, for thickness, width, length, and ρ density, respectively, the equation that describes the net force

Fig. 6 Wilhelmy plate

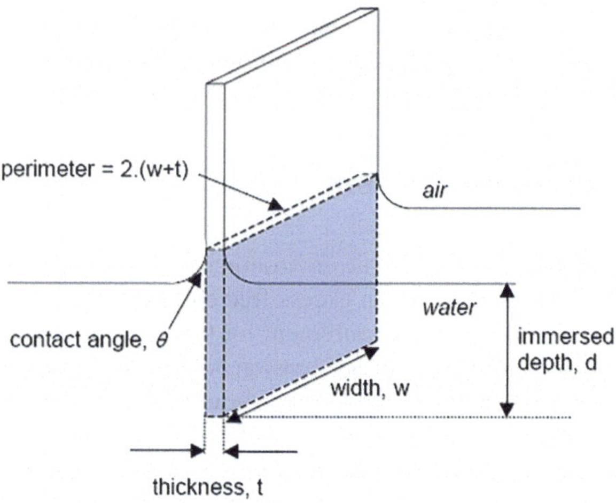

downward, F, is as follows:

$$\text{Force} = (\rho_P lwt).g - (\rho_L dwt).g + 2.(w + t).(ST).\text{Cos}\theta$$
$$(\text{Force} = \text{Weight}--\text{upthrust} + \text{surface tension})$$

where,

ρ_P–Plate density

ρ_L–Liquid density

ST–Sub-phase surface tension

θ–The angle of contact of the liquid to the plate

g–Acceleration due to gravity.

Typically, the surface pressure is set to zero before any measurements are taken, allowing us to remove the weight element from the equation above and, as a result,

$$\text{Force} = -(\rho_L dwt).g + 2.(w + t).(ST).\text{Cos}\theta$$
$$(\text{Force} = --\text{upthrust} + \text{surface tension})$$

Again, we can easily eliminate the upthrust term, since the balance consistently maintains the plate at the same level, no matter what the surface tension. So the force on a plate is presented by,

$$\text{Force} = 2.(w + t).(ST).\text{Cos}\theta$$

The filter paper employed in our investigation guarantees that there is no contact angle between liquid and plate. Thus, we have

$$\text{Force} = 2.(w + t).(ST)$$

$$\Rightarrow \quad \text{ST}_{\text{Plate}} = \frac{Force}{2(w + t)}$$

In the present case, "the surface pressure and surface tension are represented as mN/m because the perimeter is represented in m and the force is expressed in mN.

3.2 Surface Pressure Versus Average Area Per Molecule Isotherm: Langmuir Monolayer at Air–Water Interface

The most important measure of an amphiphilic material's monolayer qualities is its surface pressure as a function of the water surface area that every molecule might occupy. This is done while maintaining a steady temperature. Surface pressure variation with molecule area is named pressure-area isotherm ($\pi - A$ isotherm) or simply isotherm. When the monolayer is compressed at a constant rate (5 mm/min), an isotherm is often recorded.

To create a diluted solution, a small amount of an appropriate LB-compatible substance is first dissolved in a highly volatile and water-insoluble solvent (such as chloroform). Now, using a tiny syringe as depicted in Fig. 7a, a few microliters of this solution are applied very slowly at the air–water interface (on the water surface), giving" the volatile solvent enough time to evaporate.

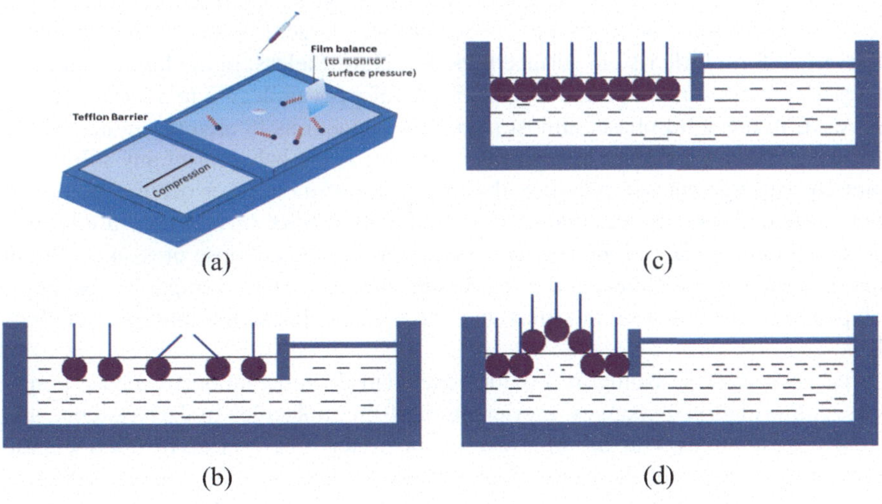

Fig. 7 **a** Gaseous phase of Langmuir monolayer, **b** Liquid phase of the Langmuir monolayer, **c** Solid phase of the Langmuir monolayer, **d** Collapse phase of the Langmuir monolayer

The dispersed molecules then continue to drift aimlessly around the 2-D water's surface, behaving as though they were molecules of gas in the space of 3-D. The molecules are very loosely packed in what is known as the gas phase. Although recent X-ray diffraction studies provide much more detailed information concerning the structure of different monolayer phases [46], the terminology suggested by Harkins [47] has been used in this study (to monitor surface pressure).

As seen in Fig. 7b, every molecule in this phase of gas takes up a considerable molecular area (A).

The random movement of molecules on the surface of water now tends to approach while the LB moveable barrier trough is permitted to move slowly thanks to a computer-controlled mechanism, as seen in Fig. 7c. We refer to this state as the liquid phase. Figure 7d illustrates the observation of a well-organized 2-dimensional arrangement of molecules upon additional compression of the barrier. This phase, referred to as the solid phase, is when the monolayer behaves like a well-organized, two-dimensional crystalline form.

This two-dimensional organized molecular assembly of suitable molecules at the interface of air and water is called a Langmuir monolayer. This Langmuir monolayer can also be formed on another suitable liquid surface such as benzene.

Further compression of the barrier leads to the total collapse of this monolayer due to mechanical instability [48] as shown in Fig. 7d. In this position, the organization of the molecules is destroyed and multilayers and aggregates of molecules have been formed on the surface of the water, which is shown in Fig. 7d.

The surface pressure vs. area per molecule (π–A) curve is produced by plotting the records of surface pressure (π) and area per molecule (A). Figure 8 displays an ideal isotherm curve with clear phase separations.

In this isotherm curve, the gaseous phase, liquid phase, solid phase, and collapse region are shown distinctly. In the gaseous phase, a large drop in area per molecule is observed for a little rise in surface pressure. Nevertheless, in the liquid phase, an increase in large surface pressure reduces the area per molecule to a lesser degree. However, in the solid phase, almost negligible reductions in area per molecule are seen by increasing large surface pressure and this is the behavior of any solid-state material. At large surface pressure, the collapsing of monolayer occurs and at this point, a sudden decrease of area per molecule with a fall of surface pressure occurs. The temperature, pH and the barrier speed all affect the collapse pressure. Certain materials show two different collapse pressure situations. For example, isotherms of polypeptides first collapse into bilayers, which then collapse into micro-crystallites [49].

The pressure-area isotherm provides excellent documentation for phase transitions, conformational changes, reorientation of the molecules in 2-D systems, and stability of monolayer at the interface of air–water. There is a close association between the features of the isotherm and the molecular form. In general, isotherms are highly sensitive to molecule nature, its orientation, composition of the mixtures, etc. Gaines [32].

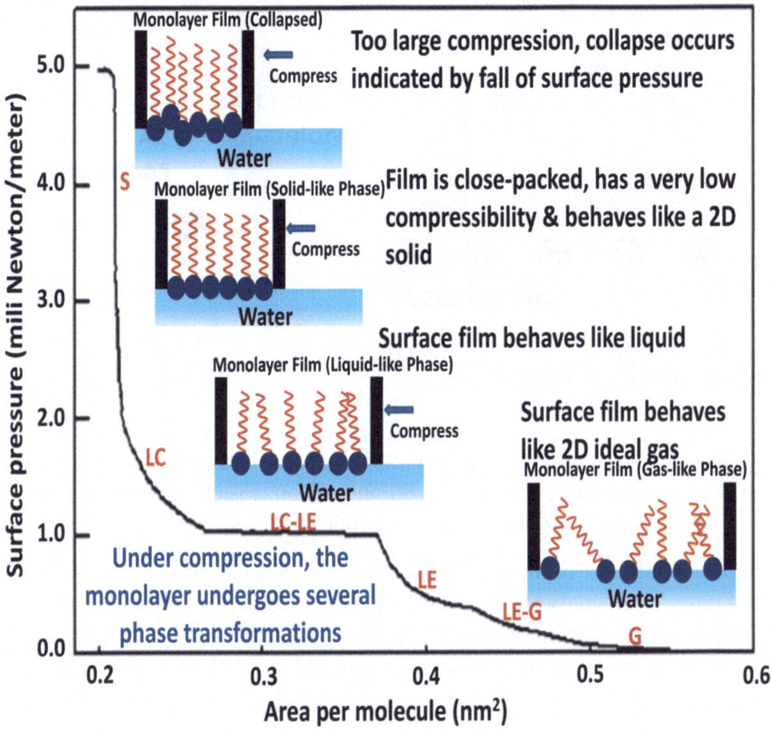

Fig. 8 Surface pressure versus area per molecule isotherm

4 Hybrid Monolayer

As already mentioned, present interest lies in the production of ultrathin films of non-amphiphilic organic molecules with the help of various supporting matrices such as long-chain fatty acid namely SA, or a polymer namely PMMA. In these situations, the sample along with the matrix molecules are first made into solutions in a highly volatile, water-insoluble solvent, like chloroform. Using a tiny syringe, these two solutions are distributed at the interface of water after being combined in a predetermined ratio.

The molecules are arranged in an orderly fashion and create a highly stable monolayer at the interface of air–water when the solvent molecules evaporate. If the barrier is compressed gently, this monolayer can be easily moved to the solid substrate to make films of LB. The non-amphiphilic organic molecules were ideally not suitable for the creation of the Langmuir monolayer. These sample molecules may become enmeshed in the matrix molecules during incorporation, forming a stable "Langmuir monolayer at an interface between air and water.

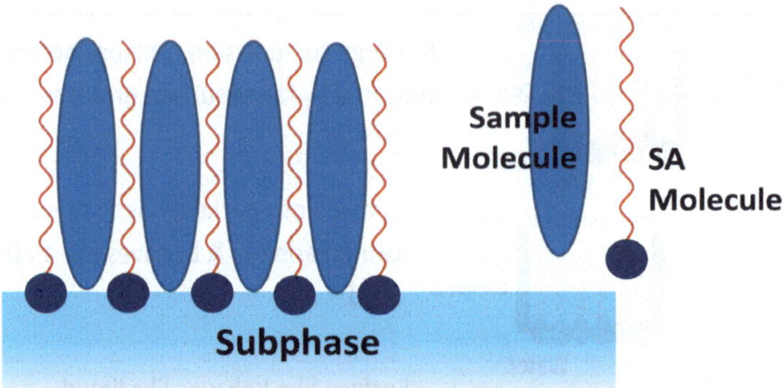

Fig. 9 A mixed (hybrid) monolayer

The resulting monolayer, dubbed mixed Langmuir monolayer, is easily transferable to solid substrates like glass, quartz, and so on. Figure 9 depicts a schematic illustration of the hybrid monolayer's molecular structure.

In this regard, it should be noted that the π–A isotherms of pure components and binary mixtures were analyzed to establish the thermodynamic behavior of mixing several 2-component systems [50]. By applying the additivity rule, one may determine the excess area of mixing (A_E) as a function of mixture composition, molecular weight, and surface pressure, obtaining a measure of non-ideality [32, 51, 52]

$$A_E = A_{12} - A_{av.}$$

$$A_{av.} = A_1N_1 + A_2N_2$$

where, supposing mole fraction additivity, $A_{av.}$ indicates the computed average area per molecule of mixture and, A_{12} represents the mixed monolayer actual area per molecule.A_1 and A_2 are thearea per molecule (or monomer) of every single component at a particular surface pressure, N_1 and N_2 are their corresponding mole fractions.

The binary components miscibility of mixed monolayer could be researched by plotting the graph between the actual area data (A_{12}) per molecule and molefraction of the sample in a mixed monolayer having an SA or PMMA at a particular surface pressure. If this plot exactly coincides with that of the theoretical curve as seen by the rule of additivity, $A_{av.} = A_1N_1 + A_2N_2$ (which is a straight line), then we can say that either complete mixing or total demixing of constituent molecules of the monolayer has occurred. This suggests that given a molecule binary mixture that shows perfect miscibility or total immiscibility, $A_E = 0$.

We can conclude that a repulsive kind of interaction between the matrix and the sample molecules has happened if the experimental curve exhibits any positive deviation. This could result in complete phase separation and the creation of aggregates.

The positive deviation indicates a strong attractive interaction between the unlike molecules and almost no interactions between the like molecules.

Again if a negative deviation with respect to the ideality curve is observed then we can say that an attractive type of interaction has occurred between the binary components of mixed monolayer, which indicates complete immiscibility of the constituent molecules. Clusters and aggregates may form when similar molecules have strong attractive interactions with one another.

It is significant to remember that in bulk binary miscible mixtures, if 2 liquids are ideally miscible, it typically indicates that they are the same in both physical and chemical aspects, meaning that their intermolecular forces are the same and the solute's presence has no effect on the solvent molecules. But total immiscibility indicates that there is very little contact between dissimilar molecules and a strong, attractive interaction between like molecules.

The nature of complete mixing or demixing of component molecules in the mixed film is shown by Steady State Fluorescence Spectroscopic and UV–Vis Absorption studies, even though we could not say with certainty whether complete mixing and demixing occurs in monolayer based on this additivity characteristics study. The collapse pressure versus molefraction plot is also useful to get an overview of the nature of interactions involved in the various components of the mixed monolayer.

If the collapse pressure is observed to be independent of the molefraction i.e., if the plot is almost parallel to the X-axis (along which the molefraction is plotted) then we can say completely demixing of binary components has occurred which might lead to the development of aggregates in LB films.

The sample and matrix molecules in the mixed monolayer are immiscible, if the collapse pressures for all mole fractions are almost constant, independent of mole fractions, and do not match the ideality (solid line) curve. Aggregates of sample molecules in the mixed films" may result from the immiscibility or total demixing.

5 Forces Governing the Monolayer

The phase behavior of the floating monolayer of an LB-compatible material is governed by the sum of attractive and repulsive forces acting on it. Various forces, which govern the phase behavior of the monolayer, is vander Waal's force, force due to change in entropy during the compression, forces resulting from various interactions such as interactions between polar head groups, ion-ion, ion–dipole, and dipole–dipole, etc. Israelachvili was the the first researcher who described the origin of these forces [53]. All forces rely on both the molecular structure of the molecule under analysis and the temp at which the experiment is carried out. As the temperature rises, the constituent molecules of the monolayer experience an increase in thermal movement, leading to its expansion.

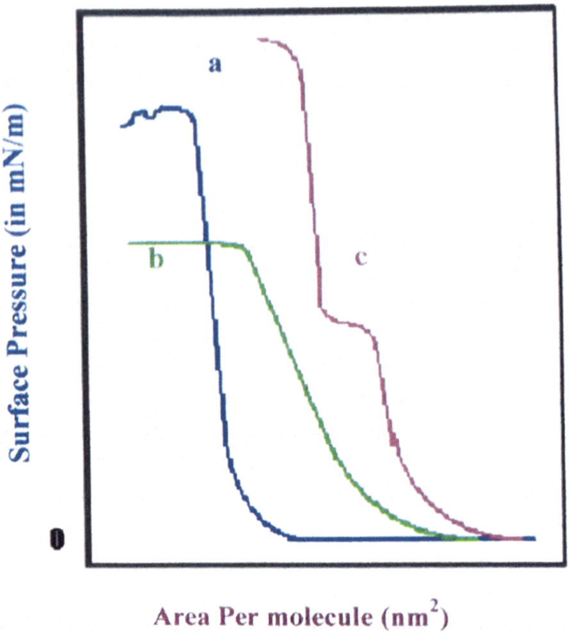

Fig. 10 Pressure area isotherms of a series of fatty acids

5.1 Vander Waal's Force

It is usually inversely related to the 6th power of distance among the molecules in contact, meaning it is a short-range force. The condensed state's existence in the monolayer is justified by this attractive force. This force changes with the 5th power of distance for long linear hydrocarbon chains, as per quantum mechanics [46]. Additionally, for saturated hydrocarbons, this force is dependent on the carbon atoms quantity in the chain of hydrocarbon. When double bonds are added to the hydrocarbon chain, it is discovered that the forces of cohesion significantly decrease.

The surface pressures v/s area per molecule isotherms for a series of fatty acids are depicted in Fig. 10.

5.2 Electrostatic Interactions

Dipole–dipole interaction, which mostly takes place in the monolayer, is the interaction among the polar head groups in a distributed monolayer. Nonetheless, ion–dipole and ion-ion interactions could also occur if the dispersed monolayer becomes partially ionized [54]. While the ion–dipole and dipole–dipole interactions change as second and third powers of separation, correspondingly, the interaction among the 2 ions is inversely proportional to separation among the ions.

According to Marsden and Schulman [55], a mostly packed monolayer for a mixture of sodium stearate and stearic acid was observed for a 1:1 ratio of soap to acid.

In this context, it should be noted that the dissociation at the interface differs energetically from that in bulk.

6 Monolayer Stability

The spread molecule monolayer at the interface of air–water is regarded as supersaturated and not in thermodynamical equilibrium. If the monolayer is compressed further after the attainment of the solid state, the collapsing of the monolayer takes place. It is also found that the collapse pressure depends on the speed at which the monolayer is being compressed. The highest pressure of the surface at which the monolayer is in the thermodynamical equilibrium having the stable bulk phase is known as equilibrium surface pressure (ESP).

Liquid surfactant monolayers are observed to collapse at pressures closer to their ESP whereas solid surfactant monolayers may be compressed to surface pressures well beyond ESP. Evaporation of molecules [56], dissolution into subphase [57], reorganization, re-orientation, and the collapse of the three-dimensional bulk crystals [58] are basic causes for the instability of the monolayer.

7 Transfer of Langmuir Monolayer onto the Solid Substrate

The exceptionally well-organized molecular assembly that exists at the interface of air–water could be transferred to a solid substrate to produce LB films. In addition to being a great model for researching the ways in which molecules interact with one another, the Langmuir monolayer is perhaps most recognized for its capacity to generate monolayers or multilayers onto solid substrates with precisely controlled thicknesses [59]. In LB films, it is possible to deposit not only a single monolayer but also numerous monolayers onto a solid substrate, producing multilayered LB films. In addition, the deposition of a single monolayer is possible. These are molecular-dimension films that are extremely well-organized and perfect. The process of production has a significant impact on the characteristics of these films. The Langmuir film from the interface of air–water was transferred onto the solid substrate by dipping the substrate vertically into the spread monolayer at a stable surface pressure. This resulted in the production of a well-ordered mono- or multi-layered film on a solid substrate. The molecular architecture of the films can be controlled using the LB approach. This implies that films can be made to have certain, desired qualities that are adapted to fit the needs of the intended users. This method was originally shown by Blodgett and Langmuir [27], and the films that follow their names are referred to as LB films.

Different architectures of LB films are observed depending on the substrate surface and the type of monolayer. Additionally, there are three distinct LB film deposition processes, which are described below.

7.1 Y-Type Deposition

The most popular method for depositing a Langmuir monolayer onto a solid substrate is Y-type deposition. The sort of substrate used determines the kind of interactions involved in this kind of deposition. If a hydrophilic substrate is used then the hydrophilic part of the substrate interacts with the hydrophilic heads of the surfactants. The schematic diagram for the said deposition is shown in Fig. 11.

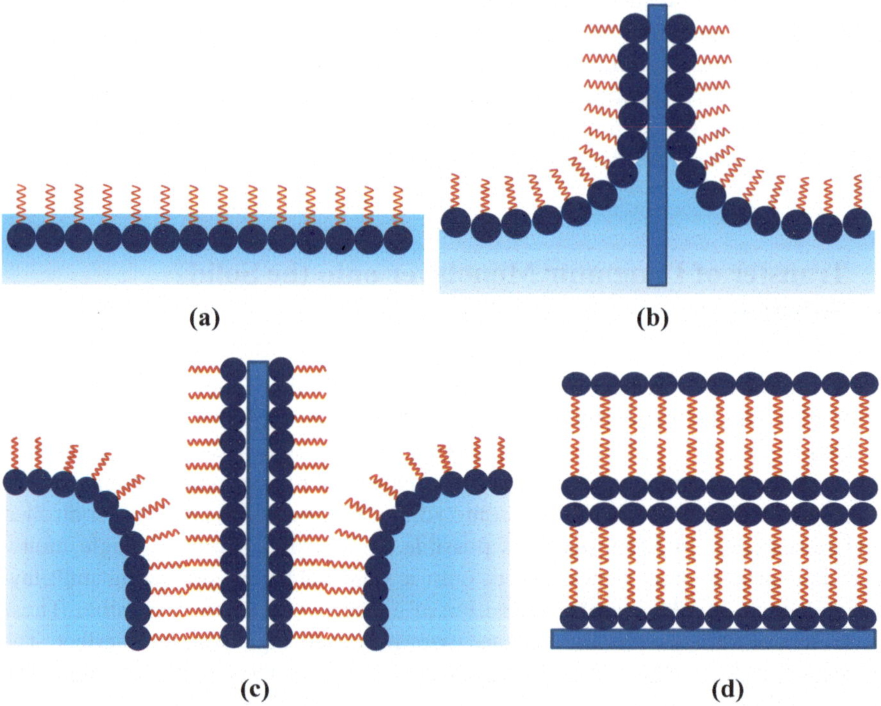

Fig. 11 Y-type deposition

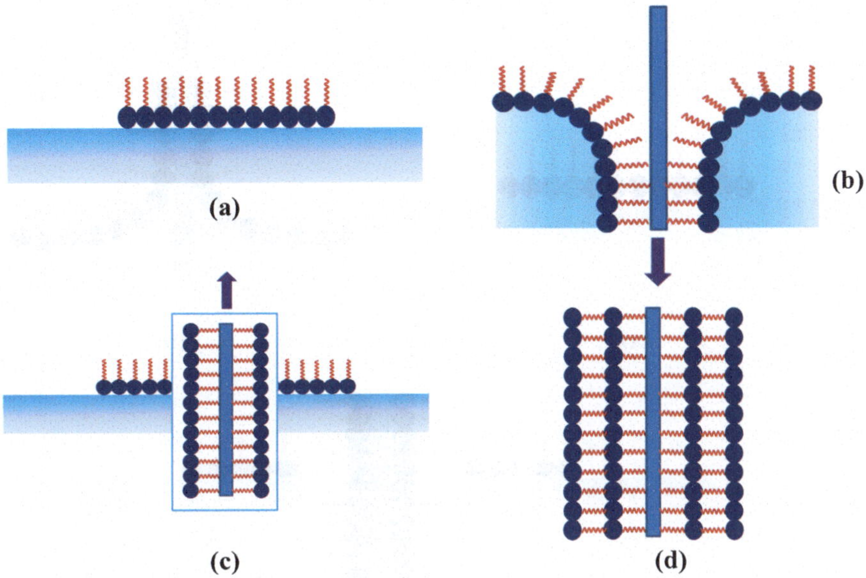

Fig. 12 X-type deposition

7.2 X-Type Deposition

X-type deposition is less common than Y-type deposition. The surfactants' hydrophobic tails interact with the hydrophobic substrate. The X-type deposition scheme is shown in the Fig. 12.

7.3 Z-Type Deposition

Z-type deposition is also less used than the Y-type deposition. The substrate is hydrophilic and interacts with the hydrophilic heads of the surfactants. The Z-type deposition scheme is shown in Fig. 13.

7.4 Langmuir–Schaefer (LS) Method

Irving Langmuir and V. J. Schaefer jointly introduced another method of thin film deposition, which is known as LS's method. In this method, LB multilayered structures of suitable materials can be obtained by the horizontal lifting technique [60]. The Japanese first invented this method of thin film fabrication nearly a thousand

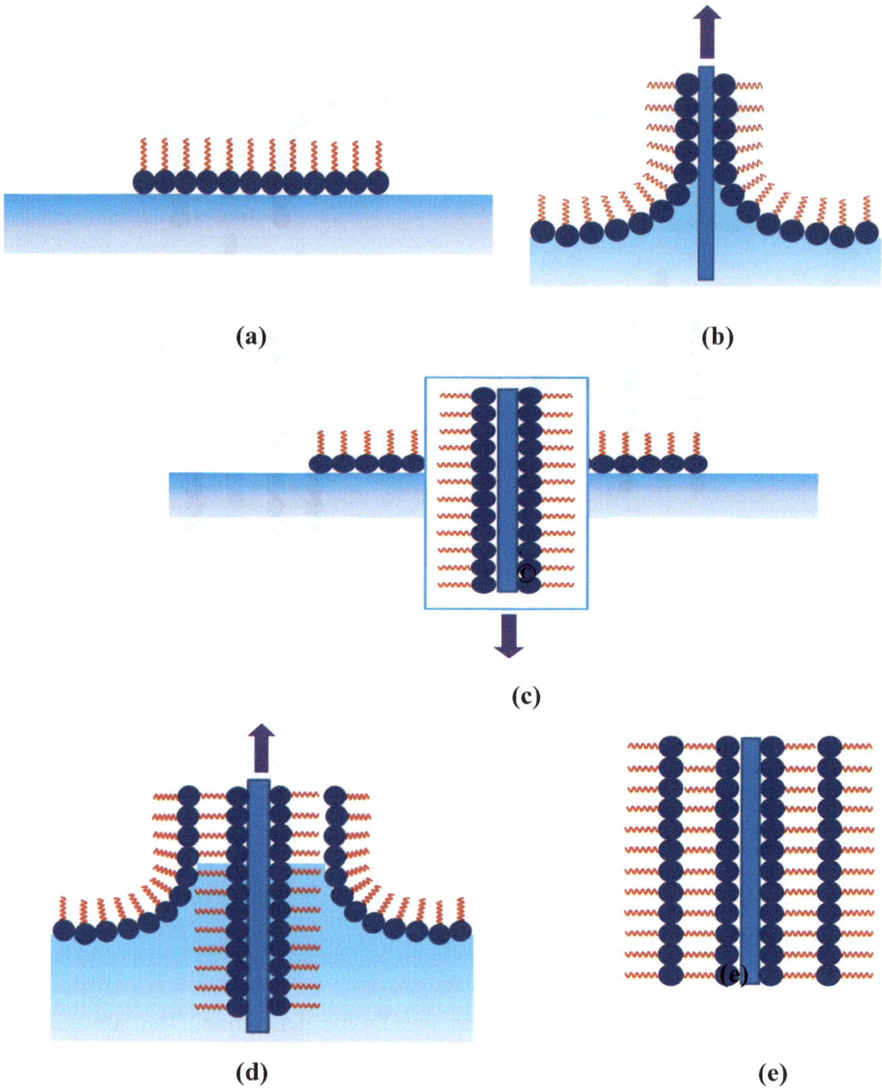

Fig. 13 Z-type deposition

years ago and called it Sumi Nagashi [61] but Langmuir and Schaefer were the first who give a systematic explanation.

The deposition of very hard films, or two-dimensional solid regions in the pressure-area (π–A) isotherm, is a very valuable application of this approach. There are two essential processes in this film deposition method:

1. At the contact between the air and the water, the formation of a compact monolayer of an appropriate substance and

2. Horizontal lifting of compressed monolayer on the substrate.

At the air–water interface, a compressed monolayer of suitable LB-compatible material is initially created, as can be seen in Fig. 14a. This step is performed in the beginning. After that, a substrate like quartz is dipped horizontally on top of the monolayer in a way that is slow and smooth in a manner that is similar to what is seen in Fig. 14b. Now when the substrate is lifted from the interface of air–water, depending on the substrate nature used, a monolayer is transferred onto it. Here, the deposition is of X-type. The monolayer transferred in this way is shown in Fig. 14c. Although this process of thin film deposition is simple it contains some limitations. For instance, the nature of the film created by this technique is found to depend on the speed with which the substrate touches the monolayer. However, the monolayers of polymeric amphiphilic molecules are very much suitable for deposition by this technique because of their high viscous force. The LS method can be very useful in the near future if one can eliminate the minor limitations that it possesses. It is because; this particular technique has the following advantages:

1. The horizontal deposition rate is not reduced with the increase in film viscosity. Hence, thermally stable polymeric films can be prepared by this technique.
2. Secondly, the formation of non-centrosymmetric X-type multilayer films of suitable LB materials is possible, which leads to various technological applications in molecular electronic devices.
3. The third vital benefit of this method is the likelihood of constructing a highly closed-pack, ordered three-dimensional molecular organization commonly known as a superlattice that exhibits new photophysical and electrical characteristics. These changes in the photophysical and electrical properties are highly important both from fundamental as well as the application perspective.

8 Aggregation in Langmuir Blodgett (LB) Films

Aggregation is the most general phenomenon for a broad range of aromatic compounds and dyes, which is commonly observed in crystals and organized media like LB films. In a close-packed crystalline structure, the mother chromophore is used to orient the molecule in a particular direction. This orientation leads to a strong coupling between the interacting dipoles. Hence, different alterations in the spectral, photophysical, and morphological behaviors are observed. In the late 1960s, H. Kuhn and his co-workers showed that highly organized monolayers might be formed when long-chain surfactant dyes such as cyanine dyes are mixed with a long-chain fatty acid. Using an extended dipole model, they also investigated the aggregation for various configurations of the clustered dye molecules. The orientation and distance of the dye molecules are related to the spectrum energy changes between the aggregate and monomer by use of this extended dipole model [62, 63]. There are three types of aggregation viz. J, H, and I-types of aggregation. In this particular model, they

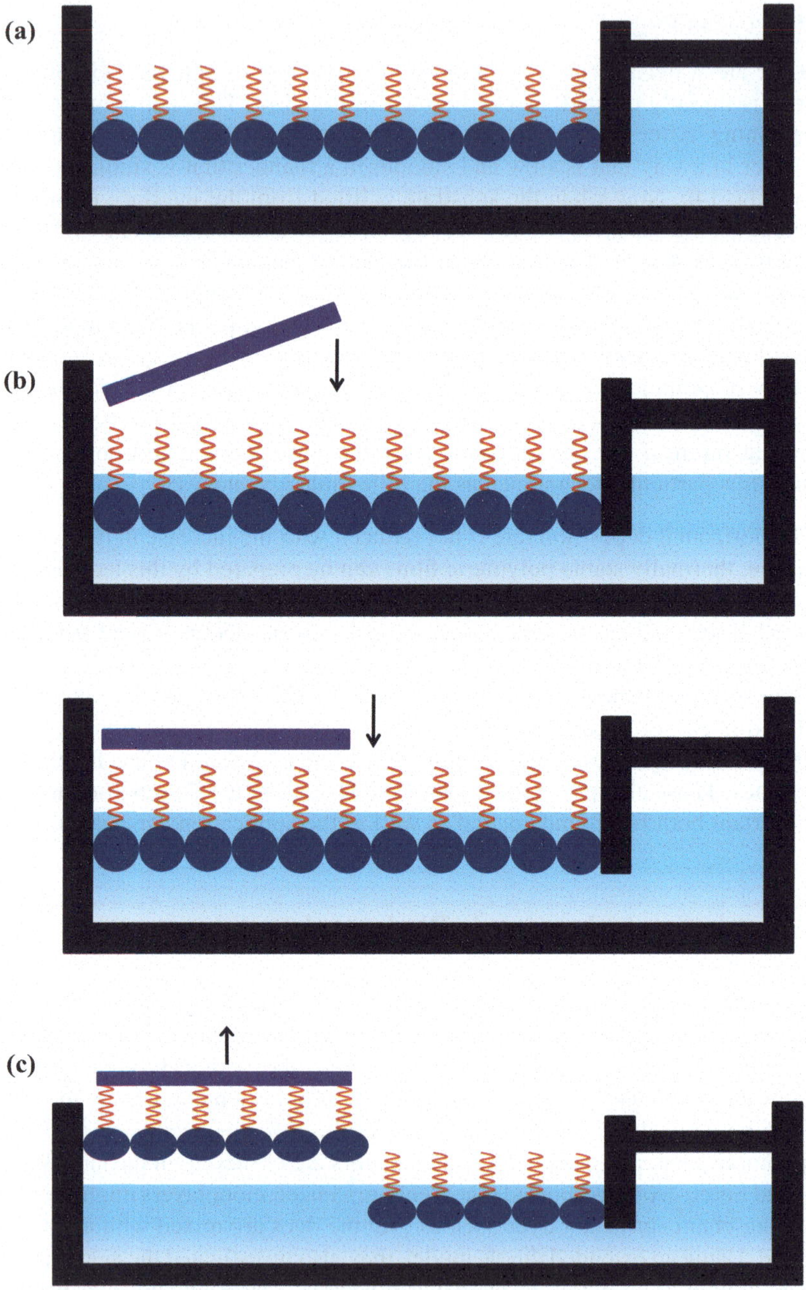

Fig. 14 **a** Langmuir monolayer formation at the contact between the air and the water, **b** Vertical immersion of substrate, **c** Horizontal lifting of monolayer onto a solid substrate

related the dye monomer by a one-dimensional electron gas model for a branched π-electron system. The J- and H-type of aggregations are two typical examples of such types of aggregation. The band shifts in the J-and H-type aggregates were calculated by considering the interaction among a molecule pair in the aggregates.

For example, the J-aggregates of cyanine dyes show a distinctive acute peak of absorption named the J-band, which is red-shifted through the monomer band and shifted towards the shorter wavelength area [64, 65]. Because of their high extinction coefficient, short radiative lifespan resonance fluorescence, and huge optical non-linearities, J-aggregates show great promise as spectral sensitizers in photography and non-linear optical materials. The J-aggregate is mainly comprised of a large array of closely packed molecules, the transition dipole moments of those are aligned head to tail in a brickwork arrangement. The exciton band is formed due to the strong interactions between two interacting dipoles.

According to the exciton model given by Kasha et al. [66, 67] the splitting of exciton (excited state species) in the excited states is mainly accountable for the appearance of the strong spectral shift observed in the absorption spectrum of a composite molecule. A triplet-state exciton may also result as a consequence of the splitting of the exciton in the excited state.

A composite molecule's parallel transition dipoles produce the exciton energy level diagram that is depicted in Fig. 15a. The double arrow in the figure represents the polarization axis for a molecular electronic transition that is being studied, and the ovals in the figure depict the molecular profile. In the right side of the figure, the exciton states E′ and E″ are represented by the vector diagram. From such a diagram one can easily get valuable information regarding (1) the energy of exciton splitting E, and (2) the transition dipole moment which gives a transition probability between the ground state E′ and excited state E″ of the molecule.

Lowering of energy (E is negative) can occur when the interacting dipoles are aligned out-of-phase i.e., in that case, E is situated lower in comparison to the "vander Waal's" displaced situations of the component molecules. But when the interacting dipoles are in phase then due to the repulsive nature of interaction E is positive and hence E″ is displaced upwards from the origin. The transition from the ground state to the excited state E″ is justified by the transition dipole moment of the composite molecule, which is equal to the sum of the individual transition dipole moments of the component molecules.

The exciton energy level of a composite double molecule when the transitions between two dipoles occur when they are interacting along a line is also shown in Fig. 15b.

Figure 15c also depicts co-planar generated transition dipoles in a composite double molecule that contribute to the exciton energy map.

The line connecting the centers of the interacting molecules (α), which is 0 in Fig. 15a and 90° in Fig. 15b, and the polarization axes are continuously varied in this general instance.

The expression for the energy required to split the exciton band can be written as,

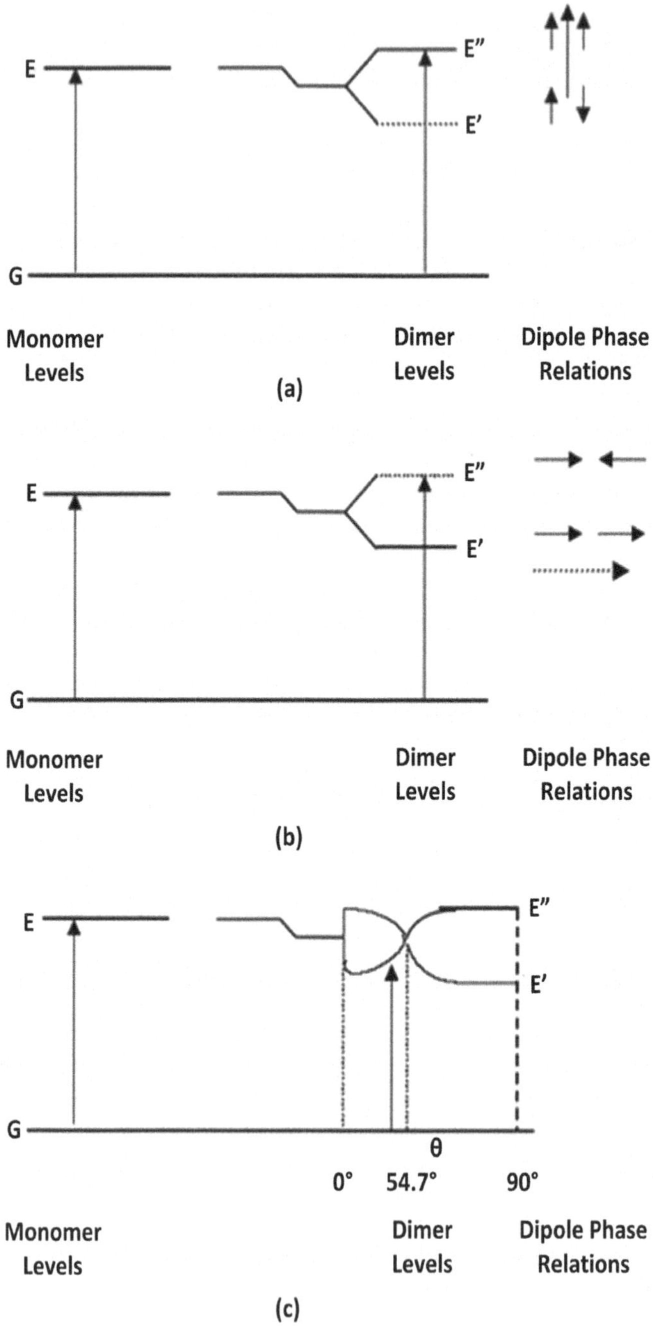

Fig. 15 Formation of exciton band

$$\Delta E = \frac{2M^2}{r^3}\left(1 - 3\cos^2\alpha\right)$$

Here, M is represented as the transition dipole moment and α shows the angle formed by the dipole moment vector connecting the two dipoles' centers with r. An exciton band could arise as a result of the 2 dipoles' dipole-to-dipole interaction. The exciton band that has just formed could be located energetically above & below the band of monomeric.

Whether there are any transitions from the state ground to the excited state or not depends on the angle α between the two dipoles. For the angle falling between the $0^{\circ} < \alpha < 54.7^{\circ}$, When compared to the energy level of the monomeric band, the exciton band is found to be lower. Because of this, a change in the absorption spectra occurs in the lower energy area, which is referred to as a red shift. The aggregates in discussion are referred to as J-type aggregates, and they are the ones responsible for such a change [62, 63].

Again, if the angle formed by the two dipoles is within the range of $54.7^{\circ} < \alpha < 90^{\circ}$, the newly created exciton band will have an energy level that is greater than that of the monomeric band.

This causes a shift in the absorption spectra that is usually referred to as the 'blue shift' in the higher energy area. H-type aggregation is the sort of aggregation that causes a shift like this in the absorption spectra of composite molecules. If the angle among 2 interacting dipoles is between 54.7 and 60°, then the related aggregation is called an Intermediate or I-type of aggregation. This type of aggregation only occurs in the most exceptional cases [66, 67]. The formation of I-type aggregates was reported in various molecules such as octadecyl rhodamines, spiropyrans, and various derivatives of anthracene (9-phenyl anthracene) [42].

9 Characterizations of LB Films

9.1 Physical and Mechanical Properties

The physical properties of such ultrathin films are of fundamental importance to make them fit for technological applications as these ultrathin films are sometimes found to be very unstable, fragile, and soft. For this, scientists have to investigate the behaviors of these films to check their stability as time passes.

The actual performance and the possibility of practical applications of these ultrathin films are found to depend largely on some of their mechanical properties. The precise measurements of the mechanical properties of films are, therefore, crucial for their design to be highly effective. As such, a number of mechanical properties like elastic modulus, indentation hardness, film-substrate adhesion, and internal (residual) stresses can be estimated to get an idea about the internal bonding and architecture of the thin films.

9.2 *Photophysical/Spectroscopic Characterizations*

After the formation of LB films, photophysical characterizations of these films are generally investigated to assess the molecular mechanisms involved during the formation of films. For this, traditional spectroscopic methods viz. (i) UV–Vis absorption spectroscopy, (ii) Steady-state fluorescence spectroscopy, (iii) Excitation spectroscopy, (iv) FTIR spectroscopy, etc. are generally employed to analyze the molecular level structure and to probe molecular organization.

There are some reports regarding the presence of dimmer, higher order n-mers, excimers, etc. in LB films and the existence of such an aggregated species in LB films is generally verified in the light of spectroscopic methods.

9.3 *Structural Characterizations*

To get some idea about the nature of the surface morphology of crystalline domains and their architecture as well as to get an extent of roughness in LB films, the traditional structural analysis viz. (1) AFM (Atomic Force Microscopy), (2) SEM, (3) XRD (X-ray crystallography) (4) Raman Spectroscopy, (5) STM ("Scanning Transmission Microscopy"), (6) BAM ("Brewster Angle Microscopy"), (7) TEM ("Transmission Electron Microscopy"), etc. May be very much useful.

9.4 *Measurement of Thickness*

A spectroscopic ellipsometer is generally employed to estimate the film thickness at room temperature. In such cases, the successful growth of LB films after every alternate deposition can be evidenced by variation of thickness with a number of layers, and in general an exponential growth with a number of layers can be observed.

10 Application Perspective and Future Prospects of LB Films

The future projections of LB thin films are promising, with possible applications in several fields.

LB films could be utilized in the development of next-generation electronic devices, like organic field-effect transistors, OLEDs ("Organic Light-Emitting Diodes"), and flexible electronics, due to their precise control over film thickness and properties [68]. They also have the potential for use in photovoltaic devices and sensors, enabling improved light absorption and charge transport properties [69].

These films can be widely used as templates to create nanoscale structures, facilitating advancements in nanotechnology, including the fabrication of nanoparticles and nanowires [70]. They are also being explored for drug delivery systems. They may also be used in piezoelectric devices to produce biosensors and tissue engineering due to their capability to incorporate biomolecules, mimic the biomembranes, and control surface properties [71]. To modify surfaces for various applications, such as anti-fouling coatings, super-hydrophobic surfaces, and biocompatible interfaces LB films can be used [72]. These ultrathin films may play an essential role in enhancing the energy storage devices performance like batteries [73] and super-capacitors [74] by enhancing electrode materials. Continued research on LB films may lead to the advancement of novel thin-film devices for displays, photodetectors, and other technologies. The versatility and precision of LB film fabrication make it a valuable technique for materials science and engineering, suggesting a bright future for its applications in diverse fields.

Acknowledgements Authors are grateful to both Professor S. A. Hussain and Professor Debajyoti Bhattacharjee of Thin Film and Nano Science Laboratory, Department of Physics, Tripura University, Suryamaninagar for providing constant help, support, and encouragement during the preparation of this review article.

References

1. Roberts, G.G.: Langmuir-Blodgett Films. Plenum Press, New York (1990)
2. Peterson, I.R.: Langmuir–Blodgett dilms. J. Phys. D Appl. Phys. **23**, 379–395 (1990)
3. Ulman, A.: An Introduction to Ultrathin Organic Films: From Langmuir-Blodgett Films to Self-Assemblies. Academic Press, New York (1991)
4. Tredgold, R.H.: Order in Organic Thin Films. Cambridge University Press, UK (1994)
5. Petty, M.C.: Langmuir-Blodgett Films: An Introduction. Cambridge University Press, Cambridge, UK (1996)
6. de Moura, F., Trisic, M.: Molecular dynamics simulation of a perylene-derivative Langmuir film. J. Phys. Chem. B **109**, 4032–4041 (2005)
7. Roberts, G.G., Petty, M.C., Baker, S., Fowler, M.T., Thomas, N.J.: Electronic devices incorporating stable phthalocyanine Langmuir–Blodgett films. Thin Solid Films **132**, 113–123 (1985)
8. Rece, T.J., Bucharme, S., Sorokin, A., Poulsen, V.M.: Nonvolatile memory element based on a ferroelectric polymer Langmuir–Blodgett film. Appl. Phys. Lett. **82**, 142–144 (2003)
9. Zuqiang, B., Kezhi, W., Linpei, J., Chunnui, H.: Colloids Surf. A Physicochemical Eng. Asp. **257–258**, 67–70 (2005)
10. Han, K., Wang, Q., Tang, G., Li, H., Sheng, X., Huang, Z.: Effect of subphase divalent ions on optical second harmonic properties of hemicyanine Langmuir–Blodgett films interleaved with compatible spacer. Thin Solid Films **476**, 152 (2005)
11. Mayya, K.M., Jain, N., Goleb, A., Langevin, D., Sastry, M.: Time-dependent complexation of glucose-reduced gold nanoparticles with octadecylamine Langmuir monolayers. J. Colloid Interface Sci. **270**, 133 (2004)
12. Natali, D., Sampietro, M., Franco, L., Bolognesic, A., Botta, C.: Mobility anisotropy in Langmuir–Blodgett deposited poly (3-methoxypentyl-tiophene)-based thin film transistors. Thin Solid Films **472**, 238 (2005)

13. Franklin, B.: Of the stilling of waves by means of oil. Philos. Trans. R. Soc. **64**, 445–451 (1774)
14. Rayleigh, L.: On the calculation of the frequency of vibration in its gravest mode with an example from hydrodynamics. Phil. Mag. **48**, 321–337 (1899)
15. Pockels, A.: Surface tension. Nature **43**, 437–439 (1891)
16. Pockels, A.: On the relative contamination of the water-surface by equal quantities of different substances. Nature **47**, 418–419 (1892)
17. Pockels, A.: Relations between the surface tension and relative contamination of water surfaces. Nature **48**, 152–154 (1893). https://doi.org/10.1038/048152a0
18. Pockels, A.: On the spreading of oil upon water. Nature **50**, 223–224 (1894)
19. Devaux, H.: Oil films on water and on mercury. Smithson. Inst. Ann. Rep., 261–273 (1913)
20. Langmuir, I.: Chemical reactions at low pressures. J. Am. Chem. Soc. **37**, 1139–1167 (1915)
21. Langmuir, I.: The constitution and fundamental properties of solids and liquids: part-I: solids. J. Am. Chem. Soc. **38**, 2221–2295 (1916)
22. Langmuir, I.: The adsorption of gases on plane surface of glass, mica and olatinum. J. Am. Chem. Soc. **40**, 1361–1403 (1918)
23. Langmuir, I.: The mechanism of the surface phenomena of flotation. Trans. Faraday Soc. **15**, 62–74 (1920)
24. Langmuir, I.: Pilgrim trust lectures, molecular layers. Proc. R. Soc. A **170**, 1–39 (1939)
25. Blodgett, K.B.: Films built by depositing successive monomolecular layers on a solid surface. J. Am. Chem. Soc. **57**, 1007–1022 (1935)
26. Blodgett, K.B.: Properties of built-up films of barium stearate. J. Phys. Chem. **41**, 975–984 (1937)
27. Blodgett, K.B., Langmuir, I.: Built-up films of barium stearate and their optical properties. Phys. Rev. **54**, 964–982 (1937)
28. Kuhn, H.: Energie übertragung in monomolekularen Schichten. Natruwiss **54**, 429–435 (1967)
29. Kuhn, H., Möbius, D.: Systems of monomolecular layers—assembling and physico-chemical behavior. Angewandte Chemie (Int. Ed) **10**, 620–637 (1971). https://doi.org/10.1002/anie.197106201
30. Kuhn, H., Möbius, D., Bucher, H.: Physical Methods of Chemistry, vol. 1. In: Weissberger, A., Rossiter, B. (eds.) Part 3B Chap. 7. Wiley, New York (1972)
31. Kuhn, H.: Functionalized monolayer assembly manipulation. Thin Solid Films **99**, 1–16 (1983). https://doi.org/10.1016/0040-6090(83)90352-8
32. Gaines, G.L.: Insoluble Monolayers at the Liquid-Gas Interfaces. Wiley Inter Science, New York (1966)
33. Chowdhury, A., Pal, A.J.: Transient electroluminescence from rubrene light-emitting diodes using double voltage pulse. J. Synth. Mat. **106**, 85–88 (1999). https://doi.org/10.1016/S0379-6779(99)00113-7
34. Wijekoon, W.M.K.P., Karna, S.P., Talapatra, G.B., Prasad, P.N.: Second-harmonic generation studies of differences in molecular orientation of Langmuir-Blodgett films fabricated by vertical and horizontal dipping techniques. J. Opt. Soc. Am. B **10**, 213–221 (1993)
35. Ehler, T.T., Noe, L.J.: Surface plasmon studies of thin silver/gold bimetallic films. Langmuir **11**, 4177–4179 (1995). https://doi.org/10.1021/LA00010A088
36. Arora, K., Chand, S., Malhotra, B.D.: Recent developments in bio-molecular electronics techniques for food pathogens. Anal. Chim. Acta **568**, 259–274 (2006). https://doi.org/10.1016/j.aca.2006.03.078
37. Frumkin, A.N., Iofa, Z.A., Gerovich, M.A.: The potential difference at the boundary water-gas. Zh. Fiz. Khim. **30**, 1455–1468 (1956). Chem. Abstr. **51**, 4173 (1957)
38. Randles, J.E.B., Schiffrin, D.J.: The temperature-dependence of the surface potential of aqueous electrolytes. Electro anal J. Chem. **10**, 480–484 (1965). https://doi.org/10.1016/0022-0728(65)80049-3
39. Collins, K.D., Washabaugh, M.W.: The Hofmeister effect and the behaviour of water at interfaces. Q. Rev. Biophys. **18**, 323–422 (1985). https://doi.org/10.1017/s0033583500005369
40. Deb, S., Biswas, S., Hussain, S.A., Bhattacharjee, D.: Spectroscopic Characterizations of the mixed Langmuir Blodgett (LB) films of 2,2′-biquinoline molecules: evidence of dimer formation. Chem. Phys. Lett. **405**, 323–329 (2005)

41. Hussain, S.A., Deb, S., Bhattacharjee, D.: Spectroscopic Characterizations of non-amphiphilic 2-(4-biphenylyl)-6-phenyl benzoxazole molecules at the air-water interface and in Langmuir-Blodgett Films. J. Lumin. **114**, 197–206 (2005)
42. Hussain, S.A., Deb, S., Biswas, S., Bhattacharjee, D.: Langmuir Blodgett Films of 9-phenyl anthracene molecules incorporated into different matrices. Spectrochim. Acta Part A **61**, 2448–2454 (2005)
43. Acharya, S., Bhattacharjee, D., Sarkar, J., Talapatra, G.B.: Spectroscopic study of non-amphiphilic 2-(4-biphenylyl)-5-(4-tert-butylphenyl)-l,3,4-oxadiazole aggregates at air–water interface and in Langmuir–Blodgett films. Chem. Phys. Lett. **393**, 1–6 (2004)
44. Acharya, S., Bhattacharjee, D., Talapatra, G.B.: Spectroscopic study of non-amphiphilic 9-phenyl fluorene assembled in Langmuir–Blodgett films in two different matrices: dimer and excimer formation. Chem. Phys. Lett. **372**, 97–103 (2003)
45. Acharya, S., Bhattacharjee, D., Talapatra, G.B.: Spectroscopic characteristics of 9-cyanoanthracene mixed with polymethyl methacrylate and stearic acid in Langmuir–Blodgett films. Chem. Phys. Lett. **352**, 429–434 (2002)
46. Salem, L.: Attractive forces between long saturated chains at short distances. J. Phys. Chem. **37**, 2100–2113 (1962)
47. Harkins, D.: The Physical Chemistry of Surface Films. Reinhold, New York (1952)
48. Maggio, B., Ahkong, Q.F., Lucy, J.A.: Poly(ethylene glycol), surface potential and cell fusion. Biochem. J. **158**, 647–650 (1976). https://doi.org/10.1042/bj1580647
49. Shah, D.O., Schulman, J.H.: Influence of calcium, cholesterol, and unsaturation on lecithin monolayers. J. Lipid Res. **8**, 215–226 (1967). https://doi.org/10.1016/S0022-2275(20)40140-3
50. Sadrazedh, N., Yu, H., Zografi, G.: Surface phase behavior of poly(tert-butyl methacrylate)−fatty acid glyceride binary mixtures as monolayers at the air−water interface. Langmuir **14**, 151–156 (1998)
51. Dörfler, H.D.: Mixing behavior of binary insoluble phospholipid monolayers. Analysis of the mixing properties of binary lecithin and Cephalin systems by application of several surface and spreading techniques. Adv. Colloid Interface Sci. **31**, 1–110 (1990)
52. Barnes, G.T.: On the calculation of excess areas of mixing in two-component monolayers. J. Colloid Interface Sci. **144**, 299–300 (1991). https://doi.org/10.1016/0021-9797(91)90262-7
53. Israelachvili, J.N.: Intermolecular and Surface Forces. Academic Press (1985)
54. Birks, B.P.: Insoluble monolayers of weakly ionising low molar mass materials and their deposition to form Langmuir-Blodgett multilayers. Adv. Colloid Interface Sci. **34**, 343–432 (1991)
55. Marsden, J., Schulman, J.H.: Molecular interaction in monolayers. Trans. Faraday Soc. **34**, 748–758 (1938). https://doi.org/10.1039/TF9383400748
56. Brooks, J.H., Alexanader A.E.: Radiation of evaporation by monolayers. In: La Mer, V.K. (ed.) Academic Press, New York (1962)
57. Ter Minassian-Saraga, L.: Recent work on spread monolayers, adsorption and desorption. J. Colloid Sci. **11**, 398–414 (1956). https://doi.org/10.1016/0095-8522(56)90157-X
58. Brooks, J.H., Alexander, A.E.: Spreading and collapse phenomena in the fatty alcohol series. J. Phys. Chem. **66**, 1851–1853 (1962). https://doi.org/10.1021/j100816a018
59. Vollhardt, D., Kato, T., Kawano, M.: Nucleation and growth of three-dimensional structures in supersaturated arachidic acid monolayers: an atomic force microscopy study. J. Phys. Chem. **100**, 4141–4147 (1996). https://doi.org/10.1021/jp9519880
60. Langmuir, I., Schaefer, V.J.: The effect of dissolved salts on insoluble monolayers. J. Am. Chem. Soc. **59**, 2400–2414. https://doi.org/10.1021/Ja01290A091
61. Messier, J., Marc, G.: Étude par résonance paramagnétique électronique de la structure de couches monomoléculaires de stéarate de cuivre. J. Phys. (Paris) **32**, 799–804 (1971). https://doi.org/10.1051/jphys:019710032010079900
62. Czikkely, V., Forsterling, H.D., Kuhn, H.: Light absorption and structure of aggregates of dye molecules. Chem. Phys. Lett. **6**, 11–14 (1970). https://doi.org/10.1016/0009-2614(70)80062-8
63. Bucher, H., Kuhn, H.: Extended dipole model for aggregates of dye molecules. Chem. Phys. Lett. **6**, 207–210 (1970). https://doi.org/10.1016/0009-2614(70)80220-2

64. Jelly, E.E.: Spectral absorption and fluorescence of dyes in the molecular state. Nature **138**, 1009–1010 (1963). https://doi.org/10.1038/1381009a0
65. Scheiber, S.: Variability of the absorption spectra of some sensitizing dyes and its cause. Angew. Chem. **49**, 563 (1936)
66. Auweraer, M., Verschuere, B., Schryver, F.C.: Absorption and fluorescence properties of Rhodamine B derivatives forming Langmuir–Blodgett films. Langmuir **4**, 583 (1988). https://doi.org/10.1021/LA00081A016
67. Miyata, A., Heard, D., Unuma, Y., Higashigaki, Y.: Three types of aggregates of spiropyran with long and short hydrophobic alkyl chains. Thin Solid Films **210–211**, 175–177 (1992). https://doi.org/10.1016/0040-6090(92)90202-M
68. Hussain, S.A., Bhattacharjee, D.: Langmuir Blodgett films and molecular electronics. Mod. Phys. Lett. B **23**, 3437–3451 (2009). https://doi.org/10.1142/S0217984909021508
69. Liu, L., Qiao, L.-X., Liu, S.-Z., Cui, D.-M., Zhang, C.-M., Zhou, Z.-J., Du, Z.-L., Wong, W.-Y.: Polyplatinayne/polyoxometalate composite Langmuir–Blodgett films: preparation, structural characterization, and potential optoelectronic applications. J. Polym. Sci. Part A Polym. Chem. **46**, 3193–3206 (2008)
70. Subramanian, A., Tiwale, N., Lee, W.-I., Nam, C.-Y.: Templating functional materials using self-assembled block copolymer thin-film for nanodevices. Front. Nanotechnol. 3, (2021). https://doi.org/10.3389/fnano.2021.766690
71. Oliviera, O.N., Caseli, L., Ariga, K.: The past and the future of Langmuir and Langmuir–Blodgett films. Chem. Rev. **122**, 6459–6516 (2022). https://doi.org/10.1021/acs.chemrev.1c00754
72. Adamson, A.W.: Physical Chemistry of Surfaces. Wiley-Interscience, New York (1990)
73. Fang, C., Yoon, I., Hubble, D., Tran, T.-N., Kostecki, R., Liu, G.: Recent applications of Langmuir–Blodgett technique in battery research. ACS Appl. Mater. Interfaces **14**, 2431–2439 (2022)
74. Mendoza, A.E.P., Schmidt, A., Cesca, E.C., Westphal, E., Serbena, J.P.M., Orth, E.S., Zarbin, A.J.G., Winnischofer, H.: Hybrid heterostructured Langmuir–Blodgett films based on graphene and triruthenium clusters as electrode for energy storage devices. JCIS Open **9**, 100080 (2023). https://doi.org/10.1016/j.jciso.2023.100080

Bioactive Glass for Biomedical Application: An Overview

Sushma Yadav, Dharmendra Yadav, Pravin Kumar, Ashishkumar Yadav, Gurudeo Nirala, and Sandeep Yadav

Abstract The term "bioceramics" describes the ceramics utilized in the reconstruction and restoration of damaged or diseased musculoskeletal system parts. Based on their biological response, bioceramics can be divided into three categories: basically bioinert materials (like zirconia and alumina), bioactive materials (like HA and bioactive glasses), and biodegradable materials (like TCP). Bioactive ceramics, including calcium phosphates, glass–ceramics, and bioactive glasses, can encourage the formation of hydroxyapatite deposits that resemble bone at their surface and offer an interface that contributes to the functional durability of the tissue. An amorphous material with an arbitrary atom configuration is called bioactive glass. The majority of bioactive glasses are mostly composed of SiO_2, Na_2O, CaO, and P_2O_5. Fast bonding to bone occurred when the bioactive glass's silica content was between 42 and 53%. For the production of silicate-, phosphate-, and borate-based bioactive

S. Yadav
Department of Mechanical Engineering, Government Polytechnic Bakshi Ka Talab, Lucknow, Uttar Pradesh 226201, India

Department of Ceramic Engineering, IIT BHU, Varanasi, Uttar Pradesh, India

D. Yadav (✉)
Department of Physics, Kisan College, Patliputra University, Patna, Bihar, India
e-mail: dharmendray.phy15@gmail.com

D. Yadav · G. Nirala
Department of Physics, IIT BHU, Varanasi, Uttar Pradesh, India

P. Kumar
Department of Chemistry, Institute of Science, Banaras Hindu University, Varanasi, Uttar Pradesh 226201, India

A. Yadav
Department of Physics, HVPS RJ College, University of Mumbai, Ghatkopar West, Mumbai 400086, India

G. Nirala
Department of Basic Science and Humanities, Government of Polytechnic Munger, Munger, Bihar, India

S. Yadav
Department of Physics, University of Allahabad, Prayagraj, Uttar Pradesh, India

U. Kumar (ed.), *Defects Engineering in Electroceramics for Energy Applications*, Engineering Materials, https://doi.org/10.1007/978-981-97-9018-0_12

glasses, several compositional modifications are made to the bioactive glasses. They have been employed, among other things, in dental root replacements, middle ear replacements, and bone scaffolds. Bioactive glasses have the ability to form a chemical link with living bone tissue, which is further enhanced by biological responses on the material's surface. Due of its low mechanical qualities, bioglass cannot be used for large load-bearing applications. These are employed in situations where mechanical strength is not required and regeneration is essential.

Keywords Bioceramics · Bioglass · Composition · Synthesis · Application

1 Introduction

Any biological or synthetic material, or combination thereof, that is utilized for the human body to enhance and restore bodily function is referred to as a biomaterial [1]. As the field of biomedical science has developed, biomaterials have gained recognition as a crucial area of study. For a biomaterial to serve its purpose effectively, it needs to be compatible with the body. Incompatible materials cause undesired or unfavorable reactions with bodily tissue and blood, which is not acceptable [2]. The ability of a biomaterial to induce a favorable host response in a clinical application is the definition of the term. Stated differently, this term designates the substance as non-toxic or detrimental to nearby tissues. Biocompatibility is a crucial characteristic for a biomaterial [3]. Biocompatibility is a term used to characterize biomaterials and refers to both a material's compatibility and ability to exist within the human body. Implantable devices intended for long-term internal use have historically raised biocompatibility concerns [4]. Three factors need to be taken into account while applying biomaterials and researching their biocompatibility. The first was that depending on the application site, a material's response may vary. Therefore, the context in which a material is utilized may have a greater bearing on its biocompatibility than its inherent qualities. Second, more and more applications required the material to interact directly with the tissues instead of being ignored by them, as is the case with inert materials. Lastly, some uses permitted the material to gradually deteriorate rather than stay in the body indefinitely. In 1987, the definition of biocompatibility changed to incorporate a material's ability to function with a host response that is acceptable under specific conditions [5].

Glass has been used by humans for thousands of years for a variety of purposes, including the use of natural glass in arrowheads and tools and the usage of drinking vessels and glass beads in ancient Egypt and Mesopotamia. Recent applications of glass include fibres for telecommunications, chemical reactions of glassware, and optical and architectural (e.g., window) applications [6]. The bioactive glasses have osteoinductive and osteoconductive properties, which provide the ability to repair damaged tissues. The bioactivity depends on the morphology because it is directly correlated with the dissolution rate. The bioactivity will be higher if the contact between the surface area of the material and the physiological fluid is greater [7].

After being implanted, the dissolution of bioactive glass takes place, releasing the ions, and the growth of a carbonated hydroxyapatite layer takes place on the surface. The network modifiers, such as sodium and calcium, produce the non-bridging Si-O bond, which reduces the network connectivity of SiO_2 and enhances the dissolution process [8]. During the dissolution process, firstly, the Si-OH bonds are formed due to the exchange of Na and Ca cations with H^+ from the biological fluid. Phosphate ions may also be released if they were initially present. The hydrolysis of Si-O-Si bonds caused by an increase in pH is followed by the formation of additional Si-OH bonds, which re-polymerize to produce a glass surface depleted in Na and Ca cations. Following the transfer of Ca^{+2} and PO_4^{-3} ions to the surface, an amorphous calcium phosphate layer is formed, which crystallizes into hydroxyapatite with the absorption of carbonate and hydroxide ions from the biological fluid [9, 10]. The bioglass has the ability to enhance the differentiation of mesenchymal stem cells, enzyme activity, osteoblast adhesion, and revascularization for use in bone tissue engineering applications. L. Hench discovered the first bioglass, which is related to bone, which not only connects directly to bone but also releases dissolution ions that promote the growth of cells at the genetic level, resulting in the development of osteogenesis [11]. Bioactive glass has a random atom structure and is amorphous due to the quick cooling of the molten ceramic. The majority of bioactive glasses are mostly composed of SiO_2, Na_2O, CaO, and P_2O_5. Bioactive glass's exact composition determines its bioactivity, osteoconductivity, and biodegradability. Bioactive glasses that have a silica content of 42–53% bond to bone quickly, while glasses with a silica content of 54–60% take 2–4 weeks to bond, and glasses with a silica content of more than 60% do not directly bond to bone [12].

To create silicate-, phosphate-, and borate-based bioactive glasses, several modifications are applied to their composition. They have been utilized in dental root replacements, middle ear replacements, bone scaffolds, and other procedures [13]. By means of a chemical reaction on their surface and subsequent biological responses, bioactive glasses have the ability to form a link with living bone tissue. At the surface of the implant, there is ion leaching and exchange that causes the glass network to dissolve and a calcium deficit to begin and worsen [14]. The drawbacks of bioactive glasses have been resolved with the development of glass–ceramics. Glass–ceramics are crystalline glasses consisting of two phases: the glassy phase and the crystalline phase. A step in the manufacturing of glass–ceramics is the use of heat treatment to a base glass to induce controlled crystallization and transform it into a glass-crystal combination. The growth of fine-grained crystalline phases is promoted by the heat treatment. Consequently, a variety of particular features, including as bioactivity, machinability, and improved mechanical properties, can be achieved by controlling the crystallization and creation of crystal phases [15]. Due to its low mechanical qualities, bioglass cannot be used in large load-bearing applications. These are employed in situations when mechanical strength is not required and regeneration is significant [16].

2 Bioactive Compositions

Figure 1 displays the Na_2O-CaO-SiO_2 ternary phase diagram with P_2O_5 at 6 weight percent. Soft tissue bonds have been observed to form between specific glasses in Region B, which is subdivided into Region A. Region D glasses break down too quickly, for example, within 10–30 days of implantation, due to inadequate network connectivity. In contrast, region B glasses are thought to be bioactive and fully adhere to the bone. The boundary between B and D is determined by the surface area to solution ratio. Glasses in region C behave like inert materials and do not disintegrate because of the region's stronger network connection and low modifier levels [17, 18].

The properties of bioglass can change based on its composition, processing methods, and intended applications. Here are some factors that can influence property changes in bioglass materials:

1. **Composition**: Altering the relative proportions of components such as silica, sodium oxide, calcium oxide, phosphoruspentaoxide, and other additives can significantly impact the properties of bioglass. Different compositions can affect bioactivity, mechanical strength, and degradation rates [19].
2. **Processing Techniques**: The method used to synthesize bioglass, whether through traditional melt-quenching, sol–gel techniques, or microwave synthesis, can influence its properties. Each method may result in variations in structure, crystallinity, and homogeneity [20]. The sol–gel derived bioglass have higher surface area and porosity which makes them more bioactive as compared to melt derived bioglass [21].
3. **Heat Treatment**: Thermal treatments can be applied to bioglass to modifiy its properties. Annealing can relieve internal stresses, while controlled heating and cooling processes can affect crystalinity and bioactivity.Thermal treatment can

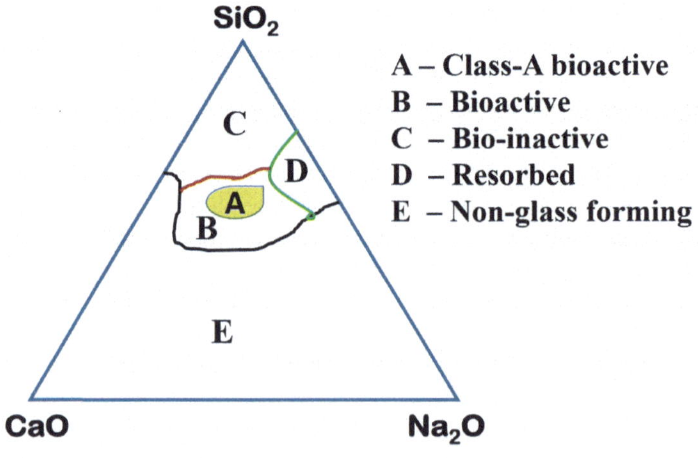

Fig. 1 Property changes of bioglass materials [17]

change the elastic properties of bioglass also at higher temperature, the bioactivity can also inhibit [22].

4. **Particle size**: The size of bioglass particles can impact its behavior. Finer particles have increased surface area, affecting degradation rates and interactions with biological tissues [23].
5. **Additives**: Introducing specific additives like strontium oxide, boron oxide, or other trace elements can impact unique properties to bioglass. Strontium containing bioglass enhance bone regeneration [24].
6. **Bioactivity**: The ability of bioglass to form a hydroxycarbonate apatite (HCA) layer on its surface when in contact with body fluid is a key indicator of its bioactivity. Changes on composition can influence the speed and extent of bioactive response [6].
7. **Mechanical strength**: Adjusting the composition and structure of bioglass can impact its mechanical strength, making it more suitable for load bearing applications in orthopedics [11].
8. **Degradation rate**: Controlled degradation is often desirable in bioglass for specific applications. Changes in composition and processing can influence how quickly the material degrades over time, releasing ions that promote tissue regeneration [25].
9. **Antimicrobial properties**: Bioglass compositions, such as S53P4 exhibit antimicrobial properties. Changes in composition can affect these properties, making the material more or less effective against bacterial action [26].

3 Bioactive Glass Composition

The basic structure of bioactive glass can be described as a silicate glass with a specific composition that includes.

3.1 *Silica (SiO_2)*

Silica is the primary glass-forming component and makes up a significant portion of the glass structure. It provides the basic structure and stability to the material.

i. **Structures and Stability**: Silica serves as the base component that gives bioglass its glassy structure and stability. It is the primary glass- forming component and provides the basic structural framework for the material [27].
ii. **Bioactivity**: Silica contributes to the bioactive nature of bioglass. When bioglass comes into contact with body fluid, the silica rich glass surface undergoes ion exchange with the surrounding tissues. This interaction initiates the formation of HA. The formation of HA on the bioglass surface promotes the integration of the material with the host bone tissue.

iii. **Biocompatibility**: Silica is biocompatible and ensures that bioglass is well tolerated by the human body, reducing the risk of adverse reactions when used in medical and dental applications.
iv. **Chemical Interaction**: It is a crucial component for the chemical reactions that occurs within the bioglass when it is implanted in the body. It facilitates the exchange of ions, which is essential for the bioactive properties of the material [25].

3.2 Calcium Oxide (CaO)

Calcium oxide is another key component of bioglass. It is responsible for the bioactivity of the material. Calcium oxide is added to the glass composition to provide bioactivity. It is one of the elements that allow the glass to bond with bone tissue. When bioglass comes into contact with the body fluid, the calcium ions in the surrounding tissues, promoting the formation of hydroxyapatite, a mineral that is found in natural bone. Its presence is critical for several reasons:

i. **Bioactivity**: Calcium oxide is a key component that gives bioglass its bioactive properties. When bioglass comes into contact with body fluid, the calcium ions (Ca^{+2}) in the glass can be released and interact with the surrounding tissues. This interaction triggers the formation of hydroxyapaptite. The formation of HA on the bioglass surface promotes the integration of the material with the host bone tissue, which is essential for applications like bone grafts and dental implants.
ii. **Biocompatibility**: Calcium oxide contributes to the overall biocompatibility of bioglass. It ensures that the material is well tolerated by the human body and reduces the risk of adverse reactions when used in medical and dental applications.
iii. **Mineralization**: It is an essential mineral for bone growth and repair. The presence of calcium ions released from bioglass promotes bone mineralization and remodeling, supporting the integration of the material with the surrounding bone tissue.
iv. **Osteoconductivity**: It enhances the osteoconductivity of the bioglass. Osteoconductivity refers to the ability of a material to support and encourage the growth of a new bone tissue, making it an important property for orthopedic and dental applications [28].

3.3 Phosphorus Pentoxide (P_2O_5)

Phosphorus pentaoxide is another important component of many bioglass formulations. It is included in bioglass for several reasons, and it plays a significant role in its bioactive properties and applications. It promotes bioactivity and plays a role in forming a layer of hydroxyapatite on the surface of the glass, which is similar to the mineral found in bone.

i. **Bioactivity**: Phosphorous is a key element for the bioactivity of the bioglass. The presence of P_2O_5 is crucial for the glass to promote the formation of Hydroxyapatite (HA), which is a mineral found in natural bone. The interaction between the phosphate ions released from the glass and calcium ion in body fluids leads to the formation of HA. This HA layer is a critical aspect of ability of bioglass to bond with bone tissue and promotes integration.
ii. **Enhanced Biocompatibility**: The inclusion of P_2O_5 contributes to the biocompatibility of bioglass. Biocompatibility is essential to ensure that the material is well- tolerated by the body and does not cause adverse reaction when implanted.
iii. **Controlling Degradation**: It influences the degradation rate of the glass when it comes into contact with body fluids. This controlled degradation can be tailored to release beneficial ions over time, which can be advantageous for tissue regeneration and drug delivery applications [29].

3.4 Sodium Oxide (Na_2O)

Sodium oxide is a common component in many bioglass formulations while it's not one of the primary bioactive components, such as silica, calcium oxide or phosphorous pentaoxide. It serves important functions in the overall composition and properties of bioglass. Sodium oxide is added to lower the melting point of the glass during the manufacturing process, making it easier to shape and work with. The role of sodium oxide in bioglass can include:

i. **Lowering melting Point**: Sodium oxide helps reduce the melting point of the glass during the manufacturing process. This makes easier to shape and process the glass into various forms, such as implant of scaffolds.
ii. **Viscosity control**: It can influence the viscosity of the glass melt, which is essential for controlling the manufacturing process and forming the glass into desired shape.
iii. **Ion Exchange**: While sodium is not the primary ion involved in the bioactivity of bioglass, it can still participate in ion exchange with body fluid to a certain extent, which contributes to the overall interaction of the glass with the surrounding tissues [30].

The specific composition of bioglass may vary depending on the intended application and the manufacturer, but these key components are common in most bioglass formulations. Some other additives are added to bioglass composition to achieve specific properties or tailor the material for different applications. These additives can vary depending on the manufacturer and the intended use of bioglass. Some Common additives and their role include.

3.5 Magnesium Oxide (MgO)

MgO is sometimes added to bioglass compositions for specific reasons related to its properties and potential benefits in biomedical applications:

i. **Controlled Degradation Rate**: Magnesium oxide can affect the degradation rate of the bioglass. The addition of MgO can help modulate the rate at which the glass dissolves in body fluids. This controlled degration is important, especially in application where the bioglass is used as a scaffold for tissue regeneration. It allows for a gradual release of ions over time, which can be beneficial for tissue healing and growth.
ii. **Stabilization of Amorphous Structure**: The addition of MgO can help stabilize the amorphous structure of the bioglass. Amorphous (non-crystalline) bioglass is desirable because it is more bioactive than crystalline materials, promoting the formation of hydroxyapatite on the surface when in contact with body fluids.
iii. **Enhanced Bioactivity**: MgO can enhance the bioactivity of bioglass, promoting the formation of a hydroxyapatite layer on its surface. This layer mimics the composition of natural bone mineral and facilitates the bonding of the bioglass with the surrounding bone tissue.
iv. **Improved Mechanical Properties**: MgO can also enhance the mechanical properties of bioglass. It can make the material more durable and resistant to fracture, which is important for applications like load bearing implants.
v. **Reduced Alkalinity**: In some cases the addition of MgO can help reduce the alkalinity of the bioglass, making it more suitable for use in sensitive biological environment [28].

3.6 Alumina

Alumina also known as aluminium oxide (Al_2O_3) is incorporated into bioglass for various medical and dental applications. Alumina is added to improve the mechanical strength of the bioglass, making it suitable for load bearing applications like orthopedic implants. It also improves the thermal stability of the glass, making it more resistant to high temperature processing or sterilization method [31].

3.7 Fluoride

Fluoride was added to some bioglass formulation for specific dental applications, especially in the context of preventing tooth decay and improving the performance of dental materials. Fluoride ions were introduced to enhance the glass ability to release fluoride over time. This can be beneficial in dental applications to help prevent tooth decay. The inclusion of fluoride serves several purposes:

i. **Dental Protection**: Fluoride is well known for its role in preventing tooth decay by strengthening tooth enamel and making it more resistant to acid attack. When incorporated into bioglass, it offers protection against dental caries.
ii. **Remineralization**: Fluoride promotes the remineralization of tooth enamel, helping to repair early-stage tooth decay or dental lesions [32].

3.8 Strontium

Strontium ions can be added to improve bone health and promotes bone formation. Strontium containing bioglasses have been investigated for their potential in orthopedic and dental applications. Strontium has been found to have a positive impact on bone health. It can promote bone formation and reduce bone resorption. This makes it beneficial for applications involving the regeneration and healing of bone tissues. It has antiosteoporotic properties, and its inclusion in bioglass can be particularly valuable in cases where bone density and strength need to be enhanced. It also influences the bone remodeling process, leading to improved bone quality and overall bone health. It is used in bioglass to promote integration of the material with host bone tissue [33].

3.9 Zinc Oxide

Zinc oxide can be added to affect the bioglass antimicrobial properties and improve its overall biocompatibility. It has inherent antimicrobial properties, and when incorporated into bioglass, it can inhibit the growth of bacteria and reduce the risk of infection at the implant site. This can be particularly valuable in dental implants and other medical applications. It also influences the mechanical properties of bioglass, potentially enhancing its strength and durability. This can be beneficial in load bearing applications such as orthopedic implants. If it used in appropriate concentrations, zinc oxide does not compromise the biocompatibity of bioglass [11, 34].

3.9.1 Boron Oxide

Boron oxide has a lowering melting point compared to some other components of bioglass, which can be useful for controlling the glass processing and forming it into various shapes or coatings. It can reduce the thermal expansion of the glass, making it more compatible with surrounding tissues and other materials. Boron oxide enhances the chemical stability of the glass [35].

3.9.2 Barium Oxide

Barium oxide can be introduced to modify the glass properties and improve its radiopacity, making it useful in dental and medical imaging applications. It is radiopaque, which means it can block X-rays and show up clearly in medical and dental imaging, such as X-rays and computed tomography (CT) scans. This property is important in situations where the visualization of implanted materials or structures is necessary, such as in dental restorations, dental fillings, or bone graft materials. The radiopacity provides by barium oxide helps clinician assess the position and integrity of the bioglass implant or material. It also provides the biocompatibity when used in appropriate concentrations [36].

3.9.3 Potassium Oxide (K2O)

It is not typically a primary component of bioglass, and it is not commonly included in the standard formulations of bioglass. Bioglass primarily consists of silica, calcium oxide, and phosphorous pentaoxide, along with other possible additive. The specific combination and concentration of additives depend on the intended use of the bioglass and the desired properties. These additives can enhance the mechanical strength, degradation rate, release of specific ions to meet the needs of various medical and dental applications [37].

4 Composition of Bioglass Used for Clinical Application

Bioglass is a type of bioactive glass with various compositions designed for specific medical and dental applications. The most well known composition is 45S5 bioglass, but there are others with distinct properties. Here are few different compositions of bioglass:

i. **45S5 Bioglass**: This is widely used composition consisting of 45% SiO_2, 24.5% Na_2O, 24.5% CaO, and 6% P_2O_5. It has ability to bond with bone and is used in bone grafts and dental applications.
ii. **13–93 Bioglass**: This composition has a higher percentage of silicon and a lower amount of sodium oxide compared to 45S5. It is often used for bone scaffold and tissue engineering [14].
iii. **S53P4 Bioglass**: S53P4 Bioglass is a specific composition of bioactive glass with unique properties and clinical applications. This composition is characterized by its relatively high concentration of phosphorous pentaoxide, which distinguishes it from other bioglass compositions. The approximate composition of S53P4 bioglass is 53% Silica and 4% P_2O_5, along with smaller amounts of other components. This higher amount of phosphorous pentaoxide, makes

it suitable for treating bone infections and defects. It has strong antimicrobial properties.

iv. **1393B3 Bioglass**: This composition includes elements such as silica, sodium oxide, calcium oxide and additional components like boron oxide and strontium oxide. This bioglass has potential to stimulate bone regeneration [11].

5 Bioactive Glass Synthesis Method

5.1 Melt Quenching

The proposed bioactive glasses constituent parts are fused together in the necessary quantities to produce the desired composition during the melt quenching process. After being cooled, the melt is powdered. It is possible to quench the molten glass in liquid nitrogen or water. Glass powder can be made by gathering and grinding frits, which are granules of different sizes. The required scale and form can be obtained by pouring the liquid mixture into molds that have been made to a certain size. Numerous types of bioactive glasses can be made with this technology [38]. The first bioactive glass was discovered by Henry L. Hench in 1970 and named 45S5, with a composition of 45% silica, 24.5% sodium oxide, 24.5% calcium oxide, and 6% phosphorus pentaoxide (all compositions in weight percent). This bioglass was developed by the traditional melt-quench method followed by quenching, in which a mixture of oxides is melted at 1300 °C [39]. The melt-quench method is applied to produce glasses with desirable dimensions. Today, more than 99 percent of the most popular glasses are manufactured using this method. The melting of raw materials such as carbonates or oxides yields a viscous liquid, followed by the rapid quenching of molten castings. For melting, the temperature ranges between 1300 and 1400 °C. After grinding the oxide precursors in a ball mill, the mixture is melted in a Pt crucible and cooled to 25 °C in order to produce an amorphous BG. A major disadvantage of this method is the evaporation of the volatile component P_2O_5 at high temperatures [40].

5.2 Sol–Gel Processing

The sol–gel approach, a low-temperature, wet chemical procedure, is the second way to make bioactive glass. In order to generate gel, the sol–gel method involves hydrolyzing and polycondensing the necessary amount of precursor solution at ambient temperature. This gel is heated and dried to create the amorphous glass structure. To allow the gel to develop, this sol is poured into a petri dish. The bonds are getting stronger as the gel ages. To create powder, the gel is dried and ground [41]. A molecule that has a core metallic ion (M) connected by an oxygen bond (O) to functional organic groups (R) is called a metal alkoxide of the M-(OR)x type. Metal alkoxides like tetraethoxysilane (TEOS) and tetramethoxysilane (TMOS) are frequently utilized as precursors for silica because of their propensity to react with water. Hydroxyl groups are substituted for the alkoxy side chain as a result of the hydrolysis reaction. When the oxygen atom in the water molecule launches a nucleophilic attack against the silicon atom, hydrolysis takes place.

Hydrolysis Reaction:

$M\text{-}(OR)_4 + 4(H_2O) \rightarrow HO\text{-}M(OR)_3 + R\text{-}OH$

R is an alkoxy functional group; an example of one would be C_2H_5OH. Clusters or branched polymeric chains are the outcome of hydrolysis, which can be adjusted by adjusting the reagent ratios. Because of an increase in the inorganic network's connectedness, viscosity rises during polycondensation.

Condensation reaction:

$(OR)_3\ M\text{-}OR + HO\text{-}/M(OR)_3 \rightarrow (OR)_3\ M\text{-}O\text{-}M(OR)_3 + ROH$

$HO\text{-}M\ (OR)_3 + HO\text{-}M(OR)_3 \rightarrow (OR)_3\ M\text{-}O\text{-}M(OR)_3 + H_2O$

As byproducts of the condensation reaction, alcohol and water are released. The gel's pores still contain water. Water is kept in the pores by the aging mechanism, which promotes solid network reprecipitation and a localized solution. As a result, interparticle necks become thicker and have higher density and strength. Usually, the aging process occurs for several hours or days at high temperatures. Small, linked holes with widths ranging from 1 to 20 nm are left behind when the monolith is separated from the pore liquid and leftover alcohols during the drying stage. Higher temperature stabilization results in additional drying, the removal of surface silanol groups, and the formation of three-dimensional silica rings [42]. By using this technique, the glass network is transformed to resemble its melt-derived counterpart while also increasing density, strength, and hardness. Calcium nitrate or calcium chloride is combined with reagents such as tri-ethyl phosphate (TEP) to create the oxides of calcium and phosphorus. Using the sol–gel process, glasses with greater uniformity and purity can be produced. Moreover, this method's entire sequence of operations is carried out at temperatures much lower than those required to create glasses by the melting process [43] (Fig. 2).

5.3 Hydrothermal Method

In the hydrothermal method, the bioactive glasses were produced at a very low temperature. The synthesis of nanomaterials using this method is highly dependent on the solubility of the reacting precursor in a water medium at the desired pressure. An autoclave is a stainless steel chamber lined with Teflon that is used for this process. In the Teflon-lined autoclave, which contains the pre-synthesised reacting precursors, water is added to the reacting precursor. The autoclave is kept at a specific temperature, creating internal pressure that dissolves the reacting precursor and starts

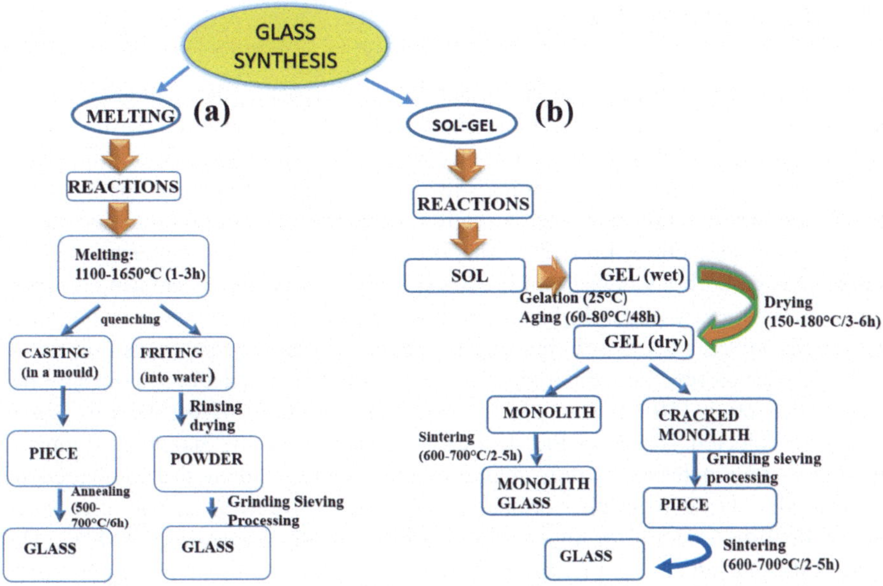

Fig. 2 Diagram depicting the synthesis of bioglass **a** Melt procedure **b** Sol–gel procedure [44]

the crystals' growth. Hydrothermal synthesis has been combined with microwaves to improve reaction kinetics. Due to the fact that the entire reaction takes place in a closed environment, it can quickly and efficiently synthesise nanomaterials while using very little energy [45]. The quaternary systems $40SiO_2$-27CaO-$27Na_2O$-$6P_2O_5$ (mol %), $60SiO_2$-30CaO-$6Na_2O$-$4P_2O_5$ (mol %), and the ternary system $58SiO_2$-33CaO-$9P_2O_5$ (wt %) were synthesised by this method [46]. A mixture of TEOS (tetraethyl orthosilicate), TEP (triethyl phosphate), $Ca(NO_3)$, and other required precursor nitrates were stirred in water. After being combined, the mixture was placed in an autoclave lined with Teflon and made of stainless steel. The hydrothermal synthesis reactor was set to 160–200 °C for 24 h. The gel-producing substance was dried at 100 °C for 24 h to produce the gel powder. The gel powder was dried and sintered at approximately 700 °C based on the results of the thermal analysis. Tuan et al. [47], synthesised bioactive glass of composition 58 wt% SiO_2, 33 wt% CaO, and 9 wt% $P_2O_5$by an acid-free hydrothermal method. For synthesis, TEOS, TEP, and calcium nitrate tetrahydrate in the required amount are stirred with water for 60 min. The mixture was transferred into a Teflon tube and autoclaved at 160 °C for 24 h. The autoclaved product was dried and sintered at 700 °C for 3 h. The glass has an amorphous structure [47].

5.4 *Microwave Synthesis*

Microwave manufacturing techniques can produce a higher quality yield in a shorter amount of time and control the fabrication procedure. In the microwave-assisted method, powders can be produced by employing an efficient and modified heat transfer throughout the entire volume [48]. In conventional heating, the vessel is heated, and heat is transferred via convection. However, microwaves transfer energy more quickly and uniformly [49]. The benefits of microwave sintering include increased sintered body density and reduced grain sizes at lower sintering temperatures, as well as significantly faster heating rates compared to conventional techniques. Moreover, microwave sintering provides mechanical properties due to the finer microstructures obtained at sintering temperatures equivalent to conventional resistance heating [50]. For microwave synthesis method, firstly the raw materials were selected based on the desired composition of the bioglass. Required amounts of raw materials were weighted and mix them to create the homogeneous mixture. Mixed raw materials were placed in a microwave heating. The microwave energy, heat the material to high temperatures, typically exceeding 1000 °C. The rapid heating results in the melting Khalid et al. synthesised bioactive E-glass fibre using the microwave method. SEM images validated the homogenous adhesion of nano-hydroxyapatite spherical particles throughout the fibres. Viability tests on mesenchymal stem cells revealed adhesion, proliferation, and growth [51]. Essien et al. [52] have successfully synthesised CaO-MgO-SiO_2 semi-crystalline bioactive glass by microwave method from rice husks and chicken eggshells, both of which are rich sources of SiO_2 and CaO, respectively. The results demonstrated that there was no aggregation of glass particles and, consequently, a sufficient surface area for the glass to react with body fluid. This was demonstrated by the rapid pH changes during immersion in SBF and the formation of apatite within three days [52].

5.5 *Flame Spray Synthesis Method*

The flame spray synthesis method allows the incorporation of nanoparticulate elements into complex materials such as BGs. For the production of inorganic nanoparticles, flame spray fabrication is a scalable and cost-effective technique. In addition, this procedure ensures a narrow particle size distribution, which contributes to the low product contamination risk. The flame spray synthesis method is one of the most efficient gas-phase-based techniques. This technique also employs metal–organic precursors to generate nanoparticles at temperatures above 1000 °C by igniting the precursors in a flame [53, 54]. The advantage of flame spray fabrication is the simple incorporation of dopants and the convenient availability of numerous nanoparticle compositions. Taubock et al. studied the effect of BG 45S5 particle size on the physical and chemical properties of composites. Experimental compositions were created using the melt-quench technique and flame spray synthesis.

The results indicated that reducing the particle size of BG to nano had a positive effect on the alkalizing potential of experimental compositions [55]. No additional energy source, such as electrically heated walls, lasers, or plasma, is required for the precursor conversion, which is an advantage compared to other gas-phase processes. By utilising oxygen over a nozzle in a tuned system, the liquid precursor is dispersed and an ignited spray is produced. To create nanoparticles, the organic components of the liquid precursor are completely ignited, while the metal components undergo oxidation. The formation of molecular nuclei via chemical reactions or condensation, followed by growth via coalescence in high-temperature regions during the duration of the process, is the fundamental principle of every gas-phase formation technique. The process generates highly chemically homogeneous nanoparticles that are combined with oxides and even salts [54].

6 Promoting the Growth of Bones

- Bioglass stimulates bone growth in two stages: first, hydroxy carbonate apatite (HCA) is formed through precipitation and dissolution; second, the HCA layer interacts with weak collagen fibrils to form a link with the bone. Cells are continuously stimulated by the dissolution products. Like traditional glass degradation mechanisms, HCA production happens in five stages [56].
- On the surface, silanol linkages (Si-OH) are formed when H^+ from the solution combines with rapidly exchanged cations (Na^+, K^+, Mg^{+2}, and Ca^{+2}) in the glass. As a result, a region rich in silica forms close to the glass surface, increasing the local pH. (Phosphate is also released at this initial stage if it is present in the glass composition.)
- At the glass-solution interface, Si-OH (silanol) is formed when OH^- ions contact with the silica glass network and break the Si–O–Si bonds, resulting in $Si(OH)_4$. This process is caused by an increase in pH.
- Si-OH groups condense in the silica-rich layer close to the glass surface.
- Migration of Ca^{+2} and PO_4^{-3} groups occurs from the solution and the bulk glass to the surface of the silica-rich layer, thus creating an amorphous $CaO\text{-}P_2O_5$ layer;
- Incorporation of hydroxyls and carbonate ions from solution and crystallisation of the $CaO\text{-}P_2O_5$ form HCA [57].

Figure 3 demonstrates the development of HCA (hydroxycarbonate apatite) on the surface of the bioactive glass. First, silanol bonds (Si-OH) form on the glass surface; second, the pH of the solution is raised to encourage the formation of a silica-rich area near the glass surface; third, silica bonds break; fourth, additional silanol forms at the glass–solution interface; fifth, a silica-rich layer precipitates; sixth, an amorphous $CaO\text{-}P_2O_5$ forms on the formed silica-rich layer; and seventh, the amorphous $CaO\text{-}P_2O_5$ layer crystallizes to form apatite [17].

The apatite layer production and bone growth surrounding the implant are influenced by the nature of the bioactive glass. A less connected network is linked to

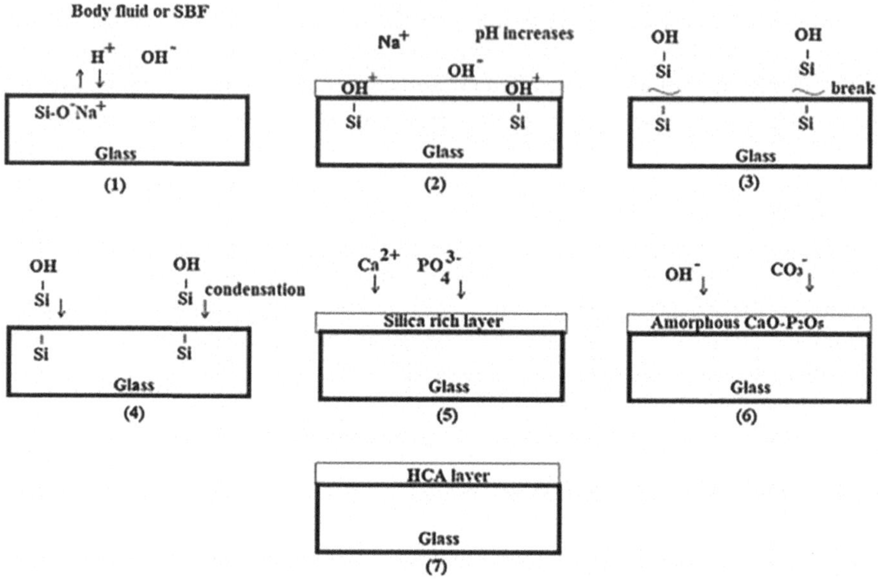

Fig. 3 Presents a schematic illustration of how bioactive glasses' surface forms hydroxycarbonate apatite (HCA)

bioactive glass with a low silica content. This accelerates the development of apatite and facilitates the breakdown of bioactive glass. Bioactivity and the activation energy of silica dissolution in glasses are frequently linked. However, the quantity of silica in glass and the connectedness of the glass network affect their bioactivity. Cations like salt and calcium speed up the rate of dissolution and boost bioactivity, whereas multivalent ions like Ti^{+4} or Al^{+3} in bioglass decrease bioactivity and solubility [57]. Because of their strongly linked network, glasses with a high silica content have reduced bioactivity and dissolution rates. Therefore, for bioactivity and dissolution, network modifiers with the ability to weaken the glass network are crucial. Bioactive glasses are perfect for cell adhesion and proliferation because they can interact with the physiological environment and promote in-vivo osteogenesis on the implant surface [17] (Fig. 4).

7 Application of Bioactive Glass

Bioactive glass, often referred to as bioglass, has a wide range of applications, primarily in the field of medicine and healthcare. Its ability to bond with biological tissues, such as bone, and promote tissue regeneration makes it valuable for various medical and dental applications. Some of the key applications of bioglass include.

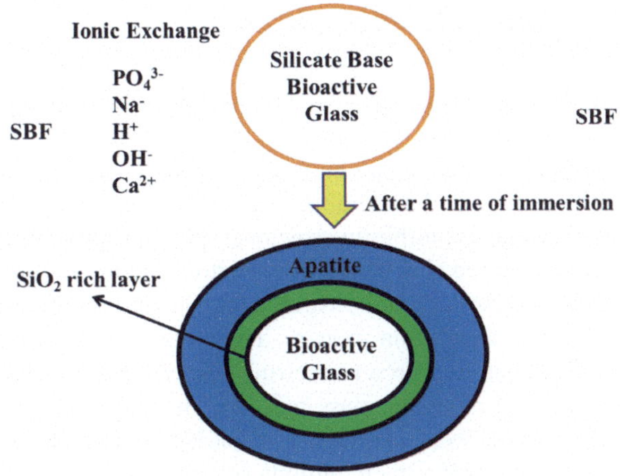

Fig. 4 Diagrammatic depiction of how silicate bioactive glasses dissolve in simulated bodily fluid [17]

7.1 *Bone Grafting*

After blood transfusions, bone grafting is the most common tissue transplant in modern medicine in the United States. Two million bone grafts are performed annually [58]. On the other hand, BGs have been used successfully as bone substitutes. In addition, BGs are the first synthetic materials capable of adhering to bone due to the formation of a biologically active layer of hydroxycarbonate apatite (HCA) over the exposed surface. In terms of structure and chemistry, this layer resembles the bone's mineral phase. Bioactive glasses are naturally brittle; they cannot be used for load-bearing components. Due to physical blending and polymer foam replication, 3D porous scaffolds with varying percentages of carbon nanotubes (CNTs) have been manufactured. The addition of 0.2% CNT to pure 13-93B1 biologically active glasses increased the compressive strength and the elastic modulus [59]. In addition, the polymer polycaprolactone (PCL) was deposited on the CNT-reinforced scaffolds using a dip-coating technique in order to modify their properties after sealing the microcracks. The polymer coating and CNT reinforcement altered the compressive strength of additively manufactured scaffolds by 98% compared to pure bioactive glass scaffolds [60]. Additionally, BAG displayed osteoinductive characteristics, similar to 45S5 BG, which stimulates osteoblastic activity through the release of its ions and the formation of apatite on its surface. Unexpectedly, the effect of this infusion surpasses that of hydroxyapatite. The dissolution products, such as P, Ca, Si, and Na, promote bone formation. In vivo studies have shown that 45S5 BG is superior to other types of bioactive ceramics in terms of its ability to repair bone [61].

7.2 Antibacterial Application

Bacterial infections have been identified as the primary clinical barrier to successful tissue regeneration. For instance, the risk of bacterial attachment and colonization in bone implants can result in implant failure or a protracted recovery [62]. Antibiotics can lose some of their antibacterial efficacy due to the constant evolution of bacteria, which results in antibiotic resistance. Clinical applications of antibacterial substances may reduce the risk of infection without inducing bacterial resistance. S53P4 and 45S5 bioglass have demonstrated antibacterial properties by increasing the pH value during the dissolution process. The metal ions are used as antibacterial agents. Ag has demonstrated broad-spectrum bactericidal activity [63]. Lu et al. developed a novel Huaxi bioactive glass ceramic. The antibacterial properties of Huaxi bioactive glass ceramic were thoroughly investigated. It has been reported that Huaxi bioactive glass ceramic effectively inhibits acid production as well as the growth of cariogenic bacteria [64].

7.3 Soft Tissues

The ability of BGs to release therapeutic ions to form an apatite layer when they are dissolved in physiological fluids accounts for their legendary success in repairing, regenerating, and replacing damaged tissue. BGs are third-generation biomaterials. A wide variety of BGs have been developed since the first one due to different glass compositions and preparation methods. Some, such as both soft and hard tissues, are very applicable to tissue engineering. The repertoire of bioactive glasses has expanded as a result of recent developments in the development of borate bioactive glass [65]. For the repair of soft tissue, borate bioactive glass advancements in expanding cell growth and its complete biodegradation are particularly effective. Yu et al. reported using fibroblast-derived sheets and silicate-based composite bioglass for the skin. It was found that the dissolution of ions stimulates fibroblasts to produce the growth factors required for wound healing and vascularization. Significant wound healing and the formation of new blood vessels were observed in vivo [66] (Fig. 5).

7.4 Dental Application

Bioglass is used in dentistry for various applications including dental fillings and dental implants. Bioglass based materials can release fluoride ions which help prevent tooth decay and promote remineralization. Bioglass can serve as a scaffold in tissue engineering, support the growth and regeneration of various tissues and organs. It is used in tissue engineering constructs for application like cartilage repair skin grafts and vascular grafts.

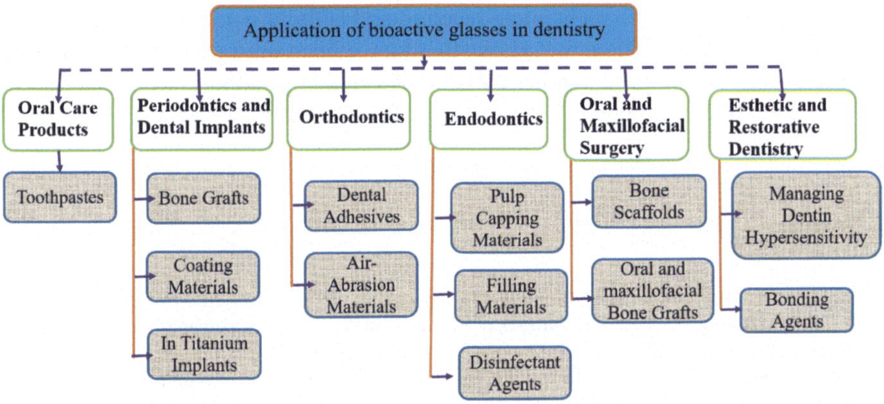

Fig. 5 Dental applications of bioglass [66]

8 Conclusion

Biomaterials are materials, either synthetic or biological, that the body uses to enhance and repair bodily functions. The ability of a material to elicit a positive host reaction in a clinical context is known as biocompatibility. It is a crucial characteristic of biomaterials and is defined by a material's compatibility and capacity to coexist with the human body. Bioglass properties can be influenced by its composition, processing methods, and intended applications. Composition, including the proportions of components like silica, sodium oxide, calcium oxide, and phosphorus pentaoxide, can affect bioactivity, mechanical strength, and degradation rates. Processing techniques, such as melt-quenching, sol–gel, or microwave synthesis, can also affect the properties of bioglass. Thermal treatments can modify properties, while particle size can affect bioglass behavior. Additives, like strontium oxide, can enhance bioglass properties. Bioactivity, mechanical strength, and degradation rate also play a role in bioglass properties. Antimicrobial properties, such as S53P4, can be affected by composition changes, making the material more or less effective against bacterial action. The creation of HCA as a result of dissolution and precipitation, as well as contact with weak collagen fibrils to create a link with the bone, are the two steps of stimulation of bone growth. The apatite layer production and bone growth surrounding the implant are influenced by the nature of the bioactive glass. A less connected network is linked to low silica content in bioactive glass, which accelerates the development of apatite and facilitates the dissolving of the glass. Bioactivity and dissolution depend on network modifiers that have the ability to erode the glass network. Bioactive glasses are perfect for cell adhesion and proliferation because they can interact with the physiological environment and promote in vivo osteogenesis on the implant surface. Bioactive glasses have osteoinductive and osteoconductive properties. Bioactivity, osteoconductivity, and biodegradability depend on the precise composition of the glass. Bioactive glasses can bond with living bone tissue through a chemical reaction and cellular reactions, but have poor mechanical properties. Bioactive glass (BGs) have been used successfully as bone substitutes and are the first synthetic materials capable of adhering to bone due to the formation of a biologically active layer of hydroxycarbonate apatite (HCA).

References

1. Narayan, R.: Encyclopedia of Biomedical Engineering. Elsevier (2018)
2. Black, J., Hastings, G. (eds.): Handbook of Biomaterial Properties. Springer Science & Business Media (2013)
3. Navarro, M., Michiardi, A., Castano, O., Planell, J.A.: Biomaterials in orthopaedics. J. R. Soc. Interface **5**(27), 1137–1158 (2008)
4. Rezaie, H.R., Bakhtiari, L., Öchsner, A.: Biomaterials and their applications. Springer International Publishing, Berlin, Germany (2015)
5. Williams, D.F.: On the mechanisms of biocompatibility. Biomaterials **29**(20), 2941–2953 (2008)
6. Hossain, S.S., Yadav, S., Majumdar, S., Krishnamurthy, S., Pyare, R., Roy, P.K.: A comparative study of physico-mechanical, bioactivity and hemolysis properties of pseudo-wollastonite and wollastonite glass-ceramic synthesized from solid wastes. Ceram. Int. **46**(1), 833–843 (2020)
7. Baino, F., Hamzehlou, S., Kargozar, S.: Bioactive glasses: where are we and where are we going? J. Funct. Biomater. **9**(1), 25 (2018)
8. González, P., Serra, J., Liste, S., Chiussi, S., León, B., Pérez-Amor, M.: Raman spectroscopic study of bioactive silica based glasses. J. Non-Cryst. Solids **320**(1–3), 92–99 (2003)
9. Vichery, C., Nedelec, J.M.: Bioactive glass nanoparticles: from synthesis to materials design for biomedical applications. Materials **9**(4), 288 (2016)
10. Hench, L.L.: Bioceramics: from concept to clinic. J. Am. Ceram. Soc. **74**(7), 1487–1510 (1991)
11. Ali, A., Singh, B.N., Yadav, S., Ershad, M., Singh, S.K., Mallick, S.P., Pyare, R.: CuO assisted borate 1393B3 glass scaffold with enhanced mechanical performance and cytocompatibility: an In vitro study. J. Mech. Behav. Biomed. Mater. **114**, 104231 (2021)
12. Hench, L.L., Thompson, I.: Twenty-first century challenges for biomaterials. J. R. Soc. Interface **7**(suppl_4), S379–91 (2010)
13. Essien, E.R., Adams, L.A.: Preparation of Mg–Zn bimetallic doped Na-containing bioceramic from sodium metasilicate. J. Porous Mater. **23**(4), 885–893 (2016)
14. Rahaman, M.N., Xiao, W., Fu, Q., Huang, W.: Review—bioactive glass implants for potential application in structural bone repair. Biomed Glas. **3**, 56–66
15. Best, S.M., Porter, A.E., Thian, E.S., Huang, J.: Bioceramics: past, present and for the future. J. Eur. Ceram. Soc. **28**(7), 1319–1327 (2008)
16. Palmero, P.: Ceramic–polymer nanocomposites for bone-tissue regeneration. Nanocomposites Musculoskelet. Tissue Regen. 331–367, (2016)
17. Henao, J., Poblano-Salas, C., Monsalve, M., Corona-Castuera, J., Barceinas-Sanchez, O.: Bioactive glass coatings manufactured by thermal spray: a status report. J. Market. Res. **8**(5), 4965–4984 (2019)
18. Cao, W., Hench, L.L.: Bioactive materials. Ceram. Int. **22**(6), 493–507 (1996)
19. Ali, A., Ershad, M., Vyas, V.K., Hira, S.K., Manna, P.P., Singh, B.N., Yadav, S., Srivastava, P., Singh, S.P., Pyare, R.: Studies on effect of CuO addition on mechanical properties and in vitro cytocompatibility in 1393 bioactive glass scaffold. Mater. Sci. Eng. C **93**, 341–355 (2018)
20. Schumacher, M., Habibovic, P., Van Rijt, S.: Mesoporous bioactive glass composition effects on degradation and bioactivity. Bioact. Mater. **6**(7), 1921–1931 (2021)
21. Letaief, N., Lucas-Girot, A., Oudadesse, H., Meleard, P., Pott, T., Jelassi, J., Dorbez-Sridi, R.: Effect of aging temperature on the structure, pore morphology and bioactivity of new sol–gel synthesized bioglass. J. Non-Cryst. Solids **402**, 194–199 (2014)

22. Begum, A.N., Rajendran, V., Ylänen, H.: Effect of thermal treatment on physical properties of bioactive glass. Mater. Chem. Phys. **96**(2–3), 409–417 (2006)
23. Yadav, D., Kumar, U., Nirala, G., Mall, A.K., Upadhyay, S.: Effect of acceptor Na^{1+} doping on the properties of perovskite $SrCeO_3$. J. Mater. Sci. Mater. Electron. **30**, 15772–15785 (2019)
24. Hasanah, L.U., Agustina, F.R., Puspita, N.F., Suherman, B., Nurosyid, F., Sandi, D.K., Iriani, Y.: Microstructure, chemical bonds, and photocatalyst activity of neodymium-doped strontium titanate ($Sr_{0.97}Nd_{0.03}TiO_3$) with 900 °C and 1000 °C sintering temperature. J. Phys. Conf. Ser. **2392**(1), 012034 (2022). (IOP Publishing)
25. Sadeghzade, S., Liu, J., Wang, H., Li, X., Cao, J., Cao, H., Tang, B., Yuan, H.: Recent advances on bioactive baghdadite ceramic for bone tissue engineering applications: 20 years of research and innovation (a review). Mater. Today Bio. **17**, 100473 (2022)
26. Prajapati, R.K., Kumar, H., Saroj, A.L.: Formation of silver particles in [PVA: CS: PEG]-$AgNO_3$ based biopolymer electrolyte membranes: structural and electric transport properties study. Phys. B **662**, 414962 (2023)
27. Yadav, D., Kumar, U., Nirala, G., Upadhyay, S.: Electrical conduction and relaxation in perovskite oxide $SrCe_{0.98}Na_{0.02}O_3$ synthesized by solid-state route. Macromol. Symp. **388**(1), 1900020 (2019)
28. Youness, R.A., Tag El-deen, D.M., Taha, M.A.: A review on calcium silicate ceramics: properties, limitations, and solutions for their use in biomedical applications. SILICON **15**(6), 2493–2505 (2023)
29. Flores-Jacobo, A., Aguilar-Reyes, E.A., León-Patiño, C.A.: Effect of dopants on the physical, mechanical, and biological properties of porous scaffolds for bone tissue engineering. Biomed. Mater. & Devices 1–22, (2022)
30. Pantulap, U., Arango-Ospina, M., Boccaccini, A.R.: Bioactive glasses incorporating less-common ions to improve biological and physical properties. J. Mater. Sci. Mater. Med. **33**, 1–41 (2022)
31. Dahinten, A.: Baghdadit-Biozemente in der Anwendung als endodontischer Funktionswerkstoff. Universität Würzburg, Ph.D. diss. (2023)
32. Nawaz, Q., de Pablos-Martin, A., e Silva, J.M., Berthold, L., Hurle, K., Jaimes, A.T., Sitarz, M., Brauer, D.S., Boccaccini, A.R.: Crystallization study of sol–gel derived 13–93 bioactive glass powder. J. Eur. Ceram. Soc. **41**(2), 1695–706 (2021)
33. Kamrani, A., Nasrabadi, M.H., Halabian, R., Ghorbani, M.: A biomimetic multi-layer scaffold with collagen and zinc doped bioglass as a skin-regeneration agent in full-thickness injuries and its effects in vitro and in vivo. Int. J. Biol. Macromol. **253**, 127163 (2023)
34. Yadav, D., Nirala, G., Kumar, U., Upadhyay, S.: Investigation on structural and optical properties of system $Sr_2Ce_{1-x}Na_xO_4$ ($0.0 \le x \le 0.10$). J. Mater. Sci. Mater. Electron. **32**, 8064–8080 (2021)
35. Sadreddini, S., Jodati, H., Evis, Z., Keskin, D.: Novel barium-doped-baghdadite incorporated PHBV-PCL composite fibrous scaffolds for bone tissue engineering. J. Mech. Behav. Biomed. Mater. 106185, (2023)
36. Han, L., Sun, J., Williams, P.A., Yang, J., Zhang, S.: Octenyl-succinylated inulins for the delivery of hydrophobic drug. Int. J. Biol. Macromol. **221**, 1112–1120 (2022)
37. Kaur, M., Singh, K.: Review on titanium and titanium based alloys as biomaterials for orthopaedic applications. Mater. Sci. Eng. C **102**, 844–862 (2019)
38. Karasu, B., Yanar, A.O., Kocak, A., Kısacık, Ö.: Bioactive glasses. El-Cezerî J. Sci. Eng. **4**(3), 436–471 (2017)
39. Yadav, S.K., Ray, S., Vyas, V.K., Prasad, S., Ali, A., Yadav, S., Majhi, M.R., Pyare, R.: Development of zirconia substituted 1393 bioactive glass for orthopaedic application. Orient. J. Chem. **33**(6), 2720 (2017)
40. Yadav, D., Nirala, G., Yadav, S., Kumar, U., Upadhyay, S., Kumar, S.: Study of structural, microstructure, impedance, and DC conductivity of Gd-doped $SrCeO_3$ ceramics. Appl. Phys. A **127**, 1–4 (2021)
41. Yadav, S., Singh, P., Pyare, R.: Synthesis, characterization, mechanical and biological properties of biocomposite based on zirconia containing 1393 bioactive glass with hydroxyapatite. Ceram. Int. **46**(8), 10442–10451 (2020)

42. Yadav, D., Kumar, P., Tripathi, A.K., Yadav, R.S., Nirala, G., Yadav, S., Yadav, A.K., and Yadav, S.: Influence of La3+ doping on structural and optical properties of SrCeO3 perovskite. Physica Scripta **99**(11), 115935 (2024).
43. Li, R., Clark, A.E., Hench, L.L.: An investigation of bioactive glass powders by sol-gel processing. J. Appl. Biomater. **2**(4), 231–239 (1991)
44. Kaur, G., Pickrell, G., Sriranganathan, N., Kumar, V., Homa, D.: Review and the state of the art: sol–gel and melt quenched bioactive glasses for tissue engineering. J. Biomed. Mater. Res. B Appl. Biomater. **104**(6), 1248–1275 (2016)
45. Durgalakshmi, D., Rakkesh, R.A., Mohanraj, J.: Graphene–metal–organic framework-modified electrochemical sensors. In: Graphene-Based Electrochemical Sensors for Biomolecules, pp. 275–296). Elsevier (2019)
46. Xuan Bui, V., Quang Vo, M., Anh Nguyen, T., Thi, B.H.: Investigation of bioactive glass-ceramic $60SiO_2$-30CaO-$10P_2O_5$ prepared by hydrothermal method. Adv. Mater. Sci. Eng. **2019**, 1–8 (2019)
47. Anh Tuan, T.V., Guseva, E., Anh Tien, N., Tan Dat, H., Vuong, B.X.: Simple and acid-free hydrothermal synthesis of bioactive glass $58SiO_2$–33CaO-$9P_2O_5$ (wt%). Crystals **11**(3), 283 (2021)
48. Farzadi, A., Solati-Hashjin, M., Bakhshi, F., Aminian, A.: Synthesis and characterization of hydroxyapatite/β-tricalcium phosphate nanocomposites using microwave irradiation. Ceram. Int. **37**(1), 65–71 (2011)
49. Chitra, S., Bargavi, P., Balakumar, S.: Effect of microwave and probe sonication processes on sol–gel-derived bioactive glass and its structural and biocompatible investigations. J. Biomed. Mater. Res. B Appl. Biomater. **108**(1), 143–155 (2020)
50. Purcar, V., Rădiţoiu, V., Nichita, C., Bălan, A., Rădiţoiu, A., Căprărescu, S., Raduly, F.M., Manea, R., Şomoghi, R., Nicolae, C.A., Raut, I.: Preparation and characterization of silica nanoparticles and of silica-gentamicin nanostructured solution obtained by microwave-assisted synthesis. Materials **14**(8), 2086 (2021)
51. Khalid, H., Suhaib, F., Zahid, S., Ahmed, S., Jamal, A., Kaleem, M., Khan, A.S.: Microwave-assisted synthesis and in vitro osteogenic analysis of novel bioactive glass fibers for biomedical and dental applications. Biomed. Mater. **14**(1), 015005 (2018)
52. Essien, E.R., Atasie, V.N., Udobang, E.U.: Microwave energy-assisted formation of bioactive CaO–MgO–SiO_2 ternary glass from bio-wastes. Bull. Mater. Sci. **39**, 989–995 (2016)
53. Strobel, L.A., Hild, N., Mohn, D., Stark, W.J., Hoppe, A., Gbureck, U., Horch, R.E., Kneser, U., Boccaccini, A.R.: Novel strontium-doped bioactive glass nanoparticles enhance proliferation and osteogenic differentiation of human bone marrow stromal cells. J. Nanopart. Res. **15**, 1–9 (2013)
54. Taygun, M.E., Boccaccini, A.R.: Nanoscaled bioactive glass particles and nanofibers. In: Bioactive Glasses, pp. 235–283. Woodhead Publishing (2018)
55. Odermatt, R., Par, M., Mohn, D., Wiedemeier, D.B., Attin, T., Tauböck, T.T.: Bioactivity and physico-chemical properties of dental composites functionalized with nano- vs. micro-sized bioactive glass. J. Clin. Med. **9**(3), 772 (2020)
56. Jones, J.R.: Review of bioactive glass: from Hench to hybrids. Actabiomaterialia **9**(1), 4457–4486 (2013)
57. O'Donnell, M.D.: Melt-derived bioactive glass. In: Bio-Glasses, pp. 13–28. Wiley, Chichester (UK) (2012)
58. Karadjian, M., Essers, C., Tsitlakidis, S., Reible, B., Moghaddam, A., Boccaccini, A.R., Westhauser, F.: Biological properties of calcium phosphate bioactive glass composite bone substitutes: current experimental evidence. Int. J. Mol. Sci. **20**(2), 305 (2019)
59. Dixit, K., Sinha, N.: Compressive strength enhancement of carbon nanotube reinforced 13–93B1 bioactive glass scaffolds. J. Nanosci. Nanotechnol. **19**(5), 2738–2746 (2019)
60. Yadav, D., Kumar, U., Upadhyay, S.: Study of structural, electrical, and photoluminescent properties of $SrCeO_3$ and Sr_2CeO_4. J. Adv. Ceram. **8**, 377–388 (2019)
61. Aparicio, C., Ginebra, M.P.: Biomineralization and biomaterials: fundamentals and applications. Woodhead Publishing (2015)

62. Yadav, S.K., Ray, S., Ershad, M.D., Vyas, V.K., Prasad, S., Ali, A., Yadav, S., Majhi, M.R., Pyare, R.: Development of zirconia substituted 1393 bioactive glass for orthopaedic application (2017)
63. Yadav, S., Ali, A., Krishnamurthy, S., Singh, P., Pyare, R.: In-vitro analysis of bioactivity, hemolysis, and mechanical properties of Zn substituted calcium zirconium silicate (baghdadite). Ceram. Int. **47**(11), 16037–16053 (2021)
64. Dai, L.L., Mei, M.L., Chu, C.H., Lo, E.C.: Antibacterial effect of a new bioactive glass on cariogenic bacteria. Arch. Oral Biol. **117**, 104833 (2020)
65. Chen, R., Li, Q., Xu, S., Han, J., Huang, P., Yu, Z., Jia, D., Liu, J., Jia, H., Shen, M., Hu, B.: Nanosized HCA-coated borate bioactive glass with improved wound healing effects on rodent model. Chem. Eng. J. **426**, 130299 (2021)
66. Jafari, N., Habashi, M.S., Hashemi, A., Shirazi, R., Tanideh, N., Tamadon, A.: Application of bioactive glasses in various dental fields. Biomater. Res. **26**(1), 31 (2022)

Bulk Metallic Glasses: Effect of Various Temperatures with Nature of Constituent Elements in Zr-Al/Ti-Ni-Cu BMGs

Patel Ram Suthar

Abstract Bulk metallic glasses (BMGs), known for their amorphous structure, exhibit a unique blend of high strength, superior magnetic properties, and excellent corrosion resistance. Produced by ultra-rapid cooling that prevents atoms from crystallizing, these materials remain metastable. Elemental substitution in BMGs can alter thermal properties such as crystallization and glass transition temperatures. Variations in thermal expansion, conductivity, and covalent radius significantly affect these changes, and using these parameters helps account for changes in their properties after altering the fraction of constituent elements.

Keywords Thermal expansion · Conductivity · Covalent radius · Glass transition temperature

1 Introduction

Bulk metallic glasses showcase outstanding physical and chemical properties, boasting high wear and corrosion resistance, alongside unique features like remarkable flexibility at elevated temperatures and exceptional strength in cold environments. Significantly, akin to specific polymer materials, they manifest a notable elastic limit approaching 2%. In comparison to their crystalline counterparts, bulk metallic glasses offer significantly increased strength, often two to three times higher. These exceptional qualities position bulk metallic glasses as an attractive choice for practical applications in the field of functional and structural materials, opening up new possibilities for advanced designs and systems. BMGs are basically categorized into two groups mainly according to the nature of their constituent elements.

P. R. Suthar (✉)
Dr. Bhimrao Ambedkar Govt. College, Sriganganagar, Rajasthan, India
e-mail: prsuthara@gmail.com

U. Kumar (ed.), *Defects Engineering in Electroceramics for Energy Applications*, Engineering Materials, https://doi.org/10.1007/978-981-97-9018-0_13

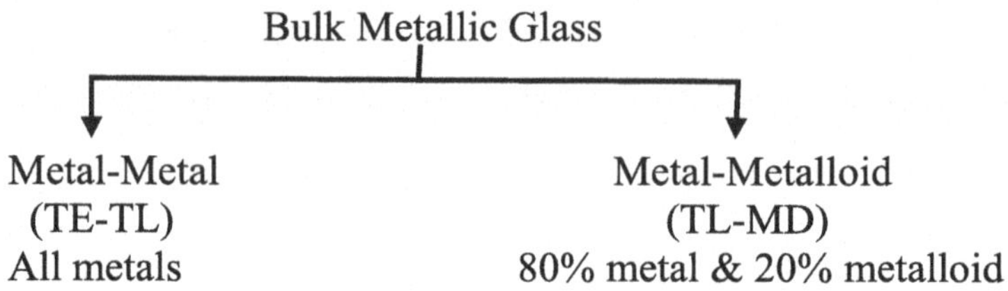

The scientific importance of understanding the intricacies of glass phenomena and the formation of bulk metallic glasses, combined with their potential for real-world applications, has piqued significant research interest. Since their initial discovery by Duwez [1] in 1960, extensive efforts have been dedicated to the development of glassy or amorphous alloys. These endeavours encompass techniques such as solid-state amorphization and rapid solidification.

The inception of what we now classify as "bulk" amorphous alloy dates back to 1969 when Chen and Turnbull [2] first identified it in ternary Pd-Cu-Si alloys [15]. The critical cooling rate [32] for these ternary bulk glass-forming alloys resides approximately at 102 K/s. Of particular note is the surge in attention within the scientific community, spurred by the emergence of novel multi-component bulk metallic glasses. Illustrative instances encompass $La_{55}Al_{25}Ni_{20}$ (first identified by Inoue et al. in 1989 [3]) and $Zr_{41.2}Ti_{13.8}Cu_{12.5}Ni_{10}Be_{22.5}$ (introduced by Peker and Johnson in 1993 [4]). These alloys can be synthesized at reduced cooling rates via direct casting from molten liquid, representing a significant stride in BMG fabrication methodologies.

A comprehensive grid system is defined by atomic percentages, as glass-forming compositions exhibit varying properties, resulting in different alloys within that framework. Applying this grid system to two distinct alloys within a single component yields approximately 1.2 trillion potential metallic glass alloys, created from 32 elemental components. It's important to note that this count excludes rare earth elements and high-cost elements like Sc, Ta, Ru, and Rh, and their inclusion would significantly expand the total number of possible metallic glass alloys beyond the 1.2 trillion estimates. BMGs employ nearly 54 different elements in their fabrication. By adhering to the laws governing the production of bulk metallic glasses, a wide array of these materials can be synthesized (Fig. 1).

To stabilize a super metallic liquid, three essential empirical component rules are required:

i. The multicomponent metallic alloy system should consist of three or more elements.
ii. There should be a substantial atomic size difference, exceeding approximately 12%, among the primary constituent elements of the metallic alloy system.
iii. The elements involved should exhibit the negative heat of mixing.

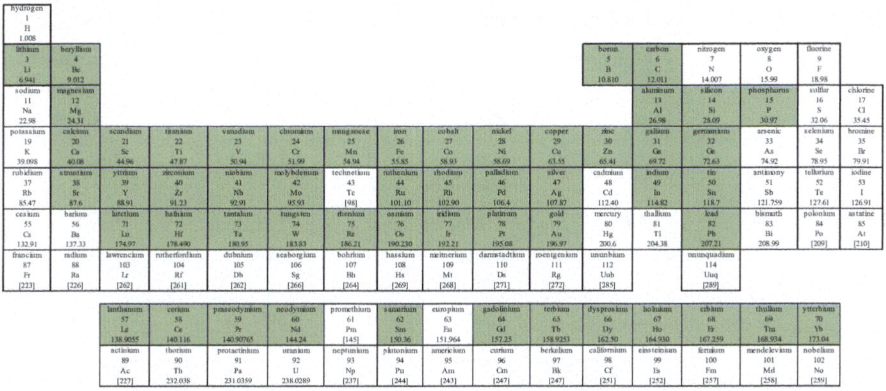

Fig. 1 The fabrication of bulk metallic glasses (BMGs) relies on the incorporation of elements from the periodic table

> Correct sentence is:
> a. Metallic glasses have low electrical resistance.
> b. The electrical resistance for metal glasses will not vary with temperature.
> c. They possess very High eddy current losses.
> d. They have low corrosion resistance.

These three guidelines to enable BMG formation instead of crystallization as given by Inoue are as follows:

I. The atomic size ratios of at least two constituents of the alloy should be $\alpha <$ 0.89 for small relative to large.
II. The value of negative heat of mixing should be large.
III. Atomic components should be several.

The simple empirical rules for the search of compositions likely to yield new metal/metalloid bulk metallic glasses [24], as given by Shen and Schwarz [5], apply to bulk ferromagnetic glasses prepared by flux melting and water quenching [6]. These rules [5] are (Table 1):

a. For the alloy to be considered, it is imperative that it comprises a minimum of three components, with two being metallic in nature.
b. The alloy's composition necessitates the presence of two or more metallic elements characterized by distinct atomic sizes and nearly negligible heats of mixing.
c. The metallic constituents of the alloy should exhibit significant negative heats of mixing with the metalloid(s).
d. The overall metalloid content of the alloy should hover around 20 atomic percent.

Table 1 The constituent elements of Zr-Al/Ti-Ni-Cu BMGs with their size ratio in atomic radius

Element	Atomic number	Atomic weight	Group	Block	Atomic radius (pm)	Size ratio*
Al	13	26.98	Main	p	143	0.89375
Cu	29	63.55	Late transition	d	128	0.8
Ni	28	58.69	Transition	d	124	0.775
Ti	22	47.86	Early transition	d	147	0.91875
Zr	40	91.22	Early transition	d	160	–

* The ratio between the atomic radius of a specific element and that of the Zr element

e. It is crucial to meticulously eliminate heteronucleants, such as crystalline oxide inclusions, by dissolving the oxides in a flux or neutralizing the oxygen through dissolution in the melt.

> The atomic radius in picometres of the elements Zr, Ti, Cu, Ni, Be, Y, and Mg, are 148, 136, 138, 121, 90, 162, and 130, respectively. Whether the alloy composition $Zr_{40}Ti_{15}Cu_{11}Ni_{11}Be_{21.5}Y_1Mg_{0.5}$ qualifies as a bulk metallic glass?

2 Advantages of Bulk Metallic Glasses

- **Elevated Strength Levels**: BMGs consistently showcase outstanding mechanical attributes, boasting high strength and hardness that render them well-suited for structural applications.
- **Effective Corrosion Resistance**: Numerous BMGs exhibit remarkable resistance to corrosion and wear, ensuring prolonged durability, particularly in challenging environmental conditions.
- **Distinctive Characteristics**: BMGs often present distinctive physical and chemical attributes absent in conventional crystalline materials, encompassing elevated elasticity, heightened glass transition temperatures, and superior electrical resistivity.
- **Versatile Manufacturing Processes**: BMGs offer flexibility in manufacturing through diverse processing techniques, including casting, molding, and rapid solidification, enhancing adaptability in production.
- **Amorphous Configuration**: The absence of a crystalline structure in BMGs results in isotropic properties, facilitating uniform performance in various directions.

3 Drawbacks of Bulk Metallic Glasses (BMGs)

- **Inherent Brittleness**: Despite their elevated strength, BMGs may exhibit relative brittleness compared to traditional metals, limiting their suitability for applications demanding extensive ductility.
- **Technical Challenges in Processing**: The fabrication of BMGs can be demanding, requiring precise control over cooling rates and alloy compositions, which may contribute to elevated production costs.
- **Constrained Alloy Combinations**: The formation of BMGs is constrained by specific criteria governing the selection of elements, limiting the spectrum of potential alloy compositions and applications.
- **Vulnerability to Crystallization**: BMGs are susceptible to crystallization upon exposure to heat, potentially leading to degradation in mechanical properties and compromised performance over time.

4 Future Perspectives of Bulk Metallic Glasses (BMGs)

- **Pioneering Structural Materials**: BMGs exhibit promise as avant-garde structural materials for aerospace, automotive, and biomedical applications, attributed to their remarkable strength-to-weight ratio and corrosion-resistant properties.
- **Functional Device Development**: Ongoing research endeavors aim to harness the unique properties of BMGs in the creation of functional devices such as sensors, actuators, and magnetic materials, capitalizing on their amorphous structure and tunable characteristics.
- **Tailoring Properties for Specific Applications**: Anticipated advancements in alloy design and processing techniques are expected to enable the customization of BMG properties, broadening their adaptability and usefulness across diverse industries.
- **Revolutionizing Manufacturing through Additive Techniques**: The exploration of additive manufacturing methods for BMGs holds the potential to revolutionize production, allowing for the creation of intricate geometries and personalized components with meticulous control over microstructure and properties.
- **Incorporation into Multifunctional Composites**: The integration of BMGs with other materials, such as polymers or ceramics, stands as a prospect for developing multifunctional composites endowed with enhanced properties, paving the way for innovative applications in engineering and technology.

5 Bulk Metallic Glass Formation

The process of glass formation is typically linked with non-metallic materials, yet it's a recent development in the realm of metals. While in most material categories, crystallization takes place as a substance is melted and subsequently cooled, there are instances where rapid cooling can circumvent crystallization, leading to the creation of an amorphous, glass-like structure as opposed to a crystalline one.

The selection of elements in the metal alloy is pivotal in enabling the creation of metallic glasses. When subjected to rapid cooling, specific combinations of elements are predisposed to give rise to metallic glasses. Prominent alloying elements found in metallic glasses encompass transition metals like iron, nickel, and cobalt, which are blended with diverse other elements like boron, silicon, or phosphorus. Thoughtful deliberation on alloy composition [29] is paramount for promoting the development of glassy structures.

At critical cooling rates, often around 1000 K/s or lower, metals have the capability to generate bulk metallic glasses (BMGs) [7]. These alloys undergo vitrification during the cooling process, preserving an amorphous atomic structure instead of crystallizing. BMGs have attracted considerable interest thanks to their distinctive blend of characteristics, ease of processing, and their amorphous structure. The properties and attributes of BMGs render them intriguing from both technological and scientific perspectives. Their distinctive blend of mechanical, magnetic, thermal, and electrical characteristics opens the door to a variety of potential applications. Their capacity to be processed in bulk dimensions represents a significant advantage. The formation of bulk metallic glasses depends on various factors, one of which is glass-forming ability [8].

What is difference between silica glasses and bulk metallic glasses?

5.1 Glass Forming Ability

Exploring and comprehending glass forming ability (GFA) holds significant importance, particularly in fields like metallurgy. The role of glass-forming ability is pivotal in influencing the evolution of metallic glasses or amorphous metals, renowned for their extraordinary properties that make them indispensable across a diverse range of applications. The critical cooling rate emerges as a vital parameter for evaluating the GFA of an alloy. A higher GFA indicates that the alloy is more inclined to assume a glassy structure during the cooling process, rather than crystallizing. GFA is associated with a material's capability to initiate and preserve its amorphous structure throughout the solidification process. In simple terms, a high GFA signifies that a material readily adopts a glassy state, while a low GFA implies a proclivity for crystallization.

Certainly, the generation of bulk metallic glasses [27] is intricately governed by numerous factors, encompassing atomic size mismatch, entropy of mixing, thermodynamic stability [33], amorphous phase stabilization [33], composition, viscosity, cooling rate, and undercooling. This ensemble of factors collectively assumes a pivotal role in delineating the glass-forming ability of an alloy and its capacity to maintain an amorphous structure throughout the solidification process.

- **Atomic Size Mismatch**: A significant disparity in atomic sizes [28] among the constituent elements within an alloy hinders crystallization and promotes the development of a metallic amorphous structure. This is often referred to as the "atomic size effect".
- **Entropy of Mixing**: High entropy of mixing, associated with a disordered arrangement of atoms in the liquid phase, hinders crystallization and can facilitate the formation of metallic amorphous structures.
- **Thermodynamic Stability**: GFA is affected by the contrast in Gibbs free energy between the amorphous and crystalline phases [26]. A reduced thermodynamic impetus for crystallization commonly fosters the retention of the metallic amorphous state.
- **Stabilization of Amorphous Phase**: The concept of amorphous phase stabilization entails preserving the non-crystalline, disordered atomic structure found in materials like metallic glasses throughout their production and application. Improving GFA contributes to the stability of the non-crystalline/amorphous phase.
- **Composition**: The composition of alloying elements plays a pivotal role in the formation of BMGs. Specific alloy compositions exhibit a greater propensity for creating BMGs compared to others.
- **Viscosity**: Elevated viscosity during the cooling process can impede the mobility of atoms, consequently restraining crystallization and encouraging the development of a metallic amorphous structure.
- **Cooling Rate**: Swift cooling rates, achieved through techniques such as splat cooling or melt spinning, facilitate the creation of metallic amorphous structures by limiting the available time for crystal growth.
- **Under Cooling**: The extent to which a liquid can be under cooled below its equilibrium melting temperature without crystallization is of paramount importance. Typically, greater under cooling tends to correlate with higher GFA. The relationship between under cooling and GFA is noteworthy.

To gain a thorough understanding and predict the compositions of bulk metallic glasses (BMGs), various theories have been developed. These theories aim to clarify the fundamental principles governing the formation of glass and provide insights into which combinations of elements are most likely to yield BMGs. Maintaining the unique characteristics of amorphous alloys requires careful operation within specified temperature ranges. This attention to temperature parameters is essential for preserving the non-crystalline structure and exceptional properties that distinguish BMGs from traditional crystalline materials. By adhering to these temperature considerations, researchers, and manufacturers can optimize the synthesis of BMGs with desired properties.

> Why is a high glass-forming ability (GFA) necessary for the synthesis of bulk metallic glasses (BMGs)?

5.2 *Temperatures for Bulk Metallic Glasses*

A wide range of temperatures is essentially both in the creation of BMGs and in the characterization of their distinct properties. BMGs often feature a broad super cooled liquid region, enabling their processing at comparatively lower temperatures than those required for traditional crystalline alloys. Processing BMGs within a temperature range hinges on the alloy composition, typically involving rapid cooling during the transition from the liquid phase to the glassy state [31] at rates approaching millions of degrees celsius per second. The various temperatures [9] for BMGs include: T_x (crystallization temperature), T_l (liquidus temperature), T_g (glass transition temperature), T_m (melting temperature), T_n (nose temperature), T_e (eutectic temperature), ΔT_x (supercooled liquid region), and $T_{rg} = T_g/T_l$ (reduced glass-transition temperature).

5.2.1 Crystallization Temperature (T_x)

The crystallization temperature, known as T_x, marks the point at which the amorphous structure of bulk metallic glass begins its transformation into a crystalline form. During this transition, atoms rearrange themselves into ordered, crystalline structures. The exact temperature at which this crystallization commences fluctuates depending on the particular composition of the bulk metallic glass. The crystallization process can influence diverse properties of bulk metallic glasses, resulting in alterations in mechanical, magnetic, electrical, and thermal characteristics.

The crystallization temperature is not fixed and depends on the unique composition of the bulk metallic amorphous alloy. Unlike crystalline metals, bulk metallic glasses lack long-range atomic order, showcasing disordered or non-crystalline atomic structures. When heated, BMGs undergo a transition from an amorphous (non-crystalline) to a structural crystalline state. Bulk metallic glasses exhibit an amorphous or glassy structure, indicating the absence of long-range order among the atoms in the material. This lack of a crystalline lattice arrangement is attained through swift cooling from the liquid state, preventing the atoms from organizing into a crystalline structure. Bulk metallic glasses typically possess a critical cooling rate, below which they maintain complete amorphousness. If the cooling rate is excessively slow, there exists a risk of crystallization. The temperature at which crystallization initiates is commonly termed the crystallization temperature [30]. The mechanical and thermal properties of bulk metallic glasses can be influenced by crystallization. To preserve the advantageous amorphous structure and capitalize on its distinctive properties, the manufacturing process often involves rapid cooling.

5.2.2 Liquidus Temperature (T_l)

The term "liquidus temperature" isn't commonly used in the context of bulk metallic glasses, but if applied, it might refer to the temperature range where these amorphous alloys exhibit characteristics akin to those of a supercooled liquid. In this state, the atomic structure resembles that of a liquid, yet it maintains a solid form due to kinetic restrictions or elevated viscosity.

Traditionally, "liquidus temperature" (T_l) signifies the point at which a material, often an alloy, transitions completely from a solid to a liquid state upon heating. However, bulk metallic glass amorphous alloys, with their disordered atomic structures, don't neatly align with the concept of a clear melting point typical in crystalline materials. Instead, they are known for their amorphous structure, lacking a pronounced melting point and typically showcasing a super cooled liquid region. As the temperature in this region rises, they slowly become more viscous and softer.

BMGs don't have a clear melting point like crystalline materials; while some researchers might colloquially mention a "liquidus" phase, the transition from a non-crystalline to a more liquid-like state in BMGs lacks the distinct temperature seen in traditional alloys. The variability in the behaviour of bulk metallic glasses concerning temperature, including any "liquidus-like" characteristics, depends on factors such as alloy composition, constituent element fractions, processing parameters, and other influencing elements.

5.2.3 Glass Transition Temperature (T_g)

Absolutely! The glass transition temperature (T_g) stands as a pivotal parameter in comprehending the behaviour of bulk metallic glasses, delineating the temperature at which these amorphous materials transition from a rigid, glassy state to a more viscous or rubbery state. This transition involves molecular mobility and relaxation, leading to changes in various properties like mechanical, electrical, and thermal.

The specific value of T_g for bulk metallic glass amorphous alloys can vary significantly based on the alloy's composition. Different compositions exhibit different glass transition behaviours, and there isn't a universal T_g value applicable to all bulk metallic glass amorphous alloys. Generally falling within the range of −50 to 600 °C or more, these values are approximate and subject to variation depending on the alloy composition/system.

Understanding T_g is critical for processing and applying bulk metallic glasses. Below T_g, the material is in a glassy, rigid state, while above T_g, it becomes more ductile, exhibiting properties typical of a liquid in a super cooled state. In essence, T+ serves as a critical parameter elucidating the characteristics of bulk metallic glasses, denoting the temperature at which they undergo a shift from a glassy, rigid state to a more pliable and elastomeric state. Bulk metallic glasses undergo a transition, transitioning from a solid-like, rigid state to a more fluid-like, viscous state at a distinct temperature referred to as the glass transition temperature (T_g). Beyond the glass transition temperature (T_g), bulk metallic glasses display a shift toward more elastic and viscous behaviour. This characteristic makes them easier to deform compared to their brittle, glassy state below T_g. The glass transition temperature assumes a vital role in the fabrication of bulk metallic glasses, exerting influence on their malleability and the degree of ease in shaping them into intricate geometries.

In summary, bulk metallic glasses are distinctive materials characterized by their amorphous structure. Preventing crystallization during cooling and understanding the glass transition temperature is essential for preserving the advantageous properties of BMGs and shaping them into useful forms during processing. Rapid cooling techniques are commonly employed to maintain the non-crystalline structure of bulk metallic glasses.

5.2.4 Melting Temperature (T_m)

The melting temperature, T_m, signifies the temperature when a solid material transforms into a liquid state from solid state. In the context of bulk metallic glasses, T_m denotes the temperature at which the amorphous structure of the alloy material turns into a liquid during the melting process.

T_m is a crucial parameter in the processing of bulk metallic glasses, indicating the temperature range within which the alloy can be effectively melted and cast into desired shapes. Unlike conventional crystalline metals and alloys, bulk metallic glasses don't have a sharp melting point. Instead, when heated, they exhibit gradual softening behaviour, marked by the glass transition temperature (T_g). Below T_g, bulk metallic glasses behave as rigid, glassy materials. As the temperature rises above T_g, they become softer and more ductile, but a distinct phase change from solid to liquid is not clearly observed.

What sets BMGs apart is their tendency to undergo crystallization at elevated temperatures rather than melting. This involves the transformation of the amorphous structure of BMGs into crystalline phases, highlighting a unique characteristic that distinguishes them from conventional materials with a defined melting point. Understanding the distinct melting characteristics of a specific BMG requires a thorough examination of the alloy's composition, employing techniques such as differential scanning calorimetry (DSC) or other thermal analysis methods. Through these approaches, valuable insights into the material's thermal behaviour, including the

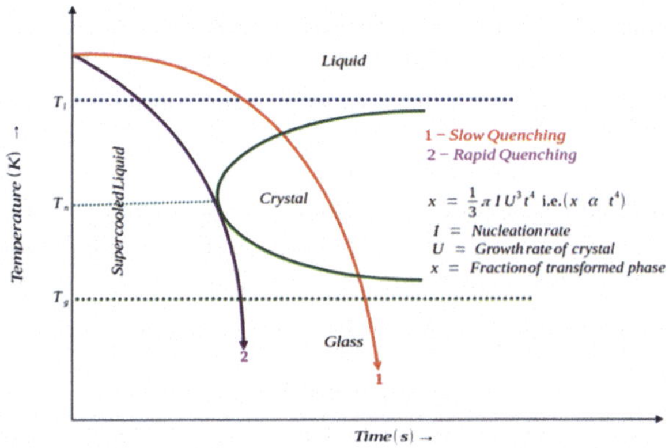

Fig. 2 Different states of matter and TTT diagram (Ref. [11])

initiation of softening and any accompanying temperature range, can be obtained, offering crucial information about the melting behaviour in BMGs.

5.2.5 Nose Temperature (T_n)

The nose temperature, T_n, assumes a prominent function within the framework of the Time–Temperature-Transformation (TTT) diagram, particularly when illustrating the crystallization behaviour of amorphous materials like bulk metallic glasses (BMGs). This diagram depicts the relationship between time and temperature concerning the transformation of the material (Fig. 2).

The T_n on the TTT curve represents the highest temperature where crystallization occurs most rapidly for BMGs. It serves as the temperature point at which the rate of the crystallization process is at its maximum. In essence, the nose temperature is a crucial parameter that helps characterize and understand the kinetics of crystallization in amorphous materials like BMGs.

5.2.6 Eutectic Temperature (T_e)

The term "eutectic temperature" doesn't typically come up in discussions about bulk metallic glass (BMG) amorphous alloys. In the expansive realm of materials science, this temperature refers to the lowest point at which a mixture of two or more elements either solidifies together or goes through complete melting. The concept is more applicable to conventional alloy systems, where a eutectic mixture forms with a specific composition and corresponding eutectic temperature.

Now, when it comes to BMGs, their amorphous and non-crystalline nature distinguishes them. They don't follow the typical phase transformations observed in crystalline materials, and their atomic arrangement is anything but regular. BMGs navigate through intricate stages of glass transition, crystallization, and various structural changes, resulting in the absence of eutectic points in their behaviour. It's a fascinating illustration of how the distinctive qualities of BMGs break away from the established norms of traditional alloy systems. In the realm of bulk metallic glasses (BMGs), the emphasis lies in attaining a sufficiently high cooling rate during solidification. This priority aims to uphold the amorphous structure and prevent crystallization, ensuring the preservation of the unique properties associated with BMGs. Rather than focusing on a eutectic temperature; the critical cooling rate [16] emerges as a parameter more relevant to BMGs. It denotes the speed at which the molten alloy needs to be cooled to prevent the emergence of crystalline phases during the solidification process.

5.2.7 Super-Cooled Liquid Region (ΔT_x)

It delineates the temperature range wherein a solid-state composition of the bulk metallic glass (BMG) manifests characteristics akin to those found in a supercooled liquid.

Mathematically expressed as the temperature difference between crystallization temperature (T_X) and glass transition temperature (T_g) [17]:

$$\Delta T_X = T_X - T_g.$$

Within the super cooled liquid region [6], the arrangement of atoms in the BMG becomes progressively disordered, drawing an analogy to the atomic arrangement in a liquid. Hindered by the high viscosity ingrained in the non-crystalline or amorphous structure of BMGs, atoms in this region are unable to move freely, in contrast to a true liquid. Despite this disorder, the BMG retains its solid-state characteristics, as the atoms are constrained by their positions.

This phenomenon in BMGs, known as the super cooled liquid region, is distinctive, underscoring their capacity to exhibit liquid-like behaviours and maintain an amorphous nature even at temperatures considerably below their conventional crystalline melting points. This unique behaviour is of paramount importance in various applications, particularly in fields where the distinctive combination of mechanical strength, low corrosion, thermal stability, and prowess in electrical and magnetic properties offered by BMGs is highly prized.

Differentiate between glass transition temperature and liquidus temperature?

5.2.8 Reduced Glass-Transition Temperature ($T_{rg} = T_g/T_l$)

T_{rg}, the reduced glass transition Temperature [18], is a key dimensionless parameter in metallic glasses, providing valuable insights into a material's glass-forming ability. In this context, T_g denotes the glass transition temperature, the juncture at which non-crystalline/amorphous material shifts from a rigid, glassy state to a more rubbery or viscous state. Conversely, T_l represents the liquidus temperature, indicating the temperature at which the material transitions into a fully liquid state.

Expressed by the formula $T_{rg} = T_l/T_g$, this ratio normalizes the values, eliminating the influence of specific temperature scales and enabling a more universal comparison of glass-forming ability across diverse materials. A higher T_{rg} value signals an enhanced glass-forming ability. Put simply, materials with a higher T_{rg} are more inclined to adopt a glassy (amorphous) structure during the cooling process, steering clear of crystallization. This parameter proves invaluable in the exploration and design of metallic glasses, where the pursuit of an amorphous structure is often a key objective to leverage the unique mechanical and physical properties of these materials.

How are metallic glasses formed?
(a) By decreasing the rate of heating
(b) By increasing the rate of cooling
(c) By decreasing the rate of cooling
(d) By rapidly heating and then cooling slowly

6 Analysis of T_g and T_x for Zr-Al-Ni-Cu BMGs

The temperature-related properties of the bulk metallic glass amorphous alloy are intricately influenced by the constituent elements and their proportions, with factors such as thermal conductance, covalent radius, and coefficient of thermal expansion playing pivotal roles in shaping these effects. The changes in crystallization temperature (T_x) and glass transition temperature (T_g) [19] observed in various Zr-Al/Ti-Ni-Cu bulk metallic glasses (BMGs) have been contingent upon the following hypotheses [9, 11].

a. In BMGs, the element designated as the base element is the one with the highest quantitative fraction within the BMG composition. However, if there is no alteration in the base element, the element with the highest quantitative fraction among the remaining elements is designated as the base element in that BMG.
b. The aforementioned points are also applicable in situations where there is a modification in the proportion of elements, yet no significant alteration occurs in the crystal structure.
c. In BMGs, when the quantitative fraction of any two elements is equal, the base element is determined by the higher number of atoms, that is, the element with the lower atomic weight.
d. Partial replacement of elements by others is necessary.

e. The proportion of the base element in the bulk metallic glass, or elements with a crystal structure akin to the base element, diminishes. Concurrently, the proportion of other elements with a crystal structure differing from that of the base element increases. In this context, the modification in glass transition or crystallization temperature exhibits an inverse correlation with any two concurrent alterations in parameters α (coefficient of thermal expansion), W (thermal conductivity), and R (covalent radius).
f. In BMG, for elements sharing the crystal structure with the base element, there is an increase in the quantitative fraction of such elements. If there is no alteration in the crystal structure, then the fraction of the base element should decrease. In such instances, variations in the glass transition or crystallization temperature exhibit an inverse relationship with corresponding similar changes in all three parameters: thermal expansion coefficient (α), thermal conductivity (W), and covalent radius (R).
g. The proportion of the base element in the bulk metallic glass, or elements with a crystal structure akin to the base element, either decreases or remains constant. Concurrently, the proportion of other elements with a crystal structure distinct from that of the base element increases or remains constant. In such scenarios, alterations in the glass transition or crystallization temperature align with analogous changes in all three parameters: thermal expansion coefficient, thermal conductivity, and covalent radius.

6.1 Effective Value of A, W, and R

In Table 2, a comprehensive array of data, including crystal structure, thermal conductivity (W/(m.K)), covalent radius (pm), and the coefficient of thermal expansion (μm/(m.K)), is presented for the constituent elements of Zr-Al/Ti-Ni-Cu bulk metallic glasses (BMGs).

In the BMG alloy $Zr_{55}Al_{8.9}Ni_{7.3}Cu_{28.8}$, the coefficients of thermal expansion (α) for Zr, Al, Ni, and Cu are 5.7, 23.1, 13.4, and 16.5 μm/(m.K) respectively. The

Table 2 The constituent elements of Zr-Al/Ti-Ni-Cu BMGs with their standard values, each characterized by their values such as crystal structure, parameters: thermal expansion coefficient, thermal conductivity, covalent radius, and glass transition temperature

Element	Crystal structure	Coefficient of thermal expansion (α)	Thermal conductivity (W)	Covalent radius (R)	Glass transition temperature (T_g)
Al	fcc	23.1	237	118	349 [12]
Cu	fcc	16.5	401	138	634 [12]
Ni	fcc	13.4	90.9	121	721 [12]
Ti	hcp	8.6	21.9	136	776 [12]
Zr	hcp	5.7	22.6	148	678 [12]

quantitative fraction (per hundred) of these elements in the alloy is 0.55, 0.089, 0.073, and 0.288 respectively.

By using the formula:

$$x = \sum_i f_i x_i$$

The effective value of parameter the coefficient of thermal expansion (α) in BMG $Zr_{55}Al_{8.9}Ni_{7.3}Cu_{28.8}$ is:

$$\alpha = \sum_i f_i \alpha_i$$

$\alpha = f_{Zr}\ \alpha_{Zr} + f_{Al}\ \alpha_{Al} + f_{Ni}\ \alpha_{Ni} + f_{Cu}\ \alpha_{Cu}$

$= 0.55 \times 5.7 + 0.089 \times 23.1 + 0.073 \times 13.4 + 0.288 \times 16.5$

$= 10.9211$ μm/(m.K)

The effective value of thermal conductivity (W) for Zr, Al, Ni, and Cu is 22.6, 237, 90.9, and 401 W/(m.K) respectively.

The effective value of parameter thermal conductivity (W) in BMG $Zr_{55}Al_{8.9}Ni_{7.3}Cu_{28.8}$ is:

$W = 0.55 \times 22.6 + 0.089 \times 237 + 0.073 \times 90.9 + 0.288 \times 401$

$= 155.6467$ W/(m.K)

The effective value of covalent radius (R) for Zr, Al, Ni, and Cu is 148, 118, 121, and 138 pm respectively.

The effective value of parameter covalent radius (R) in BMG $Zr_{55}Al_{8.9}Ni_{7.3}Cu_{28.8}$ is:

$R = 0.55 \times 148 + 0.089 \times 118 + 0.073 \times 121 + 0.288 \times 138$

$= 140.479$ pm

Applying the aforementioned computational approach, the table-14.3 provides the effective values for parameters α (coefficient of thermal expansion), W (thermal conductivity), and R (covalent radius) [25] of Zr-based bulk metallic glasses (BMGs). The primary element, Zr, adopts a hexagonal close-packed (hcp) crystal structure, whereas the accompanying constituent elements exhibit a face-centred cubic (fcc) crystal structure. Within this Zr-based BMGs collection, the constituent elements are present in diverse fractional ratios. Across the BMGs series numbered from 1 to 9, both W and R values demonstrate a declining trend, whereas α value consistently displays an upward trajectory.

The data pertaining to crystallization temperature and glass transition temperature for diverse Zr-Al-Ni-Cu-based Bulk metallic glasses (BMGs) is detailed in Table 3. Zirconium (Zr) serves as the foundational element in this particular group, holding prominence due to its higher fractional presence among the constituent elements in the BMG composition. The presence of Zr in a higher fraction is a defining characteristic that influences the overall behaviour of the BMG, particularly in terms of its thermal and structural properties. In the bulk metallic glasses (BMGs), specifically in $Zr_{55}Al_{8.9}Ni_{7.3}Cu_{28.8}$ and $Zr_{54.5}Al_{9.6}Ni_{8.4}Cu_{27.5}$ compositions, a discernible trend emerges as the coefficient of thermal expansion (α) exhibits a sequential increase from 10.9211 to 10.9872 μm/(m.K).

Simultaneously, the thermal conductance (W) displays a contrasting pattern, with values decreasing from 155.6467 to 152.9796 W/(m.K), while the covalent radius (R) undergoes a consistent decrease from 140.479 to 140.102 pm. This behaviour aligns with the hypothesis postulated as "In Bulk Metallic Glass (BMG), the quantitative fraction of the base element or elements sharing the same crystal structure as the base element decreases, while the proportion of other elements with a crystal structure distinct from that of the base element increases. In such instances, the alteration in glass transition or crystallization temperature exhibits an inverse relationship with any two identical changes in α, W, and R."

Applying this hypothesis to the glass transition or crystallization temperatures [20] of the two BMGs, it is deduced that the glass transition or crystallization temperature [21] of $Zr_{54.5}Al_{9.6}Ni_{8.4}Cu_{27.5}$ must be greater than that of $Zr_{55}Al_{8.9}Ni_{7.3}Cu_{28.8}$. This deduction is substantiated by the available data, where the glass transition and crystallization temperatures for $Zr_{54.5}Al_{9.6}Ni_{8.4}Cu_{27.5}$ are measured at 765.2 K and 769.5 K, surpassing those of $Zr_{55}Al_{8.9}Ni_{7.3}Cu_{28.8}$ at 684.6 K and 691.3 K, respectively. Hence, the experimental results validate the hypothesized relationship between the physical properties and thermal behaviour of these BMGs (Fig. 3).

In Table 3, the quantitative analysis reveals a descending trend in the ratio of Zr within the bulk metallic glasses (BMG) across s. nos. 1 to 9. Notably, the foundational element's crystal structure is hexagonal close-packed (hcp), while the accompanying constituents exhibit crystal structures distinct from hcp. Examining the calculated values, the coefficient of thermal expansion (α) in BMGs, listed in S. Nos. 1–9 in Table 3, follows a diminishing sequence. Conversely, the calculated values

Table 3 The value of the α, W, and R for Zr-Al-Ni-Cu BMGs (Refs. [9, 11, 13])

S. No.	BMG	T_g K	T_x K	α μm/(m.K)	W W/(m.K)	R pm
1	$Zr_{55}Al_{8.9}Ni_{7.3}Cu_{28.8}$	684.6	765.2	10.921	155.647	140.479
2	$Zr_{54.5}Al_{9.6}Ni_{8.4}Cu_{27.5}$	691.3	769.5	10.987	152.980	140.102
3	$Zr_{54}Al_{10.2}Ni_{9.4}Cu_{26.4}$	693.6	771.2	11.050	150.787	139.762
4	$Zr_{53.5}Al_{10.9}Ni_{10.6}Cu_{25}$	704.5	775.9	11.113	147.809	139.368
5	$Zr_{53}Al_{11.6}Ni_{11.7}Cu_{23.7}$	711.4	778.7	11.179	145.142	138.991
6	$Zr_{52.5}Al_{12.2}Ni_{12.6}Cu_{22.7}$	712.8	783.8	11.245	143.259	138.668
7	$Zr_{52}Al_{12.9}Ni_{13.8}Cu_{21.3}$	714.6	789.8	11.308	140.282	138.274
8	$Zr_{51.5}Al_{13.6}Ni_{14.9}Cu_{20}$	723.5	796.8	11.374	137.615	137.897
9	$Zr_{51}Al_{14.2}Ni_{15.9}Cu_{18.9}$	725.7	791.6	11.436	135.422	137.557

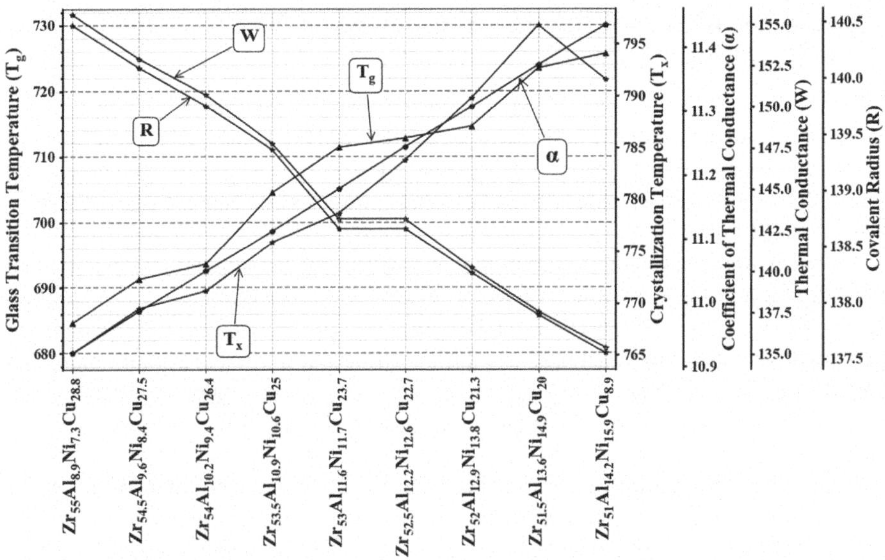

Fig. 3 The Value of T_g, T_x, α, W, and R for Zr-Al-Ni-Cu BMGs group combinations

for thermal conductance and covalent radius demonstrate an ascending pattern. In accordance with hypothesis i, positing a relationship between changes in α, thermal conductance, and covalent radius, the anticipated behaviour is further supported by the expected progression of glass transition and crystallization temperature values.

Remarkably, the experimental data affirms the hypothesis, as both glass transition and crystallization temperatures exhibit a consistent rise from serial number 1–8 in the BMGs of Table 3. Consequently, the observed experimental trends align with the predicted outcomes, substantiating the validity of hypothesis i in the context of the studied BMGs. The consistent trend of increasing glass transition temperatures persists in the case of BMGs numbered 8 and 9 as outlined in Table 3. However, it is noteworthy that the crystallization temperature of BMG $Zr_{51}Al_{14.2}Ni_{15.9}Cu_{18.9}$ (S. No. 9 in Table 3) deviates from the anticipated sequence, exhibiting a lower value compared to BMG $Zr_{51.5}Al_{13.6}Ni_{14.9}Cu_{20}$ (S. No. 8 in Table 3). Consequently, for these particular BMGs, the observed crystallization temperature results contradict the predictions set forth by the hypothesis.

7 Analysis of Crystallization Temperature (T_x) for Zr-Ti-Ni-Cu BMGs

The temperature at which crystallization occurs for BMGs $Ti_{20}Zr_{20}Cu_{60-x}Ni_x$ (where x = 10, 20, 30, and 40) is provided in Table 4. These values were obtained at heating rates of 10, 20, 30, and 40 K min^{-1} and exhibit variations corresponding to different alloy compositions.

Copper (Cu) serves as the foundational element in the subsequent group due to its higher fraction in the constituent elements of BMG combinations $Ti_{20}Zr_{20}Cu_{50}Ni_{10}$ and $Ti_{20}Zr_{20}Cu_{40}Ni_{20}$. Meanwhile, Nickel (Ni) takes precedence as the base element in the subsequent group, attributed to its higher fraction in the constituent elements of BMG $Ti_{20}Zr_{20}Cu_{20}Ni_{40}$. In BMG $Ti_{20}Zr_{20}Cu_{30}Ni_{30}$, both copper and nickel share an equal quantitative fraction, yet nickel surpasses copper in the number of atoms (Fig. 4).

Employing the specified calculation method, the effective values of α, W, and R for BMGs $Ti_{20}Zr_{20}Cu_{60-x}Ni_x$ (x = 10, 20, 30, and 40) are detailed in Table 5. Remarkably, in the BMG group combinations labeled as $Ti_{20}Zr_{20}Cu_{50}Ni_{10}$ and $Ti_{20}Zr_{20}Cu_{40}Ni_{20}$ with serial numbers 1 and 2, the proportion of the base element Cu displays a consistent decreasing trend. Despite this reduction in the order of Cu content, these combinations maintain a fcc crystal structure. In this structural maintenance, nickel (Ni) seamlessly substitutes some fraction of copper, and notably, this substitution does not alter the electron valence group. Additionally, it is noteworthy that both elements Ti and Zr share an identical fraction in these alloy combinations.

Comparatively, there is a notable and abrupt increase in the crystallization temperature (T_x) within the BMG $Ti_{20}Zr_{20}Cu_{20}Ni_{40}$ when contrasted with the BMG $Ti_{20}Zr_{20}Cu_{30}Ni_{30}$. This temperature shift is accompanied by a consistent and similar trend in the alterations of the coefficients of thermal expansion (α), thermal conductivity (W), and covalent radius (R) across all BMGs. The designation of Copper (Cu) as the base element in BMG group combinations $Ti_{20}Zr_{20}Cu_{50}Ni_{10}$ and $Ti_{20}Zr_{20}Cu_{20}Ni_{40}$ contrasts with Ni assuming this role in BMGs $Ti_{20}Zr_{20}Cu_{30}Ni_{30}$ and $Ti_{20}Zr_{20}Cu_{20}Ni_{40}$. The crystal structure of constituent elements Zr and Ti is

Table 4 The value of crystallization temperature (T_x) for Ti-Zr-Cu-Ni BMGs group combinations at different heating rate. (Ref. [14])

S. No.	BMG	T_x (K)§	T_x (K)*	T_x (K)$^{\#}$	T_x (K)^{ǂ}
1	$Ti_{20}Zr_{20}Cu_{50}Ni_{10}$	755	771	775	780
2	$Ti_{20}Zr_{20}Cu_{40}Ni_{20}$	756	768	770	772
3	$Ti_{20}Zr_{20}Cu_{30}Ni_{30}$	757	764	767	771
4	$Ti_{20}Zr_{20}Cu_{20}Ni_{40}$	778	785	791	795

§—at heating rate 10 K min^{-1}
*—at heating rate 20 K min^{-1}
#—at heating rate 30 K min^{-1}
ǂ—at heating rate 40 K min^{-1}

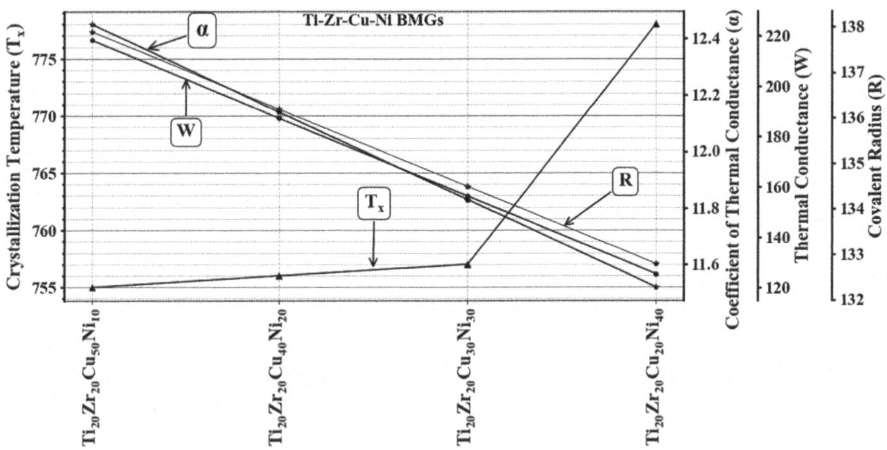

Fig. 4 Graphical presentation of T_X, α, W, and R values for various combinations within the BMGs $Ti_{20}Zr_{20}Cu_{60\text{-}x}Ni_x$ (x = 10, 20, 30 and 40) at heating rate 10 K min^{-1}

Table 5 The value of T_X, α, W, and R for BMGs $Ti_{20}Zr_{20}Cu_{60\text{-}x}Ni_x$ (x = 10, 20, 30 and 40) at heating rate 10 K min^{-1}. (Refs. [9, 11, 14])

S. No.	BMG	T_X (K)*	α {μm/(m.K)}	W {W/(m.K)}	R {pm}
1	$Ti_{20}Zr_{20}Cu_{50}Ni_{10}$	755	12.45	218.49	137.9
2	$Ti_{20}Zr_{20}Cu_{40}Ni_{20}$	756	12.14	187.48	136.2
3	$Ti_{20}Zr_{20}Cu_{30}Ni_{30}$	757	11.83	156.47	134.5
4	$Ti_{20}Zr_{20}Cu_{20}Ni_{40}$	778	11.52	125.46	132.8

* —at heating rate 10 K min^{-1}

hexagonal close-packed (hcp), while Ni and Cu exhibit a fcc arrangement. The effective values of α, W, and R, along with the crystallization temperature at heating rates of 10 K min^{-1} for BMG group combinations $Ti_{20}Zr_{20}Cu_{60\text{-}x}Ni_x$ (where x = 10, 20, 30, and 40), are effectively listed in Table 5. Notably, the proportion of the base element Cu displays a decreasing order in the BMG group combinations with serial numbers 1 and 2, specifically $Ti_{20}Zr_{20}Cu_{50}Ni_{10}$ and $Ti_{20}Zr_{20}Cu_{40}Ni_{20}$. Despite this decreasing trend, the base element Cu maintains a fcc crystal structure, experiencing partial substitution by Ni within these BMGs.

The effective values of α, W, and R for BMG group combinations $Ti_{20}Zr_{20}Cu_{60\text{-}x}Ni_x$ (x = 10, 20, 30, and 40) follow a decreasing order. Hypothesis f, "Regarding the base element of the BMG or elements sharing the same crystal structure as the base element, there is a decrease in the quantitative fraction of such elements. In this scenario, the alteration in glass transition or crystallization temperature demonstrates an inverse relationship with any two identical changes in α, W, and R." positing that for the base element of the BMG or elements with the same crystal structure as the base element, there is a decrease in their quantitative fraction, holds significance.

In this context, variations in glass transition or crystallization temperature demonstrate a reverse correlation with corresponding changes in all three coefficients of thermal expansion, thermal conductivity, and covalent radius. The trend suggested by hypothesis 'f' aligns with the crystallization temperature data in the BMG group combinations $Ti_{20}Zr_{20}Cu_{60-x}Ni_x$ (where x = 10, 20, and 30), displaying an ascending order with increasing values of T_x—755 K, 756 K, 757 K, and 778 K for x = 10, 20, 30, and 40, respectively. Within the BMG group combinations $Ti_{20}Zr_{20}Cu_{60-x}Ni_x$ (x = 30 and 40), the base element Ni experiences an increase while maintaining a fcc crystal structure. Copper (Cu) is introduced to replace a portion of Ni within the base element. Notably, the values of the crystallization temperature exhibit a significant upturn, suggesting that this pronounced increase may be attributed to the replacement of the base element from Cu to Ni.

In the realm of BMGs, specifically $Ti_{20}Zr_{20}Cu_{60-x}Ni_x$ (x = 10, 20, and 30), a discernible trend emerges as the quantitative fraction of the base element Cu experiences a consistent decrease while steadfastly maintaining a fcc crystal structure. The introduction of Ni serves the purpose of substituting a proportion of Cu within the alloy. Remarkably, the effective values of α, W, and R for the BMG group combinations $Ti_{20}Zr_{20}Cu_{60-x}Ni_x$ (x = 10, 20, 30, and 40) consistently exhibit a decreasing order.

Contrary to the expectations set by hypothesis 'f', which posits that the crystallization temperature of BMG group combinations $Ti_{20}Zr_{20}Cu_{60-x}Ni_x$ (where x = 10, 20, and 30) should increase with rising values of x, the available data at a heating rate of 20 K min^{-1} presents a contradiction. The crystallization temperature values of 771 K, 768 K, and 764 K for x = 10, 20, and 30, respectively, diverge from the anticipated ascending order. This divergence suggests that factors beyond those considered in hypothesis 'f' may influence the crystallization behaviour in this specific context.

In the case of BMG group combinations $Ti_{20}Zr_{20}Cu_{60-x}Ni_x$ (x = 30 and 40), an intriguing shift occurs as the base element Ni undergoes an increase while maintaining an fcc crystal structure. The introduction of Cu serves to substitute a portion of Ni within the alloy. Impressively, the effective values of α, W, and R for BMG group combinations $Ti_{20}Zr_{20}Cu_{60-x}Ni_x$, ranging from x = 40 to 30, consistently follow an increasing order.

In alignment with hypothesis 'f', the crystallization temperature of BMG group combinations $Ti_{20}Zr_{20}Cu_{60-x}Ni_x$ (x = 30 and 40) is anticipated to exhibit a decreasing order with the diminishing value of x. The available data supports this hypothesis, revealing crystallization temperatures of 764 K and 791 K for x = 30 and 40, respectively, under a heating rate of 20 K min^{-1}. This alignment with an increasing order provides affirmation for hypothesis 'f', suggesting a complex interplay of factors influencing the crystallization behaviour. Furthermore, consistent results across heating rates of 30 and 40 K per minute [22] reinforce the earlier observations, enhancing the credibility of the stated findings.

8 Conclusion

The endeavour to make quantitative adjustments to the constituent elements of bulk metallic glasses (BMGs) carries considerable potential, as it opens avenues for influencing key material properties like thermal expansion coefficient (α), thermal conductivity (W), and covalent radius (R). These subtle modifications possess the capability to instigate meaningful changes in fundamental characteristics such as crystallization temperature (T_x), glass transition temperature (T_g), Poisson's ratio (ν), and elastic moduli, thereby expanding the range of possibilities for customizing and optimizing the performance of these groundbreaking materials. The critical cooling rate plays a pivotal role in influencing both the crystallization and glass transition temperature of materials. Additionally, the extent of alterations in these properties following modifications in the quantitative fraction of constituent elements remains an unanswered question.

References

1. Klement, W., Willens, R.H., Duwez, P.: Non-crystalline structure in solidified gold-silicon alloys. Nature **187**, 869–870 (1960)
2. Chen, H.S., Turnbull, D.: Formation, stability, and structure of palladium-silicon based alloy glasses. Acta Metall. **17**(8), 1021–1031 (1969). ISSN 0001-6160
3. Inoue, A., Zhang, T., Masumoto, T.: Al–La–Ni amorphous alloys with a wide supercooled liquid region. Mater. Trans. JIM **30**(12), 965–972 (1989)
4. Peker, A., Johnson, W.: A highly processable metallic glass: $Zr_{41.2}Ti_{13.8}Cu_{12.5}Ni_{10.0}Be_{22.5}$. Appl. Phys. Lett. **63**, 2342–2344 (1993)
5. Shen, T.D., Schwarz, R.B.: Bulk ferromagnetic glasses prepared by flux melting and water quenching. Appl. Phys. Lett. **75**(1), 49 (1999)
6. Suryanarayana, C., Inoue, A.: Iron-based bulk metallic glasses. Int. Mater. Rev. **58**(3), 131–166 (2013). https://doi.org/10.1179/1743280412Y.0000000007
7. Li, Y., Zhao, S., Liu, Y., Gong, P., Schroers, J.: How many bulk metallic glasses are there? ACS Comb. Sci. **19**(11), 687–693 (2017)
8. Cahn, R.W.; Greer, A.L.: Chapter 19-metastable states of alloys. Encyclopedia of materials: science and technology. In: Buschow, K.H.J., Cahn, R.W., Flemings, M.C., Ilschner, B., Kramer, E.J., Mahajan, S., Veyssiere, P. (eds.) Oxford, Pergamon, vol. 5, pp. 7235–7241 (1996)
9. Suthar, P.R.: Bulk metallic glasses: general note with effect of fraction change in constituent elements on various temperatures. Eur. Chem. Bull. **12**(8), 9171–9187 (2023)
10. Senkov, O.N.: Correlation between fragility and glass-forming ability of metallic alloys. Phys. Rev. B **76**(10), 104202 (2007)
11. Suthar, P.R.: Poisson ratio of BMGs Zr-Ti-Al-Ni-Cu-Be with variation in fraction of constituent elements. Eur. Chem. Bull. **12**(8), 9131–9141 (2023)
12. Gangopadhyay, A.K., Kelton, K.F.: Prediction of elemental glass-transition temperatures of metals from thermophysical properties of liquids. J. Non-Cryst. Solids: X **2**, 100016 (2019)
13. Cai, A., Ding, D., Liu, Y., Wu, H., An, W., Zhou, G., Luo, Y., Peng, Y.: A series of Zr-based bulk metallic glasses with room temperature plasticity. Materials **9**(5), 373 (2016)
14. Rao, P.R., Bhatnagar, A.K., Majumdar, B.: Crystallization and kinetics studies of $Ti_{20}Zr_{20}Cu_{60-x}Ni_x$ (x = 10, 20, 30 and 40) metallic glasses. Bull. Mater. Sci. **43**, 36 (2020)
15. Zhang, L.-C.: Chapter 24 crystallization behavior and control of amorphous alloys. IntechOpen (2012)

16. Louzguine-Luzgin, D.V., Setyawan, A.D., Kato, H., Inoue, A.: Thermal conductivity of an alloy in relation to the observed cooling rate and glass-forming ability. Phil. Mag. **87**(12), 1845–1854 (2007)
17. Codrean, C., Buzdugan, D., Opriş, C., Muntean, R., Ambruş, S.: Corrosion behavior of bulk amorphous steels with low carbon content. Mater. Today Proc. (2022)
18. Cai, A., Ding, D., Liu, Y., Wu, H., An, W., Zhou, G., Luo, Y., Peng, Y.: A series of Zr-based bulk metallic glasses with room temperature plasticity. Materials **9**, 408 (2016). https://doi.org/10.3390/ma9060408
19. Curtis, B., Francis, C., Kmiec, S., Martin, S.W.: Investigation of the short range order structures in sodium thioborosilicate mixed glass former glasses. J. Non-Cryst. Solids **521**, 119456 (2019)
20. Cao, H., et al.: Computational thermodynamics to identify Zr-Ti-Ni-Cu-Al alloys with high glassforming ability. Acta Mater. **54**(11), 2975–2982 (2006)
21. Zhuang, Y.X.: Effects of relaxation on glass transition and crystallization of ZrTiCuNiBe bulk metallic glass. J. Appl. Phys. **88**, 6535 (2000)
22. Van de Moortèle §, B., Epicier, T., Soubeyroux, J.L., Pelletier, J.M.: Compositions of different phases appearing during devitrification of $Zr_{46.75}Ti_{8.25}Cu_{7.5}Ni_{10}Be_{27.5}$ bulk metallic glass. Philos. Mag. Lett. **84**(4), 245–256 (2004)
23. Chisholm, C.R., Haile, S.M.: Structure and thermal behavior of the new superprotonic conductor Cs (HSO)(HPO). Acta Crystallogr. Sect. B Struct. Sci. **55**, 1075–1083 (1999)
24. Shen, T.D.: Bulk ferromagnetic glasses prepared by flux melting and water quenching. Appl. Phys. Lett. **75**, 49–51 (1999)
25. Lee, F.T., Hayes, P.: Mechanisms of the reduction of zinc ferrites in H/N gas mixtures. Miner. Process. Extr. Metall. Rev. **27**, 205–213 (2006)
26. Nagase, T., Sanda, T., Nino, A., Qin, W., Yasuda, H., Mori, H., Umakoshi, Y., Szpunar, J.A.: MeV electron irradiation induced crystallization in metallic glasses: atomic structure, crystallization mechanism, and stability of an amorphous phase under the irradiation. J. Non-Cryst. Solids **358**, 2080–2086 (2012)
27. Dolan, M.D., Dave, N.C., Ilyushechkin, A., Morpeth, L.D., McLennan, K.: Composition and operation of hydrogen-selective amorphous alloy membranes. J. Membr. Sci. **285**, 30–55 (2006). https://doi.org/10.1016/j.memsci.2006.09.014
28. Suryanarayana.: Glass-forming ability of alloys. In: Bulk Metallic Glasses (2010)
29. Biswas, K., Gurao, N.P., Maiti, T., Mishra, R.S.: High entropy materials. In: Springer Science and Business Media LLC (2022)
30. Schwarz, R.B.: Bulk amorphous alloys. In: Intermetallic Compounds-Principles and Practice (2002)
31. Faupel, F.: Diffusion in metallic glasses and supercooled melts. Rev. Mod. Phys. **75**, 121–142 (2003)
32. Lu, Z.P.: A new approach to understanding and measuring glass formation in bulk amorphous materials. Intermetallics **12**, 1249–1253 (2004)
33. Tang, C., Zhou, H.: Chapter 3 thermodynamics and the glass forming ability of alloys. IntechOpen (2011)
34. Zhuang, Y.X.: Effects of relaxation on glass transition and crystallization of ZrTiCuNiBe bulk metallic glass. J. Appl. Phys. **87**, 1234–1256 (2000)

Microwave Dielectric Properties of Electroceramics

Priyambada Mallick, Snehasish Mishra, Nusrat Fatma, Dipan Ku Das, Srikanta Moharana, and Santosh Kumar Satpathy

Abstract Due to the ever-increasing demand for advanced wireless communication devices with the fast-growing world population and technological advancements, researchers have drawn their interest toward high-performance, efficient microwave and wireless communication devices. Electroceramics are advanced materials with unique electrical, dielectric, magnetic, and mechanical properties, making them suitable for advanced wireless telecommunication devices, radar technologies, and other electronic devices. The microwave dielectric properties, including dielectric constant, tangent loss, quality factor, resonant frequency, temperature, and mechanical stability of these materials, play a crucial role in the effective operation of a material to store and transmit microwave energy at microwave frequencies within advanced microwave systems, as well as in other areas like sensors, resonators, capacitors, filters, and antennas, etc. However, factors such as fabrication techniques, crystal structure, grain size, and composition can highly affect these properties.

Understanding and optimising the microwave dielectric characteristics of electroceramics necessitates extensive material characterisation, which frequently includes impedance spectroscopy, resonant cavity methods, and transmission line studies. Researchers and engineers attempt to adapt these qualities to specific applications, aiming for a balance between a high dielectric constant for energy storage and a low dielectric loss for minimal energy dissipation.

In this book chapter, electroceramic materials with outstanding microwave dielectric characteristics, such as perovskite materials and other related compounds, are critical for constructing compact and efficient devices, including resonators, filters, and antennas. Furthermore, producing temperature-stable electroceramics is crucial for consistent performance in various environmental circumstances.

This chapter will cover a broader research topic, including synthesising, characterising, and using electroceramic materials with optimised microwave dielectric characteristics. The relationship of microstructure, sintering processes, and microwave properties is also investigated. This chapter will also include additional information about specific electroceramic materials such as $BaTiO_3$, $SrTiO_3$, $CaTiO_3$, $MgTiO_3$,

P. Mallick · S. Mishra · N. Fatma · D. K. Das · S. Moharana · S. K. Satpathy (✉)
School of Applied Sciences, Centurion University of Technology and Management, Jatni, Odisha, India
e-mail: santosh.satpathy@cutm.ac.in

U. Kumar (ed.), *Defects Engineering in Electroceramics for Energy Applications*, Engineering Materials, https://doi.org/10.1007/978-981-97-9018-0_14

and other related compounds such as Lead Zirconate Titanate, Lanthanum Gallium Silicate, $CaNdAlO_4$, and the methods used for characterisation, as well as potential applications. Recent breakthroughs in microwave dielectrics for advanced wireless telecommunication devices will be explored in this context, as will their defects. Possible solutions and accomplishments will be identified.

1 Introduction

For microwave applications, dielectric characteristics in electroceramics are crucial. The microwave dielectric properties including dielectric constant, tangent loss, quality factor, resonant frequency, temperature, and mechanical stability of these materials play a crucial role in the effective operation of a material to store and transmit microwave energy at microwave frequencies within advanced microwave systems, as well as in other areas like sensors, resonators, capacitors, filters, and antennas, etc. But these properties can be highly affected by factors such as fabrication techniques, crystal structure, grain size, and composition [1, 2].

The dielectric constant and dielectric loss are the two parameters that define the characteristics of microwave dielectrics. Dielectric loss of a material is its capacity to release electrical energy as heat, whereas its dielectric constant is its capacity to store electrical energy.

Beside these two, numerous parameters including the temperature, the composition and microstructure of the electroceramic, affect the microwave dielectric characteristics. Through the ions' electronic polarizability, the electroceramic composition [3] influences both the dielectric constant and dielectric loss [4]. The dielectric constant and dielectric loss are influenced by the grain size and defect presence in the electroceramic microstructure. Due to the electroceramic's thermal expansion, temperature has an impact on both the dielectric constant and dielectric loss.

For microwave applications, electroceramics with high dielectric constants and low dielectric losses are preferred. Some examples of electroceramics with high dielectric constants and low dielectric losses are given in Fig. 1.

One can customise its microwave dielectric characteristics by varying the electroceramic composition, microstructure, and temperature. This makes it possible to create electroceramics with particular qualities for microwave applications.

2 Synthesis of Electroceramic Materials

Electroceramic materials are ceramic materials with distinct optical, magnetic, and electrical properties. They are employed in many fields, such as sensors, energy conversion, electronics, and telecommunications. Electroceramic materials can be synthesized in a variety of ways. The most popular techniques consist of.

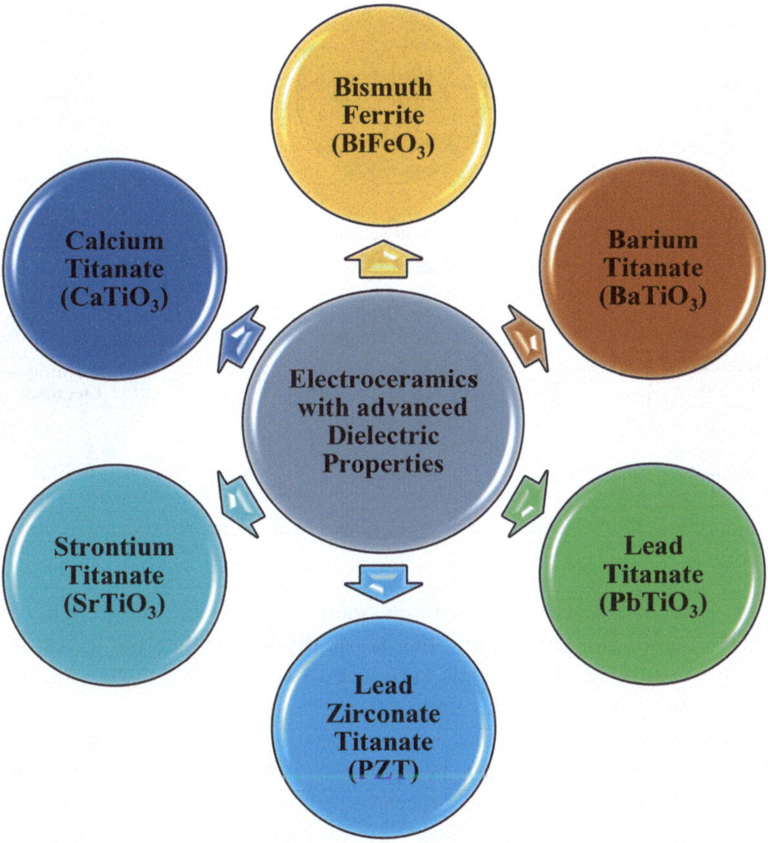

Fig. 1 Electroceramics with high dielectric constants and low dielectric losses

2.1 Solid-State Reaction Technique

Solid-state reaction synthesis (SSRS) is a flexible and popular way to create electroceramic materials. A combination of solid beginning elements must react at a high temperature to create a new solid with the appropriate structure and properties. The flow chart of Solid state raction technique is given in Fig. 2.

The following steps are commonly included in the SSRS process:

Weighing and combining the raw ingredients: The basic ingredients are incorporated in the stoichiometric ratios needed to create the intended result. To guarantee a homogenous composition, they are then adequately combined.

Calcination: A temperature lower than the melting point of any component is used for calcining the combined starting materials. This stage facilitates the creation of intermediary steps and eliminates any volatile contaminants.

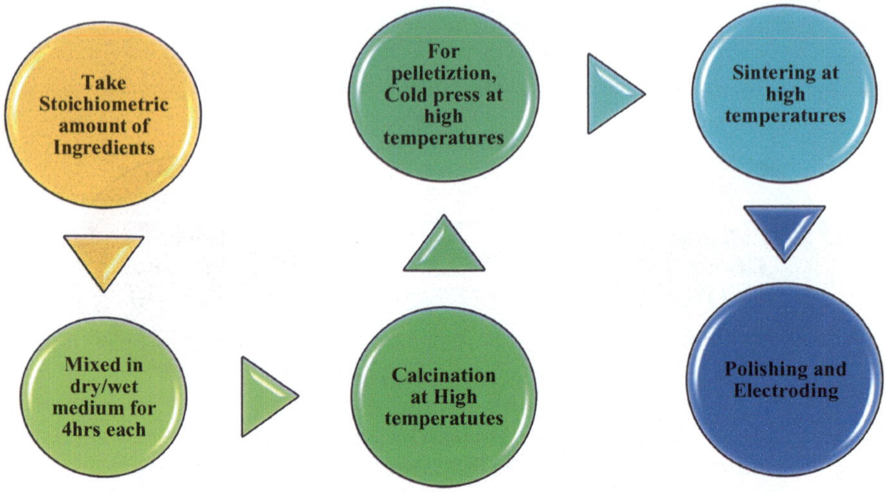

Fig. 2 Various steps for the synthesis of electroceramics by using the Solid state reaction method

Sintering: After the powder has been calcined, it is formed into pellets or other shapes and sintered at a temperature that is usually higher than the lowest melting point of the constituents. The intended product phase is created during the sintering process when the powder particles unite.

Perovskites [5, 6], ferroelectrics [7], and piezoelectric are just a few of the many electroceramic materials that may be synthesized using the straightforward and affordable SSRS method. The SSRS process is usually sluggish to create high-quality products and necessitates precise control over the reaction temperature and duration.

2.2 *Solution Based Method*

One flexible and scalable method for creating electroceramic materials is solution-based synthesis. It entails creating a precursor solution with the required ingredients, depositing the solution onto a substrate, and then turning the key into a ceramic film. The resulting electroceramic material's composition, microstructure, and characteristics can be controlled by adjusting the solution chemistry and heat conversion process. Compared to alternative manufacturing techniques, solution-based synthesis of electroceramic materials has several benefits. These include:

Scalability: Large amounts of electroceramic materials may be produced using solution-based synthesis, which qualifies it for commercial use.

Versatility: A large variety of electroceramic materials with various compositions and microstructures can be created by solution-based synthesis.

Controllability: The final electroceramic material's characteristics can be manipulated by adjusting the solution chemistry and thermal conversion procedure.

One effective method for creating electroceramic materials for a range of uses is solution-based synthesis. These uses include:

Capacitors: Capacitors are made of electroceramic materials with high dielectric constants and breakdown strengths to store electrical energy.

Piezoelectric devices: Ultrasonic transducers, actuators, and sensors employ electroceramic materials possessing piezoelectric characteristics.

Solid oxide fuel cells use electrolytes made of ceramic materials with high ionic conductivities.

Membranes: Gas separation, water filtration, and oxygen generation are all accomplished using electroceramic membranes.

There are a variety of solution-based synthesis methods for electroceramic materials, including.

2.3 *Sol–Gel Technique*

Now a day the sol–gel technique is the most adaptable method of fabrication of inorganic compounds. In this method low temperature is employed during the synthesis and also a pure phase of the material is obtained [8]. This method is extensively used for the preparation of oxides like $BiFeO_3$. This method of preparation is very simple and cheap. The crystal structure of the compound can be obtained more simply by using this method of preparation [9, 10]. To create a colloidal sol, metal alkoxides or other precursors containing metals are hydrolysed and then condensed. After that, the sol is applied to a substrate and allowed to dry to create a gel. After that, the gel is heated by calcination and sintering to make a ceramic film. The flowchart for the preparation of Bismuth ferrite by Sol–gel process is shown in Fig. 3.

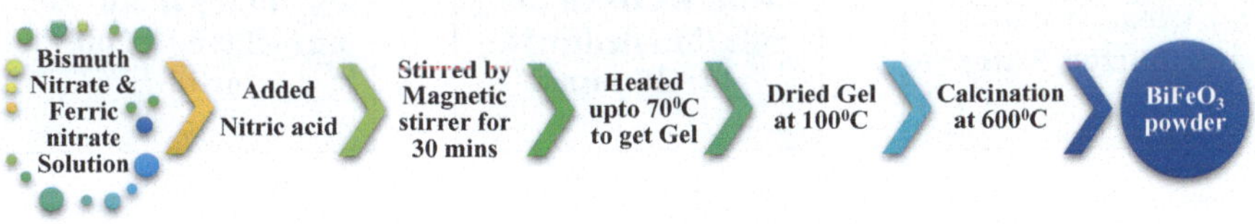

Fig. 3 Various steps for the synthesis of Bismuth ferrite by using the Sol–gel method

2.4 Hydrothermal Process

The process of synthesising materials at high temperatures and pressures in a sealed vessel with a solvent, like water, is called hydrothermal synthesis. The synthesis of electroceramic materials, a type of material with unique electrical characteristics, is frequently accomplished by this procedure [11]. Metal salts are dissolved in a solvent and heated under high pressure in a sealed autoclave in the hydrothermal synthesis process. The high temperature and pressure form the crystalline electroceramic materials more easily. The different steps for the preparation of Bismuth Ferrite using hydrothermal method are given in Fig. 4.

In order to create electroceramic materials hydrothermally, the following procedures are usually followed:

Precursor solution preparation: This step involves combining metal oxides, hydroxides, or salts with other ingredients to create the electroceramic material.

Filling a stainless-steel autoclave: An autoclave is a sealed vessel that can tolerate high pressure and heat. The precursor solution is placed into the autoclave.

Heating: An elevated temperature, usually between 100 and 300 degrees Celsius, is achieved inside the autoclave.

Autoclave pressurization: The autoclave is pressurized to a high pressure, usually in the 100–500 atmospheres range.

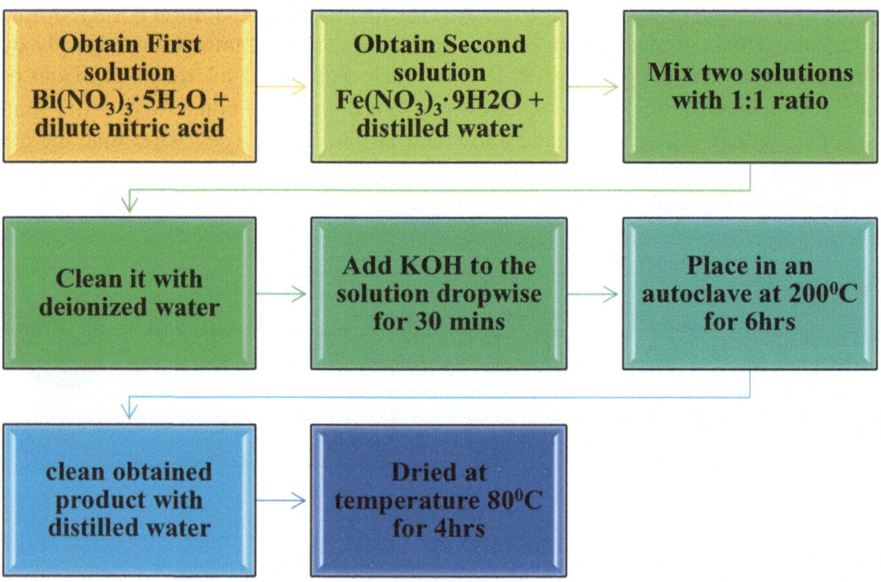

Fig. 4 Flow chart for the synthesis of Bismuth ferrite by using hydrothermal method

High temperature and pressure holding: The autoclave is kept at these temperatures for a certain amount, usually one to twenty-four hours.

Autoclave cooling: The autoclave is brought down to room temperature.

Product collection: The electroceramic material is taken out of the autoclave.

Compared to alternative synthesis techniques, including solid-state reaction and sol–gel processing, the hydrothermal synthesis of electroceramic materials offers several advantages. These benefits consist of:

- High production and purity of the product
- Homogeneous morphology and fine particle size
- Low pressure and temperature during processing
- Easy and repeatable procedure.

Many different kinds of materials have been created via the hydrothermal synthesis of electroceramic materials, such as

- Titanate of barium ($BaTiO_3$) [11]
- Titanate of lead zirconate ($Pb (Zr, Ti)O_3$) [12–14]
- Lithium Niobate ($LiNbO_3$) [15]
- $BiFeO_3$, or bismuth ferrite [16]
- Potassium Sodium Niobate (KNN) [17, 18]
- Sodium Niobate ($NaNbO_3$).

These materials are used in various applications, including Capacitors, Transducers, Actuators, Sensors, and Optical devices. The hydrothermal synthesis of electroceramic materials is a powerful tool for synthesising high-quality materials with a wide range of applications.

3 Properties of Electroceramics for Microwave Applications

3.1 Dielectric Constant

One of the most crucial characteristics of electroceramics for microwave applications is the dielectric constant. It is a measurement of the material's susceptibility to electric field polarization. A high dielectric constant for microwave applications is preferred since it raises the material's capacitance, which increases the microwave device's resonance frequency [19].

The material's composition, crystal structure, and microstructure are some variables that influence an electroceramic's dielectric constant. Lead zirconate titanate (PZT), strontium titanate ($SrTiO_3$), and barium titanate ($BaTiO_3$) are a few of the most widely utilized electroceramics in microwave applications [20]. The dielectric constants of these materials are between 100 and 1000, which is significantly greater than the dielectric constant of open space.

By doping the material with additional elements or altering the preparation conditions, it is possible to modify the dielectric constant and lose the tangent of electroceramics. For instance, strontium or calcium doping $BaTiO_3$ [21, 22] can lower the loss tangent without appreciably changing the dielectric constant. Electroceramics possessing high dielectric constants and low-loss tangents find extensive usage in microwave-related applications such as antennas, resonators, and filters. Oscillators and microwave amplifiers also use these materials.

3.2 Dielectric Strength

The highest electric field an insulating material can tolerate before degrading and conducting electricity is its dielectric strength. This characteristic is crucial because electroceramics are frequently exposed to high electric fields in microwave applications.

The material's composition, grain size, and defect presence are some variables that affect electroceramics' dielectric strength. Higher dielectric constant electroceramics typically have lower dielectric strengths. Breakdown can occur at a lower electric field because a material with a greater dielectric constant can retain more electrical energy [19, 20].

The dielectric strength of the electroceramic is also influenced by its grain size. Because they include fewer flaws, grains with smaller diameters typically have better dielectric strengths. Because faults might serve as areas where breakdown first occurs, more occasional imperfections can improve dielectric strength.

Electroceramics' dielectric strength can be determined in a number of ways. The usage of a parallel plate capacitor is one typical technique. A voltage is applied while the electroceramic is positioned between the capacitor's plates. The material is broken down and made to conduct electricity by progressively increasing the voltage. The breakdown voltage is then used to compute the dielectric strength.

Using a sphere-plane capacitor is another popular technique for determining the dielectric strength of electroceramics. This method involves applying a voltage and placing a spherical electrode on the electroceramic surface. The material is broken down and made to conduct electricity by progressively increasing the voltage. The breakdown voltage is then used to compute the dielectric strength. For microwave applications, electroceramics' dielectric strength is a crucial characteristic. Electroceramics with the appropriate qualities for particular applications can be designed by knowing the variables that affect dielectric strength.

3.3 Loss Tangent

For microwave applications, electroceramics' loss tangent property is crucial. It expresses a material's capacity to transform electrical energy into heat and is defined as the ratio of the dielectric loss to the dielectric constant. For microwave applications, a low-loss tangent is preferred since it reduces the material's power loss [23].

Low-loss tangent electroceramics are frequently utilized for resonators, filters, and substrates in microwave technology. Compact component design is made possible by the high dielectric constant of these materials.

Numerous factors, such as the material's composition, microstructure, and manufacturing conditions, influence the loss tangent of electroceramics. For instance, doping with trace amounts of calcium or strontium can lower the loss tangent of barium titanate. The size of the material's grains also affects the loss tangent; finer grains usually have smaller loss tangents.

Numerous methods, such as the dielectric resonator method, the impedance analyzer method, and the transmission line approach, can be used to estimate the loss tangent of electroceramics. The most popular technique is the dielectric resonator method, which measures the quality factor and resonant frequency of a dielectric resonator constructed from the material of interest.

For microwave applications, electroceramics' loss tangent property is crucial. Researchers can create new electroceramic materials with even more minor loss tangents by comprehending the aspects that influence them, which will make it possible to design microwave components that are more effective and compact.

3.4 Breakdown Voltage

The highest electric field given to an electroceramic material before it experiences electrical breakdown is known as the breakdown voltage. Since electroceramics are frequently employed in high-power devices like microwave filters, circulators, and isolators, this is an essential feature for microwave applications.

The breakdown voltage of electroceramics for microwave applications typically ranges from 10 to 100 kV/cm [24]. However, some electroceramics, such as barium titanate ($BaTiO_3$), have breakdown voltages as high as 500 kV/cm [25].

The breakdown voltage of an electroceramic depends on several factors, including:

Composition: Its chemical makeup greatly influences an electroceramic's breakdown voltage. For instance, lower breakdown voltages are typically found in electroceramics with a significant concentration of flaws or impurities.

Microstructure: An electroceramic's breakdown voltage is also influenced by its microstructure, which includes porosity and grain size. Greater breakdown voltages are typically found in electroceramics with smaller holes and finer grain sizes.

Processing parameters: The processing parameters employed to create an electroceramic's breakdown voltage may also be impacted. For instance, higher-temperature sintered electroceramics typically have greater breakdown voltages.

3.5 Temperature Stability

For microwave applications, electroceramics' temperature stability is a crucial characteristic. Many microwave devices use electroceramics, including filters, resonators, and amplifiers. The electroceramics in these devices are exposed to temperatures ranging from below room temperature to hundreds of degrees Celsius. Stable dielectric constant and other microwave properties are critical for the electroceramics over this temperature range. Numerous parameters, such as the crystal structure, material composition, and impurity presence, affect the temperature stability of electroceramics [26]. Compared to electroceramics with other crystal structures, those with a perovskite crystal structure are often more temperature stable. This is due to the extremely stiff and resilient perovskite crystal structure.

The temperature stability is also influenced by the electroceramic material's composition. Higher concentrations of barium or strontium in electroceramics tend to be more temperature stable than lesser concentrations in electroceramics. The reason for this is that compared to the other factors commonly present in electroceramics, barium and strontium have greater ionic radii. The crystal structure is stabilized in part by the greater ionic radius [27].

Impurities may also have an impact on the temperature stability of electroceramics. The crystal structure can be upset by impurities, altering the dielectric constant and other microwave characteristics. Reducing the impact of contaminants on thermal stability requires using high-purity materials.

A number of electroceramics that have good temperature stability for microwave applications have been developed. Some of the most common electroceramics used in microwave devices include:

- Alumina (Al_2O_3)
- Barium titanate ($BaTiO_3$) [11]
- Calcium titanate ($CaTiO_3$) [28]
- Magnesium titanate ($MgTiO_3$)
- Strontium titanate ($SrTiO_3$) [29–31].

These electroceramics have a perovskite crystal structure and a high concentration of barium or strontium. They also have a high purity, contributing to their excellent temperature stability.

3.6 Frequency Dependence

A class of ceramic materials with distinct electrical and dielectric characteristics is known as electroceramics. They are extensively used in resonators, antennas, filters, and microwave applications. For microwave applications, the dielectric constant of electroceramics is one of the most significant frequency dependency characteristics.

How easily an electric field may polarize a material is measured by its dielectric constant. This number is dimensionless and is commonly represented by the symbol ε. A material's dielectric constant varies according to the frequency of the applied electric field, which means it is frequency-dependent. The dielectric constant of electroceramics used in microwave applications must be high and comparatively constant across a broad frequency range. This is due to the fact that microwaves are high-frequency electromagnetic waves that can be effectively reflected, transmitted, and absorbed by electroceramics with high and constant dielectric constants [32].

Microwave engineers can select the proper material composition, microstructure, and crystal structure to modify the frequency dependence of the dielectric constant and dielectric loss tangent of electroceramics. As a result, they can create electroceramics with the ideal characteristics for particular microwave applications.

Examples of electroceramics that are frequently employed in microwave applications include the following:

$BaTiO_3$, or barium titanate, is a perovskite oxide ceramic with a low dielectric loss tangent and a high dielectric constant [11]. It is extensively utilized in antennas, resonators, and microwave filters. Calcium copper titanate is another perovskite oxide ceramic with a high dielectric constant and low dielectric loss tangent ($CaCu_3Ti_4O_{12}$) [33]. It works especially well in high-temperature microwave applications.

For microwave applications, an electroceramic should have a low dielectric loss tangent. A very high dielectric loss tangent may cause the electroceramic to overheat and lose some of its functionality.

The dielectric loss tangent and the dielectric constant of electroceramics are frequency-dependent and depend on several aspects, such as the crystal structure, microstructure, and composition of the material. For instance, materials containing a higher concentration of polarizable ions generally have a higher dielectric constant for electroceramics. Higher degrees of crystallinity and more ordered microstructures are typically associated with lower dielectric loss tangents in electroceramics.

3.7 Specific Dielectric Materials

Because dielectric materials can store and release electrical energy in an electric field, they are used in microwave applications. Certain electroceramic materials that are frequently employed in microwave applications include the following:

Barium titanate, or $BaTiO_3$, is suitable for microwave applications because of its high dielectric constant and low dielectric loss. It is also not too expensive to make [11].

Strontium titanate ($SrTiO_3$): Compared to $BaTiO_3$, $SrTiO_3$ has a lower dielectric constant but a lower dielectric loss. Because of this, it is a wise option for applications requiring high efficiency [29, 30].

Lead zirconate titanate (PZT): PZT is a viable option for applications requiring both energy conversion and storage because of its high dielectric constant and piezoelectric qualities [12, 13].

The precise needs of a given microwave application will determine which specific electroceramic material is most appropriate. For instance, a material with a low dielectric loss will be selected if excellent efficiency is desired. A fabric with a high dielectric constant will be chosen if high-energy storage is needed.

Apart from those mentioned above specific dielectric properties, additional features that could be significant for microwave applications include the following:

Mechanical strength: The material should withstand the mechanical stresses associated with microwave use.

Thermal conductivity: The material should have a high thermal conductivity so that heat can be dissipated efficiently.

Cost: The material should be relatively inexpensive to produce.

4 Applications

Wireless communication equipment and internet-related technologies are growing, promoting the advancement and development of microwave dielectric ceramics. Microwave dielectric ceramics are a fundamental component of the high-frequency communication sector. With the advent of 5G technology, wireless communication is expanding at a moderate rate using straight high frequency.

4.1 Ceramics Capacitor

Regarding bulk and film performance, dielectric ceramics are the best material option for energy storage devices like capacitors. Because ceramic capacitors do not rely on electric double layers or electrochemical reactions for their energy storage or charge storage mechanism, dielectric ceramics like barium titanate create conventional electrostatic capacitors rather than supercapacitors or ultracapacitors.

In 2001, EEStor Inc., a Texas-based company, began selling "electrical energy storage units" in 2007. These units are parallel plate capacitors with pure barium titanate ($BaTiO_3$) as the dielectric, used in laptops, computers, and electric vehicles, among other "utility-scale electricity storage" applications.

The barium titanate ($BaTiO_3$) [8] nanoparticle fabrication is very profitable as dielectrics are used in capacitors. $BaTiO_3$ is familiar due to its high dielectric breakdown resistance at the apex of 200 Vμ/m, and ceramics flaunt higher permittivity of nearly equal to 5000, which enables storage of the amount of energy. Barium titanate in barrier layer capacitors is doped by small amounts of electrically conducting rare earth metals, therefore creating barrier layers at the grain boundaries and between the barrier layers—the inner grain phase functions as a conducting connection connected in series.

4.1.1 Ceramic-Based Dielectric

The excellent thermal stability of dielectric ceramics, which defines their capacity and cracks under heat stress, is one of their other best qualities. Because relaxor ferroelectric ceramic material is used in energy storage devices, its slim B-H hysteresis loop has been widely studied. Due to the boosted functional electronic properties of ceramic materials, they are the best for use in storage devices and garnering energy, memory devices, sensors, and electrocaloric applications.

One feasible method for imparting various qualities to ceramics, such as mechanical, chemical, thermal stability, and flexibility, while maintaining cost-effectiveness, is to disperse a small number of ceramic nanofillers.

4.1.2 Ferroelectric Ceramic Polymer Nanocomposite

Ferroelectric ceramic-polymer composites improve the breakdown strength and flexibility of the polymers while enhancing the dielectric characteristics of the ceramic.

Barium zirconium titanate and polyvinylidene difluoride are studied and conductive to efficient energy storage in that composite because of the good intrinsic ferroelectric characteristics. The following are necessary to boost energy storage in polymer nanocomposites:

i. To improve the electrical characteristics, modify the size and form of nanoceramic
ii. A straightforward approach to repurposing materials is through new synthesis.

4.1.3 Perovskite-Type Dielectric Ceramic

Energy is stored using dielectric ceramics of the perovskite-type. Dielectric ceramics of this kind are primarily used in energy storage devices and photovoltaic solar cells. Dielectric properties are crucial for energy storage in perovskite solar cells. Dielectrics are used between plates and during the construction of capacitors to aid in energy storage. Perovskite's dielectric behavior is caused by several factors, including "material composition, sintering temperature, size of the raw materials, and synthesis processes." Materials with high dielectric constants, electrical conductivities, tangent losses, structural and magnetic properties that can be used to store energy include barium titanate ($BaTiO_3$) [7, 8], bismuth ferrite ($BiFeO_3$) [13], calcium titanate ($CaTiO_3$), and strontium titanate ($SrTiO_3$) [20].

Dielectric perovskite is formed by doping elements like Iron, Strontium, Manganese, Cobalt, etc., to store energy. At a very low temperature of 650–700 degrees centigrade, the formation of lead titanate ($PbTiO_3$) exhibits a tetragonal structure with electrical conductivity and high dielectric constant. In the synthesis of lead zirconate doping strontium, the result is that if the doping amount increases, then the dielectric constant decreases, but at the transition temperature, it attains maximum value.

4.2 Insulator

Materials with low electrical conductivity are referred to as insulators or dielectrics. By altering the dielectric constant, which measures a material's capacity to store electrical energy when subjected to an electric field, they can change the capacitance of electrical components. Based on their unique dielectric characteristics, insulators are used in the following typical applications. High-frequency devices like antennas and RF filters use ceramics with high dielectric constants. These substances help modify the resonance and impedance matching of radio frequency circuits. Ceramics with moderate dielectric constants are used as insulating substrates in electronic circuits for electrical insulation between conductive layers and supporting components on circuit boards.

In order to minimize voltage breakdown, ceramic insulators are crucial for electrical insulation in high-voltage systems, such as power transmission and distribution lines, as well as insulators on electrical poles.

The particular requirements of the application dictate the choice of ceramic material with a given dielectric constant, taking into account elements like operating frequency, temperature, and required capacitance. Because of their remarkable electrical insulating qualities and wide range of dielectric constants, ceramics are valuable materials for electrical and electronic applications.

4.3 Piezoelectric Device

Piezoelectric devices are used in sensors, actuators, and resonators. They are essentially used as transducers, transforming mechanical energy into electrical energy and vice versa. They are not usually used in applications involving material dielectric constants.

The ability of a material to store electrical energy under the influence of an electric field is measured by the dielectric constant, sometimes referred to as relative permittivity. It is essential for devices like capacitors, which modify capacitance using materials with particular dielectric constants.

Although they have unique qualities and uses, piezoelectric materials are not usually chosen because of their dielectric constant traits. Usually, non-piezoelectric materials are selected for dielectric constant applications based on their unique dielectric characteristics, which influence their ability to hold electric charge and modify the capacitance of electrical components.

4.4 Microwave Components

Materials with accurate dielectric constants are often used in microwave components such as waveguides, resonators, and antennas for optimal design and performance. In order to improve their performance in microwave systems, these components employ materials with certain dielectric constants in a variety of applications, such as radiation, resonance, impedance matching, and tuning.

Waveguide: Microwave signals are transmitted using waveguides. The phase velocity and impedance matching of the microwave signal are two ways that the dielectric constant of the material employed in the waveguide might impact the waveguide's performance. The proper dielectric constant must be selected in the materials for effective waveguide functioning.

Antennas: Microwave signals can be received or emitted by antennas. The radiation pattern, overall performance, and impedance matching of the antenna can all be impacted by the dielectric constant of materials around or close to the antenna elements. Choosing materials with particular dielectric constants allows antenna designers to maximize their designs for various uses.

Microwave filtration: Since microwave filters are in charge of isolating or prohibiting particular frequency ranges from flowing through, they are crucial parts of microwave communication systems. The features of the filter, such as bandwidth, insertion loss, and selectivity, are determined mainly by the selection of materials with various dielectric constants throughout the filter design process.

PCB substrates: Choosing the suitable dielectric material for PCB substrates is essential since it significantly affects the performance of electronic circuits. Controlling signal propagation speed, preserving constant impedance, enhancing RF and microwave circuit performance, antenna design, embedded components, high-frequency applications, miniaturization, and their use in satellite, radar, and RF systems are some of the critical applications and considerations for dielectric constant materials in PCB substrates.

1. **Propagation of signals:**

The PCB substrate material's dielectric constant affects the speed at which signals propagate. Lower dielectric constants make Faster signal transmission possible, which is vital for high-frequency applications like microwave and radio frequency circuits.

2. **Impedance regulation:**

PCB transmission line impedance is controlled mainly by the dielectric constant. Maintaining a constant impedance is crucial for maintaining signal integrity and minimising signal reflections, particularly in high-speed digital and RF applications.

3. **Satellite and RF systems:**

Customized PCB substrates with particular dielectric qualities are used in satellite communication and RF systems to meet exact electrical requirements and ensure effective signal transmission and reception. In conclusion, the dielectric constant of the material used for PCB substrates is an essential factor to consider in many different applications, especially those that involve microwave, radio frequency, and high-frequency circuits. The overall performance of electronic circuits, as well as signal propagation and impedance management, are impacted by the dielectric material selection. Choosing the appropriate dielectric material is crucial to satisfy every application's unique needs.

4.5 Dielectric Resonators

In microwave and RF engineering, dielectric resonators composed of ceramics with particular dielectric constants are essential. Among other functions, they serve as filters, antennas, duplexers, oscillators, frequency-selective surfaces, and sensors. These resonators are crucial in guaranteeing accurate control of electromagnetic waves and frequencies in these applications.

1. **Oscillating:**

For microwave oscillators to produce steady microwave signals, dielectric resonators are necessary. They are essential for microwave sources used in communication systems, radar, and related applications because they govern signal frequency and stability.

2. **Screens:**

Narrowband filters in microwave filter designs are created using dielectric resonators. Their application improves selectivity and makes it possible to reject unwanted frequencies. This is especially helpful for applications that need exact control over frequency.

3. **Switches and Reversers:**

Duplexers and diplexers use dielectric resonators to split or combine signals in transceivers. They are, therefore, necessary for effective signal management since they allow signals to be transmitted and received on the same antenna without interference. Numerous microwave and radiofrequency systems use dielectric resonators composed of ceramics with particular dielectric constants. They are necessary parts of filters, antennas, oscillators, and other devices that need to control electromagnetic waves and frequencies precisely.

4.6 Dielectric Coatings

Dielectric coatings on ceramics with certain dielectric constants find extensive use in materials engineering, electronics, and telecommunications. They are applied in capacitors, resonators, sensors, microwave components, electronic packaging, antenna design, and other areas to improve the electrical characteristics and performance of ceramics. These coatings are essential for precisely controlling signal qualities and electrical properties in a variety of systems and devices.

1. **Multilayer ceramic capacitor (MLCCS)**

Compact components with high capacitance can be made possible by applying dielectric coatings to ceramic layers in MLCCs. Because MLCCs can efficiently store electrical energy, they are essential in a wide range of electronic devices, including computers, smartphones, and consumer electronics.

2. **Dielectric-coated ceramics in sensors**

Dielectric-coated ceramics are used as transducers for sensing physical characteristics such as strain, temperature, pressure, and humidity in various sensor applications. The dielectric layer protects the underlying ceramic material and improves sensitivity simultaneously. For sensors to function accurately and consistently, this is essential. Electronic, communication, and sensing systems can all benefit from using dielectric coatings on ceramics with certain dielectric constants. These coatings improve the electrical characteristics and functionality of ceramics in various components and devices.

4.7 Dielectric Heating

Dielectric heating, which is frequently connected to appliances like microwave ovens, produces heat in dielectric materials employing alternating electric fields. Although its direct application in dielectric constant ceramics is restricted, it plays a vital role in the characterization of the behaviour of dielectric materials under controlled heating settings, as well as in ceramic production processes such as sintering and annealing.

1. **Sintering and Annealing:**

Sintering is the process of heating powdered ceramic materials to high temperatures in order to form a solid structure. This is how most dielectric constant ceramics are made. Not quite dielectric heating, but things like electrical conductivity and heat generation in ceramics are essential considerations. Precise temperature and heating rate control is necessary to obtain the appropriate ceramic characteristics.

2. **Drying:**

Ceramic components are dried using dielectric heating, which provides quick, even heating that efficiently removes moisture without raising the temperature or cracking the material.

3. **Determining the cure point:**

Ceramics with a dielectric constant have an electrical property of curie temperature. By monitoring variations in the dielectric constant with temperature, dielectric heating allows for an accurate assessment of this temperature.

4. **Announcing:**

For ceramics, annealing is a heat treatment procedure that improves structural qualities and reduces stress. Dielectric constant ceramics provide better mechanical and electrical properties because dielectric heating offers accurate temperature control during annealing. In order to ensure that dielectric stable ceramics meet precise electrical and structural requirements, dielectric heating is essential to their manufacturing, production, and quality control.

5 Conclusion

The creation of high-frequency electronic devices such as resonators, filters, antennas, and oscillators depends heavily on understanding the microwave dielectric characteristics of electroceramics. Components that function well at microwave frequencies can be designed and optimized with a sense of their features. Through intensive research and experimentation, scientists and engineers have made great strides in discovering materials with desired microwave dielectric properties, such as high dielectric constant, low dielectric loss, and stable temperature coefficients.

In conclusion, the ongoing need for faster, more dependable, and smaller electronics propels advancements in microwave dielectric characteristics of electroceramics. Ongoing research aims to develop fabrication methods, find new materials, and investigate creative applications. Electroceramics with customized microwave properties will be essential in determining the future direction of satellite technology, wireless communication, and other high-frequency applications. They will substantially contribute to advancing contemporary electronics as knowledge expands and technology advances.

References

1. Ohsato, H.: Microwave dielectrics with perovskite-type structure. Perovskite Mater.-Synth. Characterisation Prop. Appl. 281–330 (2016)
2. Tagantsev, A.K., Sherman, V.O., Astafiev, K.F., Venkatesh, J., Setter, N.: Ferroelectric materials for microwave tunable applications. J. Electroceramics **11**, 5–66 (2003)
3. Newnham, R.E.: Electroceramics. Rep. Prog. Phys.Prog. Phys. **52**(2), 123 (1989)
4. Fang, C.Y., Randal, C.A., Lanagan, M.T., Agrawal, D.K.: Microwave processing of electroceramic materials and devices. J. Electroceramics **22**, 125–130 (2009)
5. Mallick, P., Satpathy, S.K., Behera, B.: Investigation on the structural, electrical and thermistor parameters of La-doped $BiFeO_3$–$PbZrO_3$ for energy storage devices. Bull. Mater. Sci. **45**(4), 198 (2022)
6. Mallick, P., Patra, R., Mohanty, D., Satpathy, S.K.: Development and characterization of copper doped perovskite-polymer composite through high-temperature technique. Sādhanā **47**(3), 134 (2022)
7. Teresa Buscaglia, M., Harnagea, C., Dapiaggi, M., Buscaglia, V., Pignolet, A., Nanni, P.: Ferroelectric $BaTiO_3$ nanowires by a topochemical solid-state reaction. Chem. Mater. **21**(21), 5058–5065 (2009)
8. Vashisth, B.K., Bangruwa, J.S., Beniwal, A., Gairola, S.P., Kumar, A., Singh, N., Verma, V.: Modified ferroelectric/magnetic and leakage current density properties of Co and Sm co-doped bismuth ferrites. J. Alloy. Compd. **698**, 699–705 (2017)
9. Fuxue, Y.A.N., Gaoyang, Z.H.A.O., Na, S.O.N.G.: Sol-gel preparation of La-doped bismuth ferrite thin film and its low-temperature ferromagnetic and ferroelectric properties. J. Rare Earths **31**(1), 60–64 (2013)
10. Ghahfarokhi, S.E.M., Larki, M.R., Kazeminezhad, I.: The effect of Mn doped on the structural, magnetic, dielectric and optical properties of bismuth ferrite ($BiFe_{1-x}Mn_xO_3$) nanoparticles. Vacuum **173**, 109143 (2020)
11. Newalkar, B.L., Komarneni, S., Katsuki, H.: Microwave-hydrothermal synthesis and characterization of barium titanate powders. Mater. Res. Bull. **36**(13–14), 2347–2355 (2001)
12. Vaidhyanathan, B., Singh, A.P., Agrawal, D.K., Shrout, T.R., Roy, R., Ganguly, S.: Microwave effects in lead zirconium titanate synthesis: enhanced kinetics and changed mechanisms. J. Am. Ceram. Soc. **84**(6), 1197–1202 (2001)
13. Takahashi, H., Kato, K., Qiu, J., Tani, J.: Property of lead zirconate titanate actuator manufactured with microwave sintering process. Jpn. J. Appl. Phys.. J. Appl. Phys. **40**(2R), 724 (2001)
14. Manikandan, S., Aruchamy, N., Allirol, L., Girod, S., Glinsek, S., Defay, E., Granzow, T.: Dielectric relaxations of lead zirconate titanate films up to millimeter-wave frequencies. J. Am. Ceram. Soc. **105**(10), 6149–6157 (2022)
15. Holzgrafe, J., Sinclair, N., Zhu, D., Shams-Ansari, A., Colangelo, M., Hu, Y., Zhang, M., Berggren, K.K., Lončar, M.: Cavity electro-optics in thin-film lithium niobate for efficient microwave-to-optical transduction. Optica **7**(12), 1714–1720 (2020)
16. Mallick, P., Sahu, A.K., Biswal, S.K., Satpathy, S.K., Behera, B.: Investigation on structural, dielectric, thermistor parameters and negative temperature coefficient behaviour of Nd, Gd, and La-doped bismuth ferrite. Trans. Electr. Electron. Mater.Electr. Electron. Mater. **23**(5), 522–534 (2022)
17. Feizpour, M., Bafrooei, H.B., Hayati, R., Ebadzadeh, T.: Microwave-assisted synthesis and sintering of potassium sodium niobate lead-free piezoelectric ceramics. Ceram. Int. **40**(1), 871–877 (2014)
18. Gao, L., Zhou, W., Luo, F., Zhu, D., Yang, Z.: Microwave dielectric properties of potassium sodium niobate ceramics with different K/Na ratios. Ceram. Int. **42**(16), 19105–19109 (2016)
19. Sarmento, J.S., Paiva, D.V.M., de Araújo, E.V., Silva, M.A.S., Sombra, A.S.B., Mazzetto, S.E., Fechine, P.B.A.: Dielectric properties of $MCuSi_4O_{10}$ (M = Ca, Sr, Ba) electroceramic at RF and microwave frequencies. Appl. Phys. A **129**(1), 72 (2023)

20. Paiva, D.V.M., Silva, M.A.S., Sombra, A.S.B., Fechine, P.B.A.: Properties of the Sr_3MoO_6 electroceramic for RF/microwave devices. J. Alloy. Compd. **748**, 766–773 (2018)
21. Narang, S.B., Bahel, S.: Low loss dielectric ceramics for microwave applications: a review. J. Ceram. Process. Res. **11**(3), 316–321 (2010)
22. Chou, X., Wang, J., Zhao, Z., Geng, W., Zhang, W., Zhai, J.: Dielectric tunability and microwave properties of $Ba_{0.5}Sr_{0.5}TiO_3$-$BaMg_6Ti_6O_{19}$ composite ceramics for tunable microwave device applications. J. Electroceramics **26**, 185–190 (2011)
23. Nenasheva, E.A., Kanareykin, A.D., Kartenko, N.F., Dedyk, A.I., Karmanenko, S.F.: Ceramics materials based on (Ba, Sr)TiO_3 solid solutions for tunable microwave devices. J. Electroceramics **13**, 235–238 (2004)
24. Xue, S., Zhai, J., Xiao, S., Xiu, S., Shen, B.: Improved dielectric breakdown strength and suppressed energy loss in niobate glass-ceramics crystallized by microwave process. Mater. Lett. **190**, 154–156 (2017)
25. Parhi, C.C., Thirumalasetty, A.B., James, A.R., Wuppulluri, M.: Relative investigation on microwave-assisted Zr-modified $PbTiO_3$ and $BaTiO_3$ ferroelectric ceramics for energy storage application. ACS Omega **8**(41), 37752–37768 (2023)
26. Egorov, S.V., Eremeev, A.G., Kholoptsev, V.V., Plotnikov, I.V., Rybakov, K.I., Sorokin, A.A., Balabanov, S.S., Rostokina, E.Y.: Rapid microwave sintering of functional electroceramic materials. Ceram. Int. **49**(14), 24222–24228 (2023)
27. Komarneni, S., Li, Q., Stefansson, K.M., Roy, R.: Microwave-hydrothermal processing for synthesis of electroceramic powders. J. Mater. Res. **8**(12), 3176–3183 (1993)
28. Zaman, A., Uddin, S., Mehboob, N., Tirth, V., Algahtani, A., Abbas, M., Mushtaq, M., Ali, A., Sultana, F., Althubeiti, K., Ullah, K.: Structural elucidation, electronic and microwave dielectric properties of Ca (Sn_xTi_{1-x}) O_3,$(0\leq x\leq 0.8)$ lead-free ceramics. ACS Omega **7**(5), 4667–4676 (2022)
29. Tumarkin, A.V., Gaidukov, M.M., Gagarin, A.G., Samoilova, T.B., Kozyrev, A.B.: Ferroelectric strontium titanate thin films for microwave applications. Ferroelectrics **439**(1), 49–55 (2012)
30. Acikel, B., Hansen, P.J., Taylor, T.R., Nagra, A.S., Speck, J.S., York, R.A.: Tunable strontium titanate thin films for microwave devices. Integr. Ferroelectr.. Ferroelectr. **39**(1–4), 291–298 (2001)
31. Sheen, J., Wang, Y.L.: Microwave dielectric constant measurements of strontium titanate using five mixture equations. In: Joint IEEE International Symposium on Applications of Ferroelectric and Workshop on Piezoresponse Force Microscopy (ISAF/PFM), pp. 268–271. IEEE (2013)
32. Chaves, A.V., da Silva, M.A.S., Paiva, D.V.M., Mazzetto, S.E., Sombra, A.S.B., de Menezes, A.S., Fechine, P.B.A.: Electroceramic based on $Ca_{3-x}Ba_xZrSi_2O_9$ solid solutions for microwave and radiofrequency application. Ceram. Int. **49**(18), 30827–30836 (2023)
33. Li, L., Zhang, W., Chen, X.M., Zhu, H.Y.: Dielectric properties of $CaCu_3Ti_4O_{12}$, Ba $(Fe_{1/2}Nb_{1/2})$ O_3, and Sr$(Fe_{1/2}Nb_{1/2})$ O_3 giant permittivity ceramics at microwave frequencies. J. Appl. Phys. **111**(6), (2012)

Microwave Dielectric Resonator Antenna Using Electroceramics- A Perspective

Pankaj Tripathi and Preeti Kumari

Abstract The present chapter on dielectric resonator antenna (DRAs) using dielectric ceramic will draw the great attention of researcher working in the area of DRAs and ceramic dielectric used in DRAs. The chapter delivered comprehensive note at one place. Present chapter is broadly divided into two sections (i.e., DRA and DRA material). One of the section contains compact description about types of dielectric resonator antennas (DRAs), excitation techniques, modes of excitation, radiation mechanism, bandwidth and gain enhancement techniques as well as new dimension of research i.e., reconfigurable DRAs. The chapter also gives the brief idea about the parameters of ceramic dielectric materials that affecting the performance of the DRAs as well as recent development in DRAs. The other section deals with the discussion about roll of dielectric materials used for DRAs, mechanism of interaction of microwave with ceramic dielectric, parameters that affected the dielectric constant of ceramic material used for DRAs. The section also consists of exhaustive list of commercially available and laboratory synthesized dielectric ceramic as well as recently designed DRAs with commercially available and synthesized ceramic materials. For future investigations, such studies will be useful for understanding the effect on the microwave device performance on the end user side. Also, emphasis should be placed on the low cost production and eco-friendly materials, which can be recycled further, keeping in view the environmental protection aspect. This review might provide a systematic guideline to researchers/readers for design and development of novel DRAs.

Keywords Dielectric · Wideband · Antenna feed · Reconfigurable · Ceramics · Dielectric materials · Polarization · Dielectric resonator antenna · Microstructure

---

P. Tripathi (✉)
Electronics Engineering Department, Government Polytechnic Aurai, Technical Education Department, Bhadohi, Uttar Pradesh 221301, India
e-mail: pankaj.tripathi.juit@gmail.com

P. Kumari
Department of Electronics and Communication, University of Allahabad, Prayagraj 211002, India

U. Kumar (ed.), *Defects Engineering in Electroceramics for Energy Applications*, Engineering Materials, https://doi.org/10.1007/978-981-97-9018-0_15

1 Introduction

In the wireless communication and radar systems, antenna is the only interfacing device attached to both the transmitter as well as the receiver. Wideband communication systems can provide high data rates, voice biometrics, multi-lingual communication, multiple communication channels, and controlled data traffic due to availability of wide frequency spectrum. Thus, large impedance bandwidth is required for efficient wireless communication and pulse radar systems. Dielectric resonator antenna (DRA) is one of the best options to achieve wide impedance bandwidth [1]. An American physicist, R.D. Richtmyer theoretically predicted that a dielectric piece of suitable shape can function as a microwave resonator and coined the term "dielectric resonator (DR)" in 1939 [2]. Later in 1962, Okaya and Barash have worked on single crystal rutile which possesses high dielectric constant [3, 4]. They first provided the analysis of modes inside the resonator and measured its dielectric constant in the microwave frequency range. In the early 1980s, Stuart. A. Long performed an introductory experimental investigation on the resonant cylindrical dielectric cavity antenna [5].

Now-a-days, DRAs are drawing great attention for the radar and wireless communication due to their attractive advantages, such as large bandwidth, ease of excitation, high radiation efficiency, low phase noise, flexible feed arrangement, ease of integration with other active or passive microwave integrated circuit components, easily controlled characteristics and wide range of material dielectric constants. DRAs can be excited by different feeding techniques: coaxial or probe, microstrip line, slot coupling technique, coplanar waveguide and dielectric image guide. The conventional shapes of DRA: rectangular, cylindrical, hemispherical, and ring provide a typical bandwidth of around 10%, which may not be adequate for wideband applications. Bandwidth enhancement of a DRA, especially with a low profile design is a demanding task for the researchers. Different approaches are used to enhance the bandwidth of DRAs [6, 7]. Multi-element stacking is one of the approaches to enhance the bandwidth but overall antenna volume is increased in this approach. Another approach includes the merging of a hybrid radiating mode with the DRA mode. Such merger of modes occurs in dielectric resonator on patch configuration [8] and coplanar waveguide inductive slot embedded as a hybrid radiator along with the DRA [9]. By merging resonances of the RDRA and probe feed, wideband hybrid design can be obtained [10]. Although to achieve wideband impedance match, some special feeding mechanisms viz. L-shaped probe feed in cylindrical DRA [11] and tall microstrip transmission line (TML) [12] are used. The TML-DRA configuration can offer bandwidth up to a maximum of 25% with more complex construction. Several DRA geometries, such as H-shaped [13], T-shaped [14], dumb-bell shaped DRAs [15] were analyzed for wideband applications. Material selection for the DRA also plays an important role in deciding the final antenna performance. For the DRA performance to be effective, both the aspects i.e., DRA design and the DRA material selection need to be essentially considered. It is possible to get good DRA performance using proper material with simple design instead of going for complex design with improper material selection. Thus, a state-of-the art of various types of dielectric resonator materials are also discussed in the present chapter.

2 Excitation Techniques

The selection of types of excitation and its location, both play significant impact on resonant frequency, input impedance bandwidth and radiation characteristics of the DRA. General feeding methods for excitation of DRA such as co-axial probe feed, microstrip line feed, aperture coupled feed, co-planar waveguide feed and dielectric image guide feed are described below.

A. **Coaxial probe feed**: In probe excitation technique, strong coupling can be achieved by locating the probe in high E-field region of DRA. Therefore, the probe is either inserted into the dielectric resonator so that its outer surface touches the resonator everywhere or kept touching the outer surface of dielectric resonator or it is located adjacent to resonator. The height, depth of penetration and position of the probe are optimized to achieve maximum impedance bandwidth. Different types of modes can be generated by changing the location of the probe. The drilling of hole in the substrate and/or dielectric resonator is required for insertion of the probe. The advantage of probe feed is that, it can be coupled directly with 50 Ω system without using matching network. This type of feed is generally used for relatively lower frequency applications.
B. **Microstrip line feed**: In this coupling mechanism, the dielectric resonator is placed just above or proximity with the open circuited end of the microstrip line. The amount of impedance matching depends on the lateral position of dielectric resonator with respect to microstrip feed line. Dielectric constant of DRA affects significantly the degree of coupling. Strong coupling can be achieved for dielectric resonator having dielectric constant value greater than 20. This method shares the merits of using a coaxial probe, but avoids the need for drilling a hole into the dielectric. This type of feed is mainly used in the DRA integrated with MMICs without the need for a coupling slot [16].
C. **Aperture coupled feed**: In the aperture coupled feed, the feedline structure is no longer in contact with the dielectric resonator. The resonator is kept just above the slot provided in the ground plane. The slot can be fed by a microstrip coating or a waveguide. There is flexibility in the design of the feedline and the DRA in aperture couple feed technique of excitation. The aperture behaves as a magnetic current parallel to the length of the slot and excites the magnetic field in the DRA. The radiation from the microstrip feed significantly decreases and this improves the quality of polarizationas well. This coupling mechanism has the following advantages: (i) DRA can be easily integrated with the feed structure, (ii) Drilling of hole is also not required as in case of coaxial probe feed [17] and (iii) Radiating aperture is isolated from the spurious radiation from the feed. A

major drawback of the technique is that, it cannot be used for lower frequency (at and below L-band) applications because of large size of slot required.

D. **Coplanar waveguide feed**: This feed is uni-planar and thus provides good integration of the DRA with other active circuit components on the same side of substrate without the need of vias. This feeding technique is less dispersive in nature and provides low conductor losses and low radiation leakage. It is suited for millimeter wave applications because of reduced surface wave excitation in electrically thick substrates as compared to microstrip feed [17].
E. **Dielectric Image guide coupling feed**: Unlike microstrip line, the dielectric waveguide does not suffer from conductor and surface wave losses at millimeter-wave frequencies. In the DRA, particularly with low dielectric constant, the amount of coupling microstrip line is quite low. To increase the coupling, a low loss dielectric image guide is used as feed for linear DRA array [18].

3 Modes of Excitation

Simple shaped DRAs, rectangular, cylindrical and hemispherical DRAs with and without modification are commonly used for signals radiated at gigahertz frequencies. The desirable features of DRAs include high gain, wide impedance bandwidth, enhanced efficiency and application based radiation patterns. Excitation of particular mode/modes is required for the desired performance of DRAs. This is achieved through proper selection of feed type (microstrip line feed, probe feed, aperture coupling etc.) and its location. To excite particular mode/modes, knowledge of corresponding electric and magnetic field configurations is essential. The modes generally excited within different shapes of DRAs which give broadside radiation patterns over wide frequency range similar to an electric or magnetic dipole are listed in Table 1.

4 Radiation Mechanism

Radiation mechanism of antenna can be explained by symmetry breaking model (SBM) [24]. SBM states that all symmetrically connected resonators act like a filter. Filter is a two port device, which transfers energy from input port to output port. Symmetry breaks when output port is deliberately closed and energy is fed continuously through the input port to the device. The device first stores the energy upto maximum, then leaks out or radiates and therefore, the device behaves like an antenna. Thus, an antenna can be defined as a symmetric resonator, whose symmetry is broken explicitly. Radiation mechanism of DRAs can also be explained by symmetry breaking model (SBM). SBM proposes LC circuit theory for the explanation of radiation through DRA. In this model, a dielectric material mounted on ground plane act as a capacitor with finite value of capacitance and the ground plane behaves as an inductor. Symmetry breaks, when one end of dielectric material mounted on ground

Table 1 Modes of excitation in different shapes of DRA

S. No.	DRA shape	Excited mode
1	Hemispherical/conical shaped dielectric ring resonator [19]	• TE_{111} mode (Radiates like short horizontal magnetic dipole) • TM_{101} (Radiates like short electric monopole)—generally probe feed is used for excitation
2	Half- Hemispherical DRA [20]	• $HEM_{11\delta}$ like mode • TM_{101} like mode
3	Pawn-shaped HDRA [21]	• TM_{01} like mode
4	Cylindrical DRA [18]	• $TM_{01\delta}$ (Radiates like short electric monopole) • $TE_{01\delta}$ (Radiates like short magnetic monopole) • $HE_{11\delta}$ (Radiates like short horizontal magnetic dipole) • $HEM_{11\delta}$ (Generally excited using probe feed) • $HEM_{12\delta}$ (Higher order mode generally excited using aperture coupling) • $HEM_{21\delta}$ (Higher order mode) • HEM_{111} (Higher order mode generally excited using aperture coupling) • HEM_{113} (Higher order mode generally excited using aperture coupling)
5	Dielectric ring resonator	• TM_{01} (Monopole like radiation)
6	Array cylindrical antenna [22]	• $HEM_{11\delta}$ like mode
7	Rectangular DRA [23]	• $TE^x/TE^y/TE^z$ (Radiation like short magnetic dipole in x, y and z direction respectively) $TE^x_{\delta 11}$• $/TE^y_{1\delta 1}/TE^z_{11\delta}$ (Lower order mode) $TE^x_{\delta 31}$• $/TE^y_{\delta 13}/TE^z_{\delta 21}$ (Higher order mode)

plane is excited by a time varying signal through ground plane and the other end is left floating in free space, resulting in a back and forth swing of charges from capacitive to inductive elements with resonating frequency

$$f_c = 1/2\pi\sqrt{LC} \tag{1}$$

where f_c is resonating frequency, C is the capacitance of dielectric material and L is the inductance of feeding wire and ground plane. This back and forth swing of charges produces electromagnetic wave. The generated electromagnetic wave leaks out or radiates from the floating end of dielectric material and the system behaves like an antenna. It is termed as dielectric resonator antenna (DRA).

5 Bandwith Enhancement Techniques

Many radar and emerging wireless systems operate over wide frequency range. Therefore, they require broadband antennas for communication. Several methods are used for enhancement of DRA bandwidth. The bandwidth enhancement methods are broadly classified as: (i) using a combination of multi permittivity DRAs, (ii) using modified feed configuration, (iii) using modified DRA shapes and (iv) using hybrid antennas.

A. **Combination of multi permittivity DRAs**

Two or more dielectric resonators (DRs) combined together enhances the bandwidth, either because of reduction in effective dielectric constant or merging of neighboring mode excited in neighboring DRs. The foregoing concept is utilized in DRA design by providing air gap between the substrate and DR or between two DRs [25, 26], using annular DRs [23], providing notch in rectangular DRAs (RDRAs) [27], using multi-segment DRAs [28–30], co-planar parasitic DRAs [31, 32], embedded DRAs [33], staircase-shaped DRA [34, 35], perforated DRAs [36], and stepped pyramid DRAs [37] to enhance the impedance bandwidth of DRAs.

B. **Modified feed configuration**

The impedance bandwidth can also be increased without significantly affecting other properties of DRAs through modification in conventional feeding techniques or new design of feed i.e., using stub [23], probe feed with air gap [23], flat matching strips [38, 39], ring aperture feed [40–42], U-shaped aperture feed [43] or multi-mode generation through composite excitation [18].

C. **Modified DRA shapes**

Impedance bandwidth of conventional DRAs can also be improved by modifying the shape of DRs. Modified shape either generates different mode/combination of neighboring modes or reduces diffraction from the edges which results in the increase of DRA bandwidth. Some of the modified shapes are conical and inverted split conical DRAs [44, 45], tetrahedral and inverted tetrahedral DRAs [46], inverted stepped pyramidal DRAs [18], U-shaped DRA excited by elliptical conformal patch [47], fractal and CPW fed fractal DRA [48], cup-shaped inverted hemispherical DRA [49], filleted DRA [50] and uniaxial anisotropic rectangular dielectric resonator antennas [51].

D. **Hybrid Antennas**

The combination of DRs and radiating patch/slot with closer resonant frequencies provides the hybrid approach to enhance the bandwidth of DRAs. Some of the hybrid systems include: combination of circular microstrip patch and ring shaped DR fed by a slot [52], monopole antenna with ring dielectric resonator [53], combination of CPW inductive slot and rectangular DR [54], and cylindrical DR with inductive slot [55].

6 Gain Enhancement Techniques for DRAs

Gain is one of the most important parameters of Antenna. Various methods are available to enhance the gain of DRAs. Arraying of DRAs are one of the basic approaches to increase the gain of antenna, in which, gain is increased by increasing the number of DR elements. The disadvantage of this approach is that it requires significant amount of DR material. Engraving notches on two opposite side walls of DR increases the radiation from side walls as compared to its top wall for its fundamental mode resulting in enhanced gain [56]. Placement of different designs of substrates (mushroom like, strip corrugated, coaxially corrugated EBG substrate [57] and circularly polarized EBG structure [58, 59]) suppress the surface wave propagation in all radial direction resulting in the increase of antenna gain. Other efficient approaches to enhance the gain of DRAs include generation of higher order mode (by single or composite excitation), stacking of parasitic DRA on the top of driven DRA with a space [60], placement of pyramidal horn surrounding the DRA [61], using uniaxial anistropic (single crystal) dielectric materials in fundamental mode [62], optimization of place of excitation and layer thickness [63], placement of superstrate or metamaterial at particular height above the DRA [64], positioning circularly polarized DRA in circular cavity [65, 66], and using composite dielectric substrate [67].

7 Evolution of Reconfigurable DRAs

With the evolution of DRA in 1983 [68] and thereafter detailed investigation on DRAs [18, 23] have attracted antenna engineers to work for the development of these antennas. In the initial development of DRAs, different DRA structures with different types of feeding mechanisms were studied. After that, wide band antenna was developed with the excitation of multiple modes at different frequencies within the same DRA element. In some studies, the concept of conformal feed/strip is used to enhance the bandwidth further. Guha et al. (another study) compared the performance of microstrip patch antenna and DRA by keeping the feed and other structural parameters same and shows that DRAs are well ahead of microstrip patch antennas [69]. Most of the initial development of DRAs was carried out using linearly polarised antennas, but, to enhance the antenna performance such as prevailing over the effect of propagation irregularities, spin effect on satellites and ground reflection in satellite communication, the need of using circular polarisation was felt by antenna engineers or researchers.

With the advent of circular polarisation (CP) in the last decade, several single- or multi point fed CP DRAs have been investigated which can find potential application in satellite communication, GPS, Mobile communication, WLAN, WiMAX and Radar systems. A review on circularly polarized antennas with different circular polarization generation techniques are described in detail by Reena et al. [70]. DRA arrays are the most suitable configuration for the design of high gain CP-DRA. Further, rectangular DRA provides maximum radiation efficiency and therefore, the rectangular shape of DRA is the best shape for generation of circular polarization [70].

Due to the requirement of high performance, large capacity and miniaturized antennas for the development of modern wireless communication, the concept of reconfigurable antennas came into picture. Large number of reconfigurable microstrip patch antennas have been designed in the last three decades. In recent years, investigations on reconfigurable dielectric resonator antennas have also been started.

Reconfigurable antennas refer to antennas having the ability to alter one of the fundamental characteristics without disturbing/manipulating/changing other characteristics of individual radiating elements. On the basis of the type of antenna characteristics, the reconfigurable antennas are broadly categorizes as: (1) Pattern reconfigurable antennas (2) Polarization reconfigurable antennas (3) Frequency agile/ reconfigurable antennas.

A. **Pattern reconfigurable antennas**

Few investigations on pattern reconfigurable DRAs are reported in recent decades, in which radiation pattern has been altered either by changing the structure [71] and material [72] or by using phase shifters and switches [73, 74]. These antennas have the ability to reshape the main beam to provide high radiation efficiency. The DRA with dual polarization combined with pattern-reconfigurability has also been reported for multi-user multiple-input-multiple-output (MU-MIMO) applications [74]. Another study shows that changing the modes of a liquid-based DRA can provide radiation pattern reconfigurability to the antenna [75]. Reconfiguration of radiation pattern in a diversity antenna is obtained by using two DRAs. The pattern diversity of the proposed design enhances the quality and reliability of the wireless link for Digital Video Broadcasting- Handheld (DVB-H) (790–862 MHz), Wi-Fi (2.4 GHz), and WiMAX (3.5 GHz) applications [76]. The performance comparison of some of the pattern reconfigurable DRAs is given in Table 2.

B. **Polarization reconfigurable antenna**

Polarization reconfigurable DRAs provide polarization diversity in the DRAs. By generating different polarizations (linear or right handed/left handed circular) in the same frequency band, reduction in multipath fading occurs, which would give rise to the frequency reuse, thereby enhancing the system capacity and reducing the system size.

Many investigations on the microstrip reconfigurable antennas have been reported [80–96]. Three conventional methods are available to generate polarization in microstrip antennas i.e., loading of RF switches on the radiation element [80–83], introducing a reconfigurable feeding network [84–92], and using polarizers to transform the polarization type [93–95]. Due to the drawbacks of microstrip antennas, such as narrow bandwidth and low radiation efficiency, the DRAs have attracted much

Table 2 Performance comparison of pattern reconfigurable DRAs

Beam no.	−10 dB BW (%)	Reconfiguration mechanism	Switch no.	Gain (dBi)	Dielectric material	References
2	5–6 GHz (18.18%)	GaAs SPDT	2	7.21	Rogers TMM 10i $\varepsilon_r = 9.8$	[74]
2	3.75–5.37 GHz (35.5%)	Liquid-based	–	6.15	K9 glass ($\varepsilon_r = 6.85$) Liquid ethyl acetate ($\varepsilon_r = 7.1$)	[75]
4	5.14–6.02 GHz (15.77%)	PIN	4	9.74	$\varepsilon_r = 9.8$	[77]
8	5.67–5.88 (3.64%)	PIN	8	7.27	$\varepsilon_r = 10$	[78]
2	(1.25–1.9 GHz) (41.27%)	PIN	4	4.0	$\varepsilon_r = 10$	[79]

attention for the design of reconfigurable antennas. The DRAs can also provide polarization diversity. But, unfortunately, the aforementioned conventional methods of polarization reconfigurability for microstrip antennas were not found suitable for the DRAs. Thus, a new approach is required for the development of polarization reconfigurable DRAs. Recently, a liquid based polarisation reconfigurable DRA [97] and a low profile broadband polarisation reconfigurable DRA [98] have been reported. Table 3 shows the comparison of performance of polarisation reconfigurable DRAs recently reported in literature.

C. **Frequency Reconfigurable antennas**

The most exploited reconfiguration technique used in DRAs is the frequency reconfigurable technique. Different techniques have been presented in order to achieve frequency reconfigurability of DRAs, such as electronic switching [99–103], structural and mechanical changes [104], material changes [105] and variable reactive loading [101, 106]. One study for frequency agility in DRA incorporated air-filled micro-fluidic channels and used air pressure to position the slugs away and under the DRA micro-pneumatically. This provides compactness, ease of fabrication, no requirement of apparatus for external biasing field, but it is unsuitable for fast switching applications due to pneumatic control [99]. In a recent study, the frequency agility is achieved on the two port MIMO DRA with the introduction of reconfigurable parasitic slot loading in the proximity of the DRA segment/block. Both the radiating ports are orthogonal and thus, the frequency of each port can be tuned individually without affecting the frequency response of the other port. This antenna can operate in both multi-frequency mode, providing the antenna ports to independently sweep through the frequency band of interest and search the users, as well as MIMO mode to enhance the capacity whenever heavy data traffic is needed [101]. Some frequency reconfigurable DRAs are listed in Table 4.

Table 3 Performance comparison of polarisation reconfigurable DRAs

Polarisation states	Reconfiguration mechanisms	Overlapped bandwidth %		Peak gain (dBi/dBic)		Dielectric material	References
		CP	LP	CP	LP		
LHCP ↔ RHCP	Pump-switching	16.3	–	5.5	–	Ethyl acetate ($\varepsilon_r = 6.6$)	[94]
LP ↔ LHCP ↔ RHCP	p-i-n diodes	20.7	29.6	7.35	7.44	Ceramic ($\varepsilon_r = 36$)	[95]
LHCP ↔ RHCP ↔ LP($\varphi = 45^\circ$) ↔ LP($\varphi = 135^\circ$)	p-i-n diodes	20%	20%	4.0	5.0	ECCOSTOCK HIK 500F($\varepsilon_r = 20$)	[97]

Table 4 Performance of frequency reconfigurable antennas

Frequency agility technique	Switching range	Dielectric material	Applications	References
Air gap effect (Pneumatic control)	24.7% (operating frequencies: 21.4 and 16.7 GHz)	Ceramic ($\varepsilon_r = 20$)	–	[99]
Electronic switching (p-i-n diodes)	Three different frequency bands over the frequency range 3.45–6.77 GHz: 8%, 16% and 6.5%	$\varepsilon_r = 15$	IMT systems WLAN 802.11a/h/j/n	[100]
Varactor loading	Operating frequencies: 2.87, 2.57 and 2.12 GHz	$\varepsilon_r = 9.4$	Modern reconfigurable wireless systems	[106]
Parasitic slot loading (switches and varactors)	For HE mode: 630–784 MHz For TE mode: 609–818 MHz	$\varepsilon_r = 18$ $\tan\delta = 0.00015$	Dynamic spectrum access applications, MIMO applications	[101]
Electronic switching (p-i-n diodes)	1.60–2.71 GHz (Four single band modes with impedance bandwidths of 17, 11, 14 and 6%)	$\varepsilon_r = 10$	LTE/WWAN WLAN	[102]
Structural/mechanical changes (colloidal dispersions/vary columnar height)	2.5–4.5 GHz (Operating frequencies at 4.02,3.52, 3.26, 2.99, 2.87, 2.78 GHz at heights of 8, 10, 12, 16, 20 and 24 mm, respectively)	$Ba_{0.6}Sr_{0.4}TiO_3$ ($\varepsilon_r = 500$, $\tan\delta = 0.01$) Naphthenic oil ($\varepsilon_r = 2.1$, $\tan\delta = 0.001$) Effective dispersion properties ($\varepsilon_r = 8.24$, $\tan\delta = 0.00023$)	– Demonstrated—a colloidal dispersion can be utilized as a fluidic DRA	[104]
Electronic Switching (p-i-n or varactor diodes)	3–8 GHz Tuning range for p-i-n- loaded DRAs = 91%; and Varactor loaded DRAs = 55%	Eccostock HiK $\varepsilon_r = 10$, $\tan\delta = 0.001$ (Emersion and Cuming)	Top performing frequency agile antennas	[103]
Material changes (Ferrite material)	(9.4–11.4 GHz) 19%	Calcium Vanadium TTVG-1850 from Trans-Tech $\varepsilon_r = 14.8$	–	[105]

8 Parameters of Ceramic Dielectric Materials Affecting the Performance of DRAs

Some important material constraints of dielectric resonator are described here which affects the performance of DRA.

A. **Dielectric Constant/Relative Permittivity**

The dielectric constant (ε_r) values should be in the range from 5 to 30 for efficient antenna performance. In general, it should be noted that decrease in permittivity of the resonating material of the DRA can improve the radiation properties over a wider operating frequency range. Hence, low dielectric constant and low loss materials are preferred for wideband applications. The wavelength of the electromagnetic wave traversing through the dielectric material is inversely proportional to the square root of the material dielectric constant or relative permittivity. So, the high dielectric constant resonator can effectively reduce the size of the DRA for microwave frequency applications. The wavelength in the dielectric material (λ_d) can be calculated as Sebastian et al. [17]:

$$\lambda_d = \frac{\lambda_o}{\sqrt{\varepsilon_r}} \tag{2}$$

where λ_o is the wavelength in air or vacuum and ε_r is the dielectric constant of the material.

High dielectric constant materials (80–100) can be used for low profile applications when bandwidth is not a main concern. For a cylindrical and rectangular DRA, the bandwidth is about to 35–40%, and 20–40% respectively. For typical antenna application, it is recommended to use a low loss material with dielectric constant in the range 8–20 [17].

B. **Effective Dielectric Constant/Relative Permittivity**

$$\varepsilon_{eff} = \frac{H_{eff}}{\frac{h}{\varepsilon_r} + \frac{t}{\varepsilon_i} + \frac{s}{\varepsilon_s}} \tag{3}$$

where $H_{eff} = h + t + s$, h is DRA height, 't' is insert thickness, 's' is substrate thickness 'ε_r' is dielectric constant of DRA, 'ε_i' is dielectric constant of insert, and 'ε_s' is dielectric constant of substrate. Slots in DRA reduce the effective permittivity of DRA which provides wider bandwidth [18].

C. **Dissipation factor (Loss tangent)/Dielectric loss**

Dissipation factor = ε''/ε', where ε' is the real part of permittivity and ε'' is the imaginary part of permittivity. Low dielectric loss (<0.002) is generally preferred for the DRA.

D. **Temperature coefficient of resonant frequency (τ_f)**

Near zero temperature coefficient of resonant frequency. The temperature coefficient of resonant frequency decides the thermal stability of the dielectric resonator.

E. **Quality Factor**

High quality factor materials are generally preferred for the design of DRA.

F. **Sample size/Thickness**

Size of the DRA depends upon the operating frequency range and dielectric constant of the material. It is inversely proportional to ε_r by the relation $\lambda_o/\sqrt{\varepsilon_r}$, where λ_o is free space wavelength at the resonant frequency and ε_r is the dielectric constant of material.

9 Effect of Interaction of Radio-Frequency Field with Dielectric Materials

In dielectric materials, all charges are attached to specific atoms/molecules. These charges are called bound charges. By applying external electric field, positive and negative charges of neutral body displace slightly from their respective positions and form induced dipole moments whose directions align in the direction of applied external electric field. These dipoles produce their own electric field. The resultant electric field i.e., the combination of applied field and the dipole field is called the internal field. Particles having permanent dipole moment experience electrical torque in the presence of external applied field. Interaction of EM wave (composed of photons) with materials can be explained by macroscopic Maxwell's equations together with constitutive relations and boundary conditions. EM wave consists of packets of photons. When EM wave with appropriate frequency interacts with atoms of material, photonic energy is either absorbed or emitted by the atoms. When an atom emits photons, energy of the atom is decreased and energy of EM wave is increased. When the high energy photons of EM wave collide with atoms/molecules, these photons give their energy to them due to which the atoms/molecules move through the material and deposit their kinetic energy into surrounding matter.

Dielectric and magnetic polarizations occur due to permanent or induced dipole moments caused by applied field and spin moments. When applied electric field interacts with dipole moment of molecules or atoms, potential energy of molecules or atoms is converted into kinetic energy due to promotion of rotational and atomic transitions. This effect is known as Stark Effect. Rotational and atomic transitions depend upon frequency and strength of applied field. The effect which describes the interaction of magnetic field with dipole moment of molecules and atoms is known as Zeeman Effect.

A. **Material response to applied EM-field**

When EM-field is suddenly applied to a dielectric material, following fields appear that cause system to acquire non-equilibrium state for certain period of time.

1. Macroscopic field (E): (E_a–E_{dp}) or (E_a–p/3ε) i.e., average quantity in the medium and $p = \alpha E_l$, where P is induced dipole moment, α is polarizability and E_l is local field, Edp is the depolarization field, Ea is the applied electric field and ε is the absolute permittivity.
2. Depolarization field (E_{dp}): This field is generated due to particle back reaction field from charge, spin and current rearrangements. This field opposes the applied electric field.
3. Local field (E_l): This field is generated due to average EM field at a particle site because of both the applied field and field created by other sources, such as dipoles, currents, charges and spin.
4. Microscopic field (E_{micro}): It is generated due to particle interacting with the field created by discrete charges.
5. Applied electric field (E_a): It is generated due to external charges.

When EM-field (varying with time) is applied to the material, atoms, molecules, free charges, and induced/permanent dipoles and defects change their positions/ orientations according to varying EM-field but with some time lag known as relaxation time. The dynamic readjustment of atoms and molecules after interaction with EM-field is called relaxation. The time lag may be due to screening, friction, inertia and coupling. Various phenomena occur during the relaxation: i.e.,—heat conversion, photon-phonon and lattice-phonon interactions, defect diffusion and/ or higher multipole interactions. Response of material at the time of interaction with EM-field depends strongly on the material composition and lattice structure.

The crystal structure, permanent dipoles, mobility of free charges and defects contribute to the dielectric properties of material. The solids, which do not appreciably rotate or polarize due to applied EM-field have low dielectric constant values, whereas the solids which rotate or polarize appreciably according to applied EM-field have high dielectric constant values. At RF frequencies, high and low loss materials respond differently. At RF-frequencies, electronic, ionic and permanent dipole polarizations occur, so that vector dipole moment $\mu_d = (\alpha_{el} + \alpha_{ion} + \alpha_{perm})E_l$, where $\alpha_{el} = 4\pi\varepsilon R^3/3$, $\alpha_{ion} = e^2/Yd_o$, $\alpha_{perm} = |\mu_e|^2/3K_BT$. Here R is radius of ions, d_o is equilibrium separation of ions, Y is young's modulus, K_B is Boltzman constant, T is absolute temperature and μ_e is permanent dipole moment [107].

B. **Modified Clausius- Mossotti relation**

The equation $(\varepsilon_r-1)/(\varepsilon_r+2) = (N/3\varepsilon)[\alpha+|\mu_e|^2/3K_BT]$, where N is density of dipoles and ε is absolute permittivity, provides the relation between permittivity and polarizability for all types of materials (Solids, liquids and gases). The equation reveals the information that when EM-field interacts with the material, various phenomena occur at macroscopic and microscopic levels within the material causing different types of polarizations which affect the permittivity of the material.

C. **Losses during interaction of RF-field with material**

Dielectric losses occur within the polar materials because of friction caused by rotation, free charge movement and out of phase dipole coupling. Losses occur in nonpolar materials due to interaction with neighboring permanent and induced dipoles, intrinsic phonon- photon interaction in presence of EM-field and losses caused by grain morphology, defects and dislocations. In high purity crystals, enharmonic coupling with electric field and presence of defects allow photon- phonon interaction that introduces losses. Centro-symmetric dielectric crystals experience lower losses than non-centro symmetry crystals.

10 Parameters Affecting the Dielectric Constant of the Resonator

Some important material parameters which affects the dielectric constant of the resonator are described here.

A. **Refractive Index (η)**

The dielectric constant of dielectric materials measured at microwave frequencies is directly proportional to the square of the refractive index of these materials measured at optical frequencies. The relation, $\varepsilon_r = \eta^2$ is accurate only when ε_r and η are measured under same conditions. This has been found true in case of diamond ($\varepsilon_r = 5.68$, $\eta^2 = 5.85$) and germanium ($\varepsilon_r = 16$, $\eta^2 = 16.73$) [108]. But the phenomenon is not always true for all types of dielectric materials. Since the dielectric constant is frequency dependent and the possibility of getting same dielectric constant at microwave frequencies as per the aforesaid relation with refractive index is very low. In case of diamond and germanium, the same polarisation processes are getting generated at the optical and microwave frequencies. On the other hand, for other materials, the dipolar polarisation processes excited at lower frequencies do not generally happen at higher optical frequencies. The ionic and electronic polarisations mainly contribute to the permittivity at microwave frequencies.

B. **Internal Structure of Materials**

Dielectric constant (ε_r) is directly proportional to polarization (creation of resultant positive and negative charges) whose strength is only decided by crystal structure of materials. Homogeneous materials give good and uniform permittivity values throughout the volume.

C. **Temperature**

Increasing the temperature results in decrease in value of dielectric constant for high permittivity materials because orientational polarization decreases, and increase in value of dielectric constant of low permittivity materials due to generation of more charges and ions which results increase of orientational polarization.

D. **Hysteresis**

$$D = \varepsilon_0(1 + \chi_e)E = \varepsilon_0\varepsilon_r E,$$

where D is dielectric displacement, ε_o is dielectric constant of vacuum, ε_r is dielectric constant of material, χ_e is dielectric supceptibility and E is electric field. Dielectric constant (ε_r) is affected by the electric field.

E. **Density**

Density is directly proportional to dielectric constant (ε_r) of the material.

F. **Frequency range of operation (Bandwidth)**

Frequency range of operation depends upon the dielectric constant (ε_r) and dimensions of the material.

11 Dielectric Materials

Electrical insulator materials can be termed as dielectric materials. They can be polarized by an applied electric field. When a dielectric material is placed in an electric field, the positive and negative electric charges slightly shift from their average equilibrium positions causing polarization. This creates an internal electric field which reduces the overall field within the dielectric and electrical energy stored within the material reduces up to maximum, and radiation can occur from the floating end where such a material is placed on the ground plane. Dielectric ceramic ($\varepsilon_r > 0$ & $\mu_r = 1$) and ferroelectric ceramic ($\varepsilon_r > 0$ & $\mu_r > 1$) are potential ceramic materials which are used in DRAs as resonators and radiators.

Ceramic dielectrics are extensively used in the wireless communication systems and electronic devices as well. For practical applications, these materials are usually prepared by conventional ceramic fabrication routes such as, solid state route, liquid phase sintering or chemical processing. The dielectric material properties depend considerably on the chemical composition, porosity, crystal structure and imperfections in the crystal lattice. The fabrication condition also influences the material properties, e.g., calcination conditions influence the structure, secondary phase formation and dielectric characteristics. The secondary phase plays an important role in controlling the density, microstructure, grain growth and microwave dielectric properties of the ceramic resonator materials. The purity of starting material plays an important role on the sintering conditions and dielectric properties. Some common commercially available and laboratory synthesized dielectric materials are listed in Tables 5 and 6, respectively. The performance of recently designed DRAs with commercially available and synthesized ceramic materials is described in Table 7. The current state of art for the ceramic dielectric materials and the dielectric resonator antenna parameters is given in Table 8.

Table 5 Commercially available dielectric materials for DRA applications

S. No.	Dielectric material	Dielectric constant (ε_r)	Tangent loss (tanδ)
1	PTFE (Polytetrafluoroethylene or Teflon)	2.1	0.0002 @ 3 GHz
2	FR-4	4.4	0.008 @ 3 GHz
3	RT duroid	2.3	0.0012
4	Alumina (Al_2O_3)	10.2	0.0002 @ 1 GHz
5	Glass	4–10	
6	Epoxy glass PCB	5.2	
7	Wood	1.2–2.1	0.03 @ 3 GHz
8	Soil	2.44	0.0011 @ 3 GHz
9	Neoprene rubber	4	0.034 @ 3 GHz
10	Silicon	11.7–12.9	0.005 @ 1 GHz
11	GaAs	13.1	0.0016 @ 10 GHz

12 Conclusion

The intent of the chapter has been discuss about the types of dielectric resonator antennas, feeding techniques, mode/modes of excitation, radiation mechanism, bandwidth and gain enhancement techniques, parameters affecting the permittivity of dielectric ceramics used for DRAs as well as commercially and cost effective laboratory synthesized dielectric ceramics which can be used as resonators for dielectric resonator antennas. For future investigations, such studies will be useful for understanding the effect on the microwave device performance on the end user side. Also, emphasis should be placed on the low cost production and eco-friendly materials, which can be recycled further, keeping in view the environmental protection aspect. This review might provide a systematic guideline to researchers for design and development of novel DRAs.

The antennas discussed here can be used as primary feed for satellite communication. It can also be used as transmitting/receiving antenna in radars, in weapons installed in control military aircrafts, in mapping of ocean currents in harbor area, in capturing multi-spectral images to study earth science, in observing coastal and harbor traffic, in weather navigation system, in microwave radiometer to get information about various climate parameters, such as rain rate, ice morphology and ocean wind speed.

Table 6 Laboratory synthesized dielectric ceramics for DRA application

S. No.	Ceramic material	Sintering temperature (°C)	ε_r	References
Dielectric materials sintered below 1000 °C				
1	BaV_2O_6	550 °C for 1 h	11.2	[109]
2	Li_2O-1.94MgO-0.02Al_2O_3-P_2O_5	825 for 6 h	6.2	[110]
3	90 wt% $ZnNb_2O_6$-TiO_2 + 10 wt% (Li_2O-B_2O_3-SiO_2)	875	20.3	[111]
4	85 wt% $BaTi_4O_9$ + 15 wt% Li_2O-B_2O_3-SiO_2-CaO-Al_2O_3	875	26	[112]
5	$Ba_{0.5}Sr_{0.5}TiO_3$ + 10wt%PbO-BaO-B_2O_3-SiO_2 glass	875 for 2 h	27	[113]
6	$Ba_{0.5}Sr_{0.5}TiO_3$ + 8wt%PbO-BaO-B_2O_3-SiO_2 glass	900 for 2 h	25.8	[113]
7	$ZnNb_2O_6$ + 5 wt% CuO + 4B_2O_3	900	23.3	[114]
8	5Li_2O-0.583Nb_2O_5-3.248TiO_2 + 1wt%V_2O_5	920	21.5	[115]
9	$Ba_{0.5}Sr_{0.5}TiO_3$ + 5wt%PbO-BaO-B_2O_3-SiO_2 glass	925 for 2 h	13.8	[113]
10	$LiMgPO_4$	950 for 2 h	6.4	[116]
11	$(Zn_{0.8}Sn_{0.2})TiO_4$ + 10 wt% BaO-B_2O_3-SiO_2-Li_2O-CuO	950 for 4 h	17.8	[117]
12	0.83$ZnAl_2O_4$- 0.17TiO_2 + 0.2 wt% BBSZ glass	950 for 4 h	11.7	[118]
13	$Mg_3(VO_4)_2$–0.5$Ba_3(VO_4)_2$ + 0.0625 wt% Li_2CO_3	950 for 5 h	13	[119]
14	$Mg_3(VO_4)_2$–0.5$Ba(VO_4)_2$ + 0.0625 wt% Li_2O_3	950 for 5 h	12.6	[119]
15	$(Zr_{0.8}Sn_{0.2})TiO_4$ + 10wt% BaO-B_2O_3-SiO_2-Li_2O-CuO	950 for 8 h	18.4	[117]
16	$Mg_4Nb_2O_9$ + 3 wt%LiF + 6 wt% $CaTiO_3$	950 for 5 h	15.7	[120]
17	CaO-Al_2O_3-SiO_2-TiO_2-B_2O_3	950	10	[121]
18	$Ba_{0.5}Sr_{0.5}TiO_3$ + 3wt%PbO-BaO-B_2O_3-SiO_2 glass	950 for 2 h	15.0	[113]
Dielectric materials sintered above 1000 °C				
19	CeO_2-0.5WO_3-0.5TiO_2	1130	17	[122]
20	0.7$CaWO_4$–0.3$LaNbO_4$	1150	13.3	[123]
21	$MgNb_2O_6$ + 2wt% CuO	1170	19.9	[124]
22	0.96$MgTiO_3$-0.036$SrTiO_3$	1170 for 2 h	17.9	[125]
23	0.96$MgTiO_3$-0.036$SrTiO_3$ + 2wt% B_2O_3	1170 for 2 h	19	[125]
24	$Mg_{0.95}Ca_{0.05}TiO_3$ + 2 wt% B_2O_3	1200	21.2	[126]

(continued)

Table 6 (continued)

S. No.	Ceramic material	Sintering temperature (°C)	ε_r	References
25	$(Mg_{0.95}Ca_{0.05})TiO_3 + 5$ mol% B_2O_3	1200	19.6	[127]
26	$0.93(Mg_{0.6}Zn_{0.4})0.95Co_{0.05}TiO_3–0.07CaTiO_3$	1200	23.04	[128]
27	$Mg_4Ta_2O_9$	1200	14	[129]
28	$0.74CaWO_4$-$0.26TiO_2$	1250	17.5	[130]
29	$LaNbO_4$	1250	19.3	[131]
30	$La_{0.9}Nd_{0.1}NbO_4$	1250	20	[132]
31	$0.94MgTiO_3$-$0.06CaTiO_3 + 0.2$ mol% Bi_2O_3	1250	22.6	[133]
32	$0.964MgTiO_3$-$0.036SrTiO_3$	1270	20.8	[134]
33	$0.93(Mg_{0.95}Co_{0.05})TiO_3$-$0.07CaTiO_3$	1275 for 4 h	21.6	[135]
34	$0.92(Mg_{0.95}Co_{0.05})TiO_3$-$0.08CaTiO_3$	1275 for 2 h	22.1	[135]
35	$0.94(Mg_{0.95}Co_{0.05})TiO_3$-$0.06CaTiO_3$	1275 for 4 h	20.9	[135]
36	$0.95MgTiO_3$-$0.05CaTiO_3 + 0.25$ wt% CuO	1275 for 4 h	20	[136]
37	$Sm_2BaCu_{0.5}Zn_{0.5}O_5$	1280	13.4	[137]
38	$(Mg_{0.95}Zn_{0.05})TiO_3$	1300	17.1	[138]
39	$0.17Ba_5NbO_{15}–0.83BaNb_2O_6$	1300	25.2	[139]
40	$0.9Al_2O_3–0.1\ TiO_2$	1300 for 2 h	12.4	[140]
41	$Al_2O_3–TiO_2$: MnO	1300	12.4	[141]
42	$Zn(Nb_{0.35}Ta_{0.65})O_6$	1300	28	[142]
43	$0.96MgTiO_3–0.04SrTiO_3$	1300 for 4 h	20.9	[143]
44	$0.91(Mg_{0.7}Zn_{0.3})TiO_3$-$0.09CaTiO_3$	1310 for 3 h	22.5	[144]
45	$0.85(Mg_{0.95}Zn_{0.05})TiO_3$-$0.15Ca_{0.6}La_{0.8/3}TiO_3$	1320 for 3 h	25	[145]
46	$Nd(Co_{0.5}Ti_{0.5})O_3 + 0.75$ wt% B_2O_3	1320 for 4 h	27.2	[146]
47	$5MgO$-Ta_2O_5-TiO_2	1325	18	[147]
48	$Mg_{0.93}Ca_{0.07}TiO_3$	1350 for 3 h	22.15	[148]
49	$Ba(Zn_{1/3}Ta_{2/3})O_3$	1350 for 20 h	28	[149]
50	$0.89Al_2O_3$-$0.11TiO_2$-0.5 wt% ZnO	1350	12.9	[150]
51	$0.9Al_2O_3–0.1\ TiO_2$ Annealed at 1000 C	1350	12.4	[151]
52	$0.95MgTiO_3$-$0.05CaTiO_3$	1400	20	[126]
53	$0.83ZnAl_2O_4–0.17TiO_2$	1410	12.6	[151, 152]
54	$0.80ZnAl_2O_4–0.20TiO_2$	1420	14	[151]
55	$0.9Nd(Co_{0.5}Ti_{0.5})O_3–0.1SrTiO_3$	1440 for 4 h	29.3	[153]
56	$0.75MgAl_2O_4$-$0.25TiO_2$	1450	10.7	[154]
57	$Sm_3Ga_5O_{12} + TiO_2$	1450	12.33	[155]
58	$Mg_4Ta_2O_9$	1450	11.5	[129]
59	$CaYAlO_4$	1450 for 3 h	18.9	[156]

(continued)

Table 6 (continued)

S. No.	Ceramic material	Sintering temperature (°C)	ε_r	References
60	$0.95MgTiO_3$-$0.05CaTiO_3$	1450	21	[136]
61	$0.35(Al_{0.5}Ta_{0.5})O_2$–$0.65(Mg_{1/3}Ta_{2/3})O_2$	1450	26.1	[157]
62	$0.75(Al_{0.5}Ta_{0.5}O_2)$–$0.25(Ti_{0.85}Sn_{0.15})O_2$	1450 for 3 h	28	[158]
63	$0.8(Al_{0.5}Ta_{0.5})O_2$–$0.2TiO_2$	1450 for 3 h	29.4	[159]
64	$0.5MgAl_2O_4$–$0.5TiO_2$	1460	13.2	[160]
65	$0.7MgTiO_3$–$0.3MgTa_2O_6$	1460 for 3 h	23	[161]
66	$Y_3Ga_5O_{12}$	1500	11.56	[162]
67	$0.72Ba(Mg_{0.5}W_{0.5})O_3$–$0.28BaTiO_3$	1500 for 6 h	12.3	[163]
68	$0.67Ba(Mg_{0.5}W_{0.5})O_3$–$0.33BaTiO_3$	1500 for 6 h	13.1	[163]
69	BaO-33MgO-$0.35WO_3$-$0.32TiO_2$	1500	16.3	[163]
70	BaO-0.34MgO-$0.32WO_3$-$0.34TiO_2$	1500 for 12 h	14.5	[163, 164]
71	BaO-0.35MgO-$0.34WO_3$-$0.31TiO_2$	1500 for 6 h	15.4	[163]
72	BaO-0.34MgO-$0.35WO_3$-$0.31TiO_2$	1500 for 6 h	17.1	[163]
73	BaO-0.32MgO-$0.28WO_3$-$0.4TiO_2$	1500 for 6 h	18.1	[164]
74	BaO-0.32MgO-$0.26WO_3$-$0.42TiO_2$	1500 for 6 h	21.5	[164]
75	$Ba_3(Zr_{0.0645}Zn_{0.816}Ni_{0.1625}Ta_{1.957})O_9$	1510 for 24 h	29	[161]
76	$Ba(Zr_{0.09}Ni_{0.125}Zn_{0.845}Ta_{1.94})O_3$	1520 for 48 h	30	[165]
77	$0.95Ba(Zn_{0.5}Ta_{2/3})O_3$–$0.05Sr(Ga_{0.5}Ta_{0.5})O_3$	1550 for 2 h	29	[166]
78	$Ba(Zn_{1/3}Ta_{2/3})O_3$ + 1 mol%Mn	1550	30	[167]
79	$Ba(Mg_{1/3}Ta_{(2-2x)/3}W_{x/3}Ti_{x/3})O_3 (x = 0.15)$	1550 for 4 h	20	[168]
80	Al_2O_3	1550	10.05	[169]
81	$Ca(Ho_{0.5}Nb_{0.5})O_3$	1550 for 4 h	23	[170]
82	$Ba(Mg_{0.33}Ta_{0.53}Ti_{0.067}W_{0.067})O_3$	1590	23.6	[168]
83	$Ba(Mg_{0.4}Zn_{0.6})Ta_{2/3}O_3$	1600 for 4 h	27	[171]
84	$CaTa_2O_6$	1600	21.2	[172]
85	$Ba_{0.9925}(Mg_{0.33}Ta_{0.67})O_3$	1600	24.7	[173]
86	$Ba(Ma_{0.3183}Ta_{0.67})O_3$	1600	25.1	[173]
87	$Ba(Mg_{1/3}Ta_{2/3})O_3$	1640 for 20 h	24	[167]
88	$0.85Ba(Mg_{1/3}Ta_{2/3})O_3$–$0.15BaSnO_3$	1640 for 20 h	24	[174]
89	$Mg_3La_4Al_{44}O_{75}$	1680	13	[175]

Table 7 Recently designed DRAs with commercially available and synthesized ceramic materials

S. No.	Dielectric material(s)	DRA shapes (year)	Bandwidth (GHz)	Excitation mechanism	Mode	Application	References
1	Al_2O_3 ($\varepsilon_r = 9.8$)	Cylindrical DRA	1.88–2.06 (9.13%), 2.29–2.58 (11.9%), 2.9–3.93 (30.16%)	Ring-shaped patch along with an inverted L-strip (microstrip line)	$HE_{11\delta}$, $HE_{12\delta}$	WLAN (2.4 GHz) and WiMAX (2.5/3.5 GHz)	[176]
2	Al_2O_3	Cylindrical DRA	4.9–5.5 (11.5%)	Dual feed (Aperture coupling and CPW)	$HE^{x}_{11\delta}$ and $HE^{y}_{11\delta}$	MIMO applications	[177]
3	Al_2O_3	Dual segment RDRA	3.19–4 (23.14%)	Conformal strip (microstrip line)	$TE_{1\delta 1}$	Radar and satellite application (S-band)	[178]
4	Al_2O_3	Hybrid cylindrical DRA	2.24–2.56 (13.33%), 3.28–4.18 (24.13%), 5.36–5.8 (7.9%)	Annular ring shaped patch along with T-shaped printed line (Microstrip line)	$HE_{11\delta}$, $HE_{12\delta}$	WLAN (2.4/5.8 GHz) and WiMAX (2.5/3.5/5.5 GHz)	[179]
5	Al_2O_3	Stacked CDRA	3.55–10.78 (100.9%)	Annular shaped (Microstrip line)	$TE_{01\delta}$, $TM_{01\delta}$	Ultra wide band antenna application for high data rate	[180]
6	Al_2O_3	CDRA	2.42–2.9 (18.04%), 4.9–5.98 (19.85%)	Modified circular shaped aperture	$HE_{11\delta}$, $HE_{12\delta}$-like $HE_{12\delta}$	WLAN (2.4/5.2/5.8 GHz) and WiMAX (2.5/5.5 GHz)	[181]

(continued)

Table 7 (continued)

S. No.	Dielectric material(s)	DRA shapes (year)	Bandwidth (GHz)	Excitation mechanism	Mode	Application	References
7	Al_2O_3	Composite CDRA	2.25–2.79 (21.42%), 3.1–4.0 (25.35%), 5.05–5.6 (10.32%)	Modified annular ring printed line with U-shaped printed Microstrip line	$HEM_{11\delta}$, $EM_{12\delta}$	WLAN and WiMAX applications	[182]
8	Al_2O_3	Three element dual segment triangular DRA	9.27–11.67 (22.92%)	Co-axial probe feed	$TM_{10\text{-}1}$	X-band applications	[183]
9	Al_2O_3	4-Element dual segment RDRA array	3.44–3.95 GHz	Copper strip excitation (connected to microstrip line)	$TE^{y}{}_{1\delta 1}$	Radar and satellite application	[184]
10	Al_2O_3	Half split cylindrical DRA	5.1–8.3 GHz (51.2%)	Probe feed	$TM_{01\delta}$	WLAN and WiMAX	[185]
11	Al_2O_3	CDRA	1.92–2.3 GHz, 3.13–4.6 GHz, and 5.4–6.5 GHz	Modified annular ring patch feed	$HEM_{11\delta}$, $HE_{12\delta}$	WLAN/WiMAX	[186]
12	Al_2O_3	Two segment CDRA	2.3–2.5 GHz, 3.2–3.8 GHz, 4.67–7.0 GHz 8.33%; 17.14%, and 39.93%,	Microstrip line along with two ring patch	$HE_{11\delta}$, $TM_{01\delta}$	WLAN/WiMAX	[187]

(continued)

Table 7 (continued)

S. No.	Dielectric material(s)	DRA shapes (year)	Bandwidth (GHz)	Excitation mechanism	Mode	Application	References
13	Al_2O_3	Hybrid MIMO CDRA	2.24–2.38 GHz, (6.06%) 2.5–3.26 GHz (26.4%), and 4.88–7.0 GHz (35.7%) (S_{11} < -6 dB)	Two symmetric folded microstrip line feed	$HEM_{11\delta}$	LTE2500/ WLAN/WiMAX	[188]
14	Al_2O_3	CDRA	2.5–3.02, 3.76–3.86, and 4.38–4.72 GHz	Composite feed (combination of vertical strip and psi-shaped (ψ) microstrip line)	$HEM_{11\delta}$, $TM_{01\delta}$, and $HEM_{12\delta}$	WiMAX (2.5 GHz) and vehicular applications	[189]
15	Al_2O_3	Cubic DRA	$BW_{im} = 35.35\%$ $BW_{AR} = 20.62\%$	Question-mark-shaped microstrip feed	$TE_{11\delta}$	WiMAX (3.3–3.7 GHz)	[190]
16	Al_2O_3	CDRA	$BW_{im} = 40\%$ $BW_{AR} = 9.24\%$	Dual conformal strips with modified Wilkinson power divider	–	Bluetooth and Wi-Fi	[191]
17	Al_2O_3	Rectangular DRA	11.1–11.4 GHz (2.3%) 10.4–11.44 GHz (9.5%)	Microstrip line with cross slot	TE_{117}, TE_{119}, $TE_{11,11}$, $TE_{11,13}$	mm-wave and tetrahertze higher frequency bands	[192]

(continued)

Table 7 (continued)

S. No.	Dielectric material(s)	DRA shapes (year)	Bandwidth (GHz)	Excitation mechanism	Mode	Application	References
18	Al_2O_3	Square DRA	2.19–3.63 GHz (49.5%)	Microstrip-coupled cross-slot	TE_{111}	Radio frequency identification (RFID) system, Global positioning system and satellite communication/ navigation system	[193]
19	Al_2O_3	Modified cylindrical shaped dielectric resonator antenna	3.37–5.69 GHz (51.2%)	Probe feed	–	Modern wireless communication system	[194]
20	Al_2O_3	Spherical	100–110 GHz (9%)	Microstrip resonator and microstrip feed line	TE_{113}	High data-rate communication and sensing	[195]
21	$SrLa_2Mg_2W_2O_{12}$ ($\varepsilon_r = 22$)	Dual segment CDRA	9.17–10.86 GHz	Probe feed	–	X-band applications	[196]
22	$SrLa_2Mg_2W_2O_{12}$ ($\varepsilon_r = 22$)	Dual segment CDRA with meta-material	8.90–11.48 GHz (25.32%)	Probe feed	–	Radio navigation services, civil and military radar	[197]
23	Emerson & Cuming Eccostock HIK material ($\varepsilon_r = 20$)	CDRA array	2.9% Fc = 9.04 GHz	Non-resonant microstrip patch feed	$HEM_{12\delta}$	Electrically steered array applications	[198]

(continued)

Table 7 (continued)

S. No.	Dielectric material(s)	DRA shapes (year)	Bandwidth (GHz)	Excitation mechanism	Mode	Application	References
24	$Nd(Mg_{0.43}Ca_{0.07}Sn_{0.5})O_3$ ($\varepsilon r = 19.51$)	CDRA	2280–2593 MHz (12.5%), 5077–5877 MHz (14.2%)	Microstrip line feed with rectangular slot	–	Industrial, Scientific medical (ISM), High performance radio local area network (HIPERLAN), Unlicensed national information infrastructure (UNII) bands	[199]
25	Zirconium Tin Titanate ($\varepsilon_r = 38$)	Cylindrical DR which is chopped normal to its x axis dianeter or rectangular truncation in one side of resonator	BW_{im} = 112–140 MHz BW_{AR} = 28-33 MHz	Probe feed	–	–	[200]
26	Li_2O-1.94MgO-0.02Al_2O_3-P_2O_5 (LMAP) ($\varepsilon_r = 6.2$)	Composite triangular DRA	7.70–15.30 GHz (66.09%)	Probe feed	TM^z_{101}	Broadcast base-stations, radar, and satellite communications	[110]
27	K9 glass ($\varepsilon_r = 6.85$) + liquid ethyl acetate ($\varepsilon_r = 7.1$)	CDRA	3.75–5.37 GHz (35.5%)	Coaxial probe feed	$HEM_{11\delta}$ and $TM_{01\delta}$	Pattern reconfigurable design without p-i-n diode, suitable for systems which require static changes	[201]

(continued)

Table 7 (continued)

S. No.	Dielectric material(s)	DRA shapes (year)	Bandwidth (GHz)	Excitation mechanism	Mode	Application	References
28	10 wt% PBBS Glass-added Barium Strontium Titanate ($\varepsilon_r = 27$)	Dual segment CDRA	8.68–9.76 GHz (11.71%)	Aperture coupled feed	–	Radar and radio navigation, military satellite communication	[202]
29	10 wt% PBBS Glass-added Barium Strontium Titanate ($\varepsilon_r = 27$)	Dual segment RDRA	8.80–9.49 GHz	Aperture coupled feed	–	As primary feed for radar and satellite communication	[203]
30	3 wt% PBBS Glass-added Barium Strontium Titanate ($\varepsilon r = 15$)	Filleted RDRA	7.0–13.048 GHz (~60%)	Aperture coupled feed	TE^x_{211} and TE^x_{2n1}	As primary feed for parabolic reflector in radar and satellite communication	[50]
31	1 wt% $55B_2O_3$-$45Bi_2O_3$ Glass-added Magnesium cobalt titanate ($\varepsilon_r = 11.57$)	Dual segment RDRA	6.44–9.52 GHz (38.6%)	Coaxial probe feed	–	Radar and satellite communication	[204]
32	$Zn_{0.7}Mg_{0.3}TiO_3$ ($\varepsilon_r = 15.7$)	Cylindrical dielectric resonator antennas	(2.11%) Fc = 4.26 GHz	Coaxial probe feed	$HEM_{11\delta}$	Wireless communication and radar	[205]
33	$\varepsilon_r = 2.2$	Triangular DRA	4.55–5.2 (13.54%), 7.84–8.58 (8.92%)	Conformal strip connected to microstrip line	$TM_{10\delta}$, $TM_{30\delta}$	–	[206]
34	$\varepsilon_r = 12$	Four element triangular DRA	4.7–6.8 (37.61%)	Co-axial probe feed	$TM_{10\text{-}1}$	–	[207]

(continued)

Table 7 (continued)

S. No.	Dielectric material(s)	DRA shapes (year)	Bandwidth (GHz)	Excitation mechanism	Mode	Application	References
35	$\varepsilon_r = 12$	Rectangular cylindrical dielectric resonator and a metallic semi-loop	3.3–3.7 GHz (11.4%)	Modified microstrip line feed	–	Massive-MIMO array antennas	[208]
36	$\varepsilon_r = 10.2$, $\varepsilon_r = 2.2$	Cylindrical DRA array	57–64 GHz (11.5%)	Aperture–coupled feed	$HEM_{11\delta}$	Antenna-in-package design	[209]
37	$\varepsilon_r = 38$	Four-leaf-clover-shaped	2.50–2.74 GHz (9.2%)	Probe feed with T shape strip	TE_{111}	Front-end in wireless communication	[210]

Table 8 Current state of art of ceramic dielectric materials and dielectric resonant antenna parameters

Parameters	Sate of art	References
Lowest dielectric constant for ceramic	$\varepsilon_r = 3$ for $AlPO_4$ ceramic	[17]
Highest quality factor	Qf = 773 700 GHz for $0.993MgO–0.007B_2O_3$	[17]
Lowest frequency of operation	55 MHz	[211]
Lowest frequency of operation (CP-DRA)	1.5 GHz	[212]
Highest frequency of operation	94 GHz	[213]
Higher frequency of operation (CP-DRA)	60 GHz	[214]
Largest DRA	CDRA Diameter = 550 mm, height = 200 mm	[211]
Wide impedance bandwidth (LP-DRA)	126%	[215]
Wide impedance bandwidth (CP-DRA)	54%	[216]
Wide axial ratio bandwidth (CP-DRA)	50%	[216]
High gain (LP-DRA)	24.5 dBi for 529 rectangular array	[217]
High gain (CP-DRA)	12 dBi for 4 CDRA array	[218]

References

1. Ullah, U., Ali, W.F.F.W., Ain, M.F., Mahyuddin, N.M., Ahmad, Z.A.: Design of novel dielectric resonator antenna using $MgTiO_3$-$CoTiO_3$ for wideband applications. Mater. & Des. **85**, 396–403 (2015)
2. Richtmyer, R.D.: Dielectric resonators. J. Appl. Phys. **10**(6), 391 (1939)
3. Okaya, A., Barash, L.: The dielectric microwave resonator. Proc. IRE **50**(10), 2081–2092 (1962)
4. Okaya, A.: The rutile microwave resonator. Proc. IRE **48**, 1921 (1962)
5. Long, S.A., McAllister, M.W., Shen, L.C.: The resonant cylindrical dielectric cavity antenna. IEEE Trans. Antennas Propag. **31**(3), 406–412 (1983)
6. Shum, S.M., Luk, K.M.: Stacked annular-ring dielectric resonator antenna excited by axi-symmetric coaxial probe. IEEE Trans. Antennas Propag. **43**(8), 889–892 (1995)
7. Rao, Q., Denidni, T.A., Sebak, A.R.: Broadband compact stacked T-shaped DRA with equilateral-triangle cross sections. IEEE Microw. Wirel. Compon. Lett. **16**(1), 7–9 (2006)
8. Esselle, K.P., Bird, T.S.: A hybrid-resonator antenna: experimental results. IEEE Trans. Antennas Propag. **53**(2), 870–871 (2005)
9. Gao, Y., Ooi, B.L., Ewe, W.B., et al.: A compact wideband hybrid dielectric resonator antenna. IEEE Microw. Wirel. Compon. Lett. **16**(4), 227–229 (2006)
10. Zou, M., Pan, J.: Investigation of resonant modes in wideband hybrid omnidirectional rectangular dielectric resonator antenna. IEEE Trans. Antennas Propag. **63**(7), 3272–3275 (2015)
11. Kishk, A.A., Chair, R., Lee, K.R.: Broadband dielectric resonator antennas excited by L-shaped probe. IEEE Trans. Antennas Propag. **54**(8), 2182–2189 (2006)

12. Rashidian, A., Shafai, L., Klymyshyn, D.M.: Tall microstrip transmission lines for dielectric resonator antenna applications. IET Microw. Antennas Propag. **8**(2), 112–124 (2014)
13. Liang, X.L., Denidni, T.A.: H-shaped dielectric resonator antenna for wideband applications. IEEE Antennas Wirel. Propag. Lett. **7**, 163–166 (2008)
14. Gao, Y., Feng, Z., Zhang, L.: Compact asymmetrical T-shaped dielectric resonator antenna for broadband applications. IEEE Trans. Antennas Propag. **60**(3), 1611–1615 (2012)
15. Chaudhary, R.K., Srivastava, V.K., Biswas, A.: A broadband dumbbell-shaped dielectric resonator antenna. Microw. Opt. Technol. Lett. **56**(12), 2944–2947 (2014)
16. Luk, K.M., Lee, M.T., Leung, K.W., et al.: Technique for improving coupling between microstripline and dielectric resonator antenna. Electron. Lett., **35**, 357–358 (1999)
17. Sebastian, M.T., Ubic, R., Jantunen, H.: Microwave Materials and Application. Wiley & Sons (2017)
18. Petosa, A.: Dielectric Resonator Antenna Handbook, 311 p. Artech House (2007)
19. Guha, D., Gupta, B., Antar, Y.M.M.: Hybrid monopole- DRAs using hemispherical/conical shaped dielectric ring resonators: improve ultrawideband designs. IEEE Trans. Antennas Propag. **60**(1), (2012)
20. Guha, D., Antar, Y.M.M.: New half-hemispherical dielectric resonator antenna for broadband monopole- type radiation. IEEE Trans. Antennas Propag. **54**(12), (2006)
21. Guha, D., Gupta, B., Antar, Y.M.M.: New pawn-shaped dielectric ring resonator loaded hybrid monopole antenna for improved ultrawide bandwidth. IEEE Antennas Wirel. Propag. Lett. **08**, (2009)
22. Guha, D., Antar, Y.M.M.: Four element cylindrical dielectric resonator antenna for wideband monopole like radiation. IEEE Trans. Antennas Propag. **54**(9), (2006)
23. Luk, K.M., Leung, K.W. (eds.): Dielectric Resonator Antennas, 388 p. Research Studies Press (2003)
24. Sinha, D., Amaratunga, G.A.J.: Electromagnetic radiation under explicit symmetry breaking. Phys. Rev. Lett. **114**(147701), 1–7 (2015)
25. Cooper, M., et al.: Investigation of dielectric resonator antenna for L-band communications. In: Symposium on Antenna Technology and Applied Electromagnetics, Montreal, Canada, pp. 167–170 (1996)
26. Liasne, A., Gillard, R., Piton, G.: A robust slot fed dielectric resonator antenna (DRA) using an intermediate substrate. IEEE Electron. Lett. **37**(25), 1497–1498 (2001)
27. Ittipiboon, A., Petosa, A., Roscoe, D., Cuhaci, M.: An investigation of a novel broadband dielectric resonator antenna. In: IEEE Antennas and Propagation Society International Symposium. Digest, AP-S, pp. 2038–2041 (1996)
28. Petosa, A., Ittipiboon, A., Cuhaci, M., Larose, R.: Low profile phased array of dielectric resonator antennas. In: IEEE International Symposium Phased Array Systems and Technology, pp. 182–185 (1996)
29. Petosa, A., Larose, R., Ittipiboon, A., Cuhaci, M.: Active phased array of dielectric resonator antennas. In: IEEE Antennas and Propagation Symposium, Montreal, Canada, pp. 690–693 (1997)
30. Petosa, A., Larose, R., Ittipiboon, A., Cuhaci, M.: Microstrip fed array of multi-segment dielectric resonator antennas. IEEE Proc. Microw. Antennas Propag. **144**(6), 472–476 (1997)
31. Fan, Z., et al.: Parasitic coplanar three element dielectric resonator antenna sub-array. IEE Electron. Lett. **32**(9), 789–790 (1996)
32. Sangiovanni, A., Dauvignac, J.Y., Pichot, C.: Embedded dielectric resonator antenna for bandwidth enhancement. IEE Electron. Lett. **33**(25), 2090–2091 (1997)
33. Kishk, A.A.: Experimental study of broadband embedded dielectric resonator antennas excited by a narrow slot. Antennas Wirel. Prop. Lett. **4**, 79–81 (2005)
34. Chair, R., Yang, S.L., Kishk, A.A., Lee, K.F., Luk, K.M.: Aperture fed wideband circularly polarized rectangular stair shaped dielectric resonator antenna. IEEE Trans. Antenna Propag. **54**(4), 1352 (2006)
35. Chair, R., Kish, A.A., Lee, K.E.: Wideband stair shaped dielectric resonator antennas. IET Microw. Antennas Propag. **1**(2), 299–305 (2007)

36. Chair, R., Kishk, A.A., Lee, K.F.: Experimental investigation for wideband perforated dielectric resonator antenna. Electron. Lett. **42**(3), 137–139 (2006)
37. Chair, R., et al.: IEE Electron. Lett. **40**(10), 581 (2004)
38. Luk, K.M., Lee, M.T., Leung, K.W., Yung, E.K.N.: Technique for improving coupling between microstripline and dielectric resonator antenna. IEEE Electron. Lett. **35**(5), 357–358 (1999)
39. Leung, K.W., Wong, W.C., Luk, K.M., Yung, E.K.N.: Circular polarized dielectric resonator antenna excited by dual conformal strips. IEEE Electron. Lett. **36**(6), 454–455 (2000)
40. Leung, et al.: Annular slot coupled dielectric resonator antenna. IEEE Electron. Lett. **34**(13), 1275–1277 (1998)
41. Leung, K.W., Wong, W.C., Ng, H.K.: Circularly polarized slot- coupled dielectric resonator antenna with a parasitic patch. IEEE Antennas Wirel. Propag. Lett. **1**(1), 57–59 (2002)
42. Yung, K.W.N.: Annular slot- coupled dielectric resonator antenna. Electron. Lett. **34**(13), 1275–1277 (1998)
43. Li, T.W., Sun, J.S.: A wide U-shape slot fed broadband dielectric resonator antenna. In: Symposium on Antenna Technology and Applied Electromagnetics, St. Malo, France, pp. 124–125 (2005)
44. Kishk, A.A., Glisson, A.W., Yin, Y.: Microw. Opt. Technol. Lett. **29**(3), 160 (2001)
45. Kishk, A.A., Glisson, A.W., Junker, J.P.: J. Electromagn. Waves Appl. **15**(6), 777 (2001)
46. Kishk, A.A.: IEEE Trans. Antenna Propag. **51**(10), 2913 (2003)
47. Zhang, L.N., Zhong, S.S., Xu, S.D.: Broadband U-shaped dielectric resonator antenna with elliptical patch feed. Electron. Lett. **44**(16), 947–949 (2008)
48. Dhar, S., Ghatak, R., Gupta, B., Poddar, D.R.: A Wideband Minkowski fractal dielectric resonator antenna. IEEE Trans. Antenna Propag. **6**(6), 2895–2903 (2013)
49. Mukherjee, B., Patel, P., Mukherjee, J.: A novel cup-shaped inverted hemi-spherical dielectric resonator antenna for wideband applications. IEEE Antennas Wirel. Propag. Lett. **12**, 1240–1243 (2013)
50. Tripathi, P., Kumari, P., Sahu, B., Singh, R., Singh, S.P., Kumar, D.: Filleted RDRA using novel ceramic dielectric (3G-BST) with stable gain and improved bandwidth for X-band application. Ceram. Int. **44**(5), 5045–5054 (2018)
51. Fakhte, S., Oraizi, H., Matekovits, L.: High gain rectangular dielectric resonator antenna using uniaxial material at fundamental mode. IEEE Trans. Antennas Propag. **65**(1), 342–347 (2017)
52. Perron, A., Denidni, T.A., Sebak, A.R.: High gain hybrid dielectric resonator antenna for millimeter wave applications, design and implementation. IEEE Trans. Antennas Propag. **57**(10), 2882–2892 (2009)
53. Guha, D., Antar, Y.M.M., Petosa, A., Lee, D.: Improved design guidelines for the ultra wideband monopole dielectric resonator antenna. IEEE Antennas Wirel. Propag. Lett. **5**(1), 373–376 (2006)
54. Gao, Y., Popov, A.P., Ooi, B.L., Leong, M.S.: Experimental study of wideband hybrid dielectric resonator antenna on small ground plane. Electron. Lett. **42**(13), 731–733 (2006)
55. Lin, Y.F., Chen, H.M., Lin, C.H.: Compact dual band hybrid dielectric resonator antenna with radiating slot. IEEE Antenna Wirel. Propag. Lett. **8**, 6–9 (2009)
56. Fakhte, S., Oraizi, H., Matekovits, L.: Gain improvement of rectangular dielectric resonator antenna by engraving grooves on its side walls. IEEE Antennas Wirel. Propag. Lett. https://doi.org/10.1109/LAWP.2017.2702584
57. Llombart, N., Neto, A., Gerini, G., de Maagt, P.: Planar circularly symmetric EBG structures for reducing surface waves in printed antennas. IEEE Trans. Antennas Propag. **53**(10), 3210–3218 (2005)
58. Zheng, Q., Guo, C., Vandenbosch, G.A., Ding, J.: Low-profile circularly polarized array with gain enhancement and RCS reduction using polarization conversion EBG structures. IEEE Trans. Antennas Propag. **68**(3), 2440–2445 (2020). https://doi.org/10.1109/TAP.2019.2943693.
59. Coulibaly, Y., Boutayeb, H., Denidni, T.A., Talbi, L.: Gain enhancement of a dielectric resonator antenna using cylindrical electromagnetic crystal substrate. IEEE Antennas and Propagation Society International Symposium. Digest, pp. 1325–1328 (2007)

60. Petosa, A., Thirakoune, S., Ittipiboou, A.: Higher order modes in rectangular DRAs for gain enhancement. Presented at the ANTEM 2009, Banff, Canada (2009)
61. Yarge, S., Sertel, K., Volakis, J.: A directive resonator antenna using degenerate band edge crystals. IEEE Trans. Antennas Propag. **57**, 799–803 (2009)
62. Fakhte, S., Oraizi, H., Matekovits, L.: High gain rectangular dielectric resonator antenna using uniaxial material at fundamental mode. IEEE Trans. Antennas Propag. 99, (2016). https://doi.org/10.1109/TAP.2016.2627520
63. Yarga, S., Sertel, K., Volakis, J.L.: A directive resonator antenna using degenerate band edge crystals. IEEE Trans. Antennas Propag. **57**(3), (2009)
64. Antar, Y.M.M.: Antenna for wireless communications: recent advances using dielectric resonators. IET, Prog. Cicuit Devices Syst. **2**(1), 133–138 (2008)
65. Carrie, J., et al.: A K-band circularly polarized cavity backed dielectric resonator. IEEE Antennas and Propagation Symposium, Digest AP-S Baltimore, MD, pp. 734–737 (1996)
66. Carrie, J., et al.: A K-band circularly polarized dielectric resonator modeled using the transmission line matrix method. In: Progress in Antenna Technology and Applied Electromagnetic Symposium. ANTEM 96, Montreal, Canada, pp. 709–712 (1996)
67. Chattopadhyay, S., Siddiqui, J.Y., Guha, D.: Rectangular microstrip patch on composite dielectric substrate for high gain wide beam radiation patterns. IEEE Trans. Antennas Propag. **57**(10), (2009)
68. Long, S.A., McAllister, M.W., Shen, L.C.: The resonant cylindrical dielectric cavity antenna. IEEE Trans. Antennas & Propag. **31**(3), 406–412 (1983)
69. Guha, D., Kumar, C.: Microstrip patch versus dielectric resonator antennas bearing all commonly used feeds: an experimental study to choose the right element. IEEE Antennas Propag. Mag. **58**(1), 45–55 (2016)
70. Kumari, R., Gangwar, R.K.: Circularly polarized dielectric resonator antennas: design and developments. Wireless Pers. Commun. **86**, 851–886 (2016)
71. Tawk, Y., Costantine, J., Christodoulou, C.G.: A rotatable reconfigurable antenna for cognitive radio applications. Proceedings of IEEE RWS, pp. 158–161 (2011)
72. Zhu, H.L., Cheung, S.W., Yuk, T.I.: Mechanically pattern reconfigurable antenna using metasurface. IET Microw. Antennas Propag. **9**(12), 1331–1336 (2015)
73. Aboufoul, T., Alomainy, A., Parini, C.: Reconfiguring UWB monopole antenna for cognitive radio applications using GaAs FET switches. IEEE Antennas Wirel. Propag. Lett. **11**, 392–394 (2012)
74. Chen, Z., Shoaib, I., Yao, Y., Yu, J., Chen, X., Parini, C.G.: Pattern-reconfigurable dual-polarized dielectric resonator antenna. IEEE Antennas Wirel. Propag. Lett. **15**, 1273–1276 (2016)
75. Chen, Z., Wong, H.: Wideband glass and liquid cylindrical dielectric resonator antenna for pattern reconfigurable design. IEEE Trans. Antenna Propag. **65**(5), 2157–2164 (2017)
76. Huitema, L., Koubeissi, M., Mouhamadou, M., Arnaud, E., Decroze, C., Monediere, T.: Compact and multiband dielectric resonator antenna with pattern diversity for multistandard mobile handheld devices. IEEE Trans. Antennas Propag. **59**(11), 4201–4208 (2011)
77. Liu, B.J., Qiu, J.H., Wang, C.L., Li, G.Q.: Pattern reconfigurable cylindrical dielectric resonator antenna based on parasitic elements. IEEE Access **5**, 25584–25590 ((2017))
78. Zhong, L., Hong, J.-S., Zhou, H.-C.: A novel pattern-reconfigurable cylindrical dielectric resonator antenna with enhanced gain. IEEE Antennas Wirel. Propag. Lett. **15**, 1253–1256 (2016)
79. Saleem, M.K., Alkanhal, M.A.S., Sheta, A.F.: Dual strip-excited dielectric resonator antenna with parasitic strips for radiation pattern reconfigurability. Int. J. Antennas Propag. (2014). https://doi.org/10.1155/2014/865620
80. Mak, K.M., Lai, H.W., Luk, K.M., Ho, K.L.: Polarization reconfigurable circular patch antenna with a C-shape. IEEE Trans. Antennas Propag. **65**(3), 1388–1392 (2017)
81. Hu, J., Hao, Z.C., Miao, Z.W.: Design and implementation of a planar polarization-reconfigurable antenna. IEEE Antennas Wirel. Propag. Lett. **16**, 1557–1560 (2017)

82. Qin, P.Y., Weily, A.R., Guo, Y.J., Liang, C.H.: Polarization reconfigurable U-slot patch antenna. IEEE Trans. Antennas Propag. **58**(10), (2010)
83. Khidre, A., Lee, K.-F., Yang, F., Elsherbeni, A.Z.: Circular polarisation reconfigurable wideband E shaped patch antenna for wireless applications. IEEE Trans. Antennas Propag. **61**(2), 960–964 (2013)
84. Row, J.-S., Hou, M.-J.: Design of polarization diversity patch antenna based on a compact reconfigurable feeding network. IEEE Trans. Antennas Propag. **62**(10), 5349–5352 (2014)
85. Yang, W., Che, W., Jin, H., Feng, W., Xue, Q.: A polarization reconfigurable dipole antenna using polarization rotation AMC structure. IEEE Trans. Antennas Propag. **63**(12), 5305–5315 (2015)
86. Lin, W., Wong, H.: Wideband Circular polarisation reconfigurable antenna with L-shaped feeding probes. IEEE Antennas Wirel. Propag. Lett. **16**, 2114–2117 (2017)
87. Hao, Z.-C., Fan, K.-K., Wang, H.: A planar polarization reconfigurable antenna. IEEE Trans. Antennas Propag. **65**(4), 1624–1632 (2017)
88. Ji, L.Y., Qin, P.Y., Guo, Y.J., Ding, C., Fu, G., Gong, S.X.: A wideband polarisation reconfigurable antenna with partially reflective surface. IEEE Trans. Antennas Propag. **64**(10), 4534–4538 (2016)
89. Hu, J., Luo, G.Q., Hao, Z.C.: A wideband quad-polarization reconfigurable metasurface antenna. IEEE Access **6**, 6130–6137 (2018)
90. Hu, J., Hao, Z.-C., Hong, W.: Design of a wideband quad-polarization reconfigurable patch antenna array using a stacked structure. IEEE Trans. Antennas Propag. **65**(6), 3014–3023 (2017)
91. Lin, W., Wong, H.: Polarization reconfigurable aperture-fed patch antenna and array. IEEE Access **4**, 1510–1517 (2016)
92. Wu, F., Luk, K.M.: Single-port reconfigurable magneto-electric dipole antenna with quad-polarization diversity. IEEE Trans. Antennas Propag. **65**(5), 2289–2296 (2017)
93. Li, W., Xia, S., He, B., Chen, J., Shi, H., Zhang, A., Li, Z., Xu, Z.: A reconfigurable polarization converter using active metasurface and its application in horn antenna. IEEE Trans. Antennas Propag. **64**(12), 5281–5290 (2016)
94. Farzami, F., Khaledian, S., Smida, B., Erricolo, D.: Reconfigurable linear/circular polarization rectangular waveguide filtenna. IEEE Trans. Antennas Propag. **66**(1), 9–15 (2018)
95. Li, L., Li, Y., Wu, Z., Huo, F., Zhang, Y., Zhao, C.: Novel polarization-reconfigurable converter based on multilayer frequency-selective surfaces. Proc. IEEE **103**(7), 1057–1070 (2015)
96. Dhar, S., Patra, K., Ghatak, R., Gupta, B., Poddar, D.R.: Reconfigurable dielectric resonator antenna with multiple polarisation states. IET Microw. Antennas Propag. **12**(6), 895–902 (2018)
97. Chen, Z., Wong, H.: Liquid dielectric resonator antenna with circular polarization reconfigurality. IEEE Trans. Ant. Propag. **66**(1), 444–449 (2018)
98. Yang, W.W., Dong, X., Sun, W.J., Chen, J.X.: Polarization reconfigurable broadband dielectric resonator antenna with a lattice structure. IEEE Access **6**, 21212–21219 (2018)
99. Apperley, T., Okoniewski, M.: An air-gap-based frequency switching method for the dielectric resonator antenna. IEEE Antennas Wirel. Propag. Lett. **13**, 455–458 (2014)
100. Danesh, S., Rahim, S.K.A., Abedian, M., Khalily, M., Hamid, M.R.: Frequency-reconfigurable rectangular dielectric resonator antenna. IEEE Antennas Wirel. Propag. Lett. **12**, 1331–1334 (2013)
101. Yan, J.-B., Bernhard, J.T.: Implementation of a frequency-agile MIMO dielectric resonator antenna. IEEE Trans. Antennas Propag. **61**(7), 3434–3441 (2013)
102. Danesh, S., Rahim, S.K.A., Abedian, M., Hamid, M.R.: A compact frequency-reconfigurable dielectric resonator antenna for LTE/WWAN and WLAN applications. IEEE Antennas Wirel. Propag. Lett. **14**, 486–489 (2015)
103. Desjardins, J., McNamara, D.A., Thirakoune, S., Petosa, A.: Electronically frequency-reconfigurable rectangular dielectric resonator antennas. IEEE Trans. Antennas Propag. **60**(6), 2997–3002 (2012)

104. Huff, G.H., Rolando, D.L., Walters, P., McDonald, J.: A frequency reconfigurable dielectric resonator antenna using colloidal dispersions. IEEE Antennas Wirel. Propag. Lett. **9**, 288–290 (2010)
105. Petosa, A., Mongia, R.K., Cuhaci, M., Wight, J.S.: Magnetically tunable ferrite resonator antenna. Electron. Lett. **30**(13), 1021–1022 (1994)
106. Hao, C.X., Li, B., Leung, K.W., Sheng, X.Q.: Frequency-tunable differentially fed rectangular dielectric resonator antennas. IEEE Antennas Wirel. Propag. Lett. **10**, 884–887 (2011)
107. Baker-Jarvis, J., Kim, S.: The Interaction of radio frequency field with dielectric materials at macroscopic to mesoscopic scale. J. Res. Natl. Inst. Stand. Technol. **117**, (2012). https://doi.org/10.6028/jres.117.001
108. Moulson, A.J., Herbert, J.M.: Electroceramics: Materials, Properties, Applications. Wiley & Sons (2003)
109. Neelakantan, U.A., Kalathil, S.E., Ratheesh, R.: Structure and microwave dielectric properties of ultralow-temperature cofirable BaV_2O_6 ceramics. Eur. J. Inorg. Chem. 305–310, (2015)
110. Kumari, P., Tripathi, P., Sahu, B., Singh, S.P., Kumar, D.: Four-element composite triangular dielectric resonator antenna using Li_2O-1.94MgO-0.02Al_2O_3-P_2O_5 ceramic for wideband applications. J. Electron. Mater. **47**(9), 5218–5228 (2018)
111. Choi, Y.J., Park, J.H., Ko, W.J., Hwang, I.S., Park, J.G., Nahm, S.: Cofiring and shrinkage matching in low and middle permittivity dielectric compositions for a low temperature cofired ceramic system. J. Am. Ceram. Soc. **89**, 562–567 (2006)
112. Kagata, H., Kato, J.: Dielectric properties of Ca based complex perovskite at microwave frequencies. Jpn. J. Appl. Phys. **33**, 5463–5465 (1994)
113. Tripathi, P., Kumari, P., Mishra, V.K., Singh, R., Singh, S.P., Kumar, D.: Effect of PbO-B_2O_3-BaO-SiO_2 glass additive on dielectric properties of Ba0.5Sr0.5TiO_3 ceramics for radio frequency applications. J. Phys. Chem. Solids **127**, 60–67 (2019)
114. Huang, C.L., Lin, R.J., Wang, J.H.: Effect of B_2O_3 additives on sintering and microwave dielectric behaviours of CuO doped $ZnNb_2O_6$ ceramics. Jpn. J. Appl. Phys. **41**, 758–762 (2002)
115. Zheng, Q., Li, W., Shi, J.L., Guo, J.K.: Microwave dielectric properties of 5Li_2O–0.583Nb_2O_5–3.248TiO_2 ceramics with V_2O_5. J. Am. Ceram. Soc. **89**, 3305–3307 (2006)
116. Thomas, D., Pullanchiyodan, A., Sebastian, M.T.: Casting and characterization of $LiMgPO_4$ glass free LTCC tape for microwave applications. J. Eur. Ceram. Soc. **33**(1), 87–93 (2012)
117. Wang, Y.R., Wang, S.F., Wen, C.K.: Low fire $(Zr_{0.8}Sn_{0.2})TiO_4$ with glass additives. Mater. Sci. Eng. A **426**, 143–146 (2006)
118. Thomas, S., Sebastian, M.T.: Effect of B_2O_3-Bi_2O_3-SiO_2-ZnO glass on the sintering and microwave dielectric properties of 0.83$ZnAl_2O_4$–0.17TiO_2. Mater. Res. Bull. **43**, 843–851(2008)
119. Ogawa, H., Yokoi, A., Unemura, R., Kan, A.: Microwave dielectric properties of $Mg_3(VO_4)$2-x$Ba_3(VO_4)_2$ ceramics for LTCC with near zero temperature coefficient of resonant frequency. J. Eur. Ceram. Soc. **27**, 3099–3104 (2007)
120. Yokoi, A., Ogawa, H., Kan, A., Ohsato, H., Higashida, Y.: Microwave dielectric properties of $Mg_4Nb_2O_9$-3 wt% LiF ceramics prepared with $CaTiO_3$ additions. J. Eur. Ceram. Soc. **25**, 2871–2875 (2005)
121. Mc, N., Alford, N., Penn, S.J.: Sintered alumina with low dielectric loss. J. Appl. Phys. **80**, 5895–5898 (1996)
122. Anjana, P.S., Sebastian, M.T., Axelsson, A.K., Mc, N., Alford, N.: Microwave dielectric properties of CeO_2–0.5AO–0.5TiO_2 (A = Ca, Mg, Zn, Mn, Co, Ni, W) ceramics. J. Eur. Ceram. Soc. **27**, 3445–3452 (2007)
123. Kim, E.S., Kim, S.H.: Effects of structural characteristics on microwave dielectric properties of (1–x)$CaWO_4$-x$LaNbO_4$. J. Electroceram. **17**, 471–477 (2006)
124. Hsu, C.S., Huang, C.L., Tseng, J.F., Huang, C.Y.: Improved high Q microwave dielectric resonator using CuO doped $MgNb_2O_6$ ceramics. Mater. Res. Bull. **38**, 1091–1099 (2003)
125. Cho, W.W., Kakemoto, K., Ohsato, H.: Microwave dielectric properties and low temperature sintering of $MgTiO_3$-$SrTiO_3$ with B_2O_3 or CuO. Mater. Sci. Eng. B **121**, 48–53 (2005)

126. Huang, C.L., Weng, M.H.: Improved high Q values of $MgTiO_3$-$CaTiO_3$ microwave dielectric ceramics at low sintering temperatures. Mater. Res. Bull. **36**, 2741–2750 (2001)
127. Ichinose, N., Yamamoto, H.: Effect of additives on microwave dielectric properties in low temperature firing (Mg, Ca)TiO_3 based ceramics. Ferroelectrics **201**, 255–262 (1997)
128. Cha, H.J., Kang, D.H., Cho, Y.S.: Optimised microwave dielectric properties of Co- and Ca-substituted $Mg_{0.06}Zn_{0.04}TiO_3$. Mater. Res. Bull. **42**, 265–273 (2007)
129. Ogawa, H., Kan, A., Ishihara, S., Higashida, Y.: Crystal structure of corundum type $Mg_4(Nb_{2-x}Ta_x)O_9$ microwave dielectric ceramics with low dielectric loss. J. Eur. Ceram. Soc. **23**, 2485–2488 (2003)
130. Yoon, S.H., Choi, G.K., Kim, D.W., Cho, S.Y., Hong, K.S.: Mixture behaviour and microwave dielectric properties of (1–x)$CaWO_4$-xTiO_2. J. Eur. Ceram. Soc. **27**, 3087–3091 (2007)
131. Subodh, G., Sebastian, M.T.: Glass free microwave ceramics for LTCC applications. J. Am. Ceram. Soc. **90**, 2266–2268 (2007)
132. Kim, D.W., Kwon, D.K., Yoon, S.H., Hong, K.S.: Microwave dielectric properties of rare earth ortho- niobates with ferroelasticity. J. Am. Ceram. Soc. **89**, 3861–3864 (2006)
133. Huang, C.L., Yu, C.L.: Low temperature sintering and microwave dielectric properties of (1–x)$MgTiO_3$-$CaTiO_3$ ceramics bismuth addition. Jpn. J. Appl. Phys. **41**, 707–711 (2002)
134. Cho, W.W., Kakemoto, K., Ohsato, H.: High Q microwave dielectrics $SrTiO_3$ doped $MgTiO_3$ material with near zero temperature coefficient of resonant frequency. Jpn. J. Appl. Phys. **43**, 6221–6224 (2004)
135. Hwang, S.J., Kim, Y.J., Kim, H.S.: Dielectric properties of (1–x)$(Mg_{0.95}Co_{0.05})TiO_3$-x$CaTiO_3$ ceramic system at microwave frequency. Mater. Res. Bull. **37**, 2483–2490 (2002)
136. Huang, C.L., Pan, C.L., Shim, S.J.: Liquid phase sintering of $MgTiO_3$-$CaTiO_3$ microwave dielectric ceramics. Mater. Chem. Phys. **78**, 111–115 (2003)
137. Kan, A., Ogawa, H., Watanabe, M., Hatanaka, S., Ohsato, H.: Microwave dielectric properties of $Sm_2Ba(Cu_{1-x}Zn_x)O_5$ solid solutions. Jpn. J. Appl. Phys. **38**, 5629–5632 (1991)
138. Huang, C.L., Liu, S.S.: Characterisation of extremely low loss dielectrics (Mg0.95Zn0.05)TiO_3 at microwave frequency. Jpn. J. Appl. Phys. **46**, 283–285 (2007)
139. Kim, D.W., Youn, H.J., Hong, K.S., Kim, C.K.: Microwave dielectric properties of (1–x)Ba_5Nb_4O15-x$BaNb_2O_6$ mixtures. Jpn. J. Appl. Phys. **41**, 3812–3816 (2002)
140. Miyauchi, Y., Ohishi, Y., Miyake, S., Ohsato, H.: Improvement of the dielectric properties of rutile doped Al_2O_3 ceramics by annealing treatment. J. Eur. Ceram. Soc. **26**, 2093–2096 (2006)
141. Kono, M., Takagi, H., Tatekawa, T., Tamura, H.: High Q dielectric resonator material with low dielectric constant for millimeterwave application. J. Eur. Ceram. Soc. **26**, 2093–2096 (2006)
142. Zhang, Y.C., Yue, Z.X., Qi, X., Li, B., Gui, Z., Li, L.T.: Microwave dielectric properties of $Zn(Nb_{1-x}Ta_x)_2O_6$ ceramics. Mater. Lett. **58**, 1392–1395 (2004)
143. Huang, C.L., Wang, J.J., Chang, Y.P.: Dielectric properties of low loss (1–x)$(Mg_{0.95}Zn_{0.05})TiO_3$-x$SrTiO_3$ ceramic system at microwave frequency. J. Am. Ceram. Soc. **90**, 858–862 (2007)
144. Su, H., Wu, S.: Studies on the (Mg, Zn)TiO_3-$CaTiO_3$ microwave dielectric ceramics. Mater. Lett. **59**, 2337–2341 (2005)
145. Wang, J.J., Huang, C.L., Li, P.H.: Microwave dielectric properties of (1–x)$(Mg_{0.95}Zn_{0.05})TiO_3$-$Ca_{0.6}La_{0.8/3}TiO_3$ ceramic system. Jpn. J. Appl. Phys. 45(2006) 6352–6356.
146. Tseng, C.F., Huang, C.L.: Improved microwave dielectric properties of B_2O_3 doped $Nd(Co_{1/2}Ti_{1/2})O_3$ ceramics with near zero temperature coefficient of resonant frequency. Mater. Res. Bull. **42**, 9–16 (2006)
147. Bijumon, P.V., Sebastian, M.T., Mohanan, P.: Experimental investigations and three dimensional transmission line matrix simulation of $Ca_{5-x}A_xB_2TiO_{12}$ (A = Mg, Zn, Ni and Co; B = Nb and Ta) ceramic resonators. J. Appl. Phys. **98**, 124105 (2005)
148. Cho, J.Y., Yoon, K.H., Kim, E.S.: Effect of stress on microwave dielectric properties of $Mg_{0.93}Ca_{0.07}TiO_3$-$(Ca_{0.03}Li_{0.14}Sm_{0.42})TiO_3$ ceramics. Mater. Chem. Phys. **79**, 286–288 (2003)

149. Kawashima, S., Nishida, N., Ueda, I., Ouchi, H.: $Ba(Zn_{1/3}Ta_{2/3})O_3$ ceramics with low dielectric loss at microwave frequencies. J. Am. Ceram. Soc. **66**, 421–423 (1983)
150. Ohishi, Y., Miyamauchi, Y., Kakimoto, K., Oshato, H.: Proceeding AMF (2004)
151. Surendran, K.P., Santha, N., Mohanan, P., Sebastian, M.T.: Temperature stable low permittivity ceramic dielectrics in $(1-x)ZnAl_2O_4$-$xTiO_2$ system for microwave substrate applications. Eur. Phy. J. **B41**, 301–306 (2004)
152. Surendran, K.P., Mohanan, P., Manjusha, M.V., Philip, J., Sebastian, M.T.: A low loss dielectric substrate in $ZnAl_2O_4$-TiO_2 system for microelectronic applications. J. App. Phys. **98**, 044101–044109 (2005)
153. Tseng, C.F., Huang, C.L.: Microwave dielectric properties of $(1-x)Nd(Co_{1/2}Ti_{1/2})O_3$-$xSrTiO_3$ ceramics with near zero temperature coefficient of resonant frequency. Jpn. J. Appl. Phys. **46**, 1571–1573 (2007)
154. Pullar, R.C., Farrah, S., Alford, N.: $MgWO_4$, $ZnWO_4$, $NiWO_4$, and $CoWO_4$ microwave dielectric ceramics. J. Eur. Ceram. Soc. **27**, 1059–1063 (2007)
155. Kim, J.C., Kim, M.H., Nahm, S., Paik, J.H., Kim, J.H., Lee, H.J.: Microwave dielectric properties of $Re_3Ga_5O_{12}$ (Re = Nd, Sm, Eu, Dy, Yb) ceramics and effect of TiO_2 on the microwave dielectric properties of $Sm_3Ga_5O_{12}$ ceramics. J. Eur. Ceram. Soc. **27**, 2865–2870 (2007)
156. Xiao, Y., Chen, X.M., Liu, X.Q.: Microstructure and microwave dielectric characteristics of $CaRAlO_4$ (R = Nd, Sm, Y) ceramics with tetragonal K_2NiF_4 structure. J. Am. Ceram. Soc. **87**, 2143–2146 (2004)
157. Choi, J.W., Kucheiko, S., Yoon, S.J., Jai, J., Yoon, K.H.: Microwave dielectric properties of $(1-x)(Al_{1/2}Ta_{1/2})O_{2-x}(Mg_{1/3}Ta_{2/3})O_2$ Ceramics. J. Am. Ceram. Soc. **84**, 2570–2572 (2001)
158. Choi, J.W., Ha, J.Y., Yoon, S.J., Kim, H.J., Yoon, K.H.: Microwave dielectric characteristics of $0.75(Al_{1/2}Ta_{1/2})O_2$–$0.25(Ti_{1-x}Sn_x)O_2$ ceramics. J. Eur. Ceram. Soc. **23**, 2507–2510 (2003)
159. Choi, J.W., Yoon, S., Kim, H.J., Yoon, K.H.: Microwave dielectric characteristics of $(1-x)(Al_{1/2}Ta_{1/2})O_2$-$xTiO_2$ ceramics. Jpn. J. Appl. Phys. **41**, 3804–3807 (2002)
160. Surendran, K.P., Bijumon, P.V., Mohanan, P., Sebastian, M.T.: $(1-x)MgAl_2O_4$-$xTiO_2$ dielectrics for microwave and millimeter wave applications. Appl. Phys. A **81**, 823–826 (2005)
161. Huang, C.L., Chiang, K.H.: Structures and dielectric properties of a new dielectric material system $xMgTiO_3$-$(1-x)MgTa_2O_6$ at microwave frequency. J. All. Comp. **431**, 326–330 (2007)
162. Kim, J.C., Kim, M.H., Kim, J.B., Nahm, S., Paik, J.H., Kim, J.H.: Synthesis and microwave dielectric properties of $Re_3Ga_5O_{12}$ (Re = Nd, Sm, Eu, Dy, Yb, Y) ceramics. J. Am. Ceram. Soc. **90**, 641–664 (2007)
163. Takahashi, H., Ayusawa, K., Sakamoto, N.: Microwave dielectric properties of $Ba(Mg_{1/2}W_{1/2})O_3$-$BaTiO_3$ ceramics. Jpn. J. Appl. Phys. **36**, 5597–5599 (1997)
164. Takahashi, H., Ayusawa, K., Sakamoto, N.: Ba-Mg-W-Ti-O ceramics with temperature stable low microwave loss. Jpn. J. Appl. Phys. **37**, 908–911 (1998)
165. Rong, G., Newman, N., Shaw, B., Cronin, D.: Role of Ni and Zr doping on the electrical, optical, magnetic and structural properties of barium zinc tantalite ceramics. J. Mater. Res. **14**, 4011–4019 (1999)
166. Reaney, I.M., Wise, P.L., Qazi, I., Miller, C.A., Price, T.J., Cannell, D.S., Idles, D.M., Rosienky, M.J., Moussa, S.M., Bieringer, M., Noailles, L.D., Ibberson, R.M.: Ordering and quality factor in $0.95Ba(Zn_{1/3}Ta_{2/3})O_3$–$0.05Sr(Ga_{1/2}Ta_{1/2})O_3$ production resonators. J. Eur. Ceram. Soc. **23**, 3021–3034 (2003)
167. Nomura, S.: Ceramics for microwave dielectric resonator. Ferroelectrics **49**, 61–70 (1983)
168. Surendran, K.P., Mohanan, P., Sebastian, M.T.: Microwave dielectric properties of $Ba(Mg_{1/32\text{-}2x/3})W_{x/3}Ti_{x/3})O_3$ ceramics. Mater. Res. Bull. **41**, 784–790 (2006)
169. Oshato, H., Tsunooka, T., Ohishi, Y., Miyauchi, Y., Ando, M., Kakimoto, K.: Millimeter-wave dielectric ceramics of alumina and forsterite with high quality factor and low dielectric constant. J. Korean Ceramic Soc. **40**, 350–353 (2003)
170. Khalam, L.A., Sebastian, M.T.: Low loss dielectrics in the $Ca(B'_{1/2}Nb_{1/2})O_3$ [B' = lanthanides, Y] system. J. Am. Ceram. Soc. **90**, 1467–1474 (2007)

171. Surendran, K.P., Sebastian, M.T.: Low loss dielectrics in $Ba[(Mg_{1/3}Ta_{2/3})_{1-x}Ti_x]O_3$ and $Ba[(Mg_{1-x}Zn_x)_{1/3}Ta_{2/3}]O_3$ systems. J. Mater. Res. **20**, 2919–2928 (2005)
172. Lee, H.J., Kim, I.T., Hong, K.S.: Dielectric properties of AB_2O_6 compounds at microwave frequencies (A = Ca, Mg, Mn Co, Ni, Zn and B = Nb, Ta). Jpn. J. Appl. Phys. **36**, L1318–L1320 (1997)
173. Surendran, K.P., Sebastian, M.T., Mohanan, P., Moreira, R.L., Dias, A.: Effect of stoichiometry on the microwave dielectric properties of $Ba(Mg_{1/3}Ta_{2/3})O_3$ dielectric ceramics. Chem. Mater. **17**, 142–151 (2005)
174. Matsumoto, H., Tamura, H., Wakino, K.: $Ba(MgTa)O_3$-$BaSnO_3$ high Q dielectric resonator. Jpn. J. Appl. Phys. **30**, 2347–2349 (1991)
175. Kagata, H., Saito, R., Katsumura, H.: Al_2O_3-MgO-ReO_x(Re = rare earth) based LTCC and its applications to multilayer non-shrinkage substrate for microwave devices. J. Electroceram. **13**, 277–280 (2004)
176. Sharma, A., Gangwar, R.K.: Triple band dual polarized hybrid cylindrical dielectric resonator antenna with hybrid modes excitation. Prog. Electromagn. Res. C **67**, 97–105 (2016)
177. Das, G., Sharma, A., Gangwar, R.K.: Dual feed MIMO CDRA with high isolation. Microw. Opt. Technol. Lett. **59**, 1686–1692 (2017)
178. Gupta, A., Gangwar, R.K.: Design, fabrication and measurement of dual segment rectangular dielectric resonator antenna array excited with conformal strip for S-band application. Electromagnetics **36**(4), 236–248 (2016)
179. Gangwar, R.K., Sharma, A., Gupta, M., Chaudhary, S.: Hybrid cylindrical dielectric resonator antenna with $HE_{11\delta}$ and $HE_{12\delta}$ mode excitation for wireless applications. Inter. J. RF Microw. Comput.-Aided Eng. **26**, 812–818 (2016)
180. Sharma, A., Gangwar, R.K.: Asymmetrical annular shape microstrip line fed stacked cylindrical dielectric resonator antenna for UWB. In: Proceedings of Asia-Pacific Microwave Components (2016)
181. Sharma, A., Das, G., Gangwar, R.K.: Dual band circularly polarized modified circular aperture loaded cylindrical dielectric resonator antenna for wireless application. Microw. Opt. Technol. Lett. **59**, 1562–1570 (2017)
182. Sharma, A., Gangwar, R.K.: Tri-band composite cylindrical dielectric resonator antenna with hybrid mode excitation and cross-polarization suppression. Int. J. RF Microw. Comput. Aided Eng. **27**, e21130 (2017). https://doi.org/10.1002/mmce.21130
183. Gupta, A., Gangwar, R.K., Singh, S.P.: Three element dual segment triangular dielectric resonator antenna for X-Band applications. Prog. Electromagn. Res. (PIER) C **34**, 139–150 (2013)
184. Gupta, A., Gangwar, R.K.: Theoretical and experimental investigations on rectangular dielectric resonator antenna array for radar and satellite application. Electromagnetics **37**(1), 27–39 (2017)
185. Rajan, P., Gangwar, R.K.: Probe feed half split cylindrical dielectric resonator antenna for wideband application. Int. J. Electron. Commun. (AEU) **69**, 1709–1714 (2015)
186. Sahu, N.K., Sharma, A., Gangwar, R.K.: Modified annular ring patch fed cylindrical dielectric resonator antenna for WLAN/WiMAX applications. Microw. Opt. Technol. Lett. **59**(1), 120–125 (2017)
187. Sharma, A., Gangwar, R.K.: Triple band two segment cylindrical dielectric resonator antenna with a novel microstrip feed for WLAN/WiMAX applications. Microw. Opt. Technol. Lett. **57**(11), 2649–2655 (2015)
188. Sharma, A., Das, G., Gangwar, R.K.: Dual polarized triple band hybrid MIMO cylindrical dielectric resonator antenna for LTE2500/WLAN/WiMAX applications. Int. J. RF Microw. Comput. Aided Eng. **26**(9), 763–772 (2016)
189. Sharma, A., Das, G., Ranjan, P., Sahu, N.K., Gangwar, R.K.: Novel feeding mechanism to stimulate triple radiating modes in cylindrical dielectric resonator antenna. IEEE Access **4**, 9987–9992 (2016)
190. Kumar, R., Chaudhary, R.K.: A wideband circularly polarized cubic dielectric resonator antenna excited with modified microstrip feed. IEEE Antennas Wirel. Propag. Lett. **15**, 1285–1288 (2016)

191. Kumar, R., Gangwar, R.K.: Circularly polarized cylindrical dielectric resonator antenna excited by dual conformal strips along with modified Wilkinson power divider for high gain application. Microw. Opt. Technol. Lett. **59**(4), 908–913 (2016)
192. Abdulmajid, A., Khalil, Y., Khamas, S.: Higher order mode circularly polarized two layer rectangular dielectric resonator antenna. IEEE Antennas Wirel. Propag. Lett. **17**(06), (2018)
193. Yang, M., Pan, Y.: A single fed wideband circularly polarized dielectric resonator antenna. IEEE Antennas Wirel. Propag. Lett. **17**(8), (2018)
194. Chowdhary, R., Chaudhary, R.K.: An approach to generate circular polarization in a modified cylindrical shaped dielectric resonator antenna using PMC boundary approximation. IEEE Antennas Wirel. Propag. Lett. **17**(9), (2018)
195. Ahmad, Z., and Hesselbarth J.: On-chip dual polarized dielectric resonator antenna for millimeter wave application. IEEE Antennas Wirel. Propag. Lett. **17**(10), (2018)
196. Sahu, B., Singh, R., Singh, S.P., Tripathi, P.: Compact Dual segment cylindrical dielectric resonator antenna for X-Band applications. In: 2013 International Conference on Microwave and Photonics (ICMAP), pp. 13–15 (2013). https://doi.org/10.1109/ICMAP.2013.6733497
197. Sahu, B., Tripathi, P., Singh, S.P.: Investigation on Cylindrical dielectric resonator antenna with metamaterial superstrate. Wirel. Pers. Commun. **84**, 1151–1163 (2015)
198. Rana, B., Parui, S.K.: Nonresonant microstrip patch fed dielectric resonator antenna array. IEEE Antennas Wirel. Propag. Lett. **14**, 747–750 (2015)
199. Chen, Y.C., Chang, K.C., Tsai, D.Y.: A hybrid dielectric resonator antenna based upon novel complex perovskite microwave ceramic. Ceram. Int. **39**, 8043–8048 (2013)
200. Malekabadi, S.A., Neshati, M.H., Rashed-Mohassel, J.: Circular polarized dielectric resonator antenna using a single probe feed. Prog. Electromagn. Res. C **3**, 81–94 (2008)
201. Chen, Z., Wong, H.: Wideband glass and liquid cylindrical dielectric resonator antenna for pattern reconfigurable design. IEEE Trans. Antennas Propag. **65**(5), 2157–2164 (2017)
202. Kumari, P., Tripathi, P., Sahu, B., Singh, S.P., Parkash, O., Kumar, D.: Synthesis and characterisation of novel low temperature ceramic and its implementation as substrate in dual segment CDRA. J. Phys. Chem. Solids **113**, 177–185 (2018)
203. Tripathi, P., Kumari, P., Parkash, O., Kumar, D., Sahu, B., Singh, S.P.: Design of rectangular dielectric resonator antenna using glass added barium strontium titanate. In: 2015 International Conference on Microwave and Photonics (ICMAP) (2015). https://doi.org/10.1109/icmap.2015.7408748
204. Tripathi, P., Sahu, B., Singh, S.P., Parkash, O., Kumar, D.: Preparation and characterization of liquid phase ($55B_2O_3$–$45Bi_2O_3$) sintered cobalt doped magnesium titanate for wideband stacked rectangular dielectric resonator antenna (RDRA). Ceram. Int. **41**(2), 2908–2916 (2015)
205. Gangwar, R.K., Singh, S.P., Choudhary, M., Singh, N.K., Kumar, D., Lakshmi Narayana Rao, G., James Raju, K.C.: Microwave dielectric properties of $(Zn_{1-x}Mg_x)TiO_3$ (ZMT) ceramics for dielectric resonator antenna application. J. Alloy. Compd. **509**(42), 10195–10202 (2011)
206. Gupta, A., Gangwar, R.K.: Analysis of conformal strip feed triangular dielectric resonator antenna using FDTD method. In: 2015 IEEE International Conference on Computational Electromagnetics, Hong Kong, pp. 92–94 (2015). https://doi.org/10.1109/COMPEM.2015.7052567
207. Gangwar, R.K., Ranjan, P., Aigal, A.: Four element triangular dielectric resonator antenna for wireless application. Int. J. Microw. Wirel. Technol. **9**(1), 113–119 (2017). https://doi.org/10.1017/S1759078715000860
208. Zhang, Z.Y., Wu, K.L.: A wideband dual polarized dielectric magnetoelectric dipole antenna. IEEE Trans. Antennas Propag. **66**(10), (2018)
209. Sun, Y.X., Leung, K.W.: Circularly polarized substrate integrated cylindrical dielectric resonator antenna array for 60 GHz applications. IEEE Antennas Wirel. Propag. Lett. **17**(8), (2018)
210. Tang, H., Tong, C., Chen, J.X.: Differential dual polarization filtering dielectric resonator antenna. IEEE Trans. Antennas Propag. **66**(08), (2018)

211. Kingsley, S.P., O'Keefe, S.G.: Beam steering and monopulse processing of probe fed dielectric resonator antenna. Proc. Rada Sonar Navig. **146**(3), 121–125 (1999)
212. Massie, G., Caillet, M., Clenet, M., Antar, Y.M.M.: A new wideband circularly polarized hybrid dielectric resonator antenna. IEEE Antennas Wirel. Propag. Lett. **9**, 347–350 (2010)
213. Svedin, J., Huss, L.G., Karlen, D., et al.: A micro-machined 94 GHz dielectric resonator antenna for focal plane array applications. In: IEEE International Microwave Symposium, Honolulu, Hawaii, pp. 1375–1378 (2007)
214. Perron, A., Denidni, T.A., Sebak, A.R.: Circularly polarized microstrip/elliptical dielectric ring resonator antenna for millimeter-wave applications. IEEE Antennas Wirel. Propag. Lett. **9**, 783–786 (2010)
215. Guha, D., Gupta, B., Kumar, C., Antar, Y.M.M.: Segmented hemispherical DRA: New geometry characterized and investigated in multi-element composite forms for wideband antenna applications. IEEE Trans. Antennas Propag. AP **60**, 1605–1610 (2012)
216. Khalily, M., Rahim, M.K.A., Kishk, A.A.: Planar wideband circularly polarized antenna design with rectangular ring dielectric resonator and parasitic printed loops. IEEE Antennas Wirel. Propag. Lett. **11**, 905–908 (2012)
217. Keller, M.G., Cuhaci, M., Shaker, J., Petosa, A., Ittipiboon, A., Antar, Y.M.M.: A Ka-band dielectric resonator antenna reflect array. In: European MicrowaveConference-2000, Paris, France, pp. 272–275 (2000)
218. Pang, K.K., Lo, H.Y., Leung, K.W., Luk, K.M., Yung, E.K.N.: Circularly polarized dielectric resonator antenna sub-arrays. Microw. Opt. Technol. Lett. Opt. Technol. Lett. **27**(6), 377–379 (2000)

Electroceramics-Based Materials for Sensor Technology

Durvesh Gautam, Ashwani Kumar, and Yogendra K. Gautam

Abstract In recent decades, significant strides have been made in the domain of product innovation and materials technology, especially in the field of electronic devices and sensors. There is immense potential for future materials in electronic-based sensing devices, given their exceptional electrical, magnetic, and optical properties that support the adoption of emerging technologies. In modern electronics, electroceramics is crucial in facilitating new technological advancements. This chapter briefly introduces the importance and role of electroceramic materials in sensor technology. The chapter widely explore the application of various electroceramic materials such as piezoelectric, ferroelectric, semiconductor, solid, Ionic conductors, solid electrolytes thermoelectric, etc., in diverse sensor technologies. The chapter describes the various synthesis methods for electroceramic-based sensing materials. Future perspectives of electrochemical materials for enhancing sensor technology is also highlighted in the chapter.

Keywords Sensors · Electroceramics · Sensor applications · Piezoelectric ferroelectric · Environmental considerations

1 Introduction

Ceramic materials boast an unparalleled range of electrical, magnetic, and optical qualities, making them indispensable in various electronic applications. Notably, certain ceramics, identified as rapid ion conductors, have gained prominence as excellent ionic conductors, drawing significant attention over the past two decades. Their

D. Gautam · Y. K. Gautam (✉)
Smart Materials and Sensor Laboratory, Department of Physics, Ch. Charan Singh University, Meerut, Uttar Pradesh 250004, India
e-mail: ykg.iitr@gmail.com

A. Kumar (✉)
Department of Physics, Regional Institute of Education (NCERT), Bhubaneswar, Odisha 751022, India
e-mail: 01ashraj@gmail.com

U. Kumar (ed.), *Defects Engineering in Electroceramics for Energy Applications*, Engineering Materials, https://doi.org/10.1007/978-981-97-9018-0_16

pivotal roles in sensor technologies, fuel cells, and batteries have spurred extensive research efforts. Inorganic, non-metallic, water-soluble compounds with ionic contributions to their chemical bonds make up ceramics [1]. As electronics underwent rapid progress in the latter part of the 19th century, Ceramic materials' enormous potential has been realized and used to an amazing array of electrical and microelectronic devices and applications [2].

Ni-Zn and Mn-Zn ferrites began to be used more often as transformer and coil core materials in radio receivers and telephones in the mid-1920s. Due to their special qualities, which include high dielectric permittivity, high piezoelectric constants and electromechanical coupling, high pyroelectric coefficient, and, in some cases, high optical transparency and electro-optic coefficients, ferroelectric ceramics were first used technologically after World War II [3].

Since the 1970s, monolithic multilayer capacitors have been made using high-permittivity $BaTiO_3$-based dielectrics ceramics, which were initially used for tiny plate and tube capacitors (MLCs) [4]. Furthermore, microwave ceramics are being utilized more often in wireless communication systems as filters, resonators, and mixers [5].

The widest range of individual applications can be found in piezoelectric ceramics, which are typically based on $Pb(Ti, Zr)O_3$ (abbreviated PZT) compositions. These applications include, but are not limited to, sonars, ultrasonic cleaners, buzzers, accelerometers, hydrophones, surface acoustic wave filters, delay lines, piezo transformers, ultra-precision positioners for scanning probe analysis (XPM) systems, inkjet print heads, touch controls, and ultrasonic motors, such as those used in cameras [6].

Ionic conducting ceramics, including highly acceptor-doped cubic ZrO_2, are utilized in high-temperature solid oxide fuel cells (SOFCs) and electrochemical oxygen sensors in automobiles. For the cathode of SOFCs, mixed ionic electrically conducting ceramics are used [7, 8].

A variety of ceramic materials having electrical properties are referred to as electroceramics, and they are essential in many different applications. Applications include electro-optic devices, sensors, energy storage, antennas, electronic circuits, SiC power modules, and self-cleaning membranes [9]. Notably, ferroelectric, piezoelectric, electrostrictive, and pyroelectric ceramics, particularly perovskite ceramics like lead zircon titanate (PZT), have garnered significant attention. They have had a profound impact on the electronic revolution in the past [10]. Transition temperatures of over 100 K were reached by high-temperature superconducting ceramics (HTSC) discovered in the mid-1980s [11].

In the late 1980s, the electroceramics components and materials that have been mentioned thus far were generally made using standard methods like screen-printing, tape casting, or powder compacting, which are then followed by sintering procedures. Thin film deposition methods like chemical vapor deposition (CVD), chemical solution deposition (CSD), sputtering, and pulsed laser deposition (PLD) began to gradually replace these methods [2]. The newly found HTSC, if formed as thin films, was first reported to have much higher critical current densities. In addition, the concept of ferroelectric-semiconductor memory, in the late 1960s, was revived by the use of the highly developed Si semiconductor technology. These pioneering lines led to a broad research and development effort worldwide to combine the large variety of functions offered by electroceramics thin films with the standard integrated circuit chip [12]. "Sensor-actuator" technology is another field in which electroceramics is essential. It finds use in a variety of equipment, from washing machines to aeroplanes. Hazardous gases and compounds are identified at any concentration because of electroceramics-based sensor technology [13].

1.1 Sensor Technology Overview

Sensor science and engineering are relevant to virtually all aspects of life including safety, security, surveillance, monitoring, and awareness in general. Sensors are central to industrial applications being used for process control, monitoring, and safety. Sensors are also central to medicine being used for diagnostics, monitoring, critical care, and public health. Sensors can improve the world through diagnostics in medical applications; improve the performance of energy sources like fuel cells and batteries and solar power; improve health and safety and security for people; sensors for exploring space and improve environmental monitoring [14, 15]. Fundamentally, sensors are devices that detect the feature quantity of a measurement object and convert this quantity into a readable signal, which is displayed on an instrument. Simply sensing technology is a technology that uses sensors to acquire information by detecting the physical, chemical, or biological property quantities and converting them into readable signals, as shown in Fig. 1 [16].

There are a wide variety of sensors available for practically any industrial need. For demanding, mission-critical industrial applications, sensors can help improve processes, and offer unmatched asset protection [15]. In this chapter, we embark on an exploration of the vast and transformative domain of sensor technology. Sensors, integral to diverse industries, play a pivotal role in converting physical and chemical signals into valuable data. From the intricacies of mechanical sensors, including accelerometers and strain gauges, to the nuances of optical sensors utilizing light properties, we delve into the multifaceted world of sensor applications. The

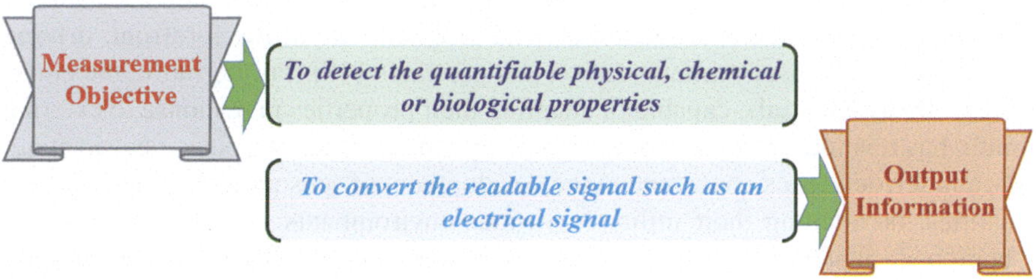

Fig. 1 Schematic diagram of sensor fundamental

chapter provides insights into the realm of chemical sensors, instrumental in environmental monitoring and industrial processes, as well as biomedical sensors that contribute to revolutionary advancements in healthcare. The basic understanding of electroceramics is not kept up with the applied technology. Because of the growing concern over environmental pollution and the safety of industrial processes, ceramic sensors are frequently used in industrial settings to monitor concentrations of various substances, such as ozone, ammonia, carbon monoxide, and solvent vapours [7]. These applications are expected to increase in the coming years. The range of sensor types and applications is expanding at a remarkable rate. The term 'electronic noses' has been introduced to describe the technology.

Revealing the fundamental principles that govern the functionality of sensors, we analyze their diverse operational mechanisms, including alterations in resistance, variations in capacitance, and the influence of piezoelectric effects. The discussion extends to the incorporation of emerging technologies such as the Internet of Things (IoT) and machine learning, enhancing sensor capabilities for the real-time analysis of data and predictive maintenance. Despite their crucial roles, sensors face challenges such as the need for miniaturization, power consumption issues, and security concerns. These challenges drive continuous research and development efforts in the field.

As we navigate through this chapter, our aim to provide readers with a comprehensive understanding of sensor technology a domain that not only shapes the present landscape of innovation but also promises to steer the future of interconnected systems and smart devices.

1.2 Importance of Advanced Materials in Sensor Technology

Within the dynamic realm of sensor technology, the importance of innovative materials serves as a cornerstone for driving innovation and efficiency. This segment explores the pivotal role that cutting-edge materials play in augmenting sensor performance, durability, and adaptability across diverse applications. Advanced materials, such as nanomaterials and engineered polymers, play a crucial role in pushing the boundaries of sensor capabilities. Nanomaterials, characterized by unique properties at the nanoscale, empower sensors to detect signals with unparalleled sensitivity [17]. Quantum dots, carbon nanotubes, and graphene are at the forefront, driving advancements in sensor miniaturization and precision [18]. Furthermore, the emergence of smart materials, capable of adapting their properties in response to external stimuli, has revolutionized sensor design. Shape memory alloys, piezoelectric materials, and ferroelectrics contribute to the development of sensors with adaptive functionalities, broadening their utility in various environments [19]. The integration of composite materials reinforces sensor robustness and resistance to challenging conditions. Sensors equipped with composite structures can endure extreme temperatures, corrosive environments, and mechanical stresses, extending their operational lifespan and reliability [20]. The significance of biocompatible materials in biomedical sensors cannot be overstated. Materials seamlessly interfacing with biological systems facilitate the creation of wearable and implantable sensors, transforming healthcare by enabling continuous monitoring and diagnostics [21]. Furthermore, the chapter delves into the role of advanced ceramics and semiconductors in sensor technology. These materials serve as the foundation for electronic and optical sensors, providing a platform for precise signal detection and transduction.

The integration of materials science and sensor innovation opens new frontiers, fostering a realm of possibilities for future technological advancements.

1.3 Role of Electroceramics in Sensor Applications

Electroceramics, categorized as advanced materials distinguished by their unique electrical properties, play a pivotal role in shaping the landscape of sensor applications [22]. This section elucidates the varied and essential functions that electroceramics fulfill in the domain of sensors, underscoring their contributions to sensitivity, stability, and responsiveness. One of the hallmark features of electroceramics is their outstanding dielectric properties. Ferroelectric materials, exemplified by lead zirconate titanate (PZT), exhibit high dielectric constants, positioning them as ideal candidates for capacitive sensors. These sensors utilize changes in capacitance to detect variations in physical parameters, spanning from pressure to humidity [23]. Another distinctive property, piezoelectricity, found in specific electroceramics, is harnessed for sensing applications, particularly in the field of acoustic and vibration sensors. Materials like piezoelectric ceramics generate electrical charges in response to mechanical stress, facilitating precise detection of sound waves and vibrations [24]. Additionally, the ferroelectric nature of electroceramics contributes to the development of pyroelectric sensors. These sensors leverage temperature changes to generate electrical signals, finding applications in infrared detectors and motion sensors [25]. Semiconducting electroceramics, including metal oxide ceramics, form the foundation of gas sensors. The interaction between target gases and the ceramic surface induces changes in conductivity, allowing for the detection of gases like carbon monoxide or nitrogen oxides in automotive exhaust systems [26].

The role of electroceramics extends into the realms of solid oxide fuel cells and electrochemical sensors [27]. Ion-conductive ceramics facilitate the movement of ions, crucial for the efficient functioning of fuel cells, while electrochemical sensors leverage the From capacitive to piezoelectric applications, these materials form the backbone of sensor technologies, driving advancements in diverse fields and contributing to the evolution of smart and responsive sensing systems.

2 Piezoelectric Materials for Sensing Applications

Piezoelectric materials have emerged as crucial components in the realm of sensing applications, demonstrating exceptional capabilities in converting mechanical stimuli into electrical signals. For piezoelectric materials in sensing applications, the direct piezoelectric effect is utilized, where there is a conversion of mechanical energy to electrical energy. The application of mechanical strain to a piezoelectric sensor induces polarization, leading to a variation in electric potential measurable across the sensor. Industries such as automotive, medical devices, Internet of Things (IoT), and energy harvesters extensively utilize piezoelectric sensors, leveraging the direct piezoelectric effect [30].

This following section explores the importance of piezoelectric materials, elucidating their fundamental principles and showcasing their diverse applications in sensing technologies.

2.1 Principles of Piezoelectricity

Sensing using piezoelectric, is the distinctive ability of specific materials to generate an electric charge when subjected to mechanical stress, as shown in Fig. 2. This phenomenon, recognized as piezoelectricity, is employed by materials such as quartz, lead zirconate titanate (PZT), and diverse ceramics. Upon experiencing mechanical deformation, these materials generate electrical charges in proportion to the applied force [31].

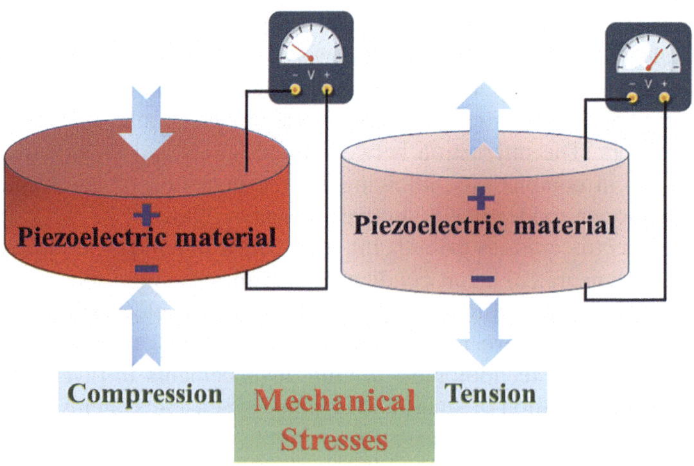

Fig. 2 Schematic of the piezoelectricity principle

2.2 Diverse Sensing Applications

Piezoelectric materials play a role in a variety of sensing technologies, spanning from accelerometers and pressure sensors to ultrasonic transducers. In accelerometers, piezoelectric crystals produce electrical signals in response to vibrations, enabling accurate motion detection. Similarly, pressure sensors employ piezoelectric elements to convert mechanical pressure into electrical signals, facilitating applications in industrial monitoring and medical devices. Additionally, touchscreens and force-sensitive devices often integrate piezoelectric materials. When pressure is applied, these materials generate electrical charges, allowing for precise and responsive touch and force sensing in electronic devices [32]. Innovatively, piezoelectric materials are also utilized for energy harvesting. The mechanical vibrations or movements in the environment can be converted into electrical energy, offering a sustainable power source for small electronic devices and sensors, especially in remote or inaccessible locations [33].

2.3 Surface Acoustic Wave (SAW) Sensors

A notable advancement in piezoelectric sensing is the utilization of Surface Acoustic Wave (SAW) devices. These sensors employ piezoelectric substrates to transmit acoustic waves across their surface. Any perturbation on the surface alters the wave characteristics, enabling the detection of changes in mass, temperature, or other environmental factors. SAW sensors have found applications in chemical sensing, environmental monitoring, and even healthcare. Surface acoustic waves (SAWs) are the guided waves that propagate along the top surface of a material with wave vectors orthogonal to the normal direction to the surface. Derived from these waves, Surface Acoustic Wave (SAW) sensors are envisioned by utilizing piezoelectric crystals to generate guided electrodynamics waves through an electromechanical coupling. This coupling, in both active and passive modes, is accomplished by incorporating inter-digitated electrode transducers (IDT) with the piezoelectric crystals. The inventive meta-designs of the periodic IDTs determine the functionality and application of SAW sensors [34].

The recent research is delving into novel piezoelectric materials, including piezoelectric polymers and composites, to push the limits of sensing capabilities. The incorporation of piezoelectric nanomaterials and the creation of flexible piezoelectric devices are expanding the horizons of sensing applications. Enhanced material composites and integration with the Internet of Things (IoT) are broadening the scope of piezoelectric sensors. Additionally, advancements in signal processing techniques offer the potential to improve the accuracy and reliability of sensor data. In summary, these innovations hold significant promise for developing more sophisticated, adaptable, and energy-efficient sensing solutions across various industries.

3 Ferroelectric Materials for Sensing Applications

Ferroelectric materials, known for their spontaneous electric polarization that can be switched by an external electric field, have become integral components in various sensing applications [35]. Ferroelectric materials contribute to non-volatile memory devices like ferroelectric random-access memory (FeRAM) and are integrated into micro-electro-mechanical systems (MEMS) for applications such as pressure sensors and microphones [36]. They are tailored for gas sensors, enabling the detection of specific gases in environmental monitoring and industrial safety [37]. In biological sensing, ferroelectric nanoparticles are utilized for their high surface area and sensitivity to environmental changes [38]. These materials are also employed in tactile sensors, haptic feedback systems, and energy harvesting devices, converting mechanical vibrations into electrical energy [33]. Additionally, ferroelectric sensors play a role in structural health monitoring [32], detecting and monitoring changes in the integrity of materials and structures in applications like civil engineering and aerospace [39]. Overall, ferroelectric materials showcase versatility and contribute significantly to advancing various sensing technologies. This section explores the significance of ferroelectric materials, their underlying principles, and their diverse applications in sensing technologies.

3.1 Principles of Ferroelectricity

Ferroelectricity, a characteristic observed in specific crystals and ceramics, is characterized by the presence of a permanent electric dipole moment that can be reversed by an external electric field [40]. The reversible polarization inherent in ferroelectric materials, exemplified by substances like lead zirconate titanate (PZT) and barium titanate ($BaTiO_3$), serves as the foundation for their integration into sensing devices. PZT, a ceramic material renowned for its exceptional ferroelectric properties, finds suitability in applications such as sensors, capacitors, and memory devices. Barium titanate ($BaTiO_3$) is a well-known ferroelectric material appreciated for its versatile properties and applications, particularly in electronics, sensors, and energy storage [41, 42].

3.2 Pyroelectric Sensors

An noteworthy application of ferroelectric materials is evident in pyroelectric sensors, where alterations in temperature lead to electric polarization [43]. Pyroelectric materials, such as lithium tantalate ($LiTaO_3$) and triglycine sulfate (TGS), are utilized in infrared detectors and motion sensors. These sensors are responsive to changes in heat, proving valuable in applications like security systems and occupancy detection [44].

3.3 Ferroelectric Capacitors in Memory Devices

Ferroelectric materials see widespread use in non-volatile memory devices, particularly in the case of ferroelectric random-access memory (FeRAM). The capacity of ferroelectric capacitors to preserve polarization without the need for a continuous power supply facilitates the creation of memory cells with rapid read and write speeds. FeRAM is applied across various electronic devices, contributing to improved data storage capabilities [45].

3.4 Strain Sensors and Energy Harvesting

Ferroelectric materials also exhibit piezoelectric properties, enabling them to operate as strain sensors. The reaction of ferroelectric materials to mechanical deformation makes them suitable for applications in monitoring structural health, identifying vibrations, and measuring forces. Additionally, research is ongoing in utilizing ferroelectric materials for energy harvesting, and converting mechanical vibrations into electrical energy for powering small electronic devices [46, 47].

As progress in materials science persists, researchers are actively exploring novel ferroelectric compounds and nanostructured materials. The merging of ferroelectric materials with emerging technologies, including the Internet of Things (IoT) and flexible electronics, presents the potential to develop more efficient and versatile sensing devices. With their diverse properties, ferroelectric materials continue to lead innovation in the realm of sensing applications.

4 Ionic Conductors and Solid Electrolytes for Sensor Applications

Ionic conductors, frequently embodied as solid electrolytes, play a vital role in diverse sensor applications, serving as a fundamental component for advanced sensing devices. Solid electrolytes have garnered widespread usage in scientific and technological realms. They prove invaluable for studying thermodynamic and kinetic phenomena and for constructing various devices, including fuel cells, batteries, sensors, and chemotronic components [48].

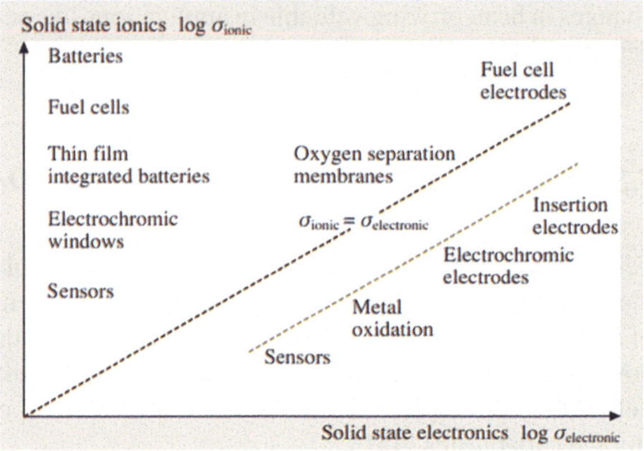

Fig. 3 Typical applications of ionic and electronic conductors as a function of the magnitude of electrical conductivity

4.1 Ionic Conductivity Fundamentals

Ionic conductors refer to materials that enable the movement of ions within their structure, providing a pathway for the flow of electric current [49]. In the realm of sensors, solid electrolytes featuring high ionic conductivity hold particular importance. Figure 3, presents the application of ionic and electronic conductors and its relevance in the context of sensor technology.

4.2 Solid Electrolyte Sensors

Solid electrolytes find extensive application in sensor technology, especially in electrochemical sensors and devices [50]. These sensors rely on the migration of ions within the solid electrolyte when exposed to specific analytes. A common approach is to use the equilibrium potential generated across a solid electrolyte given by the Nernst equation as the sensing signal. However, in some cases, stable electrode materials are not available to establish equilibrium potentials, so non-equilibrium approaches are necessary [51, 52]. For instance, solid-state oxygen sensors often employ yttria-stabilized zirconia (YSZ) as a solid electrolyte, allowing for the measurement of oxygen concentration in gases [52]. A well-known potentiometric oxygen sensor with a solid oxide electrolyte (YSZ) is the lambda sensor displayed in Fig. 4, usually employed for the detection of oxygen in exhaust gases [53].

4.3 High-Temperature Sensors

Ionic conductors are crucial in the development of high-temperature sensors, where traditional electronic conductors may not be suitable. Solid electrolytes, such as

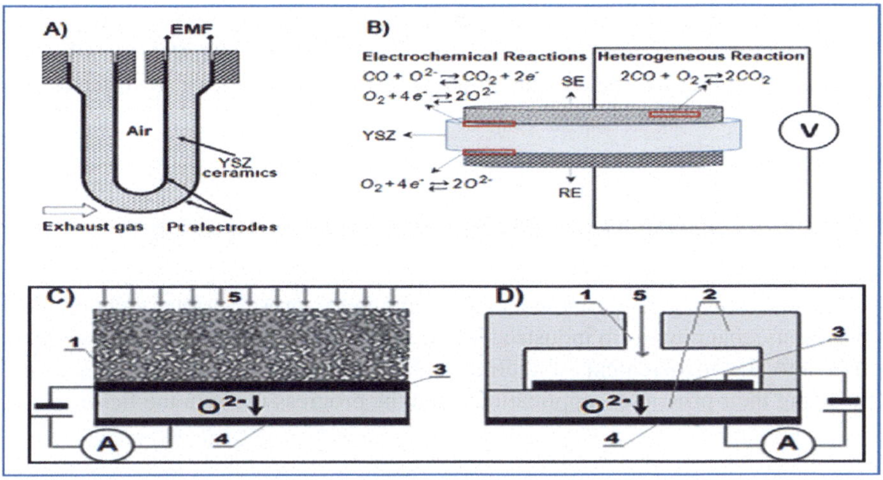

Fig. 4 An example of a potentiometric sensor used to detect oxygen in exhaust gases

doped ceria or perovskite oxides, enable the operation of sensors at elevated temperatures [54]. These sensors find applications in industries such as metallurgy and combustion processes.

4.4 *Li-Ion Batteries and Beyond*

The significance of ionic conductors goes beyond sensing applications and extends to energy storage devices, particularly lithium-ion batteries. Investigating and improving the ionic conductivity of solid electrolytes stands as a critical research focus for the advancement of battery technologies [55].

4.5 *Innovations and Emerging Trends*

Ongoing research endeavors are directed towards identifying novel materials characterized by improved ionic conductivity, mechanical stability, and chemical robustness. Developments in solid electrolytes, such as the investigation into garnet-type structures and polymer electrolytes, underscore the dynamic nature of this area. These breakthroughs set the stage for creating sensors that are more efficient, reliable, and compact, catering to a wide range of applications.

The trajectory of sensor technology is closely intertwined with continuous progress in the field of ionic conductors and solid electrolytes. Foreseen advancements encompass the incorporation of these materials into wearable sensors, implantable medical devices, and environmental monitoring systems. In the journey towards a more interconnected and sensor-rich world, the significance of ionic conductors continues to be pivotal in shaping the framework of next-generation sensing technologies.

5 Thermoelectric Materials for Temperature Sensing

Temperature sensing is a fundamental requirement across various technological applications, spanning from industrial processes to electronic devices. This segment delves into the involvement of thermoelectric materials in temperature sensing, elucidating their principles, applications, and the progress made in the field.

5.1 Thermoelectric Phenomenon

The thermoelectric effect encompasses the transformation of a temperature gradient into an electric voltage and vice versa. This phenomenon is based on the Seebeck effect, where the application of a temperature gradient results in the generation of voltage across a material [56]. Understanding this principle is crucial for the development of thermoelectric temperature sensors.

5.2 Thermocouples: Traditional Temperature Sensing

Historically, thermocouples have been the workhorses of temperature sensing. Comprising two different metal wires, they generate a voltage proportional to the temperature difference between their ends. While widely used, thermocouples have limitations in terms of accuracy and sensitivity [57].

5.3 Advanced Thermoelectric Materials

Recent progress in the field of materials science has resulted in the identification and creation of sophisticated thermoelectric materials. These materials demonstrate increased thermoelectric efficiency, enabling more precise and responsive temperature measurements. The ongoing investigation into novel compounds and nanostructured materials is broadening the potential of thermoelectric temperature sensors [58].

5.4 Semiconductor Thermoelectric Sensors

Contemporary thermoelectric temperature sensors commonly employ semiconductor materials, capitalizing on their adjustable electrical and thermal characteristics. The operation of semiconductor thermoelectric sensors involves the measurement of voltage across a diode junction, a value that changes with varying temperatures. These sensors incorporate a dual integrated circuit (IC) featuring two similar diodes and a temperature-sensitive voltage. The diodes, along with the temperature-sensitive voltage, are instrumental in temperature measurement. Semiconductor thermoelectric devices offer advantages such as faster response times, enhanced accuracy, and miniaturization, rendering them suitable for various applications, including electronics and medical devices [59].

5.5 Applications in Harsh Environments

Thermoelectric temperature sensors demonstrate superior performance in challenging environments, where conventional sensors may encounter difficulties. Their solid-state nature and adaptability to diverse materials contribute to their robustness against vibrations, corrosive substances, and extreme temperatures. Industries such as aerospace, automotive, and energy production derive advantages from the durability and resilience of thermoelectric sensors [60].

5.6 Energy Harvesting and Temperature Sensing Integration

The concept of energy harvesting, involving the conversion of ambient heat into electrical energy, intersects seamlessly with temperature sensing. In this context, thermoelectric materials serve a dual purpose by not only sensing temperature variations but also simultaneously generating electrical power [61]. This integrated approach holds significant promise, especially for applications in remote or autonomous sensor systems. By harnessing the thermoelectric effect, where temperature differences induce an electric voltage across materials, these dual-function thermoelectric materials contribute to sustainable and self-sufficient sensor systems. This innovative integration is particularly advantageous in scenarios where access to traditional power sources may be limited or impractical. The versatility of this dual-role functionality extends the potential applications to various fields, including environmental monitoring, industrial automation, and IoT devices. The ability to harvest energy from the surrounding environment while accurately sensing temperature variations makes these integrated thermoelectric sensors well-suited for challenging and remote environments. The ongoing advancements in materials science and energy harvesting technologies further propel the potential of this integrated approach, paving the way for more efficient and self-reliant sensor systems in the future.

Continual research endeavors are directed towards augmenting the efficiency and cost-effectiveness of thermoelectric materials utilized in temperature sensing applications. Innovations in this domain encompass the exploration of novel nanomaterials, advancements in manufacturing techniques, and seamless integration with emerging technologies such as the Internet of Things (IoT). Addressing challenges related to material toxicity and scalability stands as a pivotal aspect for ensuring the widespread adoption of thermoelectric temperature sensors. The ongoing efforts in these areas contribute to the evolution of temperature sensing technologies, making them more efficient, accessible, and aligned with the demands of various industries.

With the progression of technology and an increasing need for precise temperature control, thermoelectric materials remain pivotal components in the development of temperature-sensing devices.

6 Synthesis Methods for Electroceramic Based Sensing Materials

Electroceramic materials serve as a fundamental element in a multitude of sensor technologies, crucial for converting physical and chemical stimuli into measurable signals. This segment explores the various synthesis methods utilized in the creation of these advanced materials, encompassing a spectrum of techniques ranging from conventional approaches to state-of-the-art processes.

6.1 Synthesis of Traditional Ceramic

Traditional approaches to ceramic synthesis, including solid-state reactions and precipitation, have played a significant historical role in the production of sensor materials for electroceramics. Solid-state reactions entail the thorough blending of precursor powders, followed by high-temperature sintering to attain the intended phase. Conversely, precipitation methods involve the controlled chemical precipitation of components, resulting in powders well-suited for applications in sensors [62].

6.2 Sol–Gel Process

The sol–gel process provides a versatile pathway for crafting electroceramic materials with customized compositions and structures. This method entails the conversion of a sol (a colloidal suspension containing nanoscale particles) into a gel, followed by heat treatment to yield the final ceramic product. The sol–gel process facilitates control over particle size, homogeneity, and purity, rendering it highly suitable for sensor applications that require precision [63, 64].

6.3 Hydrothermal Synthesis

Hydrothermal synthesis hinges on the reaction of precursors within an aqueous solution at elevated temperatures and pressures. This technique facilitates the development of crystalline electroceramic materials characterized by heightened purity and distinctive morphologies. Particularly beneficial for the production of ferroelectric and piezoelectric materials, hydrothermal synthesis plays a crucial role in crafting components essential for sensing applications [65].

6.4 Aerosol Techniques

Synthesis techniques based on aerosols, such as spray pyrolysis and flame spray pyrolysis, present benefits in scalability and the production of nanostructured materials. These methods involve the generation of fine droplets containing precursor materials, subsequently transformed into solid particles through controlled pyrolysis. The resultant electroceramic powders showcase a high surface area, contributing to enhanced sensor performance [66].

6.5 Microwave-Assisted Synthesis

Microwave-assisted synthesis has become increasingly recognized for its capacity to swiftly and energy-efficiently produce electroceramic materials. Through microwave irradiation, uniform heating is achieved, resulting in reduced reaction times and improved crystallinity. This approach proves especially advantageous in the synthesis of intricate oxide materials employed in gas sensors, where achieving rapid response times is of paramount importance [67].

6.6 Chemical Vapor Deposition (CVD)

Chemical vapor deposition (CVD) stands out as a sophisticated method for coating substrates with thin films of electroceramic materials. In sensor applications, CVD offers meticulous regulation of film thickness and composition, facilitating the production of sensor devices with exceptional performance. The films deposited through this technique demonstrate outstanding adhesion and uniformity, essential factors contributing to the reliability of sensor coatings [68, 69].

Despite advancements in synthesis methods, challenges endure, encompassing aspects like attaining a uniform particle size distribution, managing defects, and ensuring reproducibility. Future research endeavors are set to confront these challenges with inventive approaches, including additive manufacturing and environmentally friendly synthesis methods. The exploration of sustainable, cost-effective, and scalable synthesis techniques continues to be a central focus in the ongoing advancement of electroceramics sensor materials.

The fabrication of electroceramics sensor materials involves a diverse array of methods, each presenting distinct advantages tailored to specific applications. Ranging from conventional ceramic pathways to cutting-edge aerosol and microwave-assisted techniques, the selected synthesis method plays a pivotal role in shaping the properties and performance of electroceramic materials within sensor devices.

7 Conclusion and Future Perspectives

Electroceramics-based materials for sensor technology reveals the essential role in influencing the landscape of sensing applications. From piezoelectric and ferroelectric materials to ionic conductors and thermoelectric materials, the adaptability of electroceramics has been demonstrated across diverse sensing scenarios. The discussed synthesis methods highlight the critical importance of controlled fabrication processes in determining material performance. Looking ahead, the integration of electroceramics in smart devices, the internet of things, and wearable technologies holds the promise of exciting advancements.

The future trajectory will likely involve cross-disciplinary collaborations, ethical considerations, and a commitment to environmental sustainability, crucial elements in navigating the evolving field responsibly. While challenges persist, ongoing innovation and a dedication to advancing sensor technologies are meeting them head-on. To conclude, this chapter issues a call to action, urging sustained exploration, collaboration, and ethical practices to ensure that Electroceramics continue to push the boundaries of sensor capabilities. Ultimately, this contribution aims to pave the way for a future where sensing technologies play a transformative role in various facets of our lives.

References

1. Warlimont, H.: Ceramics. In: Springer Handbooks, Springer, pp. 441–484 (2018). https://doi.org/10.1007/978-3-319-69743-7_17
2. Setter, N., Waser, R.: Electroceramic Materials (n.d.). www.elsevier.com/locate/actamat

3. Salem, M.M., Morchenko, A.T., Panina, L.V., Kostishyn, V.G., Andreev, V.G., Bibikov, S.B., Nikolaev, A.N.: Dielectric and magnetic properties of two-phase composite system: Mn-Zn or Ni-Zn ferrites in dielectric matrices. Phys. Procedia **75**, 1360–1369 (2015). https://doi.org/10.1016/j.phpro.2015.12.152
4. Randall, C.A., Newnham, R.E., Cross, L.E.: History of the First Ferroelectric Oxide, $BaTiO_3$ (n.d.)
5. Sebastian, M.T., Ubic, R., Jantunen, H.: Low-loss dielectric ceramic materials and their properties. Int. Mater. Rev. **60**, 392–412 (2015). https://doi.org/10.1179/1743280415Y.0000000007
6. Chung, C.-C.: Microstructural Evolution in Lead Zirconate Titanate (PZT) Piezoelectric Ceramics (2014). https://opencommons.uconn.edu/dissertations. (OD Dissertations)
7. Park, C.O., Akbar, S.A., Weppner, W.: Ceramic electrolytes and electrochemical sensors. J. Mater. Sci. **38**, 4639–4660 (2003). https://doi.org/10.1023/A:1027454414224
8. Marimuthu, S., Grace, A.N.: Ceramics for solid oxide fuel cells. In: Advanced Ceramics for Energy Storage, Thermoelectrics and Photonics, pp. 185–208. Elsevier (2023). https://doi.org/10.1016/B978-0-323-90761-3.00006-1
9. Chung, D.D.L., Xi, X.: A review of cement-based materials as electroceramics. Ceram. Int. **49**, 24621–24642 (2023). https://doi.org/10.1016/j.ceramint.2023.05.107
10. Baraskar, B.G., Kambale, R.C., James, A.R., Mahesh, M.L., Ramana, C.V., Kolekar, Y.D.: Ferroelectric, piezoelectric and electrostrictive properties of Sn^{4+} -modified $Ba_{0.7}Ca_{0.3}TiO_3$ lead-free electroceramics. J. Am. Ceram. Soc. **100**, 5755–5765 (2017). https://doi.org/10.1111/jace.15073.
11. Bednorz, J.G., Müller, K.A.: Possible highT c superconductivity in the Ba–La–Cu–O system. Z. Für Phys. B Condens. Matter **64**, 189–193 (1986). https://doi.org/10.1007/BF01303701
12. Atkinson, A., Moseley, P.T.: Thin film electroceramics. Appl. Surf. Sci. **65–66**, 212–219 (1993). https://doi.org/10.1016/0169-4332(93)90661-T
13. Prakash, C.: Electroceramics for sensors and actuators. In: 2012 1st International Symposium on Physics and Technology of Sensors (ISPTS-1), pp. 1–2. IEEE (2012). https://doi.org/10.1109/ISPTS.2012.6260919
14. Andrejevic, M., Burdon, M.: Defining the Sensor Society. Telev. & New Media **16**, 19–36 (2015). https://doi.org/10.1177/1527476414541552
15. Javaid, M., Haleem, A., Singh, R.P.,Rab, S., Suman, R.: Significance of sensors for industry 4.0: roles, capabilities, and applications. Sens. Int. **2**, 100110 (2021). https://doi.org/10.1016/j.sintl.2021.100110
16. Kwaaitaal, T.: The fundamentals of sensors. Sens. Actuators A Phys. **39**, 103–110 (1993). https://doi.org/10.1016/0924-4247(93)80205-U
17. Kavand, H., Nasiri, R., Herland, A.: Advanced materials and sensors for microphysiological systems: focus on electronic and electrooptical interfaces. Adv. Mater. **34**, (2022). https://doi.org/10.1002/adma.202107876
18. Li, M., Chen, T., Gooding, J.J., Liu, J.: Review of carbon and graphene quantum dots for sensing. ACS Sens. **4**, 1732–1748 (2019). https://doi.org/10.1021/acssensors.9b00514
19. Krishna Prasad, N.V., Venkata Prasad, K., Ramesh, S., Phanidhar, S.V., Venkata Ratnam, K., Janardhan, S., Manjunatha, H., Sarma, M.S.S.R.K.N., Srinivas, K.: Ceramic sensors: a mini-review of their applications. Front. Mater. **7**, (2020). https://doi.org/10.3389/fmats.2020.593342.
20. Xu, X.: Intelligent composite materials for use as sensors and actuators. In: Composite Materials, pp. 465–487. Elsevier (2021). https://doi.org/10.1016/B978-0-12-820512-9.00012-5
21. Wang, X., Tao, Y., Pan, S., Fang, X., Lou, C., Xu, Y., Wu, J., Sang, M., Lu, L., Gong, X., Luo, T., Xuan, S.: Biocompatible and breathable healthcare electronics with sensing performances and photothermal antibacterial effect for motion-detecting. Npj Flex. Electron. **6**, 95 (2022). https://doi.org/10.1038/s41528-022-00228-x
22. Kadiyala, K.G., Katari, N.K.: Ceramics for sensing application. In: Advanced Ceramic Coatings for Emerging Applications, pp. 71–84. Elsevier (2023). https://doi.org/10.1016/B978-0-323-99624-2.00013-9

23. Tiwari, B., Babu, T., Choudhary, R.N.P.: Piezoelectric lead zirconate titanate as an energy material: a review study. Mater. Today Proc. **43**, 407–412 (2021). https://doi.org/10.1016/j.matpr.2020.11.692
24. Liang, H., Hao, G., Olszewski, O.Z.: A review on vibration-based piezoelectric energy harvesting from the aspect of compliant mechanisms. Sens. Actuators A Phys. **331**, 112743 (2021). https://doi.org/10.1016/j.sna.2021.112743
25. Whatmore, R.W., Ward, S.J.: Pyroelectric infrared detectors and materials—a critical perspective. J. Appl. Phys. **133**, (2023). https://doi.org/10.1063/5.0141044
26. Mirdamadi, E.S., Haghbin Nazarpak, M., Solati-Hashjin, M.: Metal oxide-based ceramics. In: Structural Biomaterials, pp. 301–331. Elsevier (2021). https://doi.org/10.1016/B978-0-12-818831-6.00012-4
27. Bharadwaj, S.R., Varma, S., Wani, B.N.: Electroceramics for fuel cells, batteries and sensors. In: Functional Materials, pp. 639–674. Elsevier (2012). https://doi.org/10.1016/B978-0-12-385142-0.00016-7
28. Le, L.Q., Hernandez, C.H., Rodriguez, M.H., Zhu, L., Duan, C., Ding, H., O'Hayre, R.P., Sullivan, N.P.: Proton-conducting ceramic fuel cells: scale up and stack integration. J. Power. Sources **482**, 228868 (2021). https://doi.org/10.1016/j.jpowsour.2020.228868
29. Saputra, H.A.: Electrochemical sensors: basic principles, engineering, and state of the art. Mon. Für Chem.-Chem. Mon. **154**, 1083–1100 (2023). https://doi.org/10.1007/s00706-023-03113-z
30. Manjón-Sanz, A.M., Dolgos, M.R.: Applications of piezoelectrics: old and new. Chem. Mater. **30**, 8718–8726 (2018). https://doi.org/10.1021/acs.chemmater.8b03296
31. Sezer, N., Koç, M.: A comprehensive review on the state-of-the-art of piezoelectric energy harvesting. Nano Energy **80**, 105567 (2021). https://doi.org/10.1016/j.nanoen.2020.105567
32. Ju, M., Dou, Z., Li, J.-W., Qiu, X., Shen, B., Zhang, D., Yao, F.-Z., Gong, W., Wang, K.: Piezoelectric materials and sensors for structural health monitoring: fundamental aspects, current status, and future perspectives. Sensors **23**, 543 (2023). https://doi.org/10.3390/s23010543
33. Das Mahapatra, S., Mohapatra, P.C., Aria, A.I., Christie, G., Mishra, Y.K., Hofmann, S., Thakur, V.K.: Piezoelectric materials for energy harvesting and sensing applications: roadmap for future smart materials. Adv. Sci. **8**, (2021). https://doi.org/10.1002/advs.202100864
34. Mandal, D., Banerjee, S.: Surface acoustic wave (SAW) sensors: physics, materials, and applications. Sensors **22**, 820 (2022). https://doi.org/10.3390/s22030820
35. Li, S., Wang, Y., Yang, M., Miao, J., Lin, K., Li, Q., Chen, X., Deng, J., Xing, X.: Ferroelectric thin films: performance modulation and application. Mater. Adv. **3**, 5735–5752 (2022). https://doi.org/10.1039/D2MA00381C
36. Guo, R., You, L., Zhou, Y., Shiuh Lim, Z., Zou, X., Chen, L., Ramesh, R., Wang, J.: Non-volatile memory based on the ferroelectric photovoltaic effect. Nat. Commun. **4**, 1990 (2013). https://doi.org/10.1038/ncomms2990
37. Kanno, I.: Piezoelectric MEMS: ferroelectric thin films for MEMS applications. Jpn. J. Appl. Phys. **57**, 040101 (2018). https://doi.org/10.7567/JJAP.57.040101
38. Blázquez-Castro, A., García-Cabañes, A., Carrascosa, M.: Biological applications of ferroelectric materials. Appl. Phys. Rev. **5**, (2018). https://doi.org/10.1063/1.5044472
39. Aydin, A.C., Çelebi, O.: Piezoelectric materials in civil engineering applications: a review. ACS Omega **8**, 19168–19193 (2023). https://doi.org/10.1021/acsomega.3c00672
40. Liu, H.-Y., Zhang, H.-Y., Chen, X.-G., Xiong, R.-G.: Molecular design principles for ferroelectrics: ferroelectrochemistry. J. Am. Chem. Soc. **142**, 15205–15218 (2020). https://doi.org/10.1021/jacs.0c07055
41. Meng, K., Li, W., Tang, X.-G., Liu, Q.-X., Jiang, Y.-P.: A review of a good binary ferroelectric ceramic: $BaTiO_3$–$BiFeO_3$. ACS Appl. Electron. Mater. **4**, 2109–2145 (2022). https://doi.org/10.1021/acsaelm.1c00109
42. Ghoneim, M.T., Zidan, M.A., Alnassar, M.Y., Hanna, A.N., Kosel, J., Salama, K.N., Hussain, M.M.: Thin PZT-based ferroelectric capacitors on flexible silicon for nonvolatile memory applications. Adv. Electron. Mater. **1**, (2015). https://doi.org/10.1002/aelm.201500045
43. Lee, M.H., Guo, R., Bhalla, A.S.: Pyroelectric sensors. J. Electroceramics **2**, 229–242 (1998). https://doi.org/10.1023/A:1009922522642

44. Aleks, M., Jagtap, C., Kadam, V., Kolev, G., Denishev, K., Pathan, H.: An overview of microelectronic infrared pyroelectric detector. Eng. Sci. (2021). https://doi.org/10.30919/es8d535
45. Mikolajick, T., Schroeder, U., Slesazeck, S.: The past, the present, and the future of ferroelectric memories. IEEE Trans. Electron Devices **67**, 1434–1443 (2020). https://doi.org/10.1109/TED.2020.2976148
46. Yan, M., Xiao, Z., Ye, J., Yuan, X., Li, Z., Bowen, C., Zhang, Y., Zhang, D.: Porous ferroelectric materials for energy technologies: current status and future perspectives. Energy Environ. Sci. **14**, 6158–6190 (2021). https://doi.org/10.1039/D1EE03025F
47. Maurya, D., Abhijit, P.: "Dwight Viehland," Ferroelectric Materials for Energy Harvesting and Storage, 1st edn. Woodhead Publishing (2020)
48. Ye, Y., Yu, L., Lizundia, E., Zhu, Y., Chen, C., Jiang, F.: Cellulose-Based Ionic Conductor: An Emerging Material toward Sustainable Devices. Chem. Rev. **123**, 9204–9264 (2023). https://doi.org/10.1021/acs.chemrev.2c00618
49. Tuller, H.: Ionic conduction and applications. In: Springer Handbook of Electronic and Photonic Materials, pp. 213–228. Springer US, Boston, MA (2006). https://doi.org/10.1007/978-0-387-29185-7_11
50. Tombolesi, S., Zanieri, N., Bargnesi, L., Mernini, M., Lacarbonara, G., Arbizzani, C.: A sustainable gel polymer electrolyte for solid-state electrochemical devices. Polym. (Basel) **15**, 3087 (2023). https://doi.org/10.3390/polym15143087
51. Fergus, J.W.: Sensing mechanism of non-equilibrium solid-electrolyte-based chemical sensors. J. Solid State Electrochem. **15**, 971–984 (2011). https://doi.org/10.1007/s10008-010-1046-4
52. Akasaka, S., Amamoto, Y., Yuji, H., Kanno, I.: Limiting current type yttria-stabilized zirconia thin-film oxygen sensor with spiral Ta_2O_5 gas diffusion layer. Sens. Actuators B Chem. **327**, 128932 (2021). https://doi.org/10.1016/j.snb.2020.128932
53. López-Gándara, C., Ramos, F.M., Cirera, A.: YSZ-based oxygen sensors and the use of nanomaterials: a review from classical models to current trends. J Sens **2009**, 1–15 (2009). https://doi.org/10.1155/2009/258489
54. Ceramic Fuel Cells for Power Generation (n.d.)
55. Li, J., Qi, J., Jin, F., Zhang, F., Zheng, L., Tang, L., Huang, R., Xu, J., Chen, H., Liu, M., Qiu, Y., Cooper, A.I., Shen, Y., Chen, L.: Room temperature all-solid-state lithium batteries based on a soluble organic cage ionic conductor. Nat. Commun. **13**, 2031 (2022). https://doi.org/10.1038/s41467-022-29743-1
56. Tritt, T.M.: Thermoelectric materials: principles, structure, properties, and applications. In: Encyclopedia of Materials: Science and Technology, pp. 1–11. Elsevier (2002). https://doi.org/10.1016/B0-08-043152-6/01822-2
57. Deepak, J.R., Prasanna Kumar, M.. Nithishkar, M.: Review on temperature monitoring system for welding application–a case study on thermocouple array. Mater. Today Proc. (2023). https://doi.org/10.1016/j.matpr.2023.02.373
58. Park, C.R. (ed.): Advanced Thermoelectric Materials. Wiley (2019). https://doi.org/10.1002/9781119407348
59. Rettig, F., Moos, R.: Semiconducting direct thermoelectric gas sensors. In: Semiconductor Gas Sensors, pp. 261–296. Elsevier (2013). https://doi.org/10.1533/9780857098665.2.261
60. Balakrishnan, V., Phan, H.-P., Dinh, T., Dao, D., Nguyen, N.-T.: Thermal flow sensors for harsh environments. Sensors **17**, 2061 (2017). https://doi.org/10.3390/s17092061
61. Park, H., Lee, D., Park, G., Park, S., Khan, S., Kim, J., Kim, W.: Energy harvesting using thermoelectricity for IoT (Internet of Things) and E-skin sensors. J. Phys. Energy **1**, 042001 (2019). https://doi.org/10.1088/2515-7655/ab2f1e
62. Ceramic Methods. In: Essentials of Inorganic Materials Synthesis, pp. 17–21. Wiley (2015). https://doi.org/10.1002/9781118892671.ch3
63. Hench, L.L., West, J.K.: The sol-gel process. Chem. Rev. **90**, 33–72 (1990). https://doi.org/10.1021/cr00099a003
64. Dehghanghadikolaei, A., Ansary, J., Ghoreishi, R.: Sol-gel process applications: a mini-review. Proc. Nat. Res. Soc. **2**, 02008 (2018). https://doi.org/10.11605/j.pnrs.201802008

65. Sá, F.G., Silva, M.R.F., Sierra, D.L.G., Ivanov, M., Tkach, A., Vilarinho, P.M., Ferreira, P.: Comparative microwave- and conventional oven-assisted hydrothermal syntheses of $BaTiO_3$ nanoparticles for improved electroceramics. Ceram. Int. (2023). https://doi.org/10.1016/j.ceramint.2023.12.223
66. Meierhofer, F., Fritsching, U.: Synthesis of metal oxide nanoparticles in flame sprays: review on process technology, modeling, and diagnostics. Energy Fuels **35**, 5495–5537 (2021). https://doi.org/10.1021/acs.energyfuels.0c04054
67. Gawande, M.B., Shelke, S.N., Zboril, R., Varma, R.S.: Microwave-assisted chemistry: synthetic applications for rapid assembly of nanomaterials and organics. Acc. Chem. Res. **47**, 1338–1348 (2014). https://doi.org/10.1021/ar400309b
68. Sun, L., Yuan, G., Gao, L., Yang, J., Chhowalla, M., Gharahcheshmeh, M.H., Gleason, K.K., Choi, Y.S., Hong, B.H., Liu, Z.: Chemical vapour deposition. Nat. Rev. Methods Prim. **1**, 5 (2021). https://doi.org/10.1038/s43586-020-00005-y
69. Zhang, Q., Geng, D., Hu, W.: Chemical vapor deposition for few-layer two-dimensional materials. SmartMat **4**, (2023). https://doi.org/10.1002/smm2.1177

Piezoelectric, Pyroelectric, and Dielectric Properties of PZT: Nylon 11 and Graphite Doped PZT: Nylon11 Composites

Avadhesh Sharma, Minakshi Sharma, and Varsha Yadav

Abstract This paper prepared the PZT-Nylon 11 and graphite-doped PZT-Nylon 11 composites with 0–3 phase connectivity geometry using a hot press process. The samples were poled under different conditions, i.e., at different poling temperatures and fields, and characterized for their piezo, pyro, and dielectric properties. The dielectric and AC conductivity measurements were also carried out. The results showed that the piezoelectric constant and remnant polarization both increased with graphite doping. Also, the value of the d_{33} constant for the graphite-doped sample is almost three times more than the undoped sample, keeping other conditions the same. In addition, remnant polarization was increased for graphite-doped samples compared to undoped samples.

1 Introduction

Piezoelectric materials exhibit a mechanical response when subjected to an electric field and display an electrical response to mechanical inputs. Due to these unique properties, they are extensively used in sensors, transducers, electromechanical actuators, piezoelectric power supplies, and other advanced applications. These materials can be broadly classified into inorganic, organic, and composites [1]. In the inorganic category, ceramics and crystals are notable for their high dielectric constant, high piezoelectric coefficient, and strong electromechanical coupling. These materials, such as lead zirconate titanate (PZT), exhibit significant piezoelectric properties, which make them ideal for various high-performance applications. However, they

A. Sharma
School of Applied Sciences, Invertis University, Bareilly, U.P. 2443123, India

M. Sharma
School of Applied Sciences, Om Sterling Global University, Hisar, Haryana 125001, India

V. Yadav (✉)
School of Applied Sciences, Shri Venkateshwara University, Gajraula (Amroha), U.P. 244221, India
e-mail: varshayadav616@gmail.com

U. Kumar (ed.), *Defects Engineering in Electroceramics for Energy Applications*, Engineering Materials, https://doi.org/10.1007/978-981-97-9018-0_17

suffer from poor mechanical properties and relatively high acoustic impedance, which restricts their use to specific applications where these limitations can be managed. Their brittleness and high density pose additional challenges, particularly in applications requiring mechanical flexibility and low weight [2, 3]. On the other hand, the organic category includes polymers like Polyvinylidene Fluoride (PVDF). PVDF and its copolymers are attractive due to their flexibility, lower density, and ease of processing. They offer low acoustic impedance, making them suitable for acoustic matching with water or tissues. However, their piezoelectric coefficients are relatively low compared to ceramics, which limits their use in applications requiring high piezoelectric responses [4]. Piezoelectric composites, primarily ceramic and polymer composites, exhibit superior piezoelectric and dielectric properties compared to pure polymers and offer better mechanical properties than ceramics. These composites combine the advantages of both materials, providing high coupling factors, low acoustic impedance, mechanical flexibility, appropriate matching with water and human tissue, broad bandwidth, and a low mechanical quality factor [5, 6]. As a result, piezoelectric composites find applications in underwater sonar, medical diagnostic ultrasonic transducers, vibration and noise control, and morphing structures [7–9].

Piezoelectric composites can be further categorized based on the connectivity scheme of their constituents. Among these, 0–3 connectivity-type composites, consisting of ferroelectric ceramic (FC) inclusions within an extended polymer matrix, exhibit various useful physical properties [10]. These composites are widely used in modern piezoelectric technology, acoustics, and other fields. Extensive research has been conducted on different composites of PZT with various polymers such as PVDF, PVC, PVA, and copolymers. However, composites of PZT with Nylon have not been extensively reported [11]. Nylon possesses unique physical and chemical properties, such as toughness, flexibility, and resistance to chemicals, temperature, and wear [12]. Consequently, composites of PZT and Nylon are expected to offer valuable properties for applications that require these characteristics [13, 14]. We prepared flexible and freestanding films of PZT and Nylon 11 composites with 0–3 connectivity using the hot press method in this study. These films were prepared with different volume fractions of PZT and Nylon 11. Previous studies have indicated that 0–3 composites cannot be fully poled due to the screening effect of the polymer matrix, resulting in reduced overall properties and limited practical applications of the 0–3 ferroelectric composite. For example, Satish et al. [15] demonstrated that the poling behavior of the PZT phase in a PZT/PU composite doped with graphite particles was improved, enhancing the piezoelectric and pyroelectric properties of the composite. Similar findings were reported by other researchers [13–16], who observed that increasing the matrix material's electrical conductivity improved the composite's poling behavior and properties. Graphite doping has shown promising results in enhancing the electrical properties of piezoelectric composites. Including conductive graphite particles can significantly improve the poling efficiency of the ceramic phase by reducing the screening effect. This improvement is crucial for applications requiring high piezoelectric and pyroelectric responses, such as advanced sensors and actuators. The improved electrical conductivity facilitates the alignment of dipoles within the ceramic phase, leading to enhanced piezoelectric properties [15, 16].

In the present work, we characterize these PZT and Nylon 11 composites and graphite-doped PZT and Nylon 11 composites for their piezoelectric, pyroelectric, and dielectric properties. This study explores the potential advantages of these composites in various advanced applications, leveraging their unique properties and improved performance due to graphite doping. The development of these composites could lead to new and improved materials for use in fields such as biomedical engineering, aerospace, and underwater acoustics. By integrating recent advancements in the field, we aim to provide a comprehensive understanding of the potential and limitations of PZT and Nylon 11 composites, as well as graphite-doped PZT and Nylon 11 composites, in enhancing the performance of piezoelectric, pyroelectric, and dielectric applications. This research will contribute to developing more efficient and versatile materials, expanding the range of applications, and improving the performance of piezoelectric devices.

The study will focus on synthesizing these composites and their structural and morphological characterization and evaluating their piezoelectric, pyroelectric, and dielectric properties. The insights gained from this research will help optimize the composition and processing conditions to achieve the desired properties, paving the way for developing next-generation piezoelectric materials.

2 Material and Methods

2.1 *Sample Preparation*

The Nylon 11 granules obtained from Sigma-Aldrich were dissolved in DMF by heating and stirring and converted into fine powder by grinding and sheaving. The PZT powder was prepared in our laboratory using a well-known mixed oxide route. The required quantities of PZT and Nylon11 powder were mixed and ground using agate mortar for 4 h. This mixture was then placed in a die for hot pressing [17]. The mixture was first heated up to 1840 °C then pressed using the pressure of eight tons. After cooling, the samples were removed from the die, and the electrode was used with silver paste. The samples with Graphite doping were also prepared similarly by adding the required amount of Graphite powder at mixing and grinding.

2.2 *Poling*

The electrode samples were poled using an in-house developed sample holder in a silicon oil bath at different temperatures and poling voltages.

2.3 Characterization

The d_{33} constant was measured using APC-d_{33} meter. The dielectric properties were measured using an Agilent precision LCR meter (4284A) at room temperature in the 100 Hz–1 MHz range. The pyroelectric properties were determined by measuring the thermally stimulated discharge current using a Kiethley electrometer and a temperature bath at 20C/min. of rise. The P-E measurements were carried out using a Sawyer–Tower circuit and storage oscilloscope [18]. The figure of merit (FOM) was also calculated theoretically.

3 Characterization of PZT/Nylon11 (50:50) Composite

3.1 Dielectric Properties

Figure 1a and b illustrate the variation in dielectric constant and dielectric loss of a PZT/Nylon11 (50:50) composite as a function of frequency for both unpoled and poled samples at different poling fields. From these plots, it is evident that dielectric loss decreases with increasing frequency. Notably, samples poled at 4, 6, 12, and 13 kV exhibit higher dielectric losses compared to other samples. Despite this, the dielectric loss in these samples continues to decrease with increasing frequency, albeit at a slower rate and eventually approaching a constant value [17]. Additionally, a rapid increase in dielectric loss is observed as the frequency nears 10,000 Hz, which can be attributed to anomalous dielectric dispersion—a phenomenon where polarization diminishes with increasing frequency.

The dielectric loss increases with temperature due to enhanced molecular mobility and dipole relaxation in the polymer matrix. Figure 2a and b illustrate how dielectric constant and dielectric loss vary with temperature for PZT/Nylon11 (50:50)

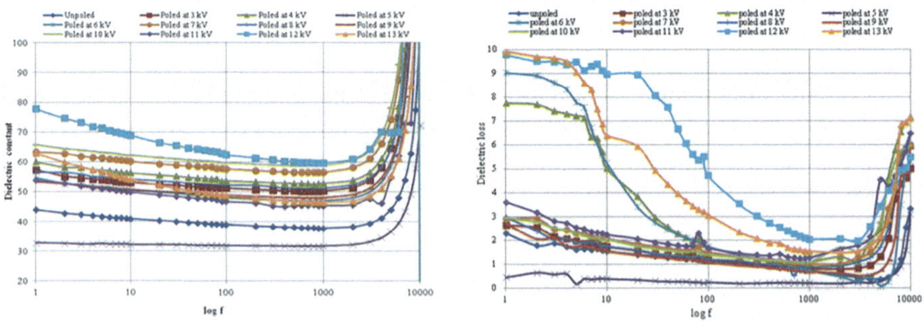

(a) (b)

Fig. 1 Variation of **a** Dielectric constant and **b** Dielectric loss with frequency of PZT/Nylon11 (50:50) composite

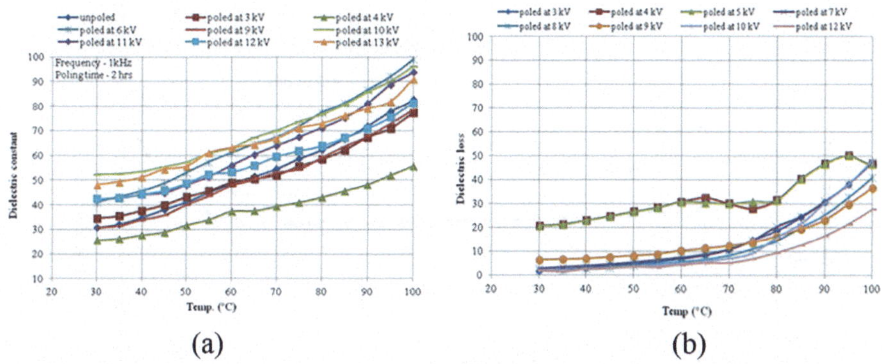

Fig. 2 Variation of **a** Dielectric constant and **b** Dielectric loss with Temperature of PZT/Nylon11 (50:50) composite

composite. Concurrently, the dielectric constant experiences a rapid rise with temperature, driven by the temperature-sensitive dielectric properties of the PZT component. Additionally, the figures reveal that all the composites exhibit a significant increase in dielectric constant with rising temperature. This is primarily because the dielectric properties of the composites are largely influenced by the PZT (lead zirconate titanate) component [18]. The PZT part of the composites is responsible for most of the observed dielectric behavior, and its properties are sensitive to temperature changes, leading to the rapid increase in dielectric constant as the temperature rises.

The dielectric constant remains nearly constant, showing a slight increase with rising frequency. At the same time, the dielectric loss is also largely independent of frequency [19], illustrating the variation of dielectric loss and dielectric constant with frequency for the PZT/Nylon11 (60:40) composite in Fig. 3. This behavior is quite similar to what is seen in the 50:50 PZT/Nylon11 composite. This suggests that the composite material retains its ability to polarize and store energy effectively over a wide range of frequencies. Similarly, the dielectric loss, which represents the energy dissipated as heat within the dielectric material under an alternating electric field, also shows minimal dependency on frequency for the 60:40 composite [20].

This implies that changes do not significantly influence the frequency of energy dissipation mechanisms in the composite. The PZT content in the composite, being higher in the 60:40 ratio compared to the 50:50 ratio, contributes to a more pronounced effect of the PZT phase on the dielectric properties. The ferroelectric nature of PZT, along with the effective poling field, ensures that the dielectric properties remain stable across a range of frequencies.

The PZT/Nylon11 (60:40) composite exhibits distinctive dielectric loss and dielectric constant behaviors when subjected to varying temperatures. Figure 4 illustrates the variation of dielectric loss and dielectric constant with frequency for the PZT/Nylon11 (60:40) composite. The behavior of the PZT/Nylon11 (60:40) composite in terms of dielectric loss and dielectric constant with temperature is similar to that observed in other PZT/Nylon11 composites, such as the 50:50

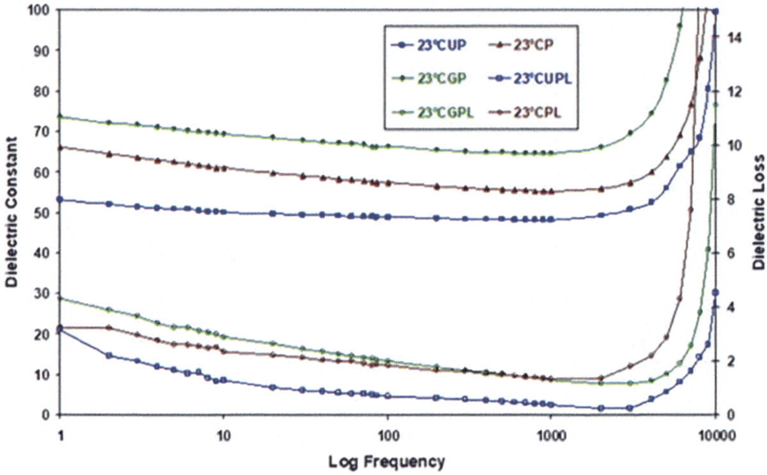

Fig. 3 Variation of **a** Dielectric loss and **b** Dielectric constant with frequency of PZT/Nylon11 (60:40) composite

ratio. As dielectric Loss is Increases with temperature due to enhanced molecular mobility, dipole relaxation, and the influence of the PZT component and dielectric constant shows a rapid increase with temperature, driven by temperature-dependent polarization, ferroelectric domain alignment, and interfacial polarization [21].

However, the higher PZT content in the 60:40 composite amplifies these effects. The composite's dielectric properties are predominantly influenced by the PZT phase, making it highly responsive to temperature variations.

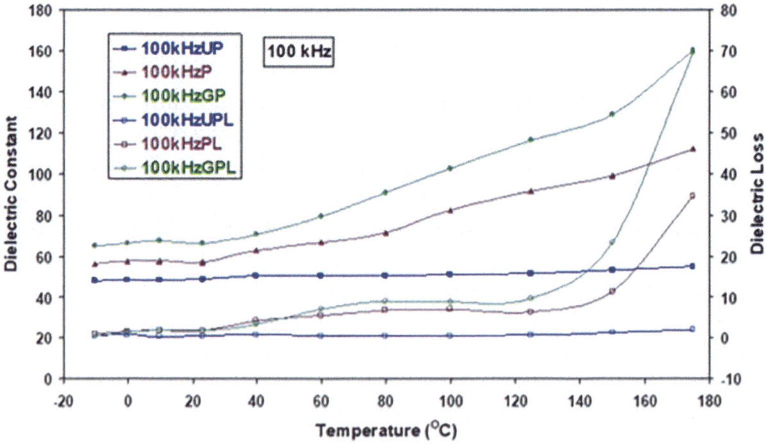

Fig. 4 Variation of **a** Dielectric loss and **b** Dielectric constant with temperature of PZT/Nylon11 (60:40) composite

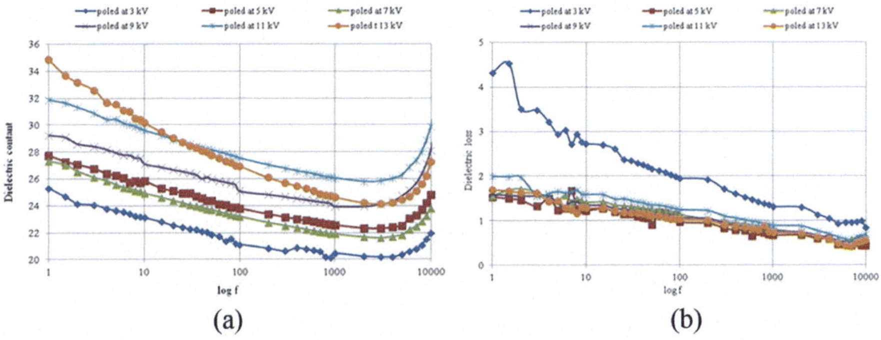

Fig. 5 Variation of **a** Dielectric constant and **b** Dielectric loss with frequency of PZT/Nylon11 (70:30) composite

Figure 5a and b present the dielectric constant and dielectric loss of the PZT/ Nylon (70:30) composite. It can be observed that both the dielectric constant and dielectric loss exhibit minimal variation with increasing frequency, maintaining a nearly constant value. This behavior can be attributed to the low ceramic fraction within the composite [22]. The reduced ceramic content leads to a smaller number of dipoles, which diminishes the effectiveness of the poling process. Consequently, most of the poling field is absorbed by the Nylon 11 matrix, resulting in the observed stability of the dielectric properties across different frequencies. This characteristic indicates that the composite's dielectric properties are predominantly influenced by the polymer matrix rather than the ceramic components [23].

3.2 Piezoelectric Properties

The piezoelectric properties of a composite are influenced by the piezoelectric characteristics of its individual phases and the degree of interconnectivity between these phases. Figure 6a and b illustrate the variation in the d33 coefficient values for the 50:50 and 60:40 composites, respectively, as the poling voltage increases. These figures clearly demonstrate the rise and fall in the d33 coefficient, allowing for an easy observation of its values at different poling voltages [24].

Figure 6a and b provide a detailed analysis of the variation in the d33 coefficient for two different PZT/Nylon composites as the poling voltage is increased. Figure 6a shows that for the 50:50 composite, the d33 coefficient initially rises from 3 to 6 kV, then decreases up to 7 kV, and subsequently rises again, reaching its maximum value at 12 kV/mm. Similarly, Fig. 6b reveals that the 60:40 composite exhibits comparable behavior, with the d33 coefficient experiencing fluctuations before achieving its peak value at 11 kV/mm.

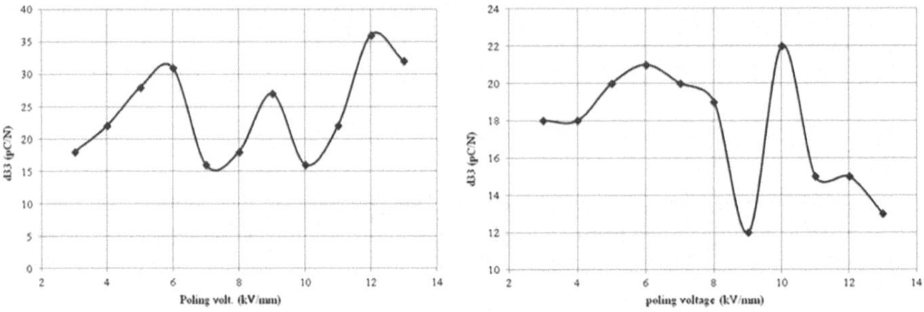

(a) (b)

Fig. 6 Variation of d33 Coefficient at various poling voltage for **a** PZT/Nylon11 (50:50) composite **b** PZT/Nylon11 (60:40) composite

This pattern of rise and fall in the d33 coefficient highlights the complex relationship between the poling voltage and the piezoelectric response of the composites. It can be observed from the figures that the d33 coefficient increases more steeply with the increase in the volume fraction of PZT, particularly for composites subjected to higher poling fields. This phenomenon can be attributed to the higher degree of alignment of the dipoles at elevated fields, which enhances the piezoelectric properties [25]. As the poling voltage increases, the dipoles within the PZT phase align more effectively, resulting in a more pronounced piezoelectric response. The observed trends underscore the importance of optimizing the poling conditions and PZT content to achieve the desired piezoelectric performance in these composites.

Figure 7a and b depict the piezoelectric figure of merit (FOM) for the 50:50 and 60:40 composites, respectively. These figures reveal distinct behaviors for each composite when subjected to increasing poling voltages. For the 50:50 composite, the FOM increases as the poling voltage rises, enhancing piezoelectric performance with higher poling fields. Conversely, the FOM for the 60:40 composite exhibits a declining trend with increasing poling voltage. This contrasting behavior highlights the different responses of the two composites under varying electric fields.

Moreover, it is noteworthy that both composites do not display any significant FOM at low poling voltages. This phenomenon can be attributed to the insufficient poling field strength at low voltages, which fails to induce adequate polarization in the materials. At these low poling fields, both the d33 and g33 coefficients, which are crucial for determining the FOM, remain almost independent of the PZT volume fraction. This lack of sufficient electric field results in minimal dipole alignment, thereby preventing the composites from achieving a measurable FOM. Consequently, the piezoelectric properties of the composites are not effectively activated until a higher poling voltage is applied, underscoring the importance of adequate poling to realize the full piezoelectric potential of these materials.

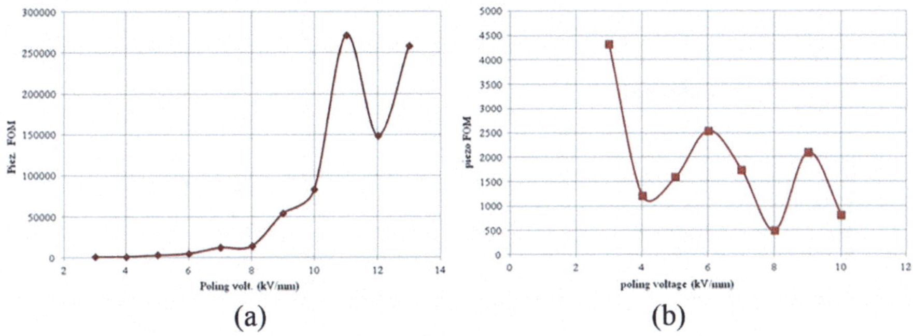

Fig. 7 Variation of piezo. Figure of matrix at various poling voltage for **a** PZT/Nylon11 (50:50) composite **b** PZT/Nylon11 (60:40) composite

3.3 Pyroelectric Properties

Figure 8 illustrates the pyroelectric coefficient of the 50:50 composite as a function of temperature. The graph shows a gradual increase in the pyroelectric coefficient up to a temperature of 65 °C. This slow rise indicates a steady enhancement in the material's ability to generate an electric charge in response to temperature changes.

However, as the temperature continues to rise and approaches 75 °C, the pyroelectric coefficient has a noticeable steep increase. This sudden surge suggests a significant enhancement in the material's pyroelectric properties at higher temperatures. An exception to this trend is observed in one sample poled at 11 kV, which does

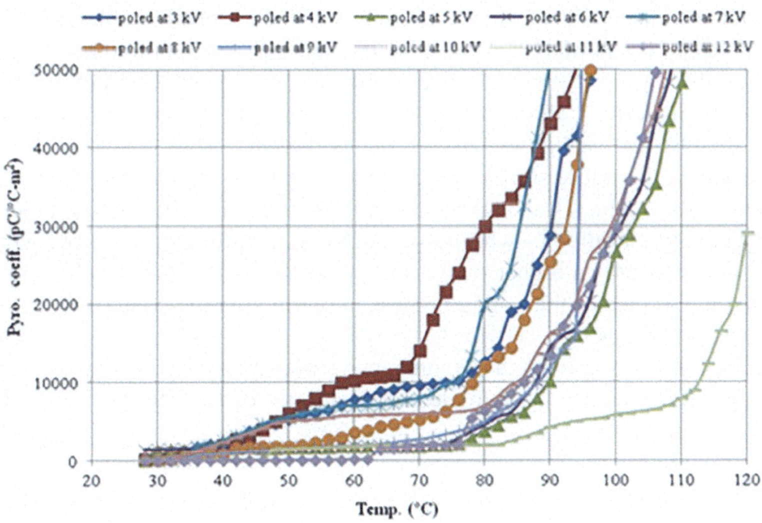

Fig. 8 Variation of pyro. Coefficient at various temperatures for PZT/Nylon11 (50:50) composite

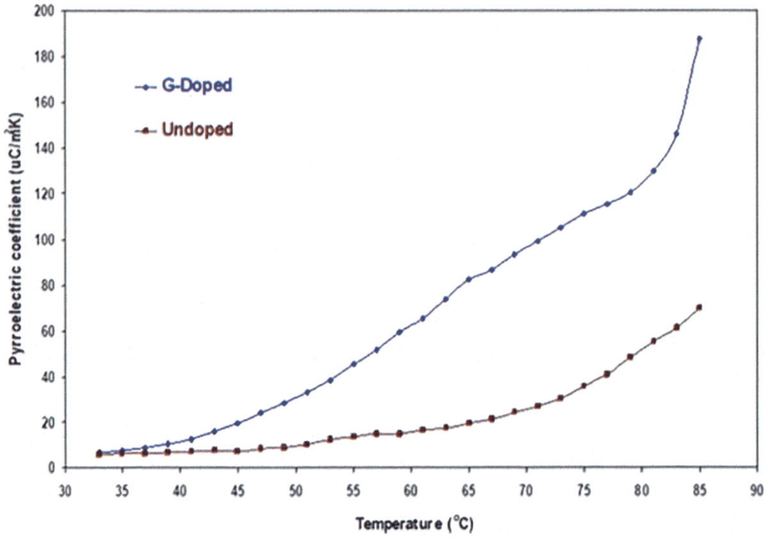

Fig. 9 Variation of pyro coefficient at various temp. of graphite doped and undoped samples PZT/Nylon11 (60:40) composite

not exhibit the same steep increase. This anomaly indicates that the poling voltage can influence the temperature dependence of the pyroelectric coefficient, potentially affecting the material's overall performance.

Figure 9 compares the pyroelectric coefficient of graphite-doped and undoped 60:40 composites. The graphite-doped samples show a significantly higher pyroelectric coefficient value of 190 $\mu C/m^2 \cdot K$, in contrast to the undoped samples, which have values between 60 and 80 $\mu C/m^2 \cdot K$. This marked difference highlights the impact of graphite doping on the pyroelectric properties. As a good conductor, Graphite enhances the material's ability to generate a charge in response to temperature changes, thereby increasing the pyroelectric coefficient.

Additionally, the figure shows that the pyroelectric coefficient increases with an increasing poling field. This observation indicates that a higher poling field leads to better alignment of dipoles within the composite, resulting in enhanced polarization and, consequently, a higher pyroelectric response [26]. This correlation underscores the importance of the poling process in optimizing the composite materials' pyroelectric properties.

Both Figs. 8 and 9 provide a comprehensive understanding of how temperature and poling field influence the pyroelectric properties of PZT/Nylon composites in different configurations.

4 Result and Discussion

The variation of dielectric loss and dielectric constant at different frequencies for poled and unpoled samples of PZT composites is analyzed to understand the impact of PZT content and graphite doping on these properties. Figure 10a and b represent the Variation of (a) dielectric constant and (b) dielectric loss at various frequencies for poled and unpoled samples for different constitutions of PZT in composites, provide insights into the frequency dependence of the dielectric constant and dielectric loss for various compositions and processing conditions.

The results indicate that the PZT content, graphite doping, and the poling process strongly influence the dielectric constant and dielectric loss. The PZT content primarily affects the magnitude of the dielectric constant and the degree of dielectric dispersion. Graphite doping enhances the dielectric constant and dielectric loss due to increased interfacial polarization and conductivity. The poling process plays a critical role in aligning the dipoles, thereby modifying the dielectric properties of the composite [27]. These findings align with previous studies, which suggest that the interplay between the matrix and filler materials, interfacial effects, and processing conditions, such as poling, governs the dielectric properties of composites. The observed dielectric constant and dielectric loss trends highlight optimizing the composite formulation and processing to achieve the desired dielectric performance for specific applications [28]. According to studies, the dielectric behavior of similar composites further supports these conclusions, indicating that the observed phenomena are consistent with established scientific understanding. For instance, the decrease in dielectric constant with reduced PZT content and the increase upon graphite doping align with previously reported mechanisms of interfacial polarization and conductive network formation [16]. Additionally, the reduced dielectric constant in poled composites corroborates the theory of phase order and interphase charge formation between the matrix and filler.

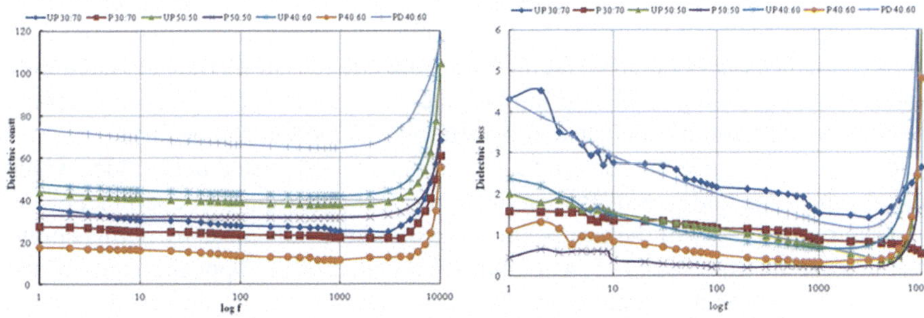

(a) (b)

Fig. 10 Variation of **a** dielectric constant and **b** dielectric loss at various frequencies for poled and unpoled samples for different constitutions of PZT in composites

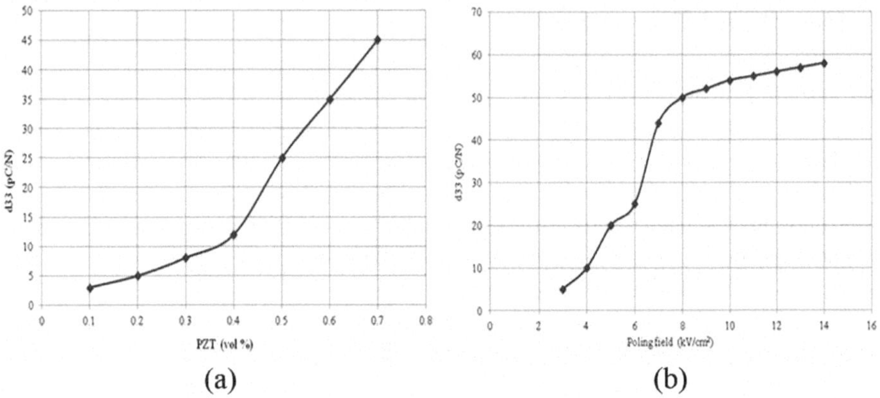

Fig. 11 Variation of d33 coefficient at **a** different PZT volume percentage in composite **b** various poling field

The d33 coefficient, a measure of the piezoelectric response in the direction of applied electric field, is influenced by several factors including the composition of the composite, the poling conditions, and the volume percentage of piezoelectric material. In PZT/Nylon composites, the interplay between the ceramic (PZT) and the polymer (Nylon) components significantly affects the piezoelectric properties. Figure 11a demonstrates the impact of varying the PZT volume percentage on the d33 coefficient of the composite and (b) represent the Variation of d33 coefficient at various poling field.

Adding PZT to the composite leads to a noticeable increase in the d33 coefficient, highlighting the crucial role of the ceramic component in enhancing the piezoelectric properties. As the volume percentage of PZT increases, the number of piezoelectric dipoles within the composite also rises, resulting in a higher d33 coefficient. This relationship indicates that the piezoelectric performance of the composite is directly proportional to the PZT content. The improved d33 coefficient with higher PZT volume percentages can be attributed to the enhanced interaction between the PZT particles and the polymer matrix, facilitating more efficient stress transfer and dipole alignment during poling. Thus, optimizing the PZT volume percentage is essential for achieving superior piezoelectric properties in PZT/Nylon composites.

The piezoelectric d33 coefficient for the composite is depicted in Fig. 11b, which illustrates how the addition of PZT significantly enhances the d33 coefficient. The piezoelectric properties of the composite are profoundly influenced by poling conditions, as evident from Fig. 11a. This influence is due to the opposite signs of the piezo-coefficients of the ceramic and polymer components. As observed, the d33 coefficient increases with an increasing poling field; however, at higher fields, the change in the d33 coefficient becomes marginal. The poling field serves as the driving force that realigns the ferroelectric domains during the poling process [29]. Therefore, a stronger poling field provides a greater driving force, promoting more effective alignment of the ferroelectric domains. Consequently, the d33 coefficient experiences

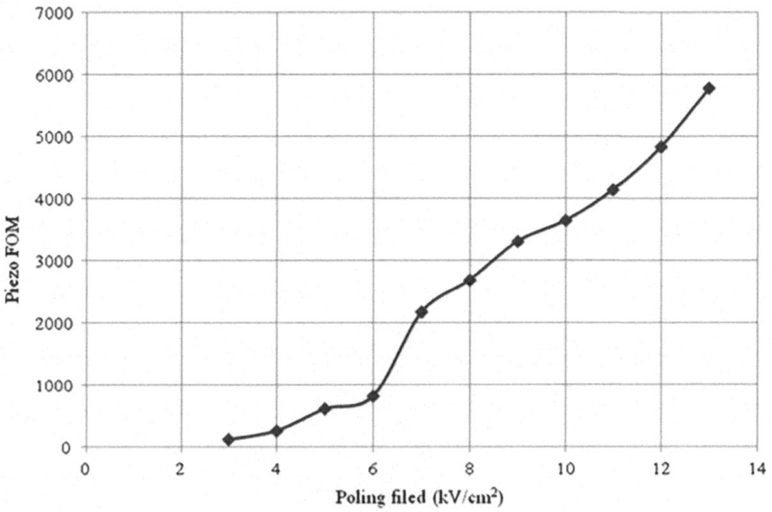

Fig. 12 Variation of Piezo. Figure of merit at various poling field

a significant increase with the rise in the poling field. Once the polarization intensity surpasses a specific threshold, the composite reaches saturation poling, causing the d33 coefficient to stabilize and exhibit minimal changes despite further increases in the poling field.

The piezoelectric figure of merit is a crucial parameter that combines the piezoelectric properties and the mechanical stability of the material, offering a comprehensive measure of the material's performance. As the poling field increases, it enhances the alignment of ferroelectric domains within the composite, thereby improving the overall piezoelectric response. Figure 12 shows Variation of Piezo. Figure of merit at various poling field.

The dielectric constant of both PZT and the polymer decreases with poling, which contributes to a higher piezoelectric figure of merit (FOM) for the composite. The inclusion of graphite in the composite leads to an increase in the d33 value, indicating an enhancement in piezoelectric performance. The presence of graphite filler facilitates the poling process, thereby improving both piezoelectric and pyroelectric activities [30]. This enhancement can be attributed to the ability of graphite particles to create continuous electric flux paths between PZT grains, which promotes more effective poling and alignment of dipoles. Consequently, the composite exhibits improved electrical properties and overall performance.

The P-E loop serves as a fundamental indicator of the ferroelectric behavior and polarization properties of these materials under varying electric fields. It's for these composites provides insights into their ferroelectric behavior, including polarization saturation, coercive field strength, and remnant polarization. This analysis is essential for comprehending how the addition of graphite impacts the ferroelectric properties of the composite, potentially altering its polarization behavior and overall performance

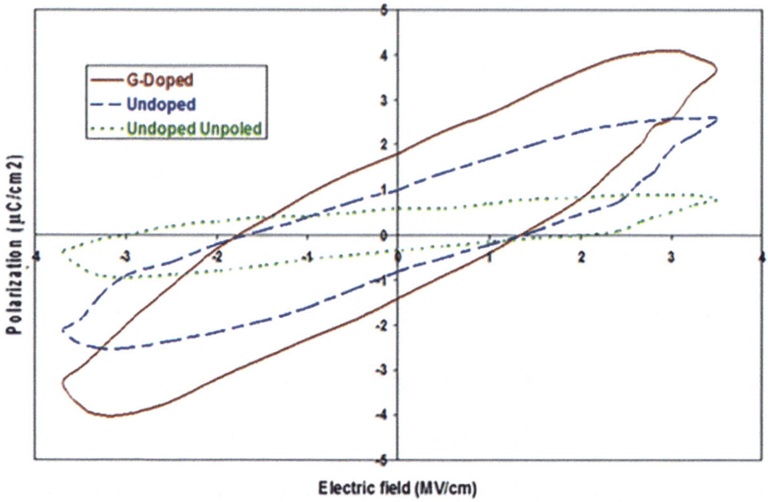

Fig. 13 P-E loop for PZT Nylon11 and graphite doped PZT Nylon 11

[31]. Figure 13 exhibits P-E Loop for PZT Nylon11 and Graphite doped PZT Nylon 11.

From the analysis of the polarization–electric field (P-E) hysteresis loop, key parameters such as remnant polarization (Pr) and coercive field (Ec) were derived and are presented in Fig. 13. These parameters are further detailed in Table 1, providing the key electrical properties of PZT/Nylon 11 composites under different conditions: undoped unpoled, undoped poled, and graphite-doped poled (1% graphite). The parameters evaluated include remnant polarization (Pr), saturation polarization (Ps), coercive field (Ec), the ratio of remnant to saturation polarization (Pr/Ps), and the piezoelectric coefficient (d33) [30].

The undoped unpoled sample exhibits the lowest values for Pr and Ps, indicating minimal polarization due to the absence of an applied poling field. The coercive field (Ec) is the highest among all samples, which suggests greater difficulty in reorienting the dipoles without poling. The piezoelectric coefficient (d33) is negligible, reflecting the lack of piezoelectric response in the unpoled state [31]. Poling significantly enhances the electrical properties of the undoped sample—both Pr and Ps

Table 1 Estimated remnant polarization (Pr) and coercive field (Ec) value from P-E hysteresis loop

Sample (PZT:Nylon11,40:60)	P_r(μC/cm^2)	P_s(μC/cm^2)	E_c(MV/cm)	P_r/P_s	d_{33}
Undoped unpoled	0.5	0.9	2.11	0.55	Nil
Undoped poled	1	2.6	1.3	0.38	12
Doped poled (Graphite 1%)	1.8	4.1	1.25	0.43	37

increase, indicating improved polarization. The coercive field decreases to 1.3 MV/cm, suggesting that the applied poling field aids in aligning dipoles [32]. The piezoelectric coefficient (d33) of 12 highlights the development of a measurable piezoelectric response due to poling. Including 1% graphite in the poled composite leads to the highest values for Pr and Ps, indicating substantial polarization. The coercive field (Ec) is slightly reduced to 1.25 MV/cm, showing that graphite doping further facilitates dipole alignment [33]. The Pr/Ps ratio improves to 0.43, suggesting a more efficient polarization process. Notably, the piezoelectric coefficient (d_{33}) increases dramatically to 37, highlighting the significant enhancement in piezoelectric response due to graphite doping.

In conclusion, the results demonstrate that poling significantly improves the piezoelectric and polarization properties of PZT/Nylon 11 composites. Furthermore, doping with graphite enhances these properties even more, leading to higher remnant and saturation polarization, a lower coercive field, and a much higher piezoelectric coefficient [26, 34]. These improvements suggest that graphite-doped PZT/Nylon 11 composites are highly promising for applications requiring enhanced piezoelectric and pyroelectric performance.

5 Conclusions

In this study, the piezoelectric composites of PZT and Nylon 11 were successfully prepared and characterized for their dielectric, piezoelectric, and pyroelectric properties. The composites exhibited notable behaviors and dependencies on the PZT content, frequency, temperature, and poling conditions. The dielectric constant and dielectric loss were found to vary significantly with changes in frequency and temperature, influenced predominantly by the PZT component and the processing conditions, including graphite doping. The analysis revealed that the dielectric properties of the composites were enhanced with the increase in PZT content and the addition of graphite due to improved interfacial polarization and conductivity.

The piezoelectric properties, characterized by the d33 coefficient, demonstrated a strong dependence on the PZT content and poling conditions. Higher PZT volume fractions and optimized poling fields resulted in improved piezoelectric performance. The study underscored the importance of poling conditions in aligning the ferroelectric domains, thereby enhancing the piezoelectric response. The piezoelectric figure of merit (FOM) also improved with increased PZT content and poling field, highlighting the potential of these composites for practical applications.

The pyroelectric properties of the composites showed a significant increase in the pyroelectric coefficient with rising temperature and poling fields. The inclusion of graphite notably enhanced the pyroelectric response, demonstrating the potential for these composites in temperature-sensing applications. The P-E hysteresis loop analysis provided insights into the ferroelectric behavior of the composites, indicating the effects of graphite doping on polarization properties.

Overall, the study confirms that PZT/Nylon 11 composites exhibit enhanced dielectric, piezoelectric, and pyroelectric properties, especially with graphite doping. These findings suggest that such composites are promising candidates for various applications, including sensors, actuators, and energy-harvesting devices. Future work may focus on optimizing the composite formulation and processing techniques to achieve even better performance and exploring new applications for these advanced materials.

References

1. Newnham, R.E.: Molecular mechanisms in smart materials. MRS Bull. **22**(5), 20–34 (1997)
2. Yu, R., Liu, Y., Gao, Z., Liu, M., Zhang, T., Zhang, H., Hu, J.: Preparation and properties of nylon 11/ethylene-vinyl alcohol/montmorillonite ternary composites. Polym. Compos. **41**(12), 5343–5354 (2020)
3. Thomas, P., Ashokbabu, A., Ravindran, R.E., Vaish, R.: Dielectric properties of nylon 11/ $CaCu_3Ti_4O_{12}$ (CCTO) nanocomposite films with high permittivity. IEEE Trans. Dielectr. Electr. Insul. **26**(2), 568–575 (2019)
4. Kumar, A., Khan, B., Yadav, V., Dixit, A., Kumar, U., Singh, M.K.: Rietveld refinement, optical, dielectric and ac conductivity studies of Ba-doped $SrSnO_3$. J. Mater. Sci. Mater. Electron. **31**(19), 16838–16848 (2020)
5. Cheema, H., Yadav, V., Maurya, R.S., Yadav, V., Kumar, A., Sharma, N., Alvi, P.A., Kumar, U.: Structural, optical and electrical properties of Mn-doped $ZnFe_2O_4$ synthesized using sol–gel method. J. Mater. Sci. Mater. Electron. **32**(18), 23578–23600 (2021)
6. Khoroshun, L.P., Maslov, B.P., Leshchenko, P.V.: Prediction of the effective properties of piezoactive composite materials. Kiev Izdatel Naukova Dumka (1989)
7. Szewczyk, P.K., Busolo, T., Kar-Narayan, S., Stachewicz, U.: Wear-resistant smart textiles using nylon-11 triboelectric yarns. ACS Appl. Mater. & Interfaces (2023)
8. Chan, H.L.W., Ng, P.K.L., Choy, C.L.: Effect of poling procedure on the properties of lead zirconate titanate/vinylidene fluoride-trifluoroethylene composites. Appl. Phys. Lett. **74**(20), 3029–3031 (1999)
9. Alam, I., Sa, K., Das, S., Subramanyam, B.V.R.S., Raiguru, J., Samanta, B., Kumar, P., Mahanandia, P.: Dielectric behavior of PZT/graphene oxide composites. Phys. Status Solidi (a) **216**(20), 1900108 (2019)
10. De Campos Fuzari, Jr., G., Orlandi, M.O., Longo, E., de Barros Melo, W. L., Sakamoto, W.K.: Effect of controlled conductivity on thermal sensing property of 0–3 pyroelectric composite. Smart Mater. Struct. **22**(2), (2013)
11. Pradhan, S.K., Kour, P., Kumar, A., Pandey, R., Kumar, P., Kar, M., Sinha, A.N.: Effect of ferroelectric filler nanoarchitectonics on the electrical and mechanical properties of the nanocomposite thick films of polyvinylidene fluoride and lanthanum-doped lead zirconate titanate in 0–3 connectivity. Appl. Phys. A **129**(11), 805 (2023)
12. Andrusiv, L.: Experimental study of reinforcing mechanical properties of nylon-11 by selective laser sintering. Phys. Sci. Forum **4**(1), 22 (2022). (MDPI.025015)
13. Lous, G.M., Cornejo, I.A., McNulty, T.F., Safari, A., Danforth, S.C.: Fabrication of piezoelectric ceramic/polymer composite transducers using fused deposition of ceramics. J. Am. Ceram. Soc. **83**(1), 124–128 (2000)
14. Cao, V.A., Lee, S., Kim, M., Alam, M.M., Park, P., Nah, J.: Output power density enhancement of triboelectric nanogenerators via ferroelectric polymer composite interfacial layers. Nano Energy **67**, 104300 (2020)
15. Sakamoto, W.K., Marin-Franch, P., Das-Gupta, D.K.: Characterization and application of PZT/PU and graphite doped PZT/PU composite. Sens. Actuators A **100**(2–3), 165–174 (2002)

16. Satish, B., Sridevi, K., Vijaya, M.S.: Study of piezoelectric and dielectric properties of ferroelectric PZT-polymer composites prepared by hot-press technique. J. Phys. D Appl. Phys. **35**(16), 2048 (2002)
17. Thomas, P., Ashokbabu, A., Vaish, R.: Structural, thermal and dielectric properties and thermal degradation kinetics of nylon 11/$CaCu_3Ti_4O_{12}$ (CCTO) nanocomposites. J. Therm. Anal. Calorim. **141**, 1123–1135 (2020)
18. Liu, X.F., Xiong, C.X., Sun, H.J., Dong, L.J., Liu, Y.: Piezoelectric and dielectric properties of PZT/PVC and graphite doped with PZT/PVC composites. Mater. Sci. Eng. B **127**(2–3), 261–266 (2006)
19. Xie, H., Hou, C., Qu, Y., Tian, H., Lu, H., Wu, J., Yang, S., Ma, Y.: N-SiOx/graphite/rGO-CNTs@ C composite with dense structure for high performance lithium-ion battery anode. J. Energy Storage **72**, 108452 (2023)
20. Gupta, S., Wang, D., Shetty, S., Meddeb, A., Dursun, S., Randall, C.A., Trolier-McKinstry, S.:. Cold sintering of PZT 2–2 composites for high frequency ultrasound transducer arrays. Actuators **10**(9), 235 (2021). (MDPI)
21. Da Costa Rodrigues, A., Bastos, I.N., Kappel, M.A.A., Nascimento, C.R., Ferreira, L.S., da Silva, A.L.N.: Micromechanical property study of nylon 11 and organoclay systems for offshore flexible pipe. Fibers Polym. **22**(11), 3172–3182 (2021)
22. Tiwari, V., Srivastava, G.: Structural, dielectric and piezoelectric properties of 0–3 PZT/PVDF composites. Ceram. Int. **41**(6), 8008–8013 (2015)
23. Sharma, M., Negi, C.M.S., Alvi, P.A., Yadav, V., Kumar, U.: Carbon nanomaterial-based composites as an active layer for optoelectronic device application: a comparative study. J. Mater. Sci. Mater. Electron. **34**(32), 2148 (2023)
24. Ernest Ravindran, R.S., Thomas, P., Renganathan, S.: Effect on dielectric, structural and thermal behaviour of $CaCu_3Ti_4O_{12}$ in a nylon 11 matrix. Bull. Mater. Sci. **42**, 1–7 (2019)
25. Wu, J., Fu, Y., Hu, G.H., Wang, S., Xiong, C.: Effect of stretching on crystalline structure, ferroelectric and piezoelectric properties of solution-cast nylon-11 films. Polymers **13**(13), 2037 (2021)
26. Dang, Z., Shen, Y., Fan, L., Cai, N., Nan, C., Zhao, S.: Dielectric properties of carbon fiber filled low-density polyethylene. J. Appl. Phys. **93**(9), 5543–5545 (2003)
27. Srinivas, S.N.S., Sharma, M., Yadav, V., Kumar, U.: Comprehensive study of $CsGeBr_3$ perovskite: optical, electronic, and thermoelectric properties. Emergent Mater. **6**(5), 1685–1696 (2023)
28. Tang, J., Liu, J., Huang, H.: Dielectric, piezoelectric and ferroelectric properties of flexible 0–3 type PZT/PVDF composites doped with graphene. J. Electron. Mater. **48**, 4033–4039 (2019)
29. Wang, J., Deng, Q., He, Y., Feng, Y., Yang, Y.: Phase structures, loss storage, damping, noise absorption, and mechanical properties of nano-graphite/lead zirconium titanate/room-temperature vulcanized silicone rubber composites. J. Mater. Eng. Perform. **32**(5), 2280–2287 (2023)
30. Sonia, Y., Agrawalb, R., Yadavc, V., Singhd, P., Singhe, S., Ranif, U., Vermaf, A.: Electronic and optical properties of novel double perovskite compound Cs2RbInI6. J. Ovonic Res. **19**(5), (2023)
31. Pal, A., Sasmal, A., Manoj, B., Rao, D.P., Haldar, A.K., Sen, S.: Enhancement in energy storage and piezoelectric performance of three phase (PZT/MWCNT/PVDF) composite. Mater. Chem. Phys. **244**, 122639 (2020)
32. Ciprari, D., Jacob, K., Tannenbaum, R.: Characterization of polymer nanocomposite interphase and its impact on mechanical properties. Macromolecules **39**(19), 6565–6573 (2006)
33. Murugaraj, P., Mainwaring, D., Mora-Huertas, N.: Dielectric enhancement in polymer-nanoparticle composites through interphase polarizability. J. Appl. Phys. **98**(5), (2005)
34. Or, Y.T., Wong, C.K., Ploss, B., Shin, F.G.: Polarization behavior of ferroelectric multilayered composite structures. J. Appl. Phys. **93**(7), 4112–4119 (2003)

Pyroelectric Properties of Electroceramics

Priyambada Mallick, Manoj Ku Panda, Subhashree Lopamudra, Dipan Ku Das, Srikanta Moharana, and Santosh Kumar Satpathy

Abstract This chapter presents an overview of pyroelectric properties, including principles such as the pyroelectric effect, i.e., the generation of electric polarization due to the change in temperature and polarization shifts, and their importance in advanced novel technologies. It also explains the effect of novel characteristics, including coefficient of pyroelectricity, thermal responsivity, and figure of merit, on the efficiency and sensitivity of pyroelectric devices. This chapter focuses on the numerous relevant and practical applications of pyroelectric materials in thermal energy harvesting from the external environment, pyroelectric sensing, imaging, wastewater treatment, and disinfection, among other things. Some electroceramic materials, including lead Zirconate Titanate (PZT), Potassium niobate ($LiNbO_3$), lithium niobate ($LiNbO_3$), Lithium Tantalate, and langasite ($La_3Ga_5SiO_{14}$), with notable pyroelectric properties, are presented to showcase the wide spectrum of pyroelectric applications. This chapter also highlights the role of material composition, crystal structure, and processing techniques in tailoring pyroelectric properties to suit specific application requirements. Finally, developments and future prospects in the pyroelectric properties of electroceramics for advanced energy-efficient devices will be explored in this context, as will defects in them. This study may provide considerable direction and motivation in the development of pyroelectric materials by providing a comprehensive conclusion and analysis of the most recent research developments.

P. Mallick · M. K. Panda · S. Lopamudra · D. K. Das · S. Moharana · S. K. Satpathy (✉)
School of Applied Sciences, Centurion University of Technology and Management, Jatani, Odisha, India
e-mail: santosh.satpathy@cutm.ac.in

P. Mallick
Basic Science and Humanities, Gandhi Institute For Technology, Bhubaneswar, India

U. Kumar (ed.), *Defects Engineering in Electroceramics for Energy Applications*, Engineering Materials, https://doi.org/10.1007/978-981-97-9018-0_18

1 Introduction

Ceramics derived from the Greek term keramous which means potter's clay or product which are obtained from clay and fired. Generally, ceramics materials are hard, brittle, resistance to heat and corrosion which is made after the process of shaping and then firing an inorganic material (clay) at high temperature (900–1200 °C range) [1]. Previously man-made ceramics were based on pottery pots, vessels, vases etc., which are mixed with other materials such as silica or by itself then they are harden and sintered in fire. Pottery gains their shape by numerous methods and its chemical stability after firing. Now a days ceramics are used in various sectors like industrial, domestics, building products, and materials developed for the use in advance ceramics engineering, such as semiconductor. The evolution of advance ceramics from pottery has expand the meaning of the word "ceramics". It is an inorganic, non-metallic materials that undergo firing and sintering process to gain their mechanical strength. Initially ceramics are used in electrical sector because of their stability in extreme weather and their high electrical resistivity such as SiC, which is a material utilized in heating unit.

1.1 Pyroelectricity

Pyroelectricity is a phenomenon that occurs when specific materials generate an electric charge or voltage in response to a change in temperature. This effect is occurred due to the rearrangement of electric dipole within the materials due to fluctuations in temperature [2, 3]. Ferroelectric materials have wide temperature dependent spontaneous polarization range so, they are expected to be strong pyroelectric. The word "pyroelectric" is derived from the Greek words "pyro," which represents fire or heat, and "electric," which represents the formation of electric charge. Based on pyroelectric, the polarization P of a material is a function of temperature T. In short circuit conditions with electrodes which are oriented normal to the polar direction, it explain the link between rate of temperature change (dT/dt), produced charge (Q), surface area of the material (A), pyroelectric current (I), and pyroelectric coefficient.

The majority of pyroelectric material that possesses a non-centrosymmetric crystal structure, which means that atomic arrangement, has no center of symmetry. These materials contain permanent electric dipoles within the structure, as the opposite charges are separated at the opposing ends. When there are fluctuations in temperature in pyroelectric materials, the atoms and molecules within the crystals lattice starts to vibrate.

Pyroelectric materials are generally hard crystals, but, soft pyroelectric can also be formed by using electrets [4]. As soon as the polarization of the materials changes, electric charges accumulate on its surfaces. A voltage difference is produced between the electrodes if the electrodes attached to its surfaces. This voltage difference can be measured and controlled for various applications. Pyroelectric materials are very

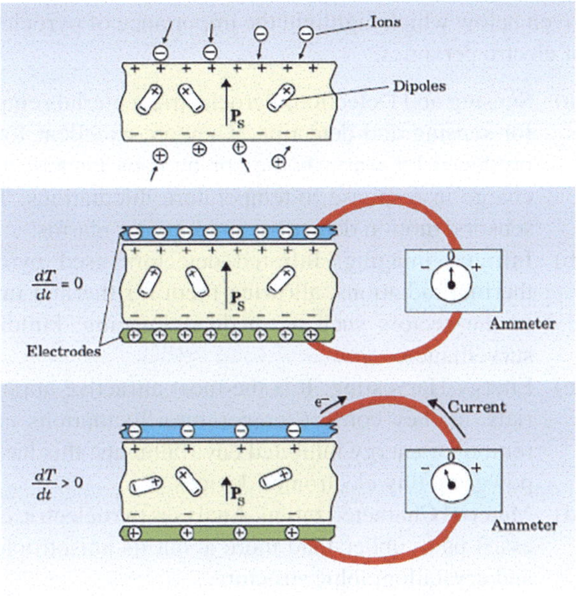

Fig. 1 Simple model of pyroelectricity from "pyroelectricity: from ancient curiosity to modern imaging tool" by Lang [5]

often used in various devices and in different applications such as, Infrared Sensors, thermal imaging, energy harvesting, and temperature monitoring etc. Pyroelectric materials are beneficials for a wide range of applications because of their special properties (Fig. 1).

1.2 Electro-Ceramics

Electro ceramics are a group of ceramics materials which are able to perform their electrical function because of their conductivity, either ionic or electronic. Ceramics are well known for their mechanical, thermal, and chemical stability, but also for their exclusive electrical, optical, and magnetic properties gained importance in several crucial technologies like communication, energy conservation and storage. Such type of materials is known as electro ceramics, which are recognized for their ability to generate an electric charge in reaction to temperature fluctuations.

1.3 Significance of Pyroelectric Properties in Electro Ceramics

The pyroelectric properties of electro ceramics confine remarkable importance due to their wide range of applications and the special capabilities. Several key aspects are given below which highlight the importance of pyroelectric properties in the domain of electro ceramics.

(a) Sensing and Detection: Pyroelectric materials employed in various applications for sensing and detection. They are excellent for detecting infrared radiation produced by warm bodies or humans because they can generate an electric charge in response to temperature fluctuations. This is particularly useful for sensors, motion detectors, and intruder alarms.
(b) Infrared imaging: Infrared detectors used pyroelectric materials to detect thermal radiations, allowing them for thermal imaging. This has applications in the sectors such as medical imaging, building diagnostics and military surveillance.
(c) Energy Harvesting: It is the most attractive applications of pyroelectric materials, as they convert temperature fluctuations into usable electric energy. In remote or energy inhibited environments, this has significant consequences for powering tiny electronics devices.
(d) Material Characterization: Analyses pyroelectric characteristics of materials can assist us to understand more about its anisotropic properties, thermal actions, and crystallographic structure.

2 Pyroelectric Effect Theory

The pyroelectric theory describes the phenomena of pyroelectricity in materials. It describes how electric polarization is generated by specific crystals and substances and eventually electric charge due to temperature fluctuations. The theory based on the principle of electric dipoles, crystal symmetry, and thermally induced changes in molecular configurations. In the absence of electric field, pyroelectric materials show spontaneous polarizations. Ferroelectric materials have spontaneous electric polarization but when an external field is applied, it can be reversed. So, all pyroelectric materials show piezoelectric properties, but reverse is not true. As a result, pyroelectric takes place below curie temperature and when the materials are heated above curie temperature, the atom come back to its stable positions (Fig. 2). So, the physical opposite of pyroelectric effect is known as electrocaloric effect.

2.1 Pyroelectric Effect: Concept and Historical Development

In minerals, pyroelectrics charges are produced due to the opposite sides of asymmetric crystals. Generally, the direction of charge's propagation is constant throughout a pyroelectric material, where in some materials the direction is diverse by an electric field. These materials are said to display ferroelectricity. Pyroelectric are usually hard crystals, however, electrets (a dielectric material that has been designed with a stable electric) can be used for generating soft pyroelectricity. The

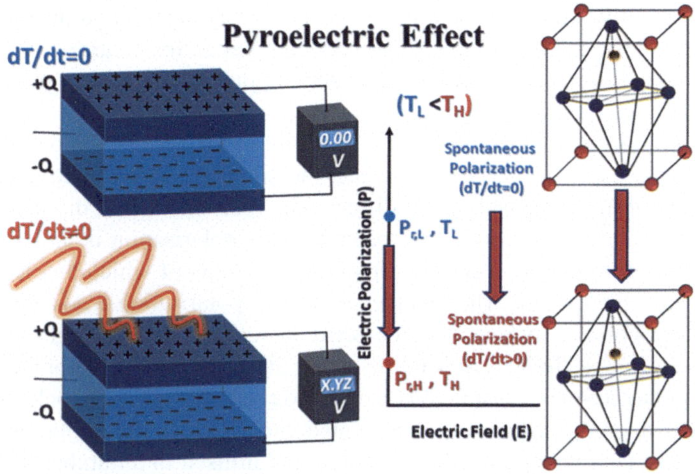

Fig. 2 Schematic of pyroelectric effect where time-dependent thermal fluctuations cause the displacement of the central atom in a non-centrosymmetric perovskite and results in an output voltage or curren [6]

change in net polarisation with respect to temperature change is used to measure pyroelectricity whereas even poor materials do not show a net dipole moment under normal circumstances.

Johann Georg Schmidt introduced pyroelectric effect in 1707, who said that," [hot] tourmaline could attract the ashes from the warm or burning coals, as the magnet does iron, but also repelling them again [after the contact]" [5]. Later in 1717 Louis Lemery discovered that small scarps of non-conducting materials were first attracted to tourmaline, but then they repelled by it once they come in contact with the stone [7]. In 19th century, pyroelectricity research gets more sophisticated. In 1824, Sir David Brewster gave its present name. In 1878 [8], Wiliam Thomson and Woldermar Voigt in 1897 [9] both contributed to the development for the theory behind pyroelectricity. Pierre Curie and Jacques Curie are two brothers, who explore pyroelectricity in 1880s, which led them to discover some of the mechanisms behind piezoelectricity [10].

2.2 Theoretical Basis of Pyroelectricity

Pyroelectric materials are typically a non-centrosymmetric crystal structure. Generally, in centrosymmetric crystal structure, the configuration of atom or ions is equal and opposite around a central point, which causes the cancellation of any net electric dipole moment. Whereas, in non-centrosymmetric crystal, the distribution of charge is not symmetric which leads to the presence of permanent electric dipoles inside the crystal lattice. In pyroelectric crystal, each electric dipole has electric dipole. The positive end of these dipoles has positively charged atoms or group of atoms, while the negative end has negatively charged atoms or group of atoms. In equilibrium condition, these dipoles are arranged randomly, therefore producing no net polarisation. In pyroelectric materials, when temperature fluctuates, the lattice structure vibrates due to thermal energy. This vibration causes the dipoles to rotate and reorient themselves. As a consequence, the arrangement of positive and negative charges within the crystal lattice differs. The reorientation of dipoles, produce a net electric polarisation within the material. This polarisation is the outcome of the alignment of dipoles in a certain direction, as results of which an electric field produce within the crystal. The change in electric polarisation leads accumulates electric charges on the surface of pyroelectric materials. These charges are mainly positive and negative charges, depending on the direction of the temperature change and initial polarisation of crystal. When electrodes are placed on pyroelectric material's surface, a voltage difference is generated due to charge accumulation between the electrodes. This voltage can be measured and utilised in a number of ways for various applications.

The pyroelectric theory explains how electric polarisation and charge accumulation results for the rearrangement of electric dipoles within a crystal lattice in response to temperature fluctuations. Pyroelectric materials are used in variety of practical applications, including temperature monitoring, motion detection, energy harvesting, and thermal imaging.

2.3 Mechanism Describing Pyroelectric Effect

In macroscopic phase, pyroelectric originates from an intrinsic ("spontaneous") electrical polarisation or from a constant electric dipole moment of the crystal's unit cell. Therefore, polarisation is altered by heating and cooling, causing electric charge to form on specific crystal faces [11]. Pyroelectric effect occurs when specific materials develop an electric charge or voltage in response to a temperature fluctuation. This effect occurs within these materials due to the asymmetric distribution of atoms or molecules within the crystal lattice. The main features of pyroelectric are the change in dipole orientation and charge separation occurs due to temperature fluctuations. The mechanism that describes pyroelectric effect are as follows:

(i) Crystal Structure and Dipole Moments

Pyroelectric materials are naturally asymmetrical crystal structure in the arrangement of positive and negative charges, resulting in permanent electric dipole moments within the crystal lattice. These dipoles can arise from several conditions, such as atom have different electronegative, ionic displacement and other crystal defects.

(ii) Thermal Excitation

When temperature fluctuates in pyroelectric materials, due to thermal energy the atoms or molecules start to vibrate and oscillate around their equilibrium positions within the crystal lattice. This vibration altered the balance of the charges temporarily within the crystal structure.

(iii) Charge in Dipole Orientation

As the temperature fluctuates, the dipole materials start to reorient due to the thermal vibration within the materials. This reorientation is because of the changes in the equilibrium positions of atoms or ions. It is ensured that the dipole reorientation is not symmetric with regard to positive and negative charges by the asymmetric crystal structure.

(iv) Surface charge Separation

When the dipole moments reorient, the charge get temporarily separated inside the crystal structure, causes the accumulation of positive and negative charges at the opposite surface of the material. Due to this separation of the charges, an electric field is generated across the materials.

(v) Electric Potential Difference and Voltage Generation

The accumulation of the charges at the surface generates an electric potential difference between both the surfaces of the pyroelectric material. If the material is electrically connected then this potential difference allows a current to flow. If the pyroelectric material is used in an open circuit, the potential difference between its surface develops a voltage. Alternatively, if the material is connected to a circuit, the created potential difference can generate an electric current.

3 Various Types of Pyroelectric Materials

The unique property of pyroelectric materials (i.e., they exhibit a temporary voltage and electric potential difference, when there is a fluctuation in temperature) makes them useful in various sector, such as detectors, sensors, and energy harvesting devices. There are various types of pyroelectric materials each having its own characteristics and applications. Some of them are given below:

1. Natural pyroelectric materials (tourmaline and quartz)
2. Synthetic Pyroelectric materials (Lithium tantalate ($LiTaO_3$) [12], Lithium Niobate ($LiNbO_3$)) [13]
3. Organic pyroelectric materials (polyvinylidene fluoride (PVDF)) [14]
4. Inorganic Pyroelectric Materials (Lead zirconate titanate (PZT) [15], Bismuth Ferrite [16], barium titanate ($BaTiO_3$) [17], Strontium Titanate,($SrTiO_3$) [18]
5. Pyroelectric Polymers (PVDF) [19]
6. Composite pyroelectric structure [20]

7. Emerging Pyroelectric Nanomaterials [21] (Nanostructured ceramics, nanoparticles, and thin film).

3.1 Natural and Synthetic Pyroelectric Materials

Natural pyroelectric materials: Natural pyroelectric materials are those minerals or crystal that exhibits pyroelectric properties in their natural occurring state [22]. These materials exhibit strong pyroelectric properties and very often use in temperature sensing device such as sensors and detectors.

Some of natural pyroelectric substances are given in Fig 3.

Some studies claim that pyroelectricity was 1st observed in quartz. Natural pyroelectric materials are limited to their intrinsic properties but still useful in niche applications.

Synthetic pyroelectric materials

Scientists and engineers have discovered a wide range of synthetic pyroelectric materials (or artificially generated materials) with improved performance. Some of the examples of artificial pyroelectric materials are as follows.

Lead Zirconate Titanate (PZT), Lithium Tantalate ($LiTaO_3$) [12], Barium Strontium Titanate (BST) [23] etc. are most commonly used synthetic pyroelectric materials. They exhibit strong pyroelectric properties, high dielectric constant and tunable properties which make them suitable for detectors in infrared radiation, sensors, transducers, and actuators. It is also used in microwave technology, as a material for phase shifters.

Fig. 3 Different types of natural pyroelectric substances

The main advantages of these materials are to perform distinct performance requirements, including sensitivity, temperature range, and response time.

3.2 Organic and Inorganic Pyroelectric Materials

Organic and inorganic materials also exhibit pyroelectric effect, where pyroelectric materials generate an electric charge when there will be a temperature change.

Organic pyroelectric materials

These are the materials usually composed of organic compounds, which contain carbon-hydrogen (C–H) bond. This category involves certain organic crystals and synthetic organic polymers [24]. They have various advantages such as flexible, lightweight, low sensitive, ease of processing and a limited temperature range. As organic pyroelectric materials are flexible, they can be mould and customised for certain applications. They are mainly use in thermal imaging device, sensors, and energy harvesting as given in Fig. 4.

Inorganic pyroelectric materials

These materials are composed of inorganic compounds, mainly crystalline structures. These materials are rigid, highly sensitive, and have a wide temperature range. In

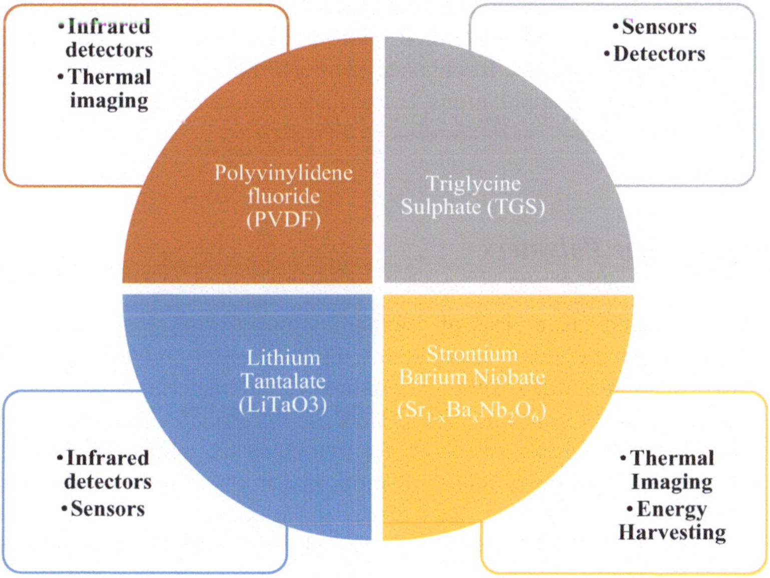

Fig. 4 Examples of organic pyroelectric materials and their uses

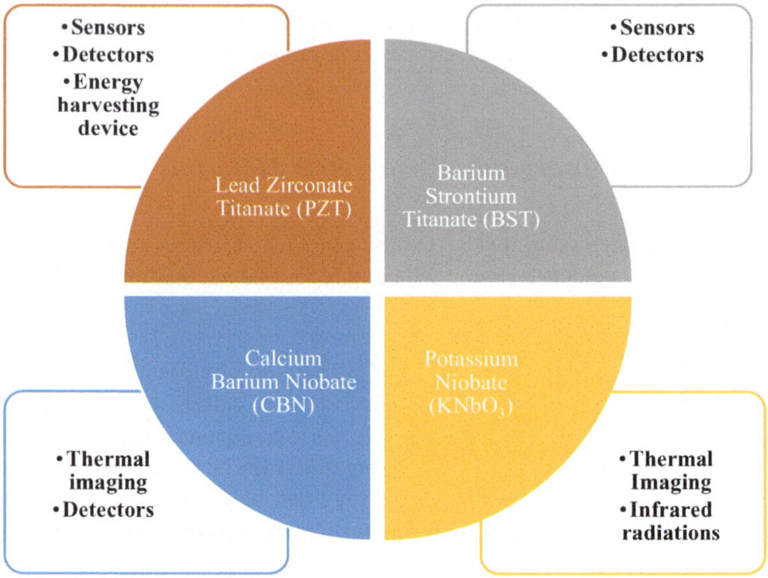

Fig. 5 Examples of inorganic pyroelectric materials and their uses

comparison to organic Pyroelectric materials, they have higher Pyroelectric coefficients and also have better thermal stability [25]. They are used in high performance infrared detectors, motion sensors, and thermal imaging cameras as shown in Fig. 5.

Both pyroelectric materials have their own benefits and applications depending on their specific requirements. To enhance the performance of pyroelectric device and expand their range of applications, researchers are exploring new materials and refining the existing ones.

3.3 Pyroelectric Polymers

Pyroelectric polymers are a class of polymers which exhibits pyroelectricity, the ability of a certain materials to generate a temporary electric charge or voltage in response to temperature fluctuations. These are organic material which consists of long repeating chains of molecular units. They are lightweight, flexible, and can be easily customized, so they are attractive for pyroelectric applications. When there are fluctuations in temperatures caused by motion of an object or heat radiation, they produce a signal which can be detected and used in various sectors [26]. They are used in energy harvesting devices, mainly low power electronic devices or sensors can be operated, which converts temperature fluctuations into electrical energy. They can also detect defects and irregularities in crystals, by measuring temperature fluctuation as given in Table 1. These polymers can be customised in various form, like thin films,

Table 1 Some examples of pyroelectric polymers

Polymer materials	Applications
Polyvinyl fluoride (PVDF)	Sensor, thermal imaging, and energy harvesting devices
Copolymer of PVDF- Trifluoro ethylene (PVDF-TrFE) [27]	Infrared detectors and pyroelectric sensors
Polyurethane	Motion sensors and energy harvesting devices
Polyvinylidene chloride (PVDC) [28]	Pyroelectric sensors and detectors

sheets and fibres. Pyroelectric polymers are often polarized during manufacturing, to improve the pyroelectric response. Because of its unique properties and potentials for various applications, it is an active area of research and development.

3.4 Composite Pyroelectric Structures

In composite pyroelectric structure, multiple pyroelectric elements are combined or layered to enhance their pyroelectric properties and functionality. These multiple pyroelectric elements are combined to form devices. Pyroelectric materials generate an electric charge when there are temperature variations [29]. This structure is customised to enhanced the performance of pyroelectric devices for various applications/purposes. They are mainly used to enhanced the sensitivity of pyroelectric devices. As we combine multiple pyroelectrics, each having its unique properties, the composite structure can detect temperature fluctuation and a extend range of frequencies. The choice of composite pyroelectric structure is to stack several pyroelectric layers on top of each other. These pyroelectric layers have different pyroelectric coefficients, which can be modified to improve the device's performance for a specific application. These devices can be customised to specific designs to respond to specific temperature changes or temperature gradients, which allow them for the applications like thermal imaging or motion detection. By integrating different frequencies of pyroelectric materials, they can detect a wide range of temperature and frequencies. This is particularly helpful in applications like infrared spectroscopy. The composite structure can be customised for powering low energy devices by utilising the energy from environment. The selection of materials in composite pyroelectric materials is very much crucial, as they have different pyroelectric coefficients, dielectric constants, and thermal properties. These composites are manufactured by the techniques such as thin film deposition, sintering, and layer by layer assembly. These processes make sure that each pyroelectric layer is properly combined to function as a single unit. They are used in various applications like motion sensors, security systems, thermal cameras, energy efficient devices, and in scientific instruments for measuring temperature fluctuations.

Table 2 Some examples of pyroelectric nanomaterials and their applications

Emerging proelectric nanomaterials	Applications
2D Materials (Graphene, transition metals dichalcogenides (TMD))	Skin attached sensors and wearable technology
Lead zirconate titanate nanowires (PZT)	Energy harvesting devices, Nano sensors, nano generators
Barium titanate nanoparticles ($BaTiO_3$)	Nanoscale heat sensors and energy scavenging systems
Ferroelectric polymer nanocomposites	Wearable electronics and flexible sensors for temperature monitoring
Quantum dots	High resolution thermal imaging and efficient energy conversion
Nanostructured thin film	Thermal cameras and noncontact temperature sensing

3.5 Emerging Proelectric Nanomaterials

These Pyroelectric materials are subsets of pyroelectric materials which are designed at nanoscale level. In recent few years they gained significant attentions for their unique properties and potentials for various applications. They are frequently use in research work as researchers explore new techniques and applications. They are very much use in energy harvesting, thermal imaging, and sensing which is given Table 2.

4 Factors Influencing Pyroelectric Properties

The electro-ceramic materials having spontaneous electric polarization which can change with temperature have pyroelectric properties which can be altered by several factors.

4.1 Composition and Doping

The composition of a material can be reformed by changing its purity in material choice, crystal structure and polarization orientation. In PZT ceramics modifying the ratio of lead, zirconium and titanium affect the pyroelectric coefficients and temperature sensitivity. Changing the stochiometric ratio in the material composition can lead to formation of secondary phase or defects which can affect the pyroelectric properties. Influencing the microstructures such as grain size or domain structure can optimize the pyroelectric properties. An alternative way to modify the pyroelectric properties is doping, which improve the polarization and change the phase transition.

For example, doping of rare earth element like lanthanum or neodymium to PZT can increase the pyroelectric property.

4.2 Crystal Structure and Symmetry

The arrangement of atoms inside a crystalline material is known as crystal structure of the materials. The strong pyroelectric behavior is shown by the materials having non-centrosymmetric crystal structures like anisotropic materials. The materials having trigonal, tetragonal, and orthorhombic crystal structures exhibit well-defined pyroelectric properties. pyroelectric materials generally lack inversion symmetry.

The materials having high symmetry like cubic crystals exhibit weak pyroelectric response whereas the materials having low symmetry exhibit most profound pyroelectricity. The reason behind this response is with temperature change lower symmetry allows in the great degree of freedom for the charge movement but in higher symmetry the cancellation of dipole happens with temperature variation.

4.3 Polarization

The inherent dipole moment inside the crystal even in the absence of electric field is known as spontaneous polarization. The change in polarization due to temperature results in pyroelectricity effect. The asymmetric arrangement of charged ions gives rise to spontaneous polarization within the crystal lattice. The materials having higher spontaneous polarization results in strong pyroelectric effect. The polarization of a pyroelectric material changes with temperature. The average dipole moment changes with changing temperature of the material. An electric charge is generated on the materials surface due to the increasing pyroelectric effect caused by temperature-dependent polarization.

The effect in which the pyroelectric results of a material depend upon the thermal conductivity of the material is known as thermal hysteresis. Due to this effect the materials show different pyroelectric behavior when exposed to heating or cooling cycles. The phase transition and domain switching of a material can influence the hysteresis loop curve of pyroelectric response. The pyroelectric response of the material can be altered by applying external electric field. The applied external electric field results in affecting temperature induced changes in polarization. The pyroelectric current is generated by the interaction of applied electric field and the pyroelectric response.

Ferroelectrics have a strong pyroelectric effect close to their phase transition temperatures and can switch polarization when an electric field is applied. As result, they are frequently the building blocks of pyroelectric EH devices. The best performances are shown by $(1 - x)Pb(Mg_{1/3}Nb_{2/3})$-$xPbTiO_3$ (PMN-PT) single crystals, but they are fragile and expensive to produce because the single crystals are grown by the Bridgman technique. The performance of polyvinylidene fluoride (PVDF) is compared to bulk ceramics, but with a much larger voltage response (up to six times that of PZT). Given that it is inexpensive, sturdy and stable under significant changes in temperature or electric field, this material appears to be very intriguing. The PZT- poly (vinylidene fluoride- hexafluoropropylene) composite, also known as PVDF-HFP, is another top-notch material. Due to its flexibility, this type of material is simple to use, but it has strong temperature dependence [30].

4.4 Temperature Change

The phase transition of electro ceramic materials associated with changes in crystal structure and spontaneous polarization which happens at specific temperature ranges. When the phase transition occurs between temperatures ranges the rapid changes in the polarization of the materials gives strong pyroelectric response. The temperature range during the phase transitions decide the direction and magnitude of pyroelectric charge. Two phase changes are seen in the undoped $PbZr_{1\text{-}y}Ti_yO_3$ system at 255 and 70 °C when $y = 0.1$. High temperature ferroelectric rhombohedral to low temperature ferroelectric rhombohedral phase transition is the lower transition, whereas the upper transition is the curie point [paraelectric to high temperature ferroelectric rhombohedral phase transition]. A key area of research has been the potential use of the high pyroelectric coefficient connected to this lower transition and the capacity to change the transition temperature [31].

The temperature at which the phase transition of materials undergoes from ferroelectric to paraelectric is called as Curie temperature. It is a critical temperature point range for ferroelectric materials like PZT and $BaTiO_3$. The spontaneous polarization of the materials disappears above curie temperature. The change in temperature with in the Curie temperature range changes the polarization significantly which results in measurable pyroelectric effect.

The measurement of response electrochemical materials pyroelectric properties are to temperature changes are known as temperature sensitivity. The temperature sensitivity is determined by the magnitude of the pyroelectric coefficient, thermal conductivity and specific heat capacity of the materials. The materials having high temperature sensitivity gives a strong pyroelectric response within a wide range of temperature. The high temperature sensitive electro chemical materials are used in precise temperature measurements. The polarization and pyroelectric charge do not depend on the current temperature but also on the thermal history of the materials. This effect on the material is called hysteresis. Within a particular temperature range the electro ceramic materials displays hysteresis in their pyroelectric response. As the materials remembers the memory of previous temperature changes, they can be used in thermal detectors or infrared sensors where hysteresis of the material is very important.

The PZN-4.5PT single crystal, with dimensions of $8 \times 4 \times 1\ mm^3$ and 8000 kg/m^3 density at 1300 °C, is subjected to a bipolar cyclic electric field. To prevent arcing, the sample's two electrodes are electrically connected to the power source and submerged in a silicon oil bath. This test was carried out using switching cycles of a bipolar triangular voltage waveform with amplitude of 2 kV/mm. The polarization starts to rapidly decelerate because the cycling field was gradually increased from zero to its maximum field value within 0.25 s. The observed shift of the polarization hysteresis loop to lower polarization values supports the existence of fatigue. Between two cases, there is a 10% remnant polarization difference [32].

The properties of electrochemical materials are influenced by the chosen temperature range of operation. The selection of the electrochemical material is depended upon the required application having temperature requirements. A process in which the external electric field is applied at a high temperature to achieve the polarization in the electro ceramic material is called as poling. A materials pyroelectric response can be affected by alignment and stability of domains which caused by the temperature range and conditions during the poling.

4.5 Thermal Conductivity

In order to provide low-power autonomous sensors, pyroelectric cells based on manufactured screen-printed PZT and commercial PVDF films are proposed as thermal energy harvesting sources. The electrical model of the cells represents them as a current source running in parallel with output impedance. The pyroelectric converted were heated and cooled by air-generated temperature changes. According to the theoretical model, generated currents and charges for temperature fluctuations from 300 to 360 k in a time period of the order of 100 s were in the range of 107A and 105C, respectively. The generated current was increased by the arrangement of the cells in parallel. Additionally, the generated current's dependance on pertinent technological parameters has been identified. After rectifying the current caused by cyclic temperature fluctuations, a 1 micro-F load capacitor was used to store the energy. It has been possible to achieve energies up to 0.5 mJ, which is sufficient to power typical autonomous sensor nodes for a measurement and transmission cycle [33].

The change in temperature results in generation of electric charge in pyroelectricity. A material gives an efficient pyroelectric response within a significant temperature gradient. The temperature gradient can be reduced by the high thermal conductivity by quickly dissipating heat. As the temperature changes this high thermal conductivity of electro ceramic materials reduce the effectiveness of pyroelectric effect. The response time of an electro ceramic material to the temperature change can be affected by thermal conductivity of the material. For faster thermal response time, materials with high thermal conductivity should be considered. The material selection for the application of electro ceramic materials depends upon the desired response time required in that application. Although in some cases the fast response is required for the material application but for the better temperature discrimination; the slower response is necessary. To match the requirement of the application the choosing of electro ceramic materials based on their thermal conductivity is necessary. In pyroelectric devices where slower thermal responses are necessary to maximize the pyroelectric effect the materials with lower thermal conductivity are preferred.

To reduce the heat dissipation and maintain a more significant temperature gradient the thermal insulation layer is added to electro ceramic devices. Thermal insulation is used for the electro ceramic materials present in the higher thermal conductivity surrounding environment to enhance its pyroelectric response. To control the temperature distribution and maintain a stable operating temperature in some applications such as heat sinks or thermally conductive materials for the electro ceramic device the heat management strategies are used. Multilayered structures of pyroelectric devices optimize the temperature gradients. These structures are helpful to achieve higher pyroelectric response in different thermal conductivities of electro ceramic materials.

5 Application of Pyroelectric Materials

Pyroelectric materials are versatile and can be used in various industries and technologies. These materials have a wide range of applications due to their unique property of generating an electric charge in response to changes in temperature. Their ability to detect and respond to temperature changes without the need for physical contact makes them valuable tool in a wide range of applications.

5.1 Pyroelectric Sensors

The devices that detect the changes in temperature and convert them into electrical signals by utilizing pyroelectric effect are known as pyroelectric sensors. The pyroelectric material exhibits a unique property in which spontaneous electric polarization that with temperature; which explains the operation principle of pyroelectric sensors.

The pyroelectric sensors are constructed using pyroelectric materials having non-centrosymmetric crystal structure. The non-centrosymmetric crystal structure means the centers of the positive and negative charges within the crystal lattice do not coincide. The pyroelectric materials having non-centrosymmetric crystal structure are:—lead zirconia titanate (PZT), lithium tantalate ($LiTaO_3$). Small and wearable electronic devices have drawn a lot of attention over the past few decades. Accordingly, a number of pyroelectric materials have been researched and used in the sensor industry. The human body continuously emits infrared rays while maintain a specific temperature. As a result, there is a temperature gradient between the human body and the environment that can be further tapped to produce electricity, especially in the winter. As a result, the pyroelectric effect allows for the conversion of the temperature variations into an electric potential between electrodes when PyNGs are incorporated into wearable technology [34].

Pyroelectric materials exhibit spontaneous polarization which means in the absence of external electric field there exist an electric dipole moment in the material. Due to the non-centrosymmetric nature of the material the displacement of charged ions within the crystal lattice helps in spontaneous polarization of the material. The lattice structure and the ion positions of a pyroelectric material shift with in the crystal lattice when the material undergoes a change in temperature, which results in changing the electric dipole moment of the material. The redistribution of electric charge within the material caused by the change in the dipole moment which results in the generation of electric voltage across the surface of pyroelectric materials. The sensor's electrodes which are placed on the opposite side of the material detect the electric voltage generated across the surface of the pyroelectric material. For any further control the voltage signals can be amplified and controlled. A thermal sensor, which is a typical application area for the pyroelectric energy conversion and can currently detect thermal signals, is used. If the conversion efficiency and total converted energy are high enough to charge electrical energy storage devices like a supercapacitor battery, it can also be used as a thermal energy harvester. Even though the PyEH concept was first proposed in the 1960s, there hasn't been much research done in this area. According to reports, in 2009, more than 50% of all consumed energy-primarily from electrical power generation and automotive systems-was wasted as heat. Additionally, a lot of heat energy is lost through electrical appliances like air conditioners, heat pumps, and refrigerators [35–38].

- Pyroelectric sensors do not need direct contact with the target object whose temperature is being measured or monitors. It can respond to temperature changes without the direct contact. Pyroelectric sensors most responsive to temperature changes that occur perpendicular to the direction of the spontaneous polarization within the materials.
- The thermal variations from specific directions can be selected by this property. The rate of change of temperature and properties of pyroelectric material are directly proportional to the magnitude of generated voltage.
- By choosing the appropriate pyroelectric material the sensitivity of sensor can be adjusted. The temperature fluctuations in any pyroelectric device can be rapidly detected by pyroelectric sensors. The pyroelectric sensors are used in motion detection for detection of high frequency response.

Ferroelectric polymers and their composites have caught the attention of the scientific community for the use in sensors and device applications due to their various functionalities, including flexibility and ease of fabrication [39]. Although polymers like PVDF have a low pyroelectric coefficient, their ease of fabrication for large-area thin films, superior stability to TGS under heat, vacuum and moisture and mechanical robustness make them ideal for the use. Despite having low heat conductivity and permittivity, PVDF has been widely used as sensors in products like burglar alarms [40].

Pyroelectric sensors have ability to generate an electric charge due to their ability to temperature changes. These sensors are used in various fields for detecting, measuring and monitoring temperature variations, motions and radiations.

- Infrared detection
- Thermal imaging
- Gas analysis
- Temperature measurement
- Non-contact switches
- Radiation detection
- Fire detection
- Environmental monitoring
- Medical devices
- Energy harvesting.

For the same PZT thickness of 100m, a 10mm square PZT cell produced an entropy variation roughly 25times greater than a 2mm square one. Although it was advantageous to quadratically reduce the volume of a pyroelectric cell with a smaller electrode area, the thickness of the cell needed to be significantly reduced in order to speed up temperature variation and strengthen the induced current and electrical signals. The time evolution of p from PZT cell at approximately 7.5 s, 11.5 s, 20 s. 30 s for the applied voltage of 40 v, 35 v, 30 v and 25 v respectively, was used to determine the threshold time. The measured capacitor's entropy variation was unaffected by PZT cell. The pyroelectric cell was better suited to serve as an entropy sensor for forecasting the ageing and failure characterization of mechanical or electrical systems because it could identify the peak value and time of the entropy generation rate [41].

5.2 *Pyroelectric Actuators*

The pyroelectric actuators working principle based on the pyroelectric effect, in which the electric charge is generated in reaction to temperature variations. They are frequently utilized in specialized applications where the creation of temperature changes is necessary for the system to function or where alternative actuation techniques are prohibitive. The pyroelectric actuators are constructed using pyroelectric materials having non-centrosymmetric crystal structure. The non-centrosymmetric crystal structure means the centers of the positive and negative charges within the crystal lattice do not coincide. The pyroelectric materials having non-centrosymmetric crystal structure are:—lead zirconia titanate (PZT), lithium tantalate ($LiTaO_3$). The lattice structure and the ion positions of a pyroelectric material shifts with in the crystal lattice when the material undergoes a change in temperature, which results in changing the electric dipole moment of the material. The redistribution of electric charge within the material caused by the change in the dipole moment which results in the generation of electric voltage across the surface of pyroelectric materials. Within the pyroelectric material an internal electric field is created by the redistribution of charges. The rate of change of temperature is proportional to the internal electric field. The pyroelectric actuators are designed in such a way that the mechanical deformation in the material can be induced by internal electric field. The mechanical deformation is helpful in designing the actuator by taking various forms such as bending, expansion or contraction for the required application. The amplitude and direction of the mechanical deformation can be precisely controlled by carefully regulating the temperature variations in the pyroelectric material. Depending on the needs of the given applications, the actuation might be either periodic or continuous. In some micro and nanoscale applications, pyroelectric actuators provide advantages in terms of precision and ease of use, but their usage is restricted to circumstances in which the generation of temperature changes is compatible with the system's requirements. Pyroelectric actuators are still being explored and developed for novel use in micro and nano-systems, especially when their special qualities can be advantageous.

- Microfluids and lab-on-a-chip devices
- Micro-scale optics
- Scanning probe microscopy
- Micro-actuators for nano-assembly
- Nano-indentation and nanomechanical testing
- Energy harvesting at micro/nano scale
- Microthermal systems
- Biomedical applications
- Nanoscale manipulation in nanorobotics.

The efficiency of PyEH can be increased by using different energy sources. Essentially, the design of the hybrid energy harvesters involves coupling two or more energy conversion mechanisms, such as the pyroelectric, piezoelectric, photothermal and thermoelectric processed, without allowing any one of them to inhibit the operation of the others. The hybrid energy harvesters that combine pyroelectric and piezoelectric systems will be the focus of this section [42]. Since all pyroelectric materials are also piezoelectric materials, combining both energy harvesting techniques has garnered a lot of interest in the scientific community. The relevant equations for the pyroelectricity and piezoelectricity also closely resemble one another because the temperature change T is equivalent to the stress in the latter's equations [35–38].

5.3 Pyroelectric Arrays

The devices which consist of an array of pyroelectric sensors which arranged in a grid or matrix are called pyroelectric arrays. The pyroelectric arrays help in capture and detect the infrared radiation pattern across the surface and the temperature variations. The working of pyroelectric arrays is depending upon the pyroelectric effect and the simultaneous measurement of multiple sensors. The pyroelectric material is the most important part of a pyroelectric array. This substance demonstrates the pyroelectric effect, which means that when its temperature changes, it produces an electric charge. Usually, the material is divided into tiny pieces, or pixels, and each pixel serves as a separate pyroelectric sensor. These pyroelectric sensor pixels are arranged in a grid or matrix to form pyroelectric arrays. Depending on the application, the arrangement may differ, but square or rectangular grids are frequently used arrangements. Localized temperature variations occur across the array's pixels as a result of incident IR radiation, which typically takes the form of heat or thermal radiation, striking the surface of the pyroelectric array. These temperature variations are brought on by variations in IR radiation intensity across the field of view. Each pyroelectric pixel produces an electric charge as a result of temperature changes brought on by the IR radiation. The rate at which the temperature changes and this charge are directly proportional. Whether the temperature is rising or falling determines the direction of the charge produced. Electronic circuity that detects the generated electric charge is connected to each individual pyroelectric sensor pixel. The readout circuitry measures the voltage signal that the charge produces across each pixel. All of the array's pixel voltage signals are read simultaneously. The array can record the spatial distribution and temporal variations of incident IR radiation or temperature changes across the field of view by examining the voltage outputs of each pixel. Depending on the particular application, it may be possible to reconstruct an image or pattern that depicts the distribution of IR radiation or temperature using the data gathered by the array. Applications like temperature profiling, motion detection or thermal imaging may benefit from this valuable information. To improve the precision and caliber of the data collected by the array, signal processing techniques may be the used. To translate voltage signals into temperature or radiometric information, this can involve noise reduction, background subtraction, and calibration.

With 128, 256 or 510 elements, the lithium tantalate chip is part of the pyroelectric linear array. A CMOS circuit processes the signals that the elements produce. The analogue circuitry, which consists of a multiplexer, an output amplifier, and a customized low-noise pre amplifier for each pixel, processes signals. The thick film substrate on which the pyroelectric chip and CMOS readout circuit are mounted is hermetically sealed inside a metal housing. An infrared-transparent window or filter lets the incident radiation pass through.

Due to their capacity to recognize changes in temperature and IR radiation patterns, pyroelectric arrays are frequently used for imaging and detection applications. These arrays allow for the capture of spatial and temporal variations in temperature or IR radiation because they are made up of numerous pyroelectric sensors arranged in a grid or matrix. Here are some typical uses for pyroelectric arrays in imaging and detection:

- Thermal imaging
- Passive infrared (PIR) motion sensors
- Gesture recognition
- Object detection and tracking
- Non-contact temperature measurement
- Flame detection
- Medical thermography
- Environmental monitoring

- Gas detection
- Spectroscopy
- Energy harvesting
- Biomedical imaging.

6 Recent Advances in Pyroelectric Research

Researchers are vigorously exploring the use of pyroelectric materials in nanoscale, to enhanced the performance of pyroelectric devices. By the use of nanomaterials, the sensitivity and response times can be improved. Research is focused on developing pyroelectric energy harvesters which can convert temperature fluctuations into electrical energy. Flexible, wearable pyroelectric sensors and energy harvesters are gaining interest as they can customised into various shapes and can merge into clothing and wearable devices. Researches are going on to create a hybrid energy harvester to capture energy from multiple sources which is formed by the integrations of pyroelectric materials with others technologies, such as thermoelectric and photovoltaic materials. These pyroelectric materials are used in various advanced sensing applications which have implements for fields like healthcare, security, and thermal monitoring. The uses of perovskites in pyroelectric materials have gained attentions for their unique pyroelectric properties with low-cost fabrication techniques. Researches are being processed to develop advanced signal processing techniques and machine learning algorithms to improve accuracy and reliability of pyroelectric sensor data. They also can be used in environmental monitoring applications to detect rapid temperature changes. A new type of pyroelectric materials was developed by Yang's group for the application of temperature sensors [43]. He developed a microwire for PyENG devices which is a transparent polydimethylsiloxane-based PZT microwire as well as a non-toxic PyENG for energy harvesting which is made up of KNbO3 nanowire. Currently, researchers are focused on developing new technique mainly materials doping, customising the nanomaterial's structure and designing new morphology. Pyroelectric materials have unique electrical properties and cutting-edge design, which can be used in energy harvesting [44, 45], temperature sensing [44, 46], wearable devices [47, 48], imaging and UV/IR detection [49, 50].

7 Challenges and Future Goals

Challenges

1. It is a challenge to develop pyroelectric materials with improved property such as wide temperature range and high pyroelectric coefficients.
2. To maximise the output with minimum energy loss is always a challenge, specifically for small scale and low temperature variations.

3. It is essential for practical applications in several industries to ensure the long-term stability and reliability of pyroelectric devices in harsh environmental conditions.
4. Reducing the manufacturing cost is a vast challenge.
5. Without affecting the performance, it is very difficult to integrate pyroelectric device into small, portable, and wearable devices. While miniaturing, manufacturing is a big challenge.
6. Developing advanced algorithms and machine learning techniques to improve data accuracy.

Goals

1. Developing improve versions of pyroelectric materials which have superior pyroelectric properties to enhanced the performance of devices like sensors and energy harvesters.
2. To expand the range of applications mainly powering low-energy devices and sensors, to encourage energy efficiency.
3. Focus on wearable and flexible pyroelectric devices which can be customised for different applications like healthcare, electronics and other applications.
4. To detect multiple environment factors (e.g., humidity, temperature etc.), a multifunctional sensor has to be created to provide more inclusive data.
5. The use of pyroelectric sensors should expand in the field of environment such as monitoring wildfire detections, monitoring air quality, and climate research.
6. Integration of pyroelectric devices with artificial intelligence (AI) to achieve automation and more accuracy to analysis data.
7. Search eco-friendly approaches for the synthesis and manufacture of pyroelectric materials to lessen environmental effects.

8 Conclusion

Pyroelectric materials are generally found in the class of electro ceramics, which has the ability to generate electric charge in response to temperature fluctuations. The materials having pyroelectric property are used in various applications like sensors, detectors, energy harvesting etc. The applications of pyroelectric materials have diverse field ranging from thermal imaging, energy harvesting and motion sensing to environmental monitoring. Research continues to develop new electro ceramics materials and improved the performance of existing materials. The materials which mainly focused are composite materials and nanomaterials. Researchers are mainly focused on the miniaturisation and integrating pyroelectric sensors and device for the portable, compact and flexible formats, like wearable devices. The materials should produce in the industrial level to ensure consistency, reliability and compatibility for different devices. Current status of the research is to improve pyroelectric properties, materials and applications. The study of pyroelectric properties in electro ceramics is a dynamic field and has potentials to improve in technology, efficiency, and sensing capabilities. Researchers and scientists are exploring the field to extend the boundaries of the achievements with the materials.

References

1. Heimann, R.B.: Classic and Advanced Ceramics: From Fundamentals to Applications. Wiley & Sons (2010)
2. Kittel, C.: Introduction to Solid State Physics. Wiley & Sons (2005)
3. Webster, J.G.: The Measurement, Instrumentation and Sensors Handbook, vol. 14. CRC Press (1998)
4. Darbaniyan, F., Dayal, K., Liu, L., Sharma, P.: Designing soft pyroelectric and electrocaloric materials using electrets. Soft Matter **15**(2), 262–277 (2019)
5. Sidney, B.: Lang: pyroelectricity: from ancient curiosity to modern imaging tool. Phys. Today **58**(8), 31–36 (2005)
6. Pandey, R., Vats, G., Yun, J., Bowen, C.R., Ho-Baillie, A.W., Seidel, J., Butler, K.T., Seok, S.I.: Mutual insight on ferroelectrics and hybrid halide perovskites: a platform for future multifunctional energy conversion. Adv. Mater. **31**(43), 1807376 (2019)
7. Roscow, J., Zhang, Y., Taylor, J., Bowen, C.R.: Porous ferroelectrics for energy harvesting applications. Eur. Phys. J. Spec. Top. **224**(14–15), 2949–2966 (2015)
8. Thomson, W.: On the thermoelastic, thermomagnetic, and pyroelectric properties of matter. Lond. Edinb. Dublin Philos. Mag. J. Sci. **5**(28), 4–27 (1878)
9. Voigt, W.: Versuch zur Bestimmung des wahren specifischen electrischen Momentes eines Turmalins. Ann. Phys. **296**(2), 368–375 (1897)
10. Curie, J., Curie, P.: Développement par compression de l'électricité polaire dans les cristaux hémièdres à faces inclines. Bull. Minér. **3**(4), 90–93 (1880)
11. Bassani, F., Liedl, G.L., Wyder, P.: Encyclopedia of Condensed Matter Physics (2005)
12. Irzaman, Siskandar, R., Nabilah, N., Aminullah, Yuliarto, B., Hamam, K.A., Alatas, H.: Application of lithium tantalate ($LiTaO_3$) films as light sensor to monitor the light status in the Arduino Uno based energy-saving automatic light prototype and passive infrared sensor. Ferroelectrics **524**(1), 44–55 (2018)
13. Poberaj, G., Hu, H., Sohler, W., Guenter, P.: Lithium niobate on insulator (LNOI) for micro-photonic devices. Laser Photonics Rev. **6**(4), 488–503 (2012)
14. Saxena, P., Shukla, P.: A comprehensive review on fundamental properties and applications of poly (vinylidene fluoride) (PVDF). Adv. Compos. Hybrid Mater. **4**, 8–26 (2021)
15. Yamada, T., Ueda, T., Kitayama, T.: Piezoelectricity of a high-content lead zirconate titanate/polymer composite. J. Appl. Phys. **53**(6), 4328–4332 (1982)
16. Mallick, P., Satpathy, S.K., Behera, B.: Investigation on the structural, electrical and thermistor parameters of La-doped $BiFeO_3$–$PbZrO_3$ for energy storage devices. Bull. Mater. Sci. **45**(4), 198 (2022)
17. Hsiao-Lin, W.: Structure and dielectric properties of perovskite–barium titanate ($BaTiO_3$). Submitt. Part. Fulfillment Course Requir. MatE **115**, 3–9 (2002)
18. Mallick, P., Satpathy, S.K.: Synthesis and characterization of zinc doped strontium titanate. Int. J. Anesth Pain Med **7**, 37 (2021)
19. Hammes, P.C.A., Regtien, P.P.L.: An integrated infrared sensor using the pyroelectric polymer PVDF. Sens. Actuators A **32**(1–3), 396–402 (1992)
20. Mallick, P., Patra, R., Mohanty, D., Satpathy, S.K.: Development and characterization of copper doped perovskite-polymer composite through high-temperature technique. Sādhanā **47**(3), 134 (2022)
21. Lingam, D., Parikh, A.R., Huang, J., Jain, A., Minary-Jolandan, M.: Nano/microscale pyroelectric energy harvesting: challenges and opportunities. Int. J. Smart Nano Mater. **4**(4), 229–245 (2013)

22. Lu, Q., Li, B., Bai, F.: Electrical effect and influence factors of tourmaline. Charact. Miner. Metals Mater. **2016**, 83–90 (2016)
23. Bunting, E.N., Shelton, G.R., Creamer, A.S.: Properties of barium-strontium titanate dielectrics. J. Am. Ceram. Soc. **30**(4), 114–125 (1947)
24. Çapan, R.: Organic pyroelectric materials for device applications. Balıkesir Üniversitesi Fen Bilimleri Enstitüsü Dergisi **12**(1), 75–90 (2010)
25. Zhang, D., Wu, H., Bowen, C.R., Yang, Y.: Recent advances in pyroelectric materials and applications. Small **17**(51), 2103960 (2021)
26. Malyshkina, O.V., Lisitsin, V.S., Dec, J., Łukasiewicz, T.: Pyroelectric and dielectric properties of calcium barium niobate single crystals. Phys. Solid State **56**, 1824–1827 (2014)
27. Furukawa, T.: Piezoelectricity and pyroelectricity in polymers. IEEE Trans. Electr. Insul. **24**(3), 375–394 (1989)
28. Ploss, B.: Dielectric nonlinearity of PVDF–TrFE copolymer. Polymer **41**(16), 6087–6093 (2000)
29. Hsieh, T.H., Ho, K.S.: Effects of thermal stability on the crystallization behavior of poly (vinylidene chloride). J. Polym. Sci. Part A: Polym. Chem. **37**(16), 3269–3276 (1999)
30. Fetisov, Y.K., Bush, A.A., Kamentsev, K.E., Srinivasan, G.: Pyroelectric effects in magnetoelectric multilayer composites. Solid State Commun. **132**(5), 319–324 (2004)
31. Dhakar, L.: Overview of energy harvesting technologies. Triboelectric Devices Power Gener. Self-Powered Sens. Appl. 9–37, (2017)
32. Shaw, C.P.: Pyroelectric properties of Mn-doped lead zirconate–lead titanate–lead magnesium niobate ceramics. J. Eur. Ceram. Soc. **22**(13), 2123–2132 (2002)
33. Mohammadi, S.: Effect of polarization fatigue on harvesting energy using pyroelectric materials. Adv. Mater. Sci. Eng. **2014**, (2014)
34. Siao, A.S., Chao, C.K., Hsiao, C.C.: Study on pyroelectric harvesters with various geometry. Sensors **15**, 19633–19648 (2015)
35. Cuadras, A., M. Gasulla., Ferrari, V.: Thermal energy harvesting through pyroelectricity. Sens. Actuators A: Phys. **158.1**, 132–139 (2010)
36. Ryu, H., Yoon, H.-J., Kim, S.-W.: Hybrid energy harvesters: toward sustainable energy harvesting. Adv. Mater. **1802898**, (2019)
37. Chen, Y., Zhang, Y., Yuan, F., Ding, F., Schmidt, O.G.: A flexible PMN-PT ribbon-based piezoelectric-pyroelectric hybrid generator for human-activity energy harvesting and monitoring. Adv. Electron. Mater **3**, 1600540 (2017)
38. You, M.H., Wang, X.X., Yan, X., Zhang, J., Song, W.Z., Yu, M., Fan, Z.Y., Ramakrishna, S., Long, Y.Z.: A self-powered flexible hybrid piezoelectric-pyroelectric nanogenerator based on non-woven nanofiber membranes. J. Mater. Chem. A **6**, 3500–3509 (2018)
39. Xie, Y., Huang, Z., Zhang, S., Li, X., Yang, W., Zhang, H., Lin, Y., He, L., Wu, B., Su, Y.: Flexible pyroelectric generators for scavenging ambient thermal energy and as self-powered thermosensors. Energy **101**, 202–210 (2016)
40. Zhang, Y., Bowen, C.R., Ghosh, S.K., Mandal, D., Khanate, H., Arafa, M., Wan, C.: Ferro electret materials and devices for energy harvesting applications. Nano Energy **57**, 118–140 (2019)
41. Zabek, D., Taylor, J., Le Boulbar, E., Bowen, C.R.: Micropatterning of flexible and free standing polyvinylidene difluoride (PVDF) films for enhanced pyroelectric energy transformation. Adv. Energy Mater. **5**, 1–6 (2015)
42. Clingman, W.H., Moore, R.G.: Application of ferroelectricity to energy conversion processes. J. Appl. Phys. **32**(4), 675–681 (1961)
43. Surmenev, R.A., Chernozem, R.V., Pariy, I.O., Surmeneva, M.A.: A review on piezo-and pyroelectric responses of flexible nano-and micropatterned polymer surfaces for biomedical sensing and energy harvesting applications. Nano Energy **79**, 105442 (2021)
44. Halim, N.A., Velayutham, T.S., Majid, W.A.: Pyroelectric, ferroelectric, piezoelectric and dielectric properties of $Na_{0.5}Bi_{0.5}TiO_3$ ceramic prepared by sol-gel method. Ceram. Int. **42**(14), 15664–15670 (2016)

45. Shen, M., Li, W., Li, M.Y., Liu, H., Xu, J., Qiu, S., Zhang, G., Lu, Z., Li, H., Jiang, S.: High room-temperature pyroelectric property in lead-free BNT-BZT ferroelectric ceramics for thermal energy harvesting. J. Eur. Ceram. Soc. **39**(5), 1810–1818 (2019)
46. Ma, N., Yang, Y.: Boosted photocurrent in ferroelectric $BaTiO_3$ materials via two dimensional planar-structured contact configurations. Nano Energy **50**, 417–424 (2018)
47. Lai, Y.X., Jiang, B.C., Liu, C.H., Chen, Z.K., Wu, C.L., Chen, J.H.: Low-alloy-correlated reversal of the precipitation sequence in Al-Mg-Si alloys. J. Alloy. Compd. **701**, 94–98 (2017)
48. Wang, X.Q., Tan, C.F., Chan, K.H., Xu, K., Hong, M., Kim, S.W., Ho, G.W.: Nanophotonic-engineered photothermal harnessing for waste heat management and pyroelectric generation. ACS Nano **11**(10), 10568–10574 (2017)
49. Yang, X., Zhang, L., Zhang, F., Huang, Y., Chen, Y.: Sulfur-infiltrated graphene-based layered porous carbon cathodes for high-performance lithium–sulfur batteries. ACS Nano **8**(5), 5208–5215 (2014)
50. Xie, M., Zabek, D., Bowen, C., Abdelmageed, M., Arafa, M.: Wind-driven pyroelectric energy harvesting device. Smart Mater. Struct. **25**(12), 125023 (2016)

Hexaferrite Composite-Based Materials: Potential Applications

Raphael Victor Barros Campos, Marcelo Antonio Santos da Silva, Juscelino Chaves Sales, Stephen Rathinaraj Benjamin, and Antonio Sergio Bezerra Sombra

Abstract Hexaferrite composites have garnered significant attention in recent years due to their unique structural characteristics, versatile synthesis methods, and diverse properties, offering promising applications in various fields. This comprehensive review explores the intricate structures of hexaferrite composites, delves into the evolving synthesis techniques, and examines their distinct properties. The discussion encompasses the magnetic, electrical, and structural features that contribute to the multifaceted functionality of hexaferrite composite materials. Furthermore, this review elucidates the potential applications of hexaferrite composites across different industries, including but not limited to telecommunications, electronics, energy storage and biomedical applications. By providing a thorough overview of the current state of research in this domain, this review aims to guide future investigations and inspire innovative approaches in the design and utilization of hexaferrite composite-based materials.

Keywords Hexaferrite composite · Antenna · Telecommunication · Dielectric · Microwave applications

R. V. B. Campos (✉) · M. A. S. da Silva · J. C. Sales · S. R. Benjamin · A. S. B. Sombra
Department of Physics, Telecommunication and Materials Science and Engineering of Laboratory (LOCEM), Federal University of Ceará (UFC), Campus PICI, Block No-930, Fortaleza, Ceará 60455-760, Brazil
e-mail: raphaelvbc@gmail.com

J. C. Sales
Departamento de Engenharia Civil, Centro de Ciências Exatas E Tecnologia, Universidade Estadual Vale Do Acaraú (UVA), Sobral, Ceará 62040-370, Brazil

S. R. Benjamin
Department of Physiology and Pharmacology, Neuroscience and Behavior Laboratory, Drug Research and Development Center (NPDM), Federal University of Ceará (UFC), Coronel Nunes de Melo 1127, Campus Porangabussu, Fortaleza, Ceará 60430-270, Brazil

U. Kumar (ed.), *Defects Engineering in Electroceramics for Energy Applications*, Engineering Materials, https://doi.org/10.1007/978-981-97-9018-0_19

1 Introduction

In recent years, there have been significant advances in fundamental, theoretical, and experimental studies on different properties of ferrites, with hexagonal crystal structures commonly known as Hexaferrites. These materials serve multiple purposes and find extensive applications in an array of devices, such as operational devices, magnetic recording media, permanent magnets, and data storage devices. Additionally, they are also utilized in microwave components and high-frequency circuits. This research has enabled the development of various microwave and mm (millimeter)-wave coatings and devices based on these materials. The growing interest in hexaferrites is due to their ferromagnetic and insulating properties at room temperature and their excellent chemical stability, mechanical hardness, wear, and corrosion resistance. These characteristics enable hexaferrites to be suitable for hard disk media, producing low noise and high Curie temperature advantages.

Hexaferrites can be classified into six distinct groups—M, U, W, X, Y, and Z—based on the Fe/O ratio and the presence of various chemical elements. This categorization arises from the specific chemical formulas and stacking sequences of the fundamental building blocks of hexaferrites. The magnetic properties of hexaferrites are closely related to these crystalline structures. There are two types of hexaferrites: uniaxial and hexaplane. Uniaxial hexaferrites have a magnetized line easy to move, while hexaplane hexaferrites are magnetized on a flat surface. These materials absorb electromagnetic radiation in the microwave frequency range and provide shielding from electromagnetic interference (EMI) [1–3]. They have diverse applications, including use in stealth military technology and healthcare. Hexaferrites exhibit outstanding microwave absorption properties, effective up to 50 GHz, while others are limited to a ferromagnetic response of only 1 GHz. Furthermore, the evolution of nanoscience and nanotechnology has played an essential role in developing microwave-absorbing materials (MAM). These materials can absorb or dissipate electromagnetic (EM) waves and transform EM energy into heat. However, the highly effective MAM features strong absorption capacity, small size, low density, and wide absorption bandwidth. The use of MAM is particularly relevant for dealing with electromagnetic pollution caused by wireless communication technologies operating in microwaves. This pollution has become a global health threat and a national defense and security priority concern. Therefore, continuous research and development of MAM has been of great importance in mitigating these problems and promoting the efficiency and safety of microwave technologies and their application in biomedical areas.

This book chapter aims to investigate hexaferrite composite-based materials, with a specific focus on their diverse applications in microwave technology, diagnostic and therapeutic biomedical interventions, and their roles in biomedical sensing and environmental applications.

1.1 EMW Absorption and Blocking Mechanisms

The impact of electromagnetic waves (EMW) on a material can be categorized into three groups: reflection, absorption, or penetration. The primary pathway for EMW attenuation is absorption, which can be influenced by various factors, such as input impedance. If the input impedance equals the free space impedance (Z_0), the EMW coupling into the material is maximum, resulting in maximum absorption. Therefore, materials with high dielectric constants are better suited for EMW absorption, while highly conductive materials are ideal for blocking EMW.

The microwave absorption properties of absorbing materials are mainly determined by the complex permittivity ϵ_r, which reflects the electric field interaction, and the complex permeability μ_r which represents the magnetic field interaction. This dynamic is demonstrated in the following equations:

$$\epsilon_r = \epsilon' - i\epsilon'' \tag{1}$$

$$\mu_r = \mu' - i\mu'' \tag{2}$$

The interaction between the electric and magnetic field with the absorbing material occurs in two distinct ways: first, there is the stored power from the external electric and magnetic field in the substance, represented by the real parts ε' and μ'. Then, energy is lost due to the external electric and magnetic field, reflected in the imaginary parts ε'' and μ'' [4]. Furthermore, it is possible to express the dielectric loss tangent ($\tan\delta\varepsilon$) as the ratio between ε'' and ε', according to Eq. (3), and in a similar way, for the magnetic loss tangent ($\tan\delta\mu$), using the Eq. (4).

$$\tan\delta_\varepsilon = \frac{\varepsilon'}{\varepsilon''} \tag{3}$$

$$\tan\delta_\mu = \frac{\mu'}{\mu''} \tag{4}$$

Reflection loss, which indicates the absorption capacity of electromagnetic (EM) waves, can be calculated using transmission line theory, as described in the following equation [5].

$$Z_1 = \sqrt{\frac{\mu_r}{\varepsilon_r}} \tan h\left[j\frac{2\pi fd}{c}\sqrt{\mu_r\varepsilon_r}\right] \tag{5}$$

$$Rl = -20\,og\left|\frac{Z_1 - Z_0}{Z_1 + Z_0}\right| \tag{6}$$

This mathematical expression is fundamental to understanding how EM waves interact and are absorbed along a transmission line. Certainly, understanding the details of this equation provides valuable insights into the efficiency and behavior of the system in question [6, 7].

Where Z_1 and Z_0 represent the impedance of the material and the impedance of the air, f is the frequency of the electromagnetic wave, d is the thickness of the absorber coating, and c is the speed of light. The reflection loss mechanism is mainly due to dielectric loss, which results primarily from loss of conduction and polarization relaxation. Ionic, electronic, dipolar and interfacial polarizations are the primary sources of polarization relaxation. According to the free electron theory, high electrical conductivity (i.e., low resistivity) will increase ε', so the loss of conduction plays a major role in the dielectric loss, and polarization relaxation is small, as indicated in Eq. (7) [8, 9].

$$\varepsilon' \approx \frac{1}{2\pi \rho f \varepsilon_0} \tag{7}$$

where ρ and ε_0 are the resistivities and free-space dielectric constant respectively. Furthermore, the magnetic loss mechanism is mainly due to natural resonance, domain wall resonance and eddy current effect [7]. Natural resonance generally occurs at lower frequencies and is related to the anisotropic domain, which can be expressed using the Eq. (8).

$$f_r = \frac{\beta H_a}{2\pi} \tag{8}$$

where f_r, H_a and β represent, respectively, the resonance frequency, the anisotropic field and the gyromagnetic proportion, it is necessary to effectively integrate dielectric and magnetic losses to achieve efficient absorption of electromagnetic waves to obtain specific impedance properties. The impedance matching property (Z) of the pattern can be determined by Eq. (9) as follows:

$$Z = \frac{Z_1}{Z_0} \tag{9}$$

When the impedance match approaches 1 or is equal to 1, the electromagnetic waves are not reflected from the surface of the absorbing substance but are directed instead into the substance. In this interval, the absolute value of RL (Reflection loss) reaches its maximum. The attenuation of electromagnetic waves inside the absorbing substance plays a crucial role in advanced MAM, as the attenuation constant (α), as defined in Eq. (10) [5], is fundamental to determining the dissipation characteristics of the substance.

$$\alpha = \sqrt{2}\pi f\left[\left[\left(\mu_r''\varepsilon_r'' - \mu_r'\varepsilon_r'\right) + \left(\mu_r''\varepsilon_r'' - \mu_r'\varepsilon_r'\right)^2 + \left(\mu_r''\varepsilon_r' - \mu_r'\varepsilon_r''\right)^2\right]^{\frac{1}{2}}\right]^{\frac{1}{2}} \tag{10}$$

The attenuation constant reveals that the main mechanisms that influence microwave absorption are dielectric and magnetic losses.

1.2 Microwave Applications

Hexagonal ferrites are a type of magnetic material that is highly versatile and has been utilized in various fields owing to its superior chemical stability, high magnetization, anti-erosion properties, and large magneto-crystalline anisotropy. The hexagonal structure of these ferrites is particularly useful in microwave absorption applications because of its high specific resistivity, strong magnetization, and permeability [10, 11]. Charanjeet et al. [10] explored the impact of double cationic (Co^{2+} and Cr^{3+}) substitutions on the magnetic, permittivity, and reflection loss behavior of M-type Ba-Sr hexaferrite. Remarkably, the study has reported a reflection loss of -52.09 dB at a frequency of 8.78 GHz, with a thickness of 5.6 mm. These significant findings demonstrate the potential for further optimizing hexaferrite materials to enable better microwave absorption performance.

Y-type barium hexaferrites substituted with La-Mg were synthesized by Roy et al. [12] and employed the sol–gel auto-combustion method. The resulting ceramics were sintered at 1050 °C for 2 h, producing a hexaferrite sample with improved shielding effectiveness. This material sample shows excellent potential for absorbing microwaves in the Ku-band frequency range, which spans from 12.5 to 17.5 GHz. The reflection loss is moderate, measuring less than −10 dB. Furthermore, hexagonal ferrites of the X-type exhibit an elevated saturation magnetization and a significant coercivity. Khan et al. [13] investigated the properties of Mg^{2+} substituted Ba_2Zn_2X hexa-ferrites using the sol–gel method for potential microwave device applications. The dielectric constant values decreased at lower frequencies but gradually increased at higher frequencies. The dielectric loss graphs also showed an increasing trend at lower frequencies but decreased at higher frequencies. AC conductivity increased with frequency and exhibited relaxation peaks at high frequencies. Multiple resonance peaks were observed at various frequencies on the reflection loss curve. Mg-doped hexagonal ferrites had the potential for microwave absorption applications. The Cole–Cole plots and complex impedance spectroscopy results revealed that grain boundaries contributed to most dielectric responses. The study found that Mg substituted Zn-X ferrite had excellent microwave absorption properties, with RL of −69.3 and −62.0 dB, suitable for absorption and super-capacitor applications. In this study, Rasool et al. [14] investigated the influence of Cr^{3+} substitution on the structural and dielectric properties of Barium-strontium-based W-type hexagonal ferrites utilized for microwave absorption. The sol–gel auto-ignition process was used to prepare the hexagonal ferrites with $Ba_{0.5}Sr_{0.5}Mg_2Cr_xFe_{16-x}O_{27}$. The outcomes of the study showed that increasing the amount of chromium decreased the dielectric permittivity, tan loss, and ac conductivity. However, the particular value of $x = 0.4$, the reflection loss value was found to be −79.62 dB at a frequency of 5.92 GHz and a thickness of 1.75 mm. Therefore, W-type hexagonal ferrites have potential uses in electronics such as microwave absorbers, high-density recording media, and multilayer chip inductors.

1.3 Diagnostic and Therapeutic Biomedical Applications

The forthcoming wave of biomedical applications will transform healthcare, integrating state-of-the-art technologies for diagnostic and therapeutic purposes. These advancements amplify precision and efficacy in medical procedures and lay the foundation for personalized and targeted treatment modalities. The convergence of diagnostic and therapeutic applications is becoming more pronounced, with pioneering methods like molecular diagnostics, advanced imaging, and precision therapies reshaping the landscape of medical interventions. From the early identification of diseases to tailored treatment plans, the cohesive integration of these next-generation technologies has significant potential to enhance patient outcomes and contribute to the progression of precision medicine. Looking ahead, the potential applications of these biomedical breakthroughs span diverse medical domains, offering inventive solutions to complex healthcare challenges and ushering in a new era of transformative and patient-centric medical practices. Ferrite materials can be magnetically excited in an external field for biomedical applications such as MRI tracers, drug delivery, and cancer treatment [15].

1.3.1 Biomedical-Sensing Applications

Magnetic Resonance Imaging (MRI)

Recent advancements in medical technology have shown remarkable potential in utilizing hexaferrite nanoparticles for high-contrast cardiovascular imaging. This innovative approach involves magnetically modulated optical coherence tomography (IC-OCT) and magnetic resonance imaging (MRI), which have proven highly effective and accurate in diagnosing cardiovascular diseases. These magnetic capabilities enable accurate imaging and contrast, improving patient outcomes and advancing medicine [16]. The research findings indicate that magnetic nanoplatelets have great potential as contrast agents for high-contrast imaging of the cardiovascular system. Due to their distinct optical and magnetic properties (below 1 mT), they can accurately locate macrophages and manipulate them in aqueous solutions. The use of BaFe-NPLTs as contrast agents in intracoronary OCT imaging has produced high-contrast images, overcoming one of the most significant limitations of current clinical applications of IC-OCT for molecular diagnosis. The results of ex vivo MRI experiments have further confirmed the potential of BaFe-NPLTs as MRI contrast agents. Recently, stabilizing polymers such as dextran can be fabricated on BHF surfaces for biomedical applications.

1.3.2 Bone Regeneration

$SrFe_{12}O_{19}$, also known as SrM, is a remarkable permanent magnetic material with excellent magnetic properties and promotes bone regeneration. Its ability to encourage the proliferation of pro-osteoblast cells and aid in the differentiation of osteoblast cells is a critical factor in its efficacy. In addition, SrM helps in bone matrix mineralization while inhibiting osteoclast cell differentiation [17]. The use of SrM MNPs in bone scaffolds can attract endogenous stem cells [18], making it a valuable tool in regenerative medicine. With its numerous benefits, SrM is essential for promoting bone growth and regeneration.

In addition, Shao and colleagues [19] developed a biogel system that utilizes the cross-linking of SiO_2 to improve the properties of SrM ceramic scaffolds bonded via 3D gel printing. The system involves using CS-Chitosan and PVA-polyvinylalcohol, which are cross-linked in situ to form a biogel and provide enhanced mechanical properties to the ceramic scaffolds. The SrM composite scaffolds have a compressive strength of 6.13 ± 2.45 MPa and a degradation rate of 4.90% after 28 days. SiO_2/CS/PVA/SrM scaffolds showed excellent biological properties for bone repair, as evidenced by the MC3T3-E1 cell proliferation method, indicating increased cell adhesion and proliferation. In addition, incorporating in situ SiO_2 improves scaffold mechanical properties and increases calcium salt deposition rates. Guo and colleagues [20] developed a scaffold comprising mesoporous bioglass and chitosan, modified with SrM nanoparticles. This scaffold effectively promoted bone regeneration and acted as a photothermal therapy agent against tumors. This research has significant implications for developing new treatments for bone-related disorders and cancer. The SrM nanoparticles possess excellent photothermal conversion capacity, resulting in the apoptosis and ablation of residual tumors when exposed to NIR laser irradiation. The combination of a magnetic field with a targeted treatment is a practical approach to treating bone deficiencies associated with tumors. Xie and colleagues [21] have proposed a novel method to address bone tumors using MCSC scaffolds. These scaffolds comprise M-type ferrite particles, mesoporous calcium silicate microspheres, and chitosan films. They are particularly useful for delivering drugs, promoting bone regeneration via osteogenic differentiation, and photothermal therapy. The MCSC/DOX (doxorubicin) scaffolds are highly effective for anti-tumor therapy. The NIR laser treatment has exhibited promising results in releasing DOX drug, both in vitro and in vivo, which is further amplified through hyperthermia ablation. The BMP-2/Smad/Runx2 pathway is crucial in encouraging the multiplication and bone-forming differentiation of hBMSCs. Additionally, higher ratios of M-type ferrite particles increase the expression of osteogenic proteins, ultimately leading to better bone formation.

1.3.3 Hyperthermia

Magnetic hyperthermia is a cutting-edge treatment modality that leverages the power of thermoseeds or magnetic nanoparticles (MNPs) to induce heat within cancer cells.

A radio frequency magnetic field generates this heat and may provide an effective means of eradicating malignant cells. It is a promising alternative to conventional cancer therapies (immuno, radio and chemotherapy) that may be of interest to the medical community and stakeholders involved in cancer care. The use of magnetic nanoparticles, particularly SrM, with high heating performance and stability, shows promise in magnetic hyperthermia [22, 23].

The utilization of hard ferromagnetic particles in hyperthermia applications has been a persistent challenge for researchers, primarily due to their higher anisotropy compared to superparamagnetic particles. Abdellahi et al. [23] introduced a hard magnetic material, SrM nanoparticles, and scaffolds for hyperthermia application. The application of an alternative magnetic field (AC) with the minimum frequency and amplitude of SrM nanoparticles composited with hydroxyapatite (HA) resulted in the achievement of high specific absorption rates. This research highlights the importance of the monodispersed HA-SrM nanocomposite, which demonstrated robust compressive strength and uniform porosity following the coating of scaffolds with gelatin. Recently, the introduction of chromium substitution in barium hexaferrite (Co_2-Y) resulted in a notable enhancement, with the material demonstrating the highest Specific Absorption Rate (SAR) at approximately 84.39 W/g, Initial Permeability (ILP) of around 1.86 nH m^2/Kg. Human osteosarcoma cells were used to examine the biocompatibility of ferrite powders. This study showed that pristine and oleic acid-coated ferrite particles are highly beneficial for proliferation and cell growth. This analysis underscores the dual advantage of chromium-substituted barium hexaferrite, showcasing its potential in magnetic hyperthermia applications and as biocompatible materials, thereby expanding their utility in diverse biomedical areas [24].

1.3.4 Drug Delivery/Carriers

$CuFe_{12}O_{19}$@HAp magnetic nanocomposites were synthesized via an ultrasound-assisted precipitation method, demonstrating superior controlled-release capabilities for the atenolol drug. The nanocomposites exhibited the highest entrapment efficiency at 63.1%, with the APTES binding to the drug attributed to the increased efficiency. The magnetic nanocomposites showed promise as effective vehicles for sustained delivery of atenolol and potential candidates for advancing controlled drug delivery systems in antihypertensive medications [25]. Recently, using Fructose as a green capping agent has led to an innovative method for synthesizing $ZnFe_{12}O_{19}$ nanostructures. The sol–gel auto-combustion process and a modified dispersive liquid–liquid microextraction method have paved the way for efficient preconcentration of BTX (Betaxolol) using hydrophobic magnetic nanostructures. The results of HPLC analysis of human blood serum samples have been impressive, showing a linear calibration curve ranging from 0.5 to 500 μg/L, with LOD and quantification limits (0.03 and 0.11 μg/L) and an exceptional recovery rate of 105.8% [26].

1.3.5 Biosensor

Hexagonal ferrite particles of the M-type, containing Barium and having a molar ratio of 4 of Fe/Ba, are highly electrocatalytic and can be substituted with transition metals. Due to their excellent electrocatalytic activity, they are well-suited for electro-oxidation methods in electroanalysis. Recently, Banguera et al. [27] used cobalt-substituted barium hexaferrites to detect acetaminophen in urine samples. A modified glassy carbon electrode can detect Acetaminophen at 0.255 μM with a linear concentration-to-current response (1–12 mM). The method had a detection limit and yielded recovery ranges of 96.6–101.0% and 93.9–98.4%. $BaFe_{12}O_{19}$ nanoparticles on a modified electrode were used for electrochemical oxidation of morphine. The electrode has a linearity of 0.06–600.0 μM and a LOD of 0.02 μM. Furthermore, the designed sensor was able to recover morphine in the morphine ampule and urine sample with a satisfactory recovery value of 103% [28].

Electrochemical methods have been widely used to study the behavior and detection of amino acids, including L-cysteine. Owing to their simplicity, sensitivity, and cost-effectiveness, the Strontium Hexaferrites (SrHFs) electrode has been optimized using cyclic voltammetry (CV) at pH 7.0. The concentration of L-cysteine at the SrHFs electrode shows a linear relationship with its anodic current. The sensor can detect L-cysteine in the concentration range of 0.5–12 μM, with a LOD of 0.166 μM and a sensitivity value of 6.638 μA/μM [29]. Additionally, a modified carbon paste electrode with strontium ferrite nanostructures was designed as an electrochemical sensor. Nanostructure modification of the electrode improved catalytic activity for tyrosine oxidation, increasing anodic peak current and decreasing oxidation overpotential. The electrode was used to measure Tyr concentration in blood and water samples using DPV (Differential Pulse Voltammetry). The anodic peak current demonstrated a linear correlation with the concentration of Tyr ranging from 0.8–300.0 μM, LOD of 0.15 μM [30]. Chen et al. [31] developed an electrochemical sensor that uses SrM NRs and a graphitic carbon nitride matrix. The sensor has a high BET surface area (325 m^2/g) and excellent electrocatalytic ability. It detects Fenitrothion in fruit samples with a detection range of 0.005 to 378.15 μM, a LOD of 0.0014 μM, and a sensitivity of 19.23 $\mu A\ \mu M^{-1}\ cm^{-2}$ by DPV method and recoveries of 97.6–99.6% (Table 1).

1.3.6 Anti-Bacterial Activity

Currently, the biomedical sector is experiencing significant interest in using particles for various applications. Notably, drug transport and antibacterial activity are among the most critical areas where these particles can demonstrate their potential. The use of ferrite nanoparticles in biomedicine is a groundbreaking development that produces an array of possibilities. It allows for the precise delivery of drugs to targeted areas and helps prevent bacterial infections. This innovation can potentially revolutionize the medical industry and bring life-changing advancements [37–40]. In a study by Sharma et al. [38] the antibacterial properties of Y-doped $BaFe_{12}O_{19}$ were

Table 1 Biosensors using Hexaferrite materials for detecting various analytes

Material used/types of ferrites/composite	Analyte	Synthesis approach	Method	Linear range	Detection limit	Sample	References
$Ba_{1.0}Co_{1.22}$ $Fe_{11.41}O_{18.11}$ $Ba_{1.14}Cu_{0.82}Fe_{11.65}O_{18.02}$	N-acetyl-*p*-aminophenol	Electro-oxidation	DPV	1–12 M	0.530 M 0.885 M	Urine	[27]
$BaFe_{12}O_{19}$	Morphine	Sol–gel auto-combustion	DPV	0.06–600.0 μM	0.02 μM	Ampule, Urine	[28]
$BaFe_{12}O_{19}$	Ammonia Gas	Solid-state reaction	DC voltage	0.2–100 ppm	0.2 ppm	–	[32]
$BaFe_{12}O_{19}$	Acetone, Ethanol	Solid-state reaction	Resistance transient	20–200 ppm	20 ppm	–	[33]
$BaFe_2O_4$-$Ba_2Fe_{12}O_{19}$	Humidity	Sol–gel auto-combustion	Relative humidity (RH)	30–90%RH	1.82 more sensitive	–	[34]
$SrFe_{12}O_{19}$	L-cysteine	Dual fuel auto-combustion	CV111	0.5–12 μM	0.166 μM	Phosphate buffer	[29]
$SrFe_{12}O_{19}$/g-C_3N_4	Fenitrothion	Hydrothermal	DPV	0.005–378.15 μM	1.4 μM	Fruits	[35]
$SrFe_{12}O_{19}$	Levofloxacin	Chemical co-precipitation	Square Wave Voltammetry (SWV)	55.2–772.8 nM	0.037 nM	Tap, lake, river water and urine	[36]

examined utilizing the citrate precursor method. The results showed that both doped and undoped samples were effective against gram-positive *S. aureus* (0.80 mg/ml) and gram-negative E. coli bacteria (0.55 mg/ml).

Bibi et al. [41] synthesized via micro-emulsion route in Pure $BaFe_{12}O_{19}$-BHF and $BaMn_{0.3}Fe_{12}O_{19}$-BMHF5 materials were tested for their antibacterial activity against *E. coli* and *S. aureus*. The samples showcase substantial antibacterial activity, effectively hindering bacterial growth. Moreover, the doped material displays significantly better antibacterial activity when compared to the undoped material. BMHF5 nanostructure is highly effective against bacteria due to oxidative stress caused by reactive species entering the cell, leading to cell death. These materials can be effectively used to combat bacterial infections. $BaFe_{12}O_{19}/CoFe_2O_4$ nanocomposites synthesized through the sol–gel method have remarkable potential, as shown in a recent study by Davarpanah et al. [42]. These nanocomposites are highly effective against bacterial and fungal pathogens, including *E. coli*, *S. aureus*, *C. albicans*, and *F. oxysporum*. Antifungal effects range within 4–64 μg/ml, and MIC and MBC values range from 256–2048 μg/ml against both bacterial pathogens. The study highlights the potential of these nanocomposites as a solution to combat various pathogens.

1.3.7 Pharmaceutical Analysis

Singhal et al. [36] developed a technology that uses M-type strontium hexagonal ferrites (SrFe) as a core. The core was modified using dopamine (Dop) and metal coating, allowing dual applications. Pharmaceutical waste materials like levofloxacin (35.6×10^{-3} min^{-1}) and sulfamethoxazole (31.3×10^{-3} min^{-1}) can be detected and degraded using electrochemical sensing and photocatalysis under visible light. The SrFe@Dop@Mn/GCE electrochemical sensor demonstrates high effectiveness in wastewater, exhibiting a linear range of 55.2–772.8 nM and a detection value of 0.037 nM for LVX detection. The designed sensor was tested on various samples and showed excellent sensing capabilities with acceptable recovery values (96.01–98.25%). Gherinasab has developed a process for rapidly developing $BaFe_{12}O_{19}$ nanoparticles using sol–gel synthesis. These nanoparticles act as a catalyst for electrochemical detection of tramadol alongside acetaminophen. The modified electrode ($BaFe_{12}O_{19}$/GSPE) can detect tramadol and acetaminophen in pharmaceutical tablets and urine samples with good linear calibration curves, LOD of 0.025 μM for tramadol and 1.5 μM for acetaminophen, and acceptable recovery values ranging from 97.1 to 104.4% [43].

1.4 Applications of Ferrites in the Environment

1.4.1 Humidity Sensor

Humidity is a critical environmental factor that affects various industries and technologies. Therefore, it is essential to detect accurate humidity levels in the air. The humidity sensors available in the market are usually made up of polymer films and ceramics. However, there has been an increased interest in studying hexa-ferrite materials for sensor applications. One of the most significant advantages of ferrites is their porosity, which is necessary for a humidity sensor. Moreover, ferrite compositions are recognised for their high resistivity that declines significantly as the surrounding air relative humidity (RH) increases. A recent study by Roy et al. [44] synthesized Y-type barium hexaferrite (Co_2-Y) ferrite $Ba_2Co_2W_xFe_{12-x}O_{22}$ using a solid-state ceramic route method. The research found that tungsten substitution within the hexaferrite increases the optical band gap linearly from 1.38 to 2.21 eV. Additionally, the study analyzed the frequency-dependent humidity-dielectric response of the material using an impedance analyzer for a wide range of frequencies (20 Hz–10 MHz) and RH variation from 20 to 95%.

In addition, Roy et al. [40] examined the synthesis of barium hexaferrite Y-type (Co_2-Y) ferrite $Ba_2Co_2W_xFe_{12-x}O_{22}$ utilizing a solid-state ceramic route method. The results indicate that tungsten substitution within the hexaferrite has a linear effect on increasing the optical band gap from 1.38 to 2.21 eV. Furthermore, the study analyzed the humidity-dielectric response of the material-employed impedance analyzer for a wide frequency range of (20 Hz–10 MHz) and RH range of 20–95%. These findings may have important implications for developing advanced materials for various applications. However, $BaFe_{12}O_{19}$ exhibits sensitivity 1.82 times higher than $BaFe_2O_4$ using the sol–gel technique (1000 °C), frequency of 1 MHz, and relative humidity of 90%. These results suggest that $BaFe_{12}O_{19}$ has potential for use in humidity-sensitive applications.

1.4.2 Photocatalytic Activity

In materials science, $SrFe_{12}O_{19}$ hexaferrite is known for its remarkable photocatalytic properties. Over the years, numerous studies have explored different techniques to improve its performance. These methods include doping with other elements, changing the synthesis conditions, and modifying the morphology of the material. The literature reveals that these approaches have shown promising results in enhancing the photocatalytic efficiency of $SrFe_{12}O_{19}$ hexaferrite. The implications of these findings are significant as they can potentially benefit various industrial processes and environmental remediation methods [45–47]. The magnetic properties of certain materials, facilitating easy recovery through external magnetic fields, produce the advantage of simple and pollution-free retrieval and reuse for photocatalysts. Several methods are available to modify the properties of materials, including ion substitution, modulation of intrinsic defects and morphological modification [48].

Ion substitution can enhance the catalytic performance of hexaferrite. Transition metal ions, including Co [49], Zn [50], Cr and Zn [51, 52], Mg [53, 54], Mn [55] or Ni [56], can be introduced to modify hexaferrite catalytic and magnetic properties. Several magnetic materials for devices and photocatalysts for remediating toxic pollutants dyes from industrial waste have been considered, including $Sr_{1-x}Nd_xFe_{12-y}Mn_yO_{19}$ co-doped nanoparticles. Using the nanoparticles effectively reduced CV dye by 90.7% in just 1.5 h, and the photocatalyst was highly cyclic stable and reusable with 1.10% losses after five successive cycles [57]. Additionally, hexaferrites doped with Yb-Zn and $CaPbFe_{12}O_{19}$ are exceptional materials for microwave, dielectric, and photocatalytic applications, specifically in removing textile contaminants. This material has exhibited a band gap (2.5–2.6 eV) and improved photocatalytic performance. The findings indicated an increased percentage degradation of methylene blue efficiency up to 96.1% when exposed to visible light irradiation for 90 min at pH 7 [58]. In addition, the compound $SrFe_{12}O_{19}$ possesses direct (1.70 eV) and indirect (0.83 eV) bandgap energy values, respectively. The compound has photocatalytic solid activity, degrading MB at a rate constant of $k = 1.36 \pm 0.03 \times 10^{-2}$/min. The photodegradation process causes the MB solution to transition from deep blue to pale white while the absorption intensity rapidly drops from 1.95 to 0.04 [59]. Polley et al. [60], they were successful in synthesizing $BaFe_{12}O_{19}$ using the co-precipitation method. Following the synthesis of the powder, MB dye was adsorbable and degraded by sunlight-induced photocatalysis. The BaM ferrite exhibited an impressive ability to degrade up to 73% of the methyl blue dye in 3 h of sunlight irradiation, owing to its calculated band gap of approximately 2.9 eV.

The ultrasonic-assisted synthesis process was recently used to create highly uniform Strontium Ferrite (SF) nanochain structures. These nanostructures were then utilized to construct an electrocatalyst, which was used to develop a nitrite sensor. The electrodes enhanced with SF exhibited impressive electroanalytic performance and electrocatalytic activity using the amperometric (*i-t*) method. The nitrite sensor detects values ranging from 20 nM to 3.6 mM with a detection value of 6.3 nM. The SF-nitrite sensor efficiently quantified nitrite from real-water samples, such as rain, tap, and river water [61]. Singhal et al. [62] developed catalysts for wastewater treatment using a combination of strontium hexaferrite, graphene (Gr), and multi-walled carbon nanotubes (MWCNT). The composites catalyze the degradation of antibiotics such as carbol fuscin, ofloxacin and levofloxacin and colorants via two oxidants (PMS, H_2O_2). Nanocomposites based on $SrFe_{12}O_{19}$ are more effective for degrading pollutants based on Fe_2O_3. PMS, as an oxidant, can degrade up to 98.1% of pollutants in 120 min, compared to only 44.6% for Fe_2O_3 nanocomposites in 180 min.

Ba-Hexa ferrite-based materials have shown promise in photocatalytic applications for degrading dyes and pollutants like Rhodamine B [63–65], phenol [66], Methyl Orange [67], and Methyl green [68]. A range of substances have been evaluated for their dielectric, structural, magnetic, and photocatalytic characteristics, including Nd and Cu co-doped barium hexaferrite Congo Red [69] and Hexahydro-1,3,5- trinitro-1,3,5 triazine (RDX) [70] tetracycline hydrochloride [71, 72], 2-Mercaptobenzothiazole [73, 74], enrofloxacin [75]. Recently, Bibi and their colleagues [76] synthesized Ba-hexaferrites doped with Cr and Co that showed promising results in degrading CV dye using photocatalytic activity with UV–Vis spectroscopy. The doped nanoparticles demonstrated a high level of degradation, breaking down 64.23% of the CV dye (60 min) under visible light irradiation (0.02927/min). The catalytic sites were activated on the material surface by solar light energy equal to its bandgap energy. Futhermore, the same group members developed Mn doped BMHF5 showed excellent Photocatalytic activity with 94.6% (CV), 92.2% (MG), and 96.5% (RhB) dyes, respectively. The bandgap of doped BMHF5 was 1.30 eV, and it exhibited better photo-degradation of all dyes under solar irradiation as compared to pure barium hexa ferrite [41]. Furthermore, Katubi and colleagues [77] conducted research on using a hydrothermal method to synthesize carbon dots and develop a Y-type hexaferrite composite to degrade MB under visible light. The study found that a composite material with a ferrite/CDs ratio of 1/0.5 exhibited significantly higher photocatalytic activity, up to 93%, compared to a single material, particularly in hydrogen peroxide. Due to its narrow optical energy band gap of 2.48 eV in the visible light spectrum, the larger flow of charge carriers on the catalyst surface, and the development of heterostructures, the nanocomposite catalyst demonstrated enhanced photocatalytic activity.

Barium hexaferrite is a catalyst demonstrating exceptional performance in degrading Methylene Blue under visible light irradiation. Through the utilization of the chemical co-precipitation method, the $BaFe_{12}O_{19}$ catalyst exhibited a remarkable ability to decolorize and mineralize the dye, achieving a rate of 70.8% and 63.7%, respectively, after 360 min. This was performed using a 3W Light Emitting Diode (LED) lamp with a 420–700 nm wavelength. It exhibits excellent catalytic activity and can degrade MB in photo-Fenton systems using visible light. These results significantly affect various industries and applications towards removing dye from wastewater [78]. Erfan et al. [79] employed ball milling to produce barium hexaferrite-based ceramic glass to absorb pollutants like Congo red and methylene blue (MB). The study revealed high adsorption capabilities, with 200 mg/g for MB and 124.5 mg/g for CR and barium hexaferrite demonstrating potential as an effective environmental adsorbent (Table 2).

1.5 Conclusion

This review highlights the potential of hexaferrite composite-based materials, which exhibit unique structures and synthesis methods that provide remarkable versatility across a range of applications. The various synthesis techniques for hexagonal ferrites provide researchers with valuable tools to tailor these materials for specific

Table 2 Application of Hexa ferrite materials in biomedical and environmental applications

Composition	Preparation technique	Temperature	Area of applications	Main results	References
Microsphere nanoparticles with $Sr_{0.5}Ba_{0.5}Dy_xSm_xFe_{8-2x}O_{19}$	Hydrothermal	500 °C	Cancer	HCT-116 and HeLa (cervical cancer cells) using MTT; no significant cytotoxic effect was observed on HEK-293 cells Induced cancer cells DNA disintegration as revealed by DAPI staining	[15]
$BaFe_{12}O_{19}$ BaFe-magnetic nanoplatelets	Hydrothermal	240 °C	Bioimaging (*magnetically modulated optical coherence tomography and MRI*) intracoronary optical coherence tomography (IC-OCT)	Visualize in macrophages ex vivo MRI experiments high potential for accurate analysis and interpretation of images of the cardiovascular system	[16]
$SrFe_{12}O_{19}$	–	–	Endogenous tissue engineering Bone scaffolds	Stimulate new bone formation	[18]
$Ba_2Co_2Fe_{11.7}Cr_{0.3}O_{22}$	Calcination	1200 °C/2 h	Hyperthermia	Saturation magnetization (*Ms*) 39.07 emu/g SAR value, 84.39 W/g, and ILP value of 1.86 nH m^2 /Kg	[24]
$BaFe_{12}O_{19}$	Ball milling	–	Water treatment	Adsorption capacity (200 mg/g 124.5 mg/g) Removal of methylene blue and Congo red from water by adsorption	[80]

(continued)

Table 2 (continued)

Composition	Preparation technique	Temperature	Area of applications	Main results	References
$Ca_{0.5}Pb_{0.4}Yb_{0.1}Zn_{1.0}Fe_{11}O_{19}$	Sol–gel auto combustion	850 °C	Photocatalyst for the removal of textile contamination	Removal of methylelne blue (10 ppm) at pH 7.0, Degradation efficiency 96.1% and time 120 min	[58]
$SrFe_{12}O_{19}$	Polymer precursor	1100 °C	Photocatalyst for degradation of organic pollutants	Degradation of methylene blue Degradation efficiency 95% and time 270 min	[59]
$CoFe_{12}O_{19}$/graphene nanocomposite	Sol–gel with natural reducing agents (carbohydrate sugar, pigments)	900 °C	Photocatalytic reduction	Degradation of methylene orange 10 ppm Degradation efficiency 75.6% and time 60 min	[81]
$BaFe_{12}O_{19} + H_2O_2$	Chemical co-precipitation		Photocatalytic reduction	Degradation of methylene blue Degradation efficiency 70.8% and time 360 min	[78]
$Ba_{1-x}Nd_xFe_{12-y}Cu_yO_{19}$	Micro-emulsion	950 °C/9 h	Photocatalytic activity	Degradation of methylene green, Degradation efficiency 92.6 % and time 60 min	[68]
$Sr_{1-x}Nd_xFe_{12-y}Mn_yO_{19}$	Micro-emulsion	950 °C/9 h	Photocatalytic activity	Degradation of crystal violet, Degradation efficiency 90.7 % and time 90 min	[57]
$Y/Ba_2Co_2W_1Fe_{11}O_{22}$	Solid-state reaction	1200 °C/2 h	Humidity sensor	60–95% RH	[44]

needs. Their exceptional properties, particularly in microwave technology, biomedical diagnostics and therapies, and environmental sensing, make hexaferrite composites highly appealing to researchers in diverse scientific disciplines. As research advances, hexaferrite composites are expected to play an increasingly significant role in driving technological innovation and scientific breakthroughs.

2 Conflicts of Interest

The authors declare no conflicts of interest.

Acknowledgements The authors would like to thank the Fundação Cearense de Apoio ao Desenvolvimento Científicoe Tecnológico, FUNCAP, Coordenação de Aperfeiçoamento de Pessoal de Nível Superior (CAPES)- and the Drug Research and Development Center (NPDM), Federal University of Ceará (UFC) in Fortaleza, Brazil.

References

1. Soroka, M., Buršík, J., Kužel, R., Prokleška, J., Aguirre, M.H.: Characterization of W-type hexaferrite thin films prepared by chemical solution deposition. Thin Solid Films **726**, 138670 (2021)
2. Phan, T.L., Tran, N., Nguyen, H.H., Yang, D.S., Dang, N.T., Lee, B.W.: Crystalline and electronic structures and magnetic properties of $BaCo_{1-x}Mn_xFe_{11}O_{19}$ hexaferrites. J. Alloys Compd. **816**, 152528 (2020)
3. View on ScienceDirect Encyclopedia of Materials
4. Green, M., Li, Y., Peng, Z., Chen, X.: Dielectric, magnetic, and microwave absorption properties of polyoxometalate-based materials. J. Magn. Magn. Mater. **497**, 165974 (2020)
5. Jia, Z., Lan, D., Lin, K., Qin, M., Kou, K., Wu, G., Wu, H.: Progress in low-frequency microwave absorbing materials. J. Mater. Sci. Mater. Electron. **29**, 17122–17136 (2018)
6. Houbi, A., Aldashevich, Z.A., Atassi, Y., Bagasharova Telmanovna, Z., Saule, M., Kubanych, K.: Microwave absorbing properties of ferrites and their composites: a review. J. Magn. Magn. Mater. **529**, 167839 (2021)
7. Duan, Y., Guan, H.: Microwave Absorbing Materials (2016)
8. Xie, P., Li, Y., Hou, Q., Sui, K., Liu, C., Fu, X., Zhang, J., Murugadoss, V., Fan, J., Wang, Y., Fan, R., Guo, Z.: Tunneling-induced negative permittivity in Ni/MnO nanocomposites by a bio-gel derived strategy. J. Mater. Chem. C Mater. **8**, 3029–3039 (2020)
9. Ren, X., Pu, X., Yin, H., Tang, Y., Yuan, H., Fan, H.: Fabrication of hierarchical PANI@W-type barium hexaferrite composites for highly efficient microwave absorption. Ceram. Int. **47**, 12122–12129 (2021)
10. Singh, J., Singh, C., Kaur, D., Narang, S.B., Jotania, R.B., Ateia, E., Kagdi, A., Joshi, R., Sombra, A.S.B., Zhou, D., Trukhanov, S., Panina, L.: Development of doped Ba–Sr hexagonal ferrites for microwave absorber applications: structural characterization, tunable thickness, absorption peaks and electromagnetic parameters. J. Alloys Compd. **855**, 157242 (2021)
11. Adersh, V., Rajan, A., Ganesanpotti, S.: $MgFe_{1.98}O_4$–$BaFe_{12}O_{19}$ magneto-dielectric composites based ferrite resonator antenna for super-high frequency applications. Ceram. Int. **48**, 24531–24539 (2022)

12. Suthar, M., Roy, P.K.: Structural, electromagnetic, and Ku-band absorption characterization of La-Mg substituted Y-type barium hexaferrite for EMI shielding application. Mater. Sci. Eng. B **283**, 115801 (2022)
13. Khan, H.A., Khan, M.A., Gulbadan, S., Ashraf, G.A., Rasool, R.T., Akhtar, M.N., Irfan, M.: Dielectric, structural and absorption properties of Mg^{2+} substituted Ba_2Zn_2X hexa-ferrites for microwave devices and applications. Ceram. Int. **49**, 39862–39874 (2023)
14. Iqbal, A., Khan, M.A., Rasool, R.T., Gulbadan, S., Algaradah, M.M., Ashraf, G.A., Alshahrani, T., Abd-Rabboh, H.S.M., Akhtar, M.N., Irfan, M.: Dielectric, reflection loss and physicochemical studies of Cr doped Ba–Sr based W type hexagonal ferrites for microwave absorption applications. Ceram. Int. **50**, 4446–4461 (2024)
15. Al-Jameel, S.S., Almessiere, M.A., Khan, F.A., Taskhandi, N., Slimani, Y., Al-Saleh, N.S., Manikandan, A., Al-Suhaimi, E.A., Baykal, A.: Synthesis, characterization, anti-cancer analysis of $Sr_{0.5}Ba_{0.5}Dy_xSm_xFe_{8-2x}O_{19}$ ($0.00 \leq x \leq 1.0$) microsphere nanocomposites. Nanomaterials **11**, 700 (2021)
16. Hu, J., Gorsak, T., Martín Rodríguez, E., Calle, D., Muñoz-Ortiz, T., Jaque, D., Fernández, N., Cussó, L., Rivero, F., Aguilar Torres, R., García Solé, J., Mertelj, A., Makovec, D., Desco, M., Lisjak, D., Alfonso, F., Sanz-Rodríguez, F., Ortgies, D.H.: Magnetic nanoplatelets for high contrast cardiovascular imaging by magnetically modulated optical coherence tomography. ChemPhotoChem **3**, 529–539 (2019)
17. Boda, S.K., Thrivikraman, G., Panigrahy, B., Sarma, D.D., Basu, B.: Competing Roles of Substrate Composition, Microstructure, and Sustained Strontium Release in Directing Osteogenic Differentiation of hMSCs (2017)
18. Wang, Q., Tang, Y., Ke, Q., Yin, W., Zhang, C., Guo, Y., Guan, J.: Magnetic lanthanum-doped hydroxyapatite/chitosan scaffolds with endogenous stem cell-recruiting and immunomodulatory properties for bone regeneration. J. Mater. Chem. B **8**, 5280–5292 (2020)
19. Duan, J., Shao, H., Wu, J., Liu, H., Wang, S., Lin, T.: Effect of SiO_2 in situ cross-linked CS/PVA on $SrFe_{12}O_{19}$ scaffolds prepared by 3D gel printing for targeting. Ceram. Int. **48**, 30307–30316 (2022)
20. Lu, J.-W., Yang, F., Ke, Q.-F., Xie, X.-T., Guo, Y.-P.: Magnetic nanoparticles modified-porous scaffolds for bone regeneration and photothermal therapy against tumors. Nanomedicine **14**, 811–822 (2018)
21. Yang, F., Lu, J., Ke, Q., Peng, X., Guo, Y., Xie, X.: Magnetic mesoporous calcium sillicate/chitosan porous scaffolds for enhanced bone regeneration and photothermal-chemotherapy of osteosarcoma. Sci. Rep. **8**, 7345 (2018)
22. Rashid, A.U., Southern, P., Darr, J.A., Awan, S., Manzoor, S.: Strontium hexaferrite ($SrFe_{12}O_{19}$) based composites for hyperthermia applications. J. Magn. Magn. Mater. **344**, 134–139 (2013)
23. Najafinezhad, A., Abdellahi, M., Saber-Samandari, S., Ghayour, H., Khandan, A.: Hydroxyapatite- M-type strontium hexaferrite: a new composite for hyperthermia applications. J. Alloys Compd. **734**, 290–300 (2018)
24. Suthar, M., Khare, D., Gangwar, A., Banerjee, S., Prasad, N.K., Dubey, A.K., Roy, P.K.: Structural, magnetic, and biocompatibility evaluations of chromium substituted barium hexaferrite (Co_2-Y) for hyperthermia application. Mater. Chem. Phys. **296**, 127348 (2023)
25. Naseri, K., Khademi, E., Mortazavi-Derazkola, S.: Introducing a new pharmaceutical agent: facile synthesis of $CuFe_{12}O_{19}$@HAp-APTES magnetic nanocomposites and its cytotoxic effect on HEK-293 cell as an efficient in vitro drug delivery system for atenolol. Arab. J. Chem. **16**, 104404 (2023)
26. Khoobi, A., Salavati-Niasari, M., Amiri, O.: Sol-gel auto-combustion synthesis and characterization of $ZnFe_{12}O_{19}$ as a nanosorbent using fructose for HPLC-UV determination of betaxolol in biological samples. J. Alloys Compd. **853**, 157393 (2021)
27. Granja-Banguera, C.P., Silgado-Cortázar, D.G., Morales-Morales, J.A.: Transition metal substituted barium hexaferrite-modified electrode: application as electrochemical sensor of acetaminophen. Molecules **27**, 1550 (2022)

28. Bagherinasab, Z., Beitollahi, H., Yousefi, M., Bagherzadeh, M., Hekmati, M.: A sensitive voltammetric morphine nanosensor based on $BaFe_{12}O_{19}$ nanoparticle-modified screen-printed electrodes. J. Iran. Chem. Soc. **17**, 717–724 (2020)
29. Lakshmikantha, J., Krishnamurthy, G., Hanumantha Nayak, R., Pari, M., Ranjitha, N., Naik, N.: Synthesis, structure, thermal, magnetic, dielectric properties of Ce^{3+} doped M-type $SrFe_{12}O_{19}$ and electrochemical determination of L-cysteine. Inorg. Chem. Commun. **146**, 110175 (2022)
30. Sheikh-Mohseni, M.A.: Electroanalysis of amino acid tyrosine by an electrochemical sensor modified by strontium ferrite nanostructure. Anal. Bioanal. Chem. Res. **6**, (2019)
31. Rajaji, U., Chinnapaiyan, S., Chen, T.W., Chen, S.M., Mani, G., Mani, V., Ali, M.A., Al-Hemaid, F.M.A., El-Shikh, M.S.: Rational construction of novel strontium hexaferrite decorated graphitic carbon nitrides for highly sensitive detection of neurotoxic organophosphate pesticide in fruits. Electrochim. Acta **371**, (2021)
32. Das, T., Das, S., Karmakar, M., Chakraborty, S., Saha, D., Pal, M.: Novel barium hexaferrite based highly selective and stable trace ammonia sensor for detection of renal disease by exhaled breath analysis. Sens. Actuators, B Chem. **325**, 128765 (2020)
33. Karmakar, M., Mondal, B., Pal, M., Mukherjee, K.: Acetone and ethanol sensing of barium hexaferrite particles: a case study considering the possibilities of non-conventional hexaferrite sensor. Sens. Actuators B Chem. **190**, 627–633 (2014)
34. Hasan, S., Azhdar, B.: Thermo-dielectric, humidito-dielectric, and humidity sensing properties of barium monoferrite and barium hexaferrite nanoparticles. Results Phys. **42**, 105962 (2022)
35. Rajaji, U., Chinnapaiyan, S., Chen, T.-W.W., Chen, S.-M.M., Mani, G., Mani, V., Ali, M.A., Al-Hemaid, F.M.A., El-Shikh, M.S.: Rational construction of novel strontium hexaferrite decorated graphitic carbon nitrides for highly sensitive detection of neurotoxic organophosphate pesticide in fruits. Electrochim. Acta **371**, 137756 (2021)
36. Kaur, P., Jyoti, Singh, S., Aggarwal, D., Nitansh, Kumar, V., Tikoo, K., Kaushik, A., Singhal, S.: Synergizing hexagonal ferrite with transition metals in core-shell-shell nanostructures (SrFe@Dop@M) as dualistic probe for detoxification and electrochemical detection of pharmaceutical drugs. Ceram. Int. **49**, 6199–6212 (2023). https://doi.org/10.1016/j.ceramint.2022.10.111
37. Thakur, A., Sharma, N., Bhatti, M., Sharma, M., Trukhanov, A.V., Trukhanov, S.V., Panina, L.V., Astapovich, K.A., Thakur, P.: Synthesis of barium ferrite nanoparticles using rhizome extract of Acorus Calamus: characterization and its efficacy against different plant phytopathogenic fungi. Nano-Struct. & Nano-Objects **24**, 100599 (2020)
38. Sharma, I., Kumari, T., Thakur, N., Sharma, P., Batoo, K.M., Verma, R.: Citrate precursor route for Y^{3+} substituted M-type Ba-hexaferrite: synthesis, the effect of doping on structural, optical, magnetic and anti-bacterial properties. Mater. Chem. Phys. **302**, 127664 (2023)
39. Djaniš, J.P., Periša, J., Boštjančič, P.H., Mihajlovski, K., Lazić, V., Dramićanin, M., Lisjak, D.: Barium hexaferrite nanoplatelets with polyphenol coatings for versatile applications as a stable, magnetic, and antimicrobial colloid. Colloids Surf. B Biointerfaces **224**, 113198 (2023)
40. Sahoo, J.K., Konar, M., Rath, J., Kumar, D., Sahoo, H.: Hexagonal strontium ferrite: cationic dye adsorption and antibacterial activity. Sep. Sci. Technol. **55**, 415–430 (2020)
41. Fatima, G., Bibi, I., Majid, F., Kamal, S., Nouren, S., Ghafoor, A., Raza, Q., Al-Mijalli, S.H., Alnafisi, N.M., Iqbal, M.: Mn-doped $BaFe_{12}O_{19}$ nanoparticles synthesis via micro-emulsion route: solar light-driven photo-catalytic degradation of CV, MG and RhB dyes and antibacterial activity. Mater. Res. Bull. **168**, 112491 (2023)
42. Davarpanah, A.M., Rahdar, A., Dastnae, M.A., Zeybek, O., Beyzaei, H.: $(1-x)BaFe_{12}O_{19}/xCoFe_2O_4$ hard/soft magnetic nanocomposites: synthesis, physical characterization, and antibacterial activities study. J. Mol. Struct. **1175**, 445–449 (2019)
43. Bagherinasab, Z., Beitollahi, H., Yousefi, M., Bagherzadeh, M., Hekmati, M.: Rapid sol gel synthesis of $BaFe_{12}O_{19}$ nanoparticles: an excellent catalytic application in the electrochemical detection of tramadol in the presence of acetaminophen. Microchem. J. **156**, 104803 (2020)
44. Suthar, M., Roy, P.K.: Evaluation of magnetic & humidity-dielectric response of tungsten substituted Y-type barium hexaferrite ($Ba_2Co_2W_xFe_{12-x}O_{22}$, $0.0 \leq x \leq 2.0$) synthesized by solid-state reaction route. Inorg. Chem Commun. **150**, 110554 (2023)

45. Elansary, M., Belaiche, M., Oulhakem, O., Alaoui, K.B., Lemine, O.M., Mouhib, Y., Iffer, E., Salameh, B., Alsmadi, A.M.: In-depth study of the photocatalytic performance of novel magnetic catalysts for efficient photocatalytic degradation of the dye orange G. Mater. Res. Bull. **170**, 112598 (2024)
46. Bavarsiha, F., Montazeri-Pour, M., Rajabi, M.: Effect of non-aqueous media on nano-crystalline $SrFe_{12}O_{19}$ particles produced by co-precipitation with metal chlorides and evaluation of their magnetic and photocatalytic properties. J. Inorg. Organomet. Polym. Mater. **30**, 2386–2396 (2020)
47. Elanthamilan, E., Elizabeth, I.B., Wang, S.-F., Lydia, I.S.: Strontium hexaferrite microspheres: synthesis, characterization and visible-light-driven photocatalytic activity towards the degradation of methylene blue dye. Opt. Mater. (Amst) **137**, 113565 (2023)
48. Elansary, M., Belaiche, M., Ahmani Ferdi, C., Iffer, E., Bsoul, I.: Novel ferrimagnetic nanomaterial $Sr_{(1-x)}La_xGd_ySm_zFe_{(12-(z+y))}O_{19}$ for recording media applications: experimental and theoretical investigations. New J. Chem. **45**, 10761–10776 (2021)
49. Mustafa, G., Bibi, I., Majid, F., Sultan, M., Taj, B., Nazeer, Z., Muhammad Umair, H., Elqahtani, Z.M., Alwadai, N., Khan, M.I., Iqbal, M.: Zn and Ni doping effect on barium hexaferrite ferroelectric, optical properties and photocatalytic activity under visible light irradiation. Phys. B Condens. Matter. **663**, (2023)
50. Muhiuddin, G., Bibi, I., Nazeer, Z., Majid, F., Kamal, S., Kausar, A., Raza, Q., Alwadai, N., Ezzine, S., Iqbal, M.: Synthesis of Ni doped barium hexaferrite by microemulsion route to enhance the visible light-driven photocatalytic degradation of crystal violet dye. Ceram. Int. **49**, 4342–4355 (2023)
51. Nazeer, Z., Bibi, I., Majid, F., Kamal, S., Ibrahim, S.M., Iqbal, M.: Energy band gap tuning of Ba_{1}-$_{x}Zn_{x}Fe_{12}$-$yCr_{Y}O_{19}$ doped with Cr and Zn ions to enhance the optical, dielectric, ferroelectric, and photocatalytic properties. Opt. Mater. (Amst) **125**, 112090 (2022)
52. Amirhosseini, Z.K., Monsef, R., Ehsanizadeh, S.A., Albahadly, W.K.Y., Majdi, H.S., Amir, A.A., Dawood, A.H., Salavati-Niasari, M.: Tailoring the photocatalytic activity of novel magnetically separable $ZnFe_{12}O_{19}$-chitosan bionanocomposites: a green preparation, structural characterization and comparative study. Int. J. Hydrog. Energy **48**, 37286–37301 (2023)
53. Irshad, A., Zulfiqar, M., Ali, H.M., Shahzadi, N., Abd El-Gawad, H.H., Chokejaroenrat, C., Sakulthaew, C., Anjum, F., Suleman, M.: Co-substituted Mg–Zn spinel nanocrystalline ferrites: synthesis, characterization and evaluation of catalytic degradation efficiency for colored and colorless compounds. Ceram. Int. **48**, (2022)
54. Bibi, I., Naz, T., Majid, F., Kamal, S., Ata, S., Almoneef, M.M., Iqbal, S., Iqbal, M.: Band gap tuning of $BaFe_{12}O_{19}$ with Mg and Mn doping to enhance solar light absorption for photocatalytic application. Int. J. Energy Res. **45**, 11193–11205 (2021)
55. Attia, E.F., Zaki, A.H., El-Dek, S.I., Farghali, A.A.: Synthesis, physicochemical properties and photocatalytic activity of nanosized Mg doped Mn ferrite. J. Mol. Liq. **231**, (2017)
56. Jadhav, S.A., Raut, A. V., Khedkar, M. V., Somvanshi, S.B., Jadhav, K.M.: Photocatalytic activity of nickel ferrite nanoparticles synthesized via sol-gel auto combustion method. Adv. Mat. Res. **1169**, (2022)
57. Bibi, F., Iqbal, S., Kalsoom, A., Jamshaid, M., Ahmed, A., Mirza, M., Qureshi, W.A.: Magnetically separable Nd and Mn co-doped $SrFe_{12}O_{19}$ hexa-ferrites nanostructures for the evaluation of structural, magnetic and photo-catalytic studies under solar irradiation for the crystal violet dye removal from the industrial wastes. Ceram. Int. **49**, 15990–16001 (2023)
58. Jamshaid, M., Nazir, M.A., Najam, T., Shah, S.S.A., Khan, H.M., Rehman, A.U.: Facile synthesis of Yb^{3+}-Zn^{2+} substituted M type hexaferrites: structural, electric and photocatalytic properties under visible light for methylene blue removal. Chem. Phys. Lett. **805**, 139939 (2022)
59. Mishra, D.D., Tan, G.: Visible photocatalytic degradation of methylene blue on magnetic $SrFe_{12}O_{19}$. J. Phys. Chem. Solids **123**, 157–161 (2018)
60. Polley, K., Kundu, R., Bera, J.: Adsorption and sunlight-induced photocatalytic degradation of methyl blue by $BaFe_{12}O_{19}$ ferrite particles synthesised through co-precipitation method. Int. J. Environ. Anal. Chem. **103**, (2023)

61. Annalakshmi, M., Balaji, R., Chen, S.-M., Chen, T.-W., Huang, Y.C.: A sensitive and high-performance electrochemical detection of nitrite in water samples based on sonochemical synthesized strontium ferrite nanochain architectures. Electrochim. Acta **360**, 136797 (2020)
62. Kaur, P., Singh, S., Renu, Jyoti, Kumar, V., Tikoo, K.B., Kaushik, A., Singhal, S.: Development of emphatic catalysts for waste water remediation via synchronized free radical and non-free radical routes with composites of strontium hexaferrite, graphene and multi-walled carbon nanotubes. Ceram. Int. **48**, 4795–4811 (2022)
63. Wang, H., Xu, Y., Jing, L., Huang, S., Zhao, Y., He, M., Xu, H., Li, H.: Novel magnetic $BaFe_{12}O_{19}$/g-C_3N_4 composites with enhanced thermocatalytic and photo-fenton activity under visible-light. J. Alloys Compd. **710**, 510–518 (2017)
64. Basha, B., Manzoor, A., Alrowaili, Z.A., Ihsan, A., Shakir, I., Al-Buriahi, M.S.: $Ba_{2-x}Ho_xSr_{2-y}Ni_yFe_{12}O_{22}$ and its composite with MXene: synthesis, characterization and enhanced visible light mediated photocatalytic activity for colored dye and pesticide. RSC Adv. **13**, 29944–29958 (2023)
65. Khan, K., Batool, Z., Manzoor, S., Ahmad, D., Aman, S., Alhashmialameer, D., Abd El-Gawad, H.H., Taha, T.A., Ashiq, M.N.: Fabrication of substituted Y-type hexaferrites/carbon dots composites for recording media and photodegradation of dye. Ceram. Int. **48**, 27550–27559 (2022)
66. Zielińska-Jurek, A., Bielan, Z., Dudziak, S., Wolak, I., Sobczak, Z., Klimczuk, T., Nowaczyk, G., Hupka, J.: Design and application of magnetic photocatalysts for water treatment. The effect of particle charge on surface functionality. Catalysts **7**, 360 (2017)
67. Sobhani-Nasab, A., Behpour, M., Rahimi-Nasrabadi, M., Ahmadi, F., Pourmasoud, S.: New method for synthesis of $BaFe_{12}O_{19}/Sm_2Ti_2O_7$ and $BaFe_{12}O_{19}/Sm_2Ti_2O_7$/Ag nano-hybrid and investigation of optical and photocatalytic properties. J. Mater. Sci. Mater. Electron. **30**, 5854–5865 (2019)
68. Bibi, F., Iqbal, S., Sabeeh, H., Saleem, T., Ahmad, B., Nadeem, M., Shakir, I., Aadil, M., Kalsoom, A.: Evaluation of structural, dielectric, magnetic and photocatalytic properties of Nd and Cu co-doped barium hexaferrite. Ceram. Int. **47**, 30911–30921 (2021)
69. Rezaei, A., Nabiyouni, G., Ghanbari, D.: Photo-catalyst and magnetic investigation of $BaFe_{12}O_{19}$–ZnO nanoparticles and nanocomposites. J. Mater. Sci. Mater. Electron. **27**, 11339–11352 (2016)
70. Raut, S.S., Adpa, S.K., Jambhale, A., Abhyankar, A.C., Kulkarni, P.S.: Enhanced photocatalytic activity of magnetic $BaFe_{12}O_{19}$ nanoplatelets than TiO_2 with emphasis on reaction kinetics, mechanism, and reusability. Ind. Eng. Chem. Res. **57**, 16192–16200 (2018)
71. Han, Y., Wang, S., Yu, X., Li, M., Yi, Z., Tang, J., Gao, H., Yang, H., Fang, L., Syed, A., Li, D.: $Ba_{0.5}Sr_{0.5}TiO_3/BaFe_{12}O_{19}$ magnetic separation photocatalysts: structural, optical, microstructure, photocatalytic activity, degradation pathway and mechanism. Ceram. Int. **49**, 37967–37982 (2023)
72. Feng, S., Xie, T., Wang, J., Yang, J., Kong, D., Liu, C., Chen, S., Yang, F., Pan, M., Yang, J., Du, H., Chen, H.: Photocatalytic activation of PMS over magnetic heterojunction photocatalyst $SrTiO_3/BaFe_{12}O_{19}$ for tetracycline ultrafast degradation. Chem. Eng. J. **470**, 143900 (2023)
73. Yan, X., Zhang, T., Meng, S., Zhou, P., Xu, Y., Wang, Y., Xie, M.: $BaFe_{12}O_{19}$/BiOBr S-scheme heterojunction magnetic nanosheets for high-efficiency photocatalytic degradation of 2-mercaptobenzothiazole. Sep. Purif. Technol. **326**, 124746 (2023)
74. Xie, M., Wang, D., Jing, L., Wei, W., Xu, Y., Xu, H., Li, H., Xie, J.: Preparation of magnetically recoverable and Z-scheme $BaFe_{12}O_{19}$/AgBr composite for degradation of 2-Mercaptobenzothiazole and Methyl orange under visible light. Appl. Surf. Sci. **521**, 146343 (2020)
75. Zhou, P., Wang, Y., Yan, X., Gan, Y., Xia, C., Xu, Y., Xie, M.: Nitrogen-defect-modified g-C_3N_4/$BaFe_{12}O_{19}$ S-scheme heterojunction photocatalyst with enhanced advanced oxidation technology synergistic photothermal degradation ability of antibiotic: insights into performance, electron transfer pathways and toxicity. Appl. Catal. B **343**, 123485 (2024)
76. Bibi, I., Muneer, M., Iqbal, M., Alwadai, N., Almuqrin, A.H., Altowyan, A.S., Alshammari, F.H., Almuslem, A.S., Slimani, Y.: Effect of doping on dielectric and optical properties of

barium hexaferrite: photocatalytic performance under solar light irradiation. Ceram. Int. **47**, 31518–31526 (2021)

77. Khan, K., Katubi, K.M., Batool, Z., Alsaiari, N.S., Manzoor, S., Aman, S., Fatima, A., Al-Buriahi, M.S., Ashiq, M.N.: Structural, electrical, and photocatalytic properties of Y-type hexaferrite/carbon dot composite. J. Mater. Sci. Mater. Electron. **34**, 1793 (2023)
78. Valero-Luna, C., Palomares-Sanchéz, S.A., Ruíz, F.: Catalytic activity of the barium hexaferrite with H_2O_2/visible light irradiation for degradation of methylene blue. Catal. Today **266**, 110–119 (2016)
79. Erfan, N.A., Abdel-Hameed, S.A.M., Mohammed, A.A.: Synthesis and characterization of barium hexaferrite ($BaFe_{12}O_{19}$) nanoparticles from iron ore waste and their possible application in water treatment. IOP Conf. Ser. Mater. Sci. Eng. **975**, 012002 (2020)
80. Ismail, I., Ibrahim, I.R., Matori, K.A., Awang, Z., Muhammad Zulkimi, M.M., Mohd Idris, F., Nazlan, R., Azis, R.S., Mohd Zaid, M.H., Rusly, S.N.A., Ertugrul, M.: Comparative study of single- and double-layer $BaFe_{12}O_{19}$-Graphite nanocomposites for electromagnetic wave absorber applications. Mater. Res. Bull. **126**, 110843 (2020)
81. Ansari, F., Soofivand, F., Salavati-Niasari, M.: Eco-friendly synthesis of cobalt hexaferrite and improvement of photocatalytic activity by preparation of carbonic-based nanocomposites for waste-water treatment. Compos. B Eng. **165**, 500–509 (2019)

Future Perspectives of Electroceramics

Dipan Kumar Das, Priyambada Mallick, Siddharth Parthasarathy, Srikanta Moharana Santosh Kumar Satpathy, and Padmaja Patnaik

Abstract Electroceramics are materials that exhibit electrical properties such as ferroelectricity, piezoelectricity, and pyroelectricity. Electroceramics play a pivotal role in modern technological advancements by serving as key components in electronic devices, energy storage systems, and sensors. This study provides a glimpse into the future perspectives of electroceramics, exploring their potential advancements and novel applications. As researchers delve deeper into the fundamental understanding of materials at the atomic and nanoscale levels, the design and synthesis of advanced electroceramics are expected to yield materials with enhanced properties and tailored functionalities. The integration of electroceramics into next-generation electronic devices such as high-performance capacitors, piezoelectric transducers, solid-state fuel cells, and memristors is anticipated to revolutionize industries and promote sustainable energy solutions. The convergence of electroceramics with emerging technologies like the Internet of Things and wearable electronics holds promise for the development of versatile and efficient devices. Electrical conductivity, dielectric behavior, thermal expansion, mechanical strength, chemical stability, etc., are among the key aspects evaluated in the context of future perspectives of electroceramics.

Exploring novel fabrication techniques and material compositions could lead to electroceramics with improved mechanical, thermal, and electrical properties. The continued advancements in additive manufacturing, nanotechnology, and thin-film deposition methods are likely to facilitate the production of intricate and miniaturized electroceramic devices, which open doors to unprecedented applications. Electroceramics are expected to significantly contribute to energy storage, battery development and supercapacitors. They are ideal for renewable energy integration and electric vehicle technology. Electroceramics could revolutionize environmental monitoring, medical diagnostics, and personalized health management through sensor technology advancements.

D. K. Das · P. Mallick · S. Parthasarathy · S. M. S. K. Satpathy (✉) · P. Patnaik (✉)
School of Applied Sciences, Centurion University of Technology and Management, Jatani, Odisha, India
e-mail: santosh.satpathy@cutm.ac.in

P. Patnaik
e-mail: padmaja.patnaik@cutm.ac.in

U. Kumar (ed.), *Defects Engineering in Electroceramics for Energy Applications*, Engineering Materials, https://doi.org/10.1007/978-981-97-9018-0_20

However, challenges such as material stability, cost-effective manufacturing, and environmental impact must be addressed alongside these promising prospects. The convergence of multidisciplinary research, innovative fabrication techniques, and growing emphasis on sustainable solutions is expected to drive the field forward. As electroceramics evolve, their transformative impact on electronics, energy, healthcare, and beyond is poised to shape a more connected, efficient, and sustainable future.

1 Introduction

Prakash (2012) discusses the importance of electroceramics in various advanced applications, such as sensors and actuators [1]. Bharadwaj (2012) highlights the role of electroceramics in fuel cells, batteries, and sensors [2]. Moulson's (1990) offers a synopsis of electroceramics' composition, characteristics, and uses [3]. Irvine (1990) focuses on the characterization of electroceramics using impedance spectroscopy [4]. Overall, these papers emphasize the diverse electrical, electromagnetic, and optical properties of electroceramics and their significance in various technological fields. Electroceramics have specific electrical, electromagnetic, or optical properties and are key components of any smart system. A powerful impedance spectroscopy technique is used to unravel the complexities of electroceramics. These materials transition from empirical study and physical demonstration into virtual design and testing.

1.1 Historical Overview

Electroceramics research and development focus on technology and device applications, encompassing inorganic materials at all scales. Applications include environment monitoring, transportation, medicine, healthcare, electronics, and communications. Surface and interface phenomena are increasingly important, driven by integration into microelectronics and MEMS technologies. Electroceramics are following conventional semiconductors in down-scaling, with nano-size effects and nanotechnology-related processes gaining importance. High permittivity Barium Titanate based piezoelectric ceramics and their polymer composites offer outstanding properties for a variety of applications in sonars, ultrasonic cleaners, micro-accelerometers, hydrophones, surface acoustic wave filters, delay lines, and micro actuators. [5].

Technological advancements and device applications in microelectronics, communications, automation, energy saving, and MEMS are the main drivers of electroceramics research. Ferroelectric ceramics have unique properties like high dielectric permittivity, piezoelectric coupling, and optical transparency. Indian organizations, including academic institutions and research laboratories, prepare, characterize, and fabricate electroceramic materials, with ongoing work and future prospects [6].

1.2 *Importance of Electroceramics*

Prakash (2012) emphasizes the significance of electroceramics in sensors and actuators, including ferrites, ferroelectrics, piezoelectrics, and pyroelectrics [1]. Bharadwaj (2012) discusses the role of electroceramics in fuel cells, batteries, and sensors, emphasizing their tailored properties for insulators, conductive ceramics, electrodes, and actuators [2]. Setter (2001) emphasizes the broad range of applications for electroceramics, including environment monitoring, transportation, medicine, health-care, electronics, and communications [5]. Rosseinsky (2009) emphasises electroceramics' considerable technical effect and the significance of academic and industrial collaboration in progressing the area [7]. Overall, the relevance of electroceramics in many technologies is emphasised in these articles, as is the continuous research and development in this sector.

Electroceramics are adaptable ceramic materials that are employed in cutting-edge civil and military applications. Ferrites, ferroelectrics, piezoelectrics, pyroelectrics, and microwave dielectrics are among them. Because their characteristics are sensitive to environmental changes, they are critical components of smart systems. With microwave ferrites and dielectrics for phase shifters, prizoelectrics for actuators, and pyroelectrics for IR detectors, processing methods is critical in regulating material performance.

Electroceramics has emerged as a critical research subject in materials science, drawing hundreds of participants from 35 countries. These seminars allow academic and industrial contact by discussing current breakthroughs and new trends in processing, characterization, and particular material families.

Point defects are crucial for electroceramics, as they influence material functionality, chemical kinetics, and preparation, conditioning, annealing, and degradation. They lead to depletion, accumulation, and inversion layers, affecting electrical and chemical effects. Understanding these effects can help design purposefully for electroceramics, such as ionically or mixed conducting oxides and halides. Nanostructured materials also affect local properties, leading to confinement effects and anomalies in ion conduction and mass transport [8].

2 Emerging Trends in Electroceramics

2.1 *Multifunctional Electroceramics*

The behaviour of capacitors, thermistors, varistors, and other electroceramic components is explained by reviewing structure–property correlations in electronic ceramic materials. The integration and miniaturisation of ceramic components have advanced quickly, resulting in the creation of multifunctional electronic packages with intricate three-dimensional circuitry. Concurrently, a vast array of intelligent sensors, transducers, and actuators are being built with composite materials to focus forces and fields. Although a number of disruptive technologies are on the horizon, the processing methods used at the moment rely on tape casting thick film techniques. As component sizes decrease to less than micrometre and nanocomposite devices become a reality, electroceramic devices should follow in the footsteps of semiconductor technology in the coming years [9].

Electroceramics are ceramic materials with specific electrical, electromagnetic, or optical properties, increasingly important in various technologies like communication, energy conversion, electronics, and sensors. Their performance depends on the complex interplay between processing, chemistry, and structure. Their development depends on ionic-covalent bonding, microstructure, and thermal processing at elevated temperatures. This chapter provides an overview of electroceramics, their synthesis, processing, electrical properties, and applications in advanced electrochemical devices [2].

2.2 *Nanoscale Electroceramics*

Kidner (2003) introduces a nested cube model (NCM) that accurately calculates the current distribution in polycrystalline ceramics at the nanoscale [10].

Krol (2002) explores the significance of interfaces in junction-based devices made of semiconducting oxides, emphasizing the necessity of nanosized structures and interface state control [11]. Alexe (2004) investigates non-traditional micro- and nanopatterning strategies for electroceramics, such as focused ion beam patterning, and bottom-up approaches, such as physical and chemical self-patterning [12]. These discoveries add to our understanding of nanoscale electroceramics and open up new options for study and development in this area.

The "bricklayer model" accurately describes the behavior of grain boundary-controlled electroceramics (BLM) in the microcrystalline domain. However, in the nanocrystalline domain, where grain boundary layers may form a large volume fraction of the overall microstructure, fundamental layer models are no longer valid. In order to treat a "nested cube model" (NCM) more precisely, this study details the creation of a pixel-based finite-difference method. This technique calculates the current distribution in polycrystalline ceramics at the point when the diameters of the grain borders and grain cores are similar. Furthermore, as the volume fraction of grain cores approaches unity (thin boundary layers), the NCM approaches layer model behavior and mimics standard effective medium treatments when the volume % of grain cores approaches zero. Consequently, the NCM can model electroceramic behavior at all grain sizes from the nanoscale to the microscale [10].

This paper examines the prospects and possible issues in the development of junction-based devices made of semiconducting oxides. It is discussed how the material properties of oxide semiconductors and conventional semiconductors differ in significant ways. A review of the research on the many types of oxide junctions, such as n–n heterojunctions, double Schottky barriers, p–n heterojunctions, and metal–semiconductor (Schottky) junctions, comes next. Future junction-based devices are anticipated to rely heavily on the utilization of nanoscale structures, with particular attention paid to the case when the space charge area is larger than the structure's dimensions. The selective activation of shallow donor or acceptor states in normally p- or n-type conducting oxides, respectively, the growth of single crystal p–n homo- and heterojunctions, and enhanced control over interface states are three crucial prerequisites for future advancements in this developing field [11].

A summary of non-traditional techniques, encompassing both top-down and bottom-up procedures, is provided for the micro- and nanopatterning of electroceramics. The top-down method looks at next-generation lithography techniques such as nanoimprint lithography, electron-beam direct writing, focused ion beam patterning, and others, while the bottom-up approach presents a number of physical and chemical self-patterning options. There is also a chapter discussing the promise and limitations of piezoresponse scanning force microscopy for ferroelectric testing of nanopatterned electroceramics [12].

2.3 *Biocompatible Electroceramics*

Shishov (2022) focuses on the development of electroconductive polymer matrices for tissue engineering, demonstrating the biocompatibility of composite materials modified with conductive additives [13]. Bharadwaj (2012) provides a broad overview of electroceramics, highlighting their diverse properties and applications in various technologies [2]. Stauss (2018) discusses the potential of biocompatible batteries for powering medical implants and devices, emphasizing the need for miniaturization and improved performance [14]. Irvine (1990) explores the characterization of electroceramics using impedance spectroscopy, showcasing its effectiveness in studying electrical properties and phenomena in these materials [4]. In summary, these papers collectively demonstrate the importance of biocompatible electroceramics in various fields, including tissue engineering, energy storage, and electronic medical implants.

Various biocompatible polymers (polyimide, chitosan, polyamide) were treated with various conductive additions to create electroconductive polymer matrices for application in tissue engineering (graphene, carbon nanotubes, polypyrrole). By employing human dermal fibroblast culture, the biocompatibility of matrices generated from composite materials was verified. Upon applying cyclic external potential, the conducting characteristics of the matrices were examined both in the dry state and in the physiological solution. It describes how the resultant composites' ion- and electron-conducting constituents affect the current flowing through the matrix and the pattern of the current sweep.

Electroceramics are ceramic materials with specific electrical, electromagnetic, or optical properties, and they are increasingly important in various technologies like communication, energy conversion, electronics, and sensors. Their performance depends on the complex interplay between processing, chemistry, and structure. Their development depends on ionic-covalent bonding, microstructure, and thermal processing at elevated temperatures. This chapter provides an overview of electroceramics, their synthesis, processing, electrical properties, and applications in advanced electrochemical devices.

Conventional electrode materials and electrolytes serve as the foundation for energy storage systems used to power electronic medical implants and sensors. These battery systems require careful encapsulation due to their toxicity, which results in large devices. These drawbacks are addressed by batteries based on biocompatible electrodes and electrolytes, which show promise as workable substitutes for powering medical implants and equipment. Giving a general overview of potential battery systems and their current state of performance is the goal of this review. A synopsis of battery topologies and their construction is also provided, with an emphasis on possible downsizing. Technological developments in biocompatible batteries are anticipated to have significant effects on environmental sensing and transient electronics, in addition to electronic medical implants and point-of-care monitoring devices.

A new conductive hydrogel method has been developed for implantable flexible electrical conductors in bioelectronics. This method involves exfoliating graphite and polymerizing zwitterionic monomers, resulting in a hydrogel with mechanical properties similar to living tissue. The hydrogel exhibits excellent adhesion, electrical conductivity, non-swelling, and high conformability in water. Its biocompatibility is confirmed through cytotoxicity tests and histological analysis. This hydrogel shows promise for next-generation bioelectronics applications [15].

2.4 Energy Harvesting and Storage

Patel (2021) investigates the effect of stress on energy storage density and electrocaloric performance in BZT ceramics, finding that stress can enhance energy harvesting and storage potential [16]. Anoufa (2013) focuses on core–shell ferroelectric ceramics and presents a theoretical approach to evaluate the possibility of storing electrical energy in such materials [17]. Zhang (2014) discusses energy harvesting based on piezoelectric Ericsson cycles in piezoceramic materials, highlighting the potential for increasing harvested energy through the coupling of electrical and mechanical excitations [18]. Overall, these papers contribute to our understanding of energy harvesting and storage in electroceramics, offering valuable insights for the development of high-performance applications in this field.

This study investigates waste energy harvesting and storage potential in $Ba_{0.85}Zr_{0.15}TiO_3$ ceramics, focusing on stress effects on energy storage density and electrocaloric performance. Polarization–electric field hysteresis loops were recorded at various temperatures and uniaxial compressive stress. Results show high energy storage density and electrocaloric temperature under different stress fields. The study contributes to understanding stress effects on BZT materials, offering high performance for energy harvesting and storage-based applications, and improving energy storage density and electrocaloric effect via stress confinement [16].

A quantitative evaluation of storing electrical energy using a ferroelectric supercapacitor built on core–shell nanoceramics was conducted. The study solved the Laplace equation for a single core–shell particle and used an effective medium approximation to obtain equations for internal electric field, permittivity, and polarization. The results were generalized to different ceramics and anisotropic particle shapes [17].

The research focuses on the use of electric generators for recycling ambient energies, particularly from mechanical vibrations. Direct piezoelectric energy harvesting leads to weak energy levels, but coupling an electric field and mechanical excitation on Ericsson-based cycles can increase the amplitude of harvested energy. A dynamic hysteresis model is developed to accurately estimate harvested energy, and theoretical predictions and experimental results are compared to validate the proposed solution [18].

2.5 *Additive Manufacturing in Electroceramics*

The papers collectively suggest that additive manufacturing, also known as 3D printing, holds great potential for the production of electroceramics. Hashemi (2020) and Browne (2020) highlight the advantages of additive manufacturing in electrochemistry, including fast prototyping, low cost, and the availability of various 3D printing technologies [19, 20]. Fieber (2019) explores the use of additive manufacturing to fabricate electrochemical double layer capacitor devices, demonstrating the flexibility and cost-effectiveness of the process [21]. Safari (2002) discusses the application of layered manufacturing techniques in producing advanced functional electroceramic components, such as piezoelectric actuators and sensors [22]. Overall, these papers indicate that additive manufacturing has the potential to revolutionize the manufacturing of electroceramics by enabling the production of complex geometries, customization, and improved performance.

Since additive manufacturing allows for quick prototyping and is relatively inexpensive, it is widely used in many fields of electrochemistry to make electrodes and devices. Additive manufacturing is also referred to as three-dimensional (3D) printing. Additive manufacturing is a very attractive technology for electrochemical applications because of the range of 3D-printing technologies accessible, such as select laser melting (SLM), stereolithography (SLA), inkjet printing, and fused deposition modelling (FDM). Specifically, during the past several years, a substantial body of research has evolved into the use of 3D printing to manufacture electrodes and devices for the conversion and storage of electrochemical energy. Even though progress has been made, a number of obstacles and problems still need to be solved before active and stable electrodes and devices for electrochemical energy conversion and storage can be 3D printed and compared to state-of-the-art technology. We will provide an outline of the rationale behind these electrochemical uses of 3D printing in this Review. Subsequently, we will talk about how different 3D printing processes and post-modification procedures affect the electrochemical performance of the electrodes and devices [20]. A hybrid additive manufacturing system was designed to fabricate supercapacitors in a single automated operation [21].

Layered Manufacturing (LM) techniques, such as Fused Deposition of Ceramics (FDC), Robocasting (RC), Micropen (MP), and Laminated Object Manufacturing (LOM), are used to create net-shape components quickly. These techniques include FDC for prototyping piezoelectric and electrostrictive actuators, multi-material actuators, piezo composite sensors, Templated Grain Growth (TGG), LOM for torsional and telescoping actuators, and Robocasting for 3–3 piezo composites and photonic band gaps structures. Micropen allows for multifunctional integrated ceramic components [22].

3 Challenges and Limitations

3.1 Material Processing and Manufacturing Challenges

Electroceramics face challenges in material processing and manufacturing. One challenge is the need for improved piezoelectric properties, which can be achieved by creating a morphotropic structure at the phase interface of the material [23]. Another challenge is the demand for cost reduction and constant threats of replacement with substitution products, which puts pressure on processed raw materials suppliers [24]. Ceramic materials offer advantages in terms of dimensional stability and quality of finish, but grinding processes can cause damage [25]. Sol–gel processing has been used to create dense insulating electroceramics in thin-layer form, avoiding the issues associated with powders [26]. Overall, there is a need for advancements in processing techniques and materials to address the challenges faced by electroceramics in manufacturing and improve their properties and cost-effectiveness [27].

The creation of a morphotropic structure at the phase interface of the material can lead to improved piezoelectric properties. The method of manufacturing the electroceramic multilayer structural element involves grinding and mixing initial materials, calculations, the addition of a binder, formation of green films, stacking and lamination of the green films, removal of the binder, and sintering of the green elements.

Nickel may be introduced into the material structure prior to calcination or may come from internal electrodes applied to the green films during further processing.

Processed raw materials suppliers face challenges in improving their margins and surviving in a market where buyers constantly demand cost reduction and threaten to replace them with substitution products.

The demand for cost reduction and threats of replacement creates pressure on processed raw materials suppliers to find ways to lower their prices and remain competitive.

This pressure may lead to a need for suppliers to reassess their materials and form routes to maintain cost-effectiveness.

Suppliers may need to experiment with new materials, shapes, and finishing procedures in order to increase the speed of production and the cost-effectiveness of their goods. A material-process combination's cost-effectiveness is determined by factors such as the cost of the process and the quantity of similar things produced. The high RD expenses, expertise, and patent laws in certain industries make it difficult for firms to mimic the process, which can provide some protection for processed raw materials suppliers.

3.2 Environmental and Sustainability Concerns

Environmental and sustainability concerns in electroceramics are driven by the need to integrate electroceramic functions into microelectronics and MEMS technologies, as well as the evolution of bulk products into materials systems. Although electrochemical methods have been studied for environmental cleanup, many of these technologies are currently at a low technical readiness level (TRL) [28]. Manufacturing technologies, including grinding tool manufacture, must also become more sustainable [29]. These considerations underline the relevance of electroceramics research and development in addressing environmental and sustainability challenges. Electroceramics are complicated materials that are employed in a variety of applications such as electronics, energy storage, and sensors.

Electroceramics production comprises numerous processes, including raw material extraction, processing, and manufacture, all of which can have an influence on the environment. Using energy-intensive methods and producing trash during production can contribute to carbon emissions and resource depletion.

Concerns about sustainability in electroceramics include the need for more efficient production methods, waste and emission reduction, and the use of ecologically friendly materials. Alternative materials and production procedures that minimize environmental consequences and increase the sustainability of electroceramics require research. Transparency in the supply chain is also vital to guarantee that raw materials are sustainably acquired and that the complete life cycle of electroceramics is taken into account in sustainability assessments [30].

3.3 Reliability and Durability Issues

Because of their wide range of applications, reliability and durability difficulties in electroceramics are critical. This field's research and development focuses on incorporating electroceramic capabilities into microelectronics and MEMS technologies, as well as enhancing thermoelectric device dependability. The use of novel characterisation techniques and nanotechnology-related procedures is becoming increasingly important in understanding the nanoscale properties of electroceramics. Furthermore, impedance spectroscopy is an effective tool for investigating the complexity of electroceramics, such as electrical inhomogeneities, electrode/electrolyte interfaces, and surface layers on glasses. Electroceramic components are developed with protective encapsulation and insulating materials to avoid electric flashover and enable safe operation to maintain dependability. Overall, electroceramics research strives to address reliability and durability challenges in order to improve performance and broaden applications [31, 32].

Because of their particular features and uses, electroceramic components are prone to reliability and durability concerns. However, using different materials for the protective encapsulation on opposing portions of the component body might pose material compatibility and long-term performance difficulties.

The selection of appropriate materials for the protective encapsulation is critical to ensuring the electroceramic component's dependability and endurance. More research and development is needed to solve these reliability and durability challenges, as well as to improve the performance of electroceramic components in diverse applications.

3.4 Cost and Scalability Challenges

Cost and scalability issues in electro ceramics are a major source of worry in the industry. Electronic ceramics for integrated circuit packaging and other applications demand a high degree of purity and a high-quality regular microstructure [33]. While ceramic materials show promise for super capacitor electrodes, their low specific surface area and surface activity limit their performance [34]. Traditional sintering processes used to shape ceramics are ineffective in generating fine shapes without micro cracks, leading to the exploration of alternative methods such as electro discharge machining [35]. The development of advanced ceramics has led to the production of various ceramic products, but their manufacturing is more challenging compared to metals and polymers [36]. The miniaturization of electronic components in the ceramic industry is approaching its technical limit, driving the need for new materials, advanced characterization tools, and innovative ceramic processing techniques [37]. 3D-printed conductive silicon carbide ceramics have high electrical conductivity.

Proton-conducting ceramic materials have potential for electrochemical devices. There are Challenges exist for commercial-scale manufacturing and Nano levels. Energy ceramics face challenges in stability under extreme redox conditions.

4 Future Applications of Electroceramics

Due to the unique electrical, dielectric, thermal, optical, and mechanical properties, electroceramics can have a wide range of future applications (Fig. 1), advancing the technology in human life.

4.1 Neuromorphic Computing

These papers collectively discuss the potential applications of electroceramics in neuromorphic computing. Schneider (2022) focuses on the use of superconductive electronics, specifically Josephson junctions, for neuromorphic computing systems [38]. Nikam (2023) explores electrochemical random access memories (ECRAMs) as a promising candidate for implementing artificial synapses in neuromorphic computing [39]. Emboras (2020) introduces opto-electronic memristors as a class of electro-optical neuromorphic hardware with potential for low-power artificial intelligence applications [40]. Kang (2022) reviews ion-driven ECRAM-based synaptic devices and their ability to emulate the role of biological synapses in neuromorphic computing systems [41]. Overall, these papers highlight the advancements and potential of electroceramics in enabling neuromorphic computing systems.

Electroceramics have shown promise in various applications, particularly in neuromorphic computing. These materials can be used to create memristive devices that mimic the synaptic behavior of biological neurons, store information non-volatilely, create sensors that provide real-time data input and feedback, and optimize signal processing. They can also be used for energy-efficient computing by switching between different polarization states. Electroceramics can be integrated into specialized hardware accelerators for tasks related to neuromorphic computing, such as neural network training and inference. They can also be used in brain-computer interfaces to enhance interaction between the human brain and artificial systems. Electroceramics can also be used in neurorobotics to make robots more adaptable to dynamic environments. They can also be used in AI applications for tasks like image and speech recognition and autonomous navigation. Lastly, they can enhance

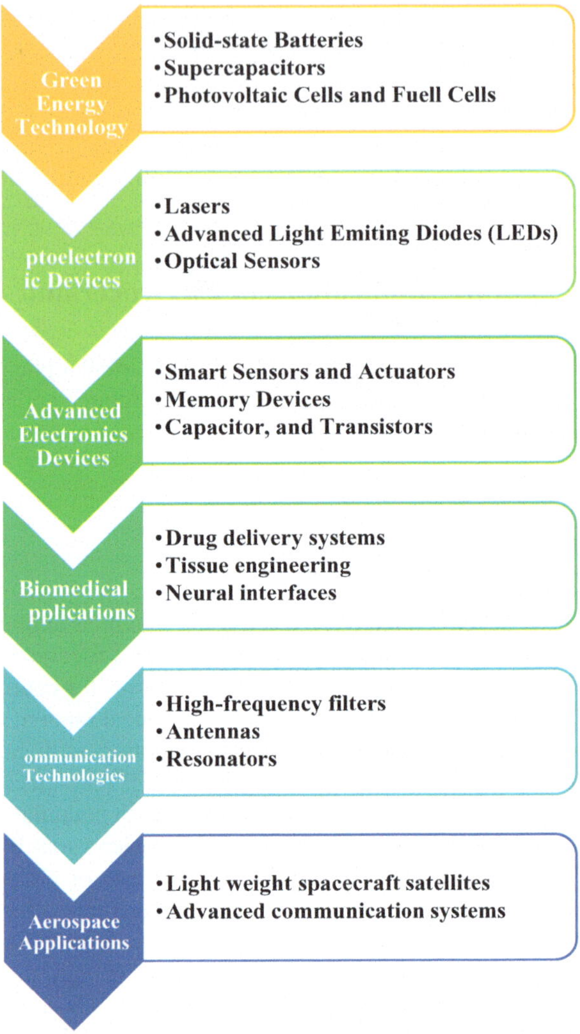

Fig. 1 Future applications of electroceramics

hardware security by implementing biologically inspired authentication and encryption mechanisms. A nonvolatile electro-chemical random-access memory is demonstrated for high-speed, low-power neuromorphic computing [42]. A plastic electrochemical neuromorphic organic device switches at low voltage and energy [43]. The fabricated three-dimensional electronic synapse exhibits the potential for low fabrication cost, high integration density, and excellent performances [44]. Spin-transfer torque mechanisms can be potentially utilized for device structures mimicking neural and synaptic functionalities [45].

4.2 Quantum Computing

The papers collectively suggest that electroceramics, specifically trapped electrons in a linear Paul trap, semiconductor quantum dots, and superconducting electrical circuits, have potential applications in quantum computing. Yu (2021) discusses the feasibility of using trapped electrons as qubits in a future quantum computer, achieving high two-qubit Bell-state fidelities [46]. Zhang (2017) highlights the advancements in semiconductor quantum dot-based quantum computing, including qubit preparation, readout, and manipulation of quantum logic gates [47]. Blais (2004) proposes an architecture using superconducting electrical circuits for quantum computing, offering strong coupling, inhibition of spontaneous emission, and entanglement of qubits [48]. Bocko (1997) discusses the prospects and challenges of implementing a quantum computer using superconducting electronics, with potential for quantum coherent computations [49].

Electroceramics have the potential to significantly contribute to the field of quantum computing, which uses quantum mechanics to perform complex calculations at speeds that classical computers cannot achieve. These materials can be used in various aspects of quantum computing, including hardware, memory, and control systems. They can be used to develop microwave resonators and capacitors for superconducting qubits, build quantum memory units, design cryogenic control components, integrate quantum sensors for precision measurements and feedback control, develop quantum transducers for quantum information conversion, contribute to quantum error correction codes and schemes, enhance quantum communication security and reliability, and enhance materials characterization for materials characterization. Electroceramics can also be integrated into specialized hardware for quantum machine learning algorithms, enabling faster and more efficient computation. However, challenges related to material quality, device integration, and control will be essential for fully harnessing the full potential of electroceramics in quantum computing.

A Cooper pair box may be used to carry out one- and two-qubit gate operations [50]. A quasi-2D set of N 108 electrons in vacuum trapped in 1D hydrogenic levels above a micron-thick helium film can be used as an easily manipulated strongly interacting set of quantum bits [51]. A quantum computer based on electrons floating over liquid helium is described [52]. By trapping electrons in vacuum within an innovative confining structure, a quantum computer can be implemented [53].

4.3 Advanced Medical Devices

The wide spectrum of uses for electroceramics, especially in healthcare and medicine, is highlighted by Setter [5]. The use of electroceramics to sensors and actuators—two crucial parts of smart systems—is the main topic of Prakash [1]. The usage of electroceramics in fuel cells, batteries, and sensors is covered in Bharadwaj (2012), with an emphasis on how their unique features are suited for certain applications. In Bharadwaj [2] an summary of electroceramics research conducted in India is given in Goel (2004), which also highlights the special qualities of ferroelectric ceramics and its prospective uses in a range of devices [6]. These studies collectively imply that electroceramics holds great promise for future improvements in medical equipment. The prospective uses of electroceramics in cutting-edge medical devices are covered in all of these articles.

Because of their special mechanical and electrical qualities, electroceramics are perfect for cutting-edge medical equipment. They can raise functionality and performance, raise patient care, and raise diagnostic potential. Ultrasound imaging, drug delivery systems, electrostimulation devices, implanted sensors, biosensors and diagnostics, wearable health monitoring, dental devices, tissue engineering, neuroprosthetics, telemedicine and remote surgery, and wearable rehabilitation devices are some of the future uses. Ultrasound transducers frequently employ piezoelectric materials like lead zirconate titanate (PZT), and implanted drug delivery devices like MEMS-based micropumps and drug reservoirs precisely regulate the release of drugs. Electroceramic components are used by electrostimulation devices, such as cochlear implants and deep brain stimulators, to transmit electrical signals to certain body parts. Electroceramic materials can also be used in biosensors and diagnostic devices, wearable health monitoring, implantable sensors, tissue engineering, neuroprosthetics, dental devices, and telemedicine. Ensuring the safety and efficacy of electroceramic-based medical devices through rigorous testing and regulatory approval is crucial for their adoption in healthcare settings. Ceramic-polymer electronic composites have been introduced for medical, telecommunication and microelectronics applications [54].

The modelling of electroceramic materials and devices is still in its infancy [55]. Impedance spectroscopy is a powerful technique for unravelling the complexities of advanced materials [4].

4.4 Green Energy Solutions

The papers collectively suggest that electroceramics have promising future applications in green energy solutions. Maletin (2014) highlights the use of electrochemical double layer capacitors (EDLCs) and hybrid devices (HDs) in hybrid transport and wind power stations [56]. The wide spectrum of uses for electroceramics is covered by Setter (2001), encompassing electronics, transportation, medical, and environmental monitoring. In Setter [5] electrochemical energy storage and conversion technologies, such as fuel cells and rechargeable batteries, are examined in Chouhan [57]. The significance of electroceramics in communication, energy storage and conversion, sensors, and other areas is emphasised by Bharadwaj [2]. Together, these publications show how electroceramics may be used to further sustainable energy solutions.

Solid oxide fuel cells (SOFCs), photovoltaic devices, thermoelectric generators, piezoelectric energy harvesting, energy storage, hydrogen production and storage, high-temperature superconductors, green electronics, environmental monitoring, and smart grid technologies are just a few of the green energy solutions that electroceramics have many potential uses in. These substances can power low-energy electronic devices and sensors, increase the efficiency of solar cells, and serve as solid electrolytes in SOFCs. They can also be utilised in high-temperature superconductors, piezoelectric energy harvesting devices, and thermoelectric generators. Proton exchange membrane electrolyzers, which produce and store hydrogen, and high-temperature superconductors, such as yttrium barium copper oxide (YBCO), which provide effective superconducting transmission lines, have more application for electroceramics. Electroceramics can also be utilised in smart grid technology and environmental monitoring equipment. However, research and development in materials science, manufacturing techniques, and device integration are crucial for achieving these applications. Electrochemical energy conversion and storage devices offer environmental and sustainable advantages over fossil fuel-based systems [58]. Aqueous rechargeable batteries and supercapacitors have exhibited high power and energy densities [59]. Electrochemical energy technologies are critical enabling technologies for renewable energy, energy management, storage, pollution control/ monitoring, and greenhouse gas reduction [60].

Energy harvest and storage to fabricate self-powered electrochemical energy storage systems has promise to power future portable and wearable electronics [61].

4.5 *Aerospace and Defence Applications*

The papers collectively suggest that electroceramics have potential future applications in aerospace and defence. Kankam (2001) discusses the widespread use of power electronics in aerospace technologies, including motor drives in aircraft and starter/generators [62]. Bashir (2018) highlights the development of electroactive polymers for aerospace applications, such as aircraft morphing and space robotics [63]. Gerada (2015) focuses on high-performance electrical machines for aerospace, emphasizing the importance of electrical machine drive systems in aircraft electrification [64]. Last but not least, Setter (2001) covers the wide range of uses for electroceramics, such as communications and electronics, as well as the new developments in the integration of electroceramic functions into nanotechnology and microelectronics [5]. All of these studies point to the potential uses of electroceramics in aerospace and defence in the future. Because of their unique characteristics, which may boost many technologies' performance, efficiency, and dependability, electroceramics are valuable materials in the aerospace and defence industries. Electronic warfare, sensors, acoustic devices, inertial navigation systems, stealth technologies, energy harvesting, thermal protection, satellite technology, directed energy weapons, cybersecurity, and unmanned aerial vehicles (UAVs) are just a few of the applications for which they can be used. In a variety of applications, these materials can improve signal processing, communication capabilities, and dependability. They can also be employed in countermeasure, spectrum management, and signal jamming applications in electronic warfare systems. Electroceramics can also be used in acoustic devices, such as underwater sonar transducers and acoustic emission sensors, for naval and underwater surveillance and communication. Further research, testing, and compliance with military and aerospace standards are necessary to ensure the safety and performance of electroceramic-based technologies in these critical sectors. The various reduced models can be further extracted from the simplified magneto-fluid dynamics equations depending on the magnitude of the magnetic Reynolds number [65]. Electric drives for thrust-vector and flight-control actuation are required for future space vehicles [66].

5 Innovations in Electroceramic Materials

5.1 High-K Dielectrics

These papers collectively discuss innovations in electroceramic materials in high-k dielectrics for energy storage applications. Wang (2021) highlights the need for high energy/power density materials in capacitors and discusses the factors that improve energy storage properties in dielectrics [67]. Valero (2021) proposes a new vision for high-k dielectric materials at the nanoscale, demonstrating the potential of ultra-thin alumina layers for improved electrical conduction and extended lifetime capacity in micro-supercapacitors [68]. Fang (2002) presents a novel UV photo-assisted chemical vapor deposition process for depositing high-k metal oxide dielectrics, demonstrating the production of good quality layers at low temperatures [69]. Clark (2014) summarizes the current status of high-k dielectrics in VLSI technology and highlights emerging applications such as resistive RAM memories and ferroelectric logic and memory devices [70].

High-k dielectric materials are crucial in semiconductor devices, particularly in advanced microelectronics and integrated circuits. They replace traditional silicon dioxide as gate dielectrics in transistors, addressing issues like leakage current and gate capacitance. Innovations in electroceramic materials have improved performance and characteristics. Hafnium-based high-k materials, such as HfO_2 and $HfSiO_4$, offer higher dielectric constants, enabling better gate control and reducing leakage current. Metal–Insulator-Metal (MIM) structures use high-k materials to create capacitors with higher capacitance and lower leakage current. Perovskite oxides, rare earth oxides, and stabilization techniques have been developed to maintain stable dielectric properties at nanoscale thicknesses. Atomic Layer Deposition (ALD) techniques have been refined for depositing high-k materials, and interface engineering techniques have been improved to reduce defects and minimize interface trap densities. Multilayer dielectric structures combine various high-k materials and insulating layers to achieve higher effective dielectric constants while maintaining low leakage current.

ALD high-k materials exhibit compelling thermal/mechanical and electrical/optical properties [71]. Dielectrics with k-values in the 9–30 range are studied as insulators between charge storage layers and control gates in flash devices [72]. The addition of a small amount of metal is an effective method to improve the dielectric properties of the high-k film [73].

Aluminates derived by mixing aluminum oxide with lanthanum and yttrium oxides have unique physical attributes for a possible application as gate dielectrics when stabilized in the rhombohedral perovskite structure [74].

5.2 Lead-Free Piezoelectric Materials

These papers collectively discuss the development and advancements in lead-free piezoelectric materials. Uchikoba (2005) highlights the ongoing efforts to reduce lead content in electroceramic devices and introduces new fibrous piezoelectric materials [75]. Lee (2012) reviews various lead-free alternatives to lead zirconate titanate (PZT) and discusses their piezoelectric properties and thermal stability [76]. Aksel (2010) focuses on lead-free piezoelectric materials for sensors and actuators, including $BaTiO_3$, $Na_{0.5}Bi_{0.5}TiO_3$, $K_{0.5}Bi_{0.5}TiO_3$, $Na_{0.5}K_{0.5}NbO_3$, and their solid solutions [77]. A review of current developments in lead-free piezoelectric materials, both organic and inorganic, and their uses in transducers, sensors, and nanogenerators is given in Wei [78]. These studies highlight the significance of creating lead-free, eco-friendly substitutes for conventional lead-based piezoelectric materials.

Environmental concerns and regulatory limitations have made the development of lead-free piezoelectric materials more crucial. Key innovations include bismuth-based materials, alkaline niobates, perovskite oxides, organic–inorganic hybrids, relaxor ferroelectrics, mechanochemical synthesis, texturing ceramics, multiferroic materials, and flexoelectric materials. These substances seek to remove harmful lead content while preserving or enhancing piezoelectric qualities. As lead-free piezoelectric materials, bismuth-containing substances like $BiFeO_3$ and $Bi_4Ti_3O_{12}$ have surfaced, and the promise of alkaline niobates has also been explored. In lead-free piezoelectric, perovskite oxides have demonstrated potential, and organic–inorganic hybrids have been created. Relaxor ferroelectrics, which have better crystalline structure and composition, have also demonstrated potential. Understanding of the structure–property correlations in lead-free piezoelectrics has increased because to advances in materials characterization, including piezoresponse force microscopy and sophisticated spectroscopic techniques.

A lead-free piezoelectric ceramic with an electric-field-induced strain comparable to typical actuator-grade PZT has been found [79].

The potassium niobate family exhibits low dielectric constants, large thickness coupling coefficient along certain non-polar directions, and low density [80]. The current development stage in lead-free piezoceramics is critically assessed [81]. The use of lead should be limited in piezoelectric materials has promoted lead-free piezoelectric ceramics [82].

5.3 Solid Oxide Fuel Cell Materials

These papers discuss various innovations in electroceramic materials for solid oxide fuel cells (SOFCs). Eguchi (1997) explores the use of ceria-based electrolytes, specifically samaria-doped ceria, which lowers the operating temperature of the cell due to its high oxide ion conductivity [83]. Huang (2001) proposes an alternate approach using Sr- and Mg-doped $LaGaO_3$ perovskite as the electrolyte and a mixed oxide-ion/electronic conductor (MIEC) as the cathode [84]. Singhal (1997) provides an overview of the ceramic materials used in different components of SOFCs and discusses the performance of cells fabricated using these materials [85]. Orera (2010) discusses new systems being investigated for application in SOFCs, including electrode materials with perovskite structures and electrolyte materials with diverse structure-types [86]. Together, these studies emphasise how crucial it is to create electroceramic materials with low operating temperatures and excellent conductivity for successful and affordable SOFCs.

Energy-converting devices known as solid oxide fuel cells (SOFCs) use high temperatures to transform chemical energy into electrical energy. Electroceramic materials are essential to their effectiveness and functionality. Yttria-Stabilized Zirconia (YSZ), Gadolinium-doped Ceria (GDC), Lanthanum Strontium Manganite (LSM), Nickel Oxides (NiO), protons-conducting electrolytes, thin-film electrolytes, proton-conducting electrolytes, mixed ionic and electronic conductors (MIECs), redox-stable perovskites, ceramic composites, infiltration techniques, additive manufacturing, advanced characterization techniques, and alternative fuels are some of the major advancements in electroceramic materials for SOFCs. These advancements have contributed to the development of more efficient, durable, and cost-effective fuel cell systems. As SOFC technology continues to evolve, research is ongoing to further enhance material properties and address challenges related to long-term reliability, thermal cycling, and manufacturing scalability.

Electrode materials in cells with ZrO_2-, CeO_2-, $LaGaO_3$-, and $La_{10}Si_6O_{27}$-based electrolytes are appraised in context of potential optimization strategies [87]. Mixed oxide-ion/electron conductors are being explored for natural gas fuel cells [88]. Oxygen-ion conducting solid electrolyte systems have been reviewed with specific emphasis on their use in solid oxide fuel cells [89]. High-temperature proton conductors can achieve high conductivity at relatively low temperatures compared to their oxygen ion-conducting counterparts [90].

5.4 Perovskite Oxides

These papers collectively provide insights into the innovations in electroceramic materials in perovskite oxides. Cao (2021) discusses the recent advances in perovskite oxides as electrode materials for supercapacitors, highlighting their unique structure and compositional flexibility [91]. Tan (2017) focuses on the use of perovskite oxides as electrode materials in nonaqueous lithium-oxygen batteries, emphasizing their tunability and potential for achieving superior battery performance [92]. Xu (2018) explores novel nanostructuring methods of perovskite electrocatalysts for energy-related applications, highlighting the importance of reducing the size of bulk perovskites to improve their electrocatalytic performance [93]. Lastly, Fu (2023) reviews recent advances in perovskite oxide electrocatalysts, with a specific focus on ordered perovskites, cation segregation, and exsolution, providing insights into the crystal structure and structure regulation of perovskite oxides [94].

Perovskite oxides are a class of electroceramic materials that exhibit several desirable properties, such as ferroelectricity, piezoelectricity, high dielectric constants, and catalytic activity. They are employed in many different domains, including as energy conversion, electronics, and catalysis. Improvements in perovskite oxide technology have led to the creation of multifunctional materials, high-K dielectric perovskites, photovoltaic perovskites, high-temperature superconductors, transparent conductive oxides, solid oxide fuel cell electrolytes, nanomaterials, strain engineering, multiferroic perovskites, and advanced characterization methods. Perovskite oxides now have a wider variety of applications thanks to these advancements, which are also accelerating the search for new components and features. Given the wide range of prospective applications and industries, perovskite oxides are a promising field for more research and development.

Since the oxygen-anion-intercalation charge storage process was proposed, perovskites have been used as an anode material for supercapacitors [95].

Supercapacitors can benefit from the use of perovskite oxides as electrode materials [91]. In the area of energy storage and conversion, perovskite oxides have shown to be quite important [96]. Perovskite-based electrodes seem to be a way to satisfy the high energy density need in the market [97].

Targeted catalysts' porous nanostructure promotes catalytic activity, which raises the installations' overall efficiency in the energy sector [98].

5.5 *2D Materials for Electroceramics*

The overview of electroceramics and its applications given in these publications does not explicitly cover advances in 2D materials electroceramic materials. The significance of electroceramics in many technologies, including their characteristics, production, processing, and applications in electrochemical devices, is covered by Bharadwaj [2]. Irvine (1990) emphasises the use of impedance spectroscopy as an effective method for examining the electrical characteristics of electroceramics and characterising them [4]. An introduction to electroceramics, covering their synthesis, characteristics, and many uses, is given by Moulson [3]. In Moulson [3] None of these studies, meanwhile, specifically address advancements in 2D electroceramic materials.

Using 2D materials, electroceramics has the potential to revolutionise several sectors, including electronics, energy storage, and sensing. These materials provide unique advantages including flexibility, tunability, and high surface-to-volume ratios. Among the significant advancements in this subject are piezoelectric materials, dielectrics, FeFETs, electrolytes, thermoelectric materials, piezotronics, magneto-electric materials, nanogenerators, and integration with other 2D materials.

The synthesis of 2D ferroelectric materials, which show spontaneous electric polarisation and may be used in sensors, non-volatile storage, and memory devices, has advanced significantly. Piezoelectric materials are employed in actuators, energy harvesting devices, and sensors because they produce electrical charge in reaction to mechanical deformation. Two-dimensional dielectrics reduce flaws and enhance dielectric characteristics.

While piezotronic devices control the parameters of electrical transmission, thermoelectric materials are investigated for their potential to transform waste heat into power. Magnetoelectric materials are employed in sensors, memory devices, and energy conversion because they can transform magnetic fields into electric signals and vice versa. Nanogenerators harvest energy from mechanical vibrations and pressure changes, used in self-powered sensors and low-power electronics.

Advanced synthesis and characterization techniques are used to control material properties and understand their structure–function relationships.

High permittivity piezoelectric ceramics and their polymer composites show unusual properties for many applications in sonars, ultrasonic cleaners, micro-accelerometers, hydrophones, surface acoustic wave filters, delay lines, and micro-actuators [6].

Layered manufacturing has been developed and utilized to provide a cost-effective route to rapidly turn new designs into net-shape components [99]. The modelling of electroceramic materials and devices is still in its infancy [55]. Electroceramics is one of the most important research areas in materials science [7].

References

1. Prakash, C.: Electroceramics for sensors and actuators. In: 2012 1st International Symposium on Physics and Technology of Sensors (ISPTS-1), pp. 1–2 (2012)
2. Bharadwaj, S.R., Varma, S., Wani, B.N.: Electroceramics for Fuel Cells, Batteries and Sensors (2012)
3. Moulson, A.J., Herbert, J.M.: Electroceramics: Materials, Properties, Applications (1990)
4. Irvine, J.T., Sinclair, D.C., West, A.R.: Electroceramics: characterization by impedance spectroscopy. Adv. Mater. **2**, 132–138 (1990)
5. Setter, N.: Electroceramics: looking ahead. J. Eur. Ceram. Soc. **21**, 1279–1293 (2001)
6. Goel, M., Sharma, S.: Electroceramics research in India. J. Electroceram. **12**, 175–179 (2004)
7. Rosseinsky, M.J.: Electroceramics. Adv. Appl. Ceram. **108**, 261–261 (2009)
8. Maier, J.: Transport in electroceramics: micro- and nano-structural aspects. J. Eur. Ceram. Soc. **24**, 1251–1257 (2004)
9. Newnham, R.E.: Electroceramics. Rep. Prog. Phys. **52**, 123–156 (1989)
10. Kidner, N.J., Ingram, B.J., Homrighaus, Z.J., Mason, T.O., Garboczi, E.J.: Impedance/dielectric spectroscopy of electroceramics in the nanograin regime. MRS Proc. **756**, 39–50 (2003)
11. Krol, R., Tuller, H.L.: Electroceramics—the role of interfaces. Solid State Ion. **150**, 167–179 (2002)

12. Alexe, M., Harnagea, C., Hesse, D.: Non-conventional micro- and nanopatterning techniques for electroceramics. J. Electroceram. **12**, 69–88 (2004)
13. Shishov, M.A., Sapurina, I.Y., Smirnova, N.V., Yudin, V.E.: Biocompatible Electroconductive Matrices for Tissue Engineering: A Comparative Study. Biointerface Research in Applied Chemistry (2022)
14. Stauss, S., Honma, I.: Biocompatible batteries—materials and chemistry, fabrication, applications, and future prospects. Bull. Chem. Soc. Jpn. **91**, 492–505 (2018)
15. Han, I.K., Song, K., Jung, S., Jo, Y., Kwon, J.I., Chung, T., Yoo, S., Jang, J., Kim, Y., Hwang, D.S., Kim, Y.S.: Electroconductive, adhesive, non-swelling, and viscoelastic hydrogels for bioelectronics. Adv. Mater. **35**, (2022)
16. Patel, S., Yadav, H., Kumar, M.: Effect of uniaxial stress on energy harvesting, storage and electrocaloric performance of BZT ceramics. J. Korean Ceram. Soc. **58**, 437–444 (2021)
17. Anoufa, M., Kiat, J.M., Kornev, I.A., Bogicevic, C.: Energy harvesting in core-shell ferroelectric ceramics theoretical approach and practical conclusions. J. Appl. Phys. **113**, 054104 (2013)
18. Zhang, B., Ducharne, B., Guyomar, D., Sebald, G.: Energy harvesting based on piezoelectric Ericsson cycles in a piezoceramic material. Eur. Phys. J. Spec. Top. **222**, 1733–1743 (2014)
19. Hashemi, S., Babic, U., Hadikhani, P., Psaltis, D.: The potentials of additive manufacturing for mass production of electrochemical energy systems. Curr. Opin. Electrochem. **20**, 54–59 (2020)
20. Browne, M.P., Redondo, E., Pumera, M.: 3D Printing for electrochemical energy applications. Chem. Rev. (2020)
21. Fieber, L., Evans, J.D., Huang, C., Grant, P.S.: Single-operation, multi-phase additive manufacture of electro-chemical double layer capacitor devices. Addit. Manuf. (2019)
22. Safari, A., Cesarano, J., Clem, P.G., Bender, B.: Fabrication of advanced functional electroceramic components by layered manufacturing (LM) methods. In: Proceedings of the 13th IEEE International Symposium on Applications of Ferroelectrics. ISAF 2002, pp. 1–6 (2002)
23. Kawaguchi, N., Akikubo, K.: Material Manufacturing Device and Material Manufacturing Method (2018)
24. Shossmann, M., Georg, K., Glazunov, A.: Ceramic Material, Method of its Manufacturing and Electroceramic Structural Element Containing Electroceramic Material (2012)
25. Harash, J., Sachdev, Cipto, Y., Joegiono.: The challenges of obtaining a competitive advantage for processed material suppliers: a conceptual study. J. Mark. Dev. Compet. **7**(1), 27–41 (2013)
26. Morrell, R., Lindsey, K.: Overv. Mater. Process. **1573**, 6–12 (1992). https://doi.org/10.1117/12.57737
27. Payne, D.A.: Powderless processing of ceramics: thin-layer electroceramics. J. Sol-Gel Sci. Technol. **2**(1),311–315 (1994). https://doi.org/10.1007/BF00486262
28. Vučić, V., Vučić, M.R.: Environmental concerns of energy chain as a reflection of sustainability. Saf. Eng. **8**(2), (2018). https://doi.org/10.7562/SE2018.8.02.08
29. Lacasa, E., Cotillas, S., Saez, C., Lobato, J., Cañizares, P., Rodrigo, M.A.: Environmental applications of electrochemical technology. What is needed to enable full-scale applications? Curr. Opin. Electrochem. **16**, 149–156 (2019). https://doi.org/10.1016/J.COELEC.2019.07.002
30. Barbara, L.: Sustainability concerns in the life cycle of bonded grinding tools. CIRP J. Manuf. Sci. Technol. **7**(3), 258–263 (2014). https://doi.org/10.1016/J.CIRPJ.2014.05.002
31. Grobbauer, P., Ott, G. and Zodl, H.: Electroceramic Component and Method of Manufacture Thereof (1997)
32. Anatychuk, I.I., Luste, O.J.: Problems of Thermoelectric Devices Reliability, pp. 632–635 (1997). https://doi.org/10.1109/ICT.1997.667609
33. Bernier, J.C.: Chemical processing for electronic ceramics: a challenge. Mater. Sci. Eng. A-Struct. Mater. Prop. Microstruct. Process. **109**, 233–241 (1989). https://doi.org/10.1016/0921-5093(89)90593-5
34. Xiaojun, Z., Xiaojun, Z., Hanbin, S., Zong-Yang, S., Martin, M.: Progress and challenges of ceramics for supercapacitors. J. Mater. **7**(6), 1198–1224 (2021). https://doi.org/10.1016/J.JMAT.2021.03.001

35. Farooqui, M.N., Patil, N.G.: A perspective on shaping of advanced ceramics by electro discharge machining. Procedia Manuf. **20**, 65–72 (2018). https://doi.org/10.1016/J.PROMFG.2018.02.009
36. Xiao, P.: "Frontiers in Ceramics" Grand Challenges, vol. 1 (2023). https://doi.org/10.3389/fceic.2023.1137377
37. Andrea, T.: Ceramics for electronics and energy: issues and opportunities. Int. J. Appl. Ceram. Technol. **10**(5), 723–730 (2013). https://doi.org/10.1111/IJAC.12148
38. Schneider, M., Toomey, E.A., Rowlands, G.E., Shainline, J.M., Tschirhart, P., Segall, K.: SuperMind: a survey of the potential of superconducting electronics for neuromorphic computing. Supercond. Sci. Technol. **35**, (2022)
39. Nikam, R.D., Lee, J., Lee, K., Hwang, H.: Exploring the cutting-edge frontiers of electrochemical random access memories (ECRAMs) for neuromorphic computing: revolutionary advances in material-to-device engineering. Small e2302593, (2023)
40. Emboras, A., Alabastri, A., Lehmann, P., Portner, K., Weilenmann, C., Ma, P., Cheng, B., Lewerenz, M., Passerini, E., Koch, U., Aeschlimann, J., Ducry, F., Leuthold, J., Luisier, M.: Opto-electronic memristors: prospects and challenges in neuromorphic computing. Appl. Phys. Lett. **117**, 230502 (2020)
41. Kang, H., Seo, J., Kim, H., Kim, H.W., Hong, E.R., Kim, N., Lee, D., Woo, J.: Ion-driven electrochemical random-access memory-based synaptic devices for neuromorphic computing systems: a mini-review. Micromachines **13**, (2022)
42. Tang, J., Bishop, D.M., Kim, S., Copel, M., Gokmen, T., Todorov, T.K., Shin, S., Lee, K., Solomon, P.M., Chan, K.K., Haensch, W.E., Rozen, J.: ECRAM as scalable synaptic cell for high-speed, low-power neuromorphic computing. 2018 IEEE International Electron Devices Meeting (IEDM), pp. 13.1.1–13.1.4 (2018)
43. Burgt, Y.V., Lubberman, E., Lubberman, E., Fuller, E.J., Keene, S.T., Faria, G.C., Faria, G.C., Agarwal, S., Marinella, M.J., Talin, A.A., Salleo, A.: A non-volatile organic electrochemical device as a low-voltage artificial synapse for neuromorphic computing. Nat. Mater. **16**(4), 414–418 (2017)
44. Gao, B., Bi, Y., Chen, H., Liu, R., Huang, P., Chen, B., Liu, L., Liu, X., Yu, S., Wong, H.-S., Kang, J.: Ultra-low-energy three-dimensional oxide-based electronic synapses for implementation of robust high-accuracy neuromorphic computation systems. ACS Nano **8**(7), 6998–7004 (2014)
45. Sengupta, A., Roy, K.: Encoding Neural and Synaptic Functionalities in Electron Spin: A Pathway to Efficient Neuromorphic Computing (2017). https://doi.org/10.48550/arXiv.1711.02235
46. Yu, Q., Alonso, A.M., Caminiti, J., Beck, K.M., Sutherland, R.T., Leibfried, D., Rodriguez, K., Dhital, M., Hemmerling, B., Haffner, H.: Feasibility study of quantum computing using trapped electrons. Phys. Rev. A (2021)
47. Zhang, X., Li, H., Wang, K., Cao, G., Guo, G.: Quantum Computation Based on Semiconductor Quantum Dots (2017)
48. Blais, A., Huang, R., Huang, R., Wallraff, A., Girvin, S.M., Schoelkopf, R.J.: Cavity quantum electrodynamics for superconducting electrical circuits: an architecture for quantum computation. Phys. Rev. A **69**, 062320 (2004)
49. Bocko, M.F., Herr, A.M., Feldman, M.J.: Prospects for quantum coherent computation using superconducting electronics. IEEE Trans. Appl. Supercond. **7**, 3638–3641 (1997)
50. Wesenberg, J.H., Ardavan, A., Briggs, G.A., Morton, J.J., Schoelkopf, R.J., Schuster, D.I., Mølmer, K.: Quantum computing with an electron spin ensemble. Phys. Rev. Lett. **103**(7), 070502 (2009)
51. Dykman, M.I., Platzman, P.M.: Quantum computing using electrons floating on liquid helium. Protein Sci. **48**, 1095–1108 (2000)
52. Karakurt, İ: Quantum computing with electrons on liquid helium. Turk. J. Phys. **27**, 383–394 (2003)
53. Ciaramicoli, G., Marzoli, I., Tombesi, P.: Scalable quantum processor with trapped electrons. Phys. Rev. Lett. **91**(1), 017901 (2003)

54. Guggilla, P., Batra, A.K.: Novel Electroceramic: Polymer Composites-Preparation, Properties and Applications (2011)
55. Waser, R.: Modeling of electroceramics—applications and prospects. J. Eur. Ceram. Soc. **19**, 655–664 (1999)
56. Maletin, Y.A., Stryzhakova, N.G., Zelinsky, S.A., Chernukhin, S.I., Tretyakov, D., Tychina, S., Drobny, D.O.: Electrochemical double layer capacitors and hybrid devices for green energy applications. Green **4**, 17–19 (2014)
57. Chouhan, N., Liu, R.: Electrochemical Technologies for Energy Storage and Conversion (2012)
58. Rodríguez-Varela, J., Alonso-Lemus, I.L., Savadogo, O., Palaniswamy, K.: Overview: current trends in green electrochemical energy conversion and storage. J. Mater. Res. **36**, 4071–4083 (2021)
59. Chang, Z., Yang, Y., Li, M., Wang, X., Wu, Y.: Green energy storage chemistries based on neutral aqueous electrolytes. J. Mater. Chem. **2**, 10739–10755 (2014)
60. Badwal, S.P., Giddey, S., Munnings, C., Bhatt, A.I., Hollenkamp, A.F.: Emerging electrochemical energy conversion and storage technologies. Front. Chem. **2**, (2014)
61. Wei, H., Cui, D., Ma, J., Chu, L., Zhao, X., Song, H., Liu, H., Liu, T., Wang, N., Guo, Z.: Energy conversion technologies towards self-powered electrochemical energy storage systems: the state of the art and perspectives. J. Mater. Chem. **5**, 1873–1894 (2017)
62. Kankam, M.D., Elbuluk, M.E.: A Survey of Power Electronics Applications in Aerospace Technologies (2001)
63. Bashir, M., Rajendran, P.: A review on electroactive polymers development for aerospace applications. J. Intell. Mater. Syst. Struct. **29**, 3681–3695 (2018)
64. Gerada, C., Galea, M., Kladas, A.G.: Electrical machines for aerospace applications. In: 2015 IEEE Workshop on Electrical Machines Design, Control and Diagnosis (WEMDCD), pp. 79–84 (2015)
65. D'Ambrosio, D., Giordano, D.D.: Electromagnetic Fluid Dynamics for Aerospace Applications. Part I: Classification and Critical Review of Physical Models (2004)
66. Ganev, E.: Selecting the best electric machines for electrical power-generation systems: high-performance solutions for aerospace more electric architectures. IEEE Electrification Mag. **2**, 13–22 (2014)
67. Wang, G., Lu, Z., Li, Y., Li, L., Ji, H., Feteira, A., Zhou, D., Wang, D., Zhang, S., Reaney, I.M.: Electroceramics for high-energy density capacitors: current status and future perspectives. Chem. Rev. **121**, 6124–6172 (2021)
68. Valero, A., Méry, A., Gaboriau, D.M., Dietrich, M., Fox, M., Chretien, J., Pauc, N., Jouan, P.Y., Gentile, P., Sadki, S.: Redefining high-k dielectric materials vision at nanoscale for energy storage: a new electrochemically active protection barrier. Electrochim. Acta **389**, 138727 (2021)
69. Fang, Q., Zhang, J., Boyd, I.W.: High-K Dielectrics by UV Photo-Assisted Chemical Vapour Deposition (2003)
70. Clark, R.D.: Emerging applications for high K materials in VLSI technology. Materials **7**, 2913–2944 (2014)
71. Gaskins, J.T., Hopkins, P.E., Merrill, D.R., Bauers, S.R., Hadland, E.C., Johnson, D.C., Koh, D., Yum, J.H., Banerjee, S., Nordell, B.J., Paquette, M.M., Caruso, A.N., Lanford, W.A., Henry, P., Ross, L.L., Li, H., Li, L., French, M., Rudolph, A.M., King, S.W.: Review—investigation and review of the thermal, mechanical, electrical, optical, and structural properties of atomic layer deposited high-k dielectrics: beryllium oxide, aluminum oxide, hafnium oxide, and aluminum nitride. ECS J. Solid State Sci. Technol. **6**, (2017)
72. Kittl, J.A., Opsomer, K., Popovici, M.I., Menou, N., Kaczer, B., Wang, X.P., Adelmann, C., Pawlak, M.A., Tomida, K., Rothschild, A., Govoreanu, B., Degraeve, R., Schaekers, M., Zahid, M., Delabie, A., Meersschaut, J., Polspoel, W., Clima, S., Pourtois, G., Knaepen, W., Detavernier, C., Afanas'ev, V.V., Blomberg, T., Pierreux, D., Swerts, J., Fischer, P.R., Maes, J.W., Manger, D., Vandervorst, W., Conard, T., Franquet, A., Favia, P., Bender, H., Brijs, B., Elshocht, S.V., Jurczak, M., Houdt, J.V., Wouters, D.J.: High-k dielectrics for future generation memory devices (Invited paper). Microelectron. Eng. **86**, 1789–1795 (2009)

73. Lin, C., Kuo, Y.: Improvement of zirconium-doped hafnium oxide high-k dielectric properties by adding molybdenum. J. Vac. Sci. & Technol. B Nanotechnol. Microelectron. Mater. Process. Meas. Phenom. **31**, 030605 (2013)
74. Shevlin, S.A., Curioni, A., Andreoni, W.: Ab initio design of high-k dielectrics: $La_{(x)}Y_{1-x}AlO_3$. Phys. Rev. Lett. **94**(14), 146401 (2005)
75. Uchikoba, F.: Study on lead-free electroceramics-piezoelectric materials focusing on crystalline orientation. In: 2005 4th International Symposium on Environmentally Conscious Design and Inverse Manufacturing, pp. 578–585 (2005)
76. Lee, H.J., Zhang, S.: Perovskite Lead-Free Piezoelectric Ceramics (2012)
77. Aksel, E., Jones, J.L.: Advances in lead-free piezoelectric materials for sensors and actuators. Sens. (Basel, Switz.) **10**, 1935–1954 (2010)
78. Wei, H., Wang, H., Yijie, X., Cui, D., Shi, Y., Dong, M., Liu, C., Ding, T., Zhang, J., Ma, Y., Wang, N., Wang, Z., Sun, Y., Wei, R., Guo, Z.: An overview of lead-free piezoelectric materials and devices. J. Mater. Chem. C **6**, 12446–12467 (2018)
79. Saito, Y., Takao, H., Tani, T., Nonoyama, T., Takatori, K., Homma, T., Nagaya, T., Nakamura, M.: Lead-free piezoceramics. Nature **432**, 84–87 (2004)
80. Maeder, M.D., Damjanovic, D., Setter, N.: Lead free piezoelectric materials. J. Electroceram. **13**, 385–392 (2004)
81. Rödel, J., Jo, W., Seifert, K.T., Anton, E., Granzow, T., Damjanovic, D.: Perspective on the development of lead-free piezoceramics. J. Am. Ceram. Soc. **92**, 1153–1177 (2009)
82. Wu, J.: Perovskite lead-free piezoelectric ceramics. J. Appl. Phys. (2020)
83. Eguchi, K.: Ceramic materials containing rare earth oxides for solid oxide fuel cell. J. Alloy. Compd. **250**, 486–491 (1997)
84. Huang, K., Wan, J., Goodenough, J.: Oxide-ion conducting ceramics for solid oxide fuel cells. J. Mater. Sci. **36**, 1093–1098 (2001)
85. Singhal, S.C.: Application of Ionic and electronic conducting ceramics in solid oxide fuel cells. ChemInform **29**, (1997)
86. Orera, A., Slater, P.R.: New chemical systems for solid oxide fuel cells. Chem. Mater. **22**, 675–690 (2010)
87. Tsipis, E.V., Kharton, V.V.: Electrode materials and reaction mechanisms in solid oxide fuel cells: a brief review. J. Solid State Electrochem. **12**, 1367–1391 (2008)
88. Goodenough, J., Huang, Y.: Alternative anode materials for solid oxide fuel cells. J. Power. Sources **173**, 1–10 (2007)
89. Badwal, S.P., Ciacchi, F.T.: Oxygen-ion conducting electrolyte materials for solid oxide fuel cells. Ionics **6**, 1–21 (2000)
90. Medvedev, D.A., Brouzgou, A., Demin, A.K., Tsiakaras, P.E.: Proton-Conducting Electrolytes for Solid Oxide Fuel Cell Applications (2017)
91. Cao, Y., Liang, J., Li, X., Yue, L., Liu, Q., Lu, S., Asiri, A.M., Hu, J., Luo, Y., Sun, X.: Recent advances in perovskite oxides as electrode materials for supercapacitors. Chem. Commun. (2021)
92. Tan, P., Liu, M., Shao, Z., Ni, M.: Recent advances in perovskite oxides as electrode materials for nonaqueous lithium–oxygen batteries. Adv. Energy Mater. **7**, (2017)
93. Xu, X., Wang, W., Zhou, W., Shao, Z., Shao, Z.: Recent advances in novel nanostructuring methods of perovskite electrocatalysts for energy-related applications. Small Methods (2018)
94. Fu, C., Ma, Q., Gao, L., Li, S.: Recent advances in perovskite oxides electrocatalysts: ordered perovskites, cations segregation and exsolution. ChemCatChem **15**, (2023)
95. Nan, H., Hu, X., Tian, H.: Recent advances in perovskite oxides for anion-intercalation supercapacitor: a review. Mater. Sci. Semicond. Process. (2019)
96. Sun, C., Alonso, J.A., Bian, J.: Recent advances in perovskite-type oxides for energy conversion and storage applications. Adv. Energy Mater. **11**, (2020)
97. Tomar, A.K., Joshi, A., Singh, G., Sharma, R.K.: Perovskite oxides as supercapacitive electrode: properties, design and recent advances. Coord. Chem. Rev. **431**, 213680 (2021)

98. Yu, J., Ran, R., Zhong, Y., Zhou, W., Ni, M., Shao, Z.: Advances in porous perovskites: synthesis and electrocatalytic performance in fuel cells and metal–air batteries. Energy & Environ. Mater. **3**, (2020)
99. Brennan, R.E., Turcu, S., Hall, A., Hagh, N.M., Safari, A.: Fabrication of electroceramic components by layered manufacturing (LM). Ferroelectrics **293**, 17–23 (2003)

SPRINGER NATURE

GPSR Compliance

The European Union's (EU) General Product Safety Regulation (GPSR) is a set of rules that requires consumer products to be safe and our obligations to ensure this.

If you have any concerns about our products, you can contact us on ProductSafety@springernature.com

In case Publisher is established outside the EU, the EU authorized representative is:

Springer Nature Customer Service Center GmbH
Europaplatz 3
69115 Heidelberg, Germany

The manufacturer's authorised representative in the EU is Springer Nature Customer Service Centre GmbH, Europaplatz 3, 69115 Heidelberg, Germany. If you have any concerns regarding our products, please contact ProductSafety@springernature.com

Printed and bound by CPI Group (UK) Ltd, Croydon, CR0 4YY
01/10/2026
02225858-0020